AF776222

Selected Titles in This Series

15 **Stephen P. Braham, Jack D. Gegenberg, and Robert J. McKellar, Editors,** The Sixth Canadian conference on general relativity and relativistic astrophysics, 1997

14 **Mourad E. H. Ismail, David R. Masson, and Mizan Rahman, Editors,** Special functions, q-series and related topics, 1997

13 **Peter A. Fillmore and James A. Mingo, Editors,** Operator algebras and their applications, 1997

12 **Dan-Virgil Voiculescu, Editor,** Free probability theory, 1997

11 **Colleen D. Cutler and Daniel T. Kaplan, Editors,** Nonlinear dynamics and time series: Building a bridge between the natural and statistical sciences, 1997

10 **Jerrold E. Marsden, George W. Patrick, and William F. Shadwick, Editors,** Integration algorithms and classical mechanics, 1996

9 **W. H. Kliemann, W. F. Langford, and N. S. Namachchivaya, Editors,** Nonlinear dynamics and stochastic mechanics, 1996

8 **Larry M. Bates and David L. Rod, Editors,** Conservative systems and quantum chaos, 1996

7 **William F. Shadwick, Perinkulam Sambamurthy Krishnaprasad, and Tudor Stefan Ratiu, Editors,** Mechanics day, 1996

6 **Anna T. Lawniczak and Raymond Kapral, Editors,** Pattern formation and lattice gas automata, 1996

5 **John Chadam, Martin Golubitsky, William Langford, and Brian Wetton, Editors,** Pattern formation: Symmetry methods and applications, 1996

4 **William F. Langford and Wayne Nagata, Editors,** Normal forms and homoclinic chaos, 1995

3 **Anthony Bloch, Editor,** Hamiltonian and gradient flows, algorithms and control, 1994

2 **K. A. Morris, Editor,** Control of flexible structures, 1993

1 **Michael J. Enos, Editor,** Dynamics and control of mechanical systems: The falling cat and related problems, 1993

Fields Institute Communications

The Fields Institute for Research in Mathematical Sciences

The Sixth Canadian Conference on General Relativity and Relativistic Astrophysics

Stephen P. Braham
Jack D. Gegenberg
Robert J. McKellar
Editors

American Mathematical Society
Providence, Rhode Island

The Fields Institute for Research in Mathematical Sciences

The Fields Institute is named in honour of the Canadian mathematician John Charles Fields (1863–1932). Fields was a visionary who received many honours for his scientific work, including election to the Royal Society of Canada in 1909 and to the Royal Society of London in 1913. Among other accomplishments in the service of the international mathematics community, Fields was responsible for establishing the world's most prestigious prize for mathematics research—the Fields Medal.

The Fields Institute for Research in Mathematical Sciences is supported by grants from the Ontario Ministry of Education and Training and the Natural Sciences and Engineering Research Council of Canada. The Institute is sponsored by McMaster University, the University of Toronto, the University of Waterloo, and York University and has affiliated universities in Ontario and across Canada.

1991 *Mathematics Subject Classification.* Primary 83–06; Secondary 85–06.

Library of Congress Cataloging-in-Publication Data

Canadian Conference on General Relativity and Relativistic Astrophysics (6th : 1995 : University of New Brunswick)

The sixth Canadian Conference on General Relativity and Relativistic Astrophysics / Stephen P. Braham, Jack D. Gegenberg, Robert J. McKellar, editors.

p. cm. — (Fields Institute communications, ISSN 1069-5265 ; 15)

"The sixth CCGRRA, held at the University of New Brunswick in Fredericton, New Brunswick on May 25–27, 1995"—Pref.

Includes bibliographical references.

ISBN 0-8218-0523-1

1. Black holes (Astronomy)—Congresses. 2. General relativity (Physics)—Congresses. 3. Relativistic astrophysics—Congresses. I. Braham, Stephen P. (Stephen Paul), 1964– . II. Gegenberg, Jack D. (Jack David), 1947– . III. McKellar, Robert J. (Robert James), 1952– . IV. Title. V. Series: Fields Institute communications ; v. 15.

QB843.B55C34 1995

523.01—dc21

97-11371

CIP

Copying and reprinting. Material in this book may be reproduced by any means for educational and scientific purposes without fee or permission with the exception of reproduction by services that collect fees for delivery of documents and provided that the customary acknowledgment of the source is given. This consent does not extend to other kinds of copying for general distribution, for advertising or promotional purposes, or for resale. Requests for permission for commercial use of material should be addressed to the Assistant to the Publisher, American Mathematical Society, P. O. Box 6248, Providence, Rhode Island 02940-6248. Requests can also be made by e-mail to `reprint-permission@ams.org`.

Excluded from these provisions is material in articles for which the author holds copyright. In such cases, requests for permission to use or reprint should be addressed directly to the author(s). (Copyright ownership is indicated in the notice in the lower right-hand corner of the first page of each article.)

© 1997 by the American Mathematical Society. All rights reserved.
The American Mathematical Society retains all rights
except those granted to the United States Government.
Printed in the United States of America.

∞ The paper used in this book is acid-free and falls within the guidelines
established to ensure permanence and durability.
This publication was prepared by the Fields Institute.
Visit the AMS home page at URL: `http://www.ams.org/`.

10 9 8 7 6 5 4 3 2 1   02 01 00 99 98 97

Contents

Preface ix

Invited Talks

Four Dimensional Black Holes and Duality in Superstring Theory 1
CLIFFORD P. BURGESS, ROBERT C. MYERS and FERNANDO QUEVEDO

Self-Similarity and Cosmology 19
ALAN A. COLEY

Field Theory on Spacetimes that are not Globally Hyperbolic 43
JOHN L. FRIEDMAN

Solving the Einstein Constraint Equations 59
JAMES ISENBERG

Covariant Double-Null Dynamics and the Internal Geometry of Aspherical Black Holes 71
WERNER ISRAEL

Charged Black Holes in Generic 2-D Dilaton Gravity 81
DOMINGO LOUIS-MARTINEZ and GABOR KUNSTATTER

Gravitational Waves from Coalescing Compact Binaries 95
ERIC POISSON

Colliding Black Holes with Linearized Gravity 117
JORGE PULLIN

Spinors, Jets, and the Einstein Equations 125
CHARLES G. TORRE

The Laws of Black Hole Mechanics in an Arbitrary Lagrangian Theory of Gravity 137
ROBERT M. WALD

Black Holes, Galaxy Formation and the Large-scale Structure of the Universe 139
MICHAEL J. WEST

Contributed Talks and Posters

When Black Holes Collide 151
PETER ANNINOS, DAVID HOBILL, EDWARD SEIDEL, LARRY SMARR and WAI-MO SUEN

A New Class of Spherically Symmetric Perfect Fluid Solutions with Kinematic Self-Similarity 157
PATRICIA M. BENOIT and ALAN A. COLEY

A Wave-Like Solution in Kaluza-Klein Theory 161
ANDREW BILLYARD and PAUL S. WESSON

Mass Inflation in a Rotating Charged Hole 167
ALFIO BONANNO

Quantum Black Holes and Hypertime 173
STEPHEN P. BRAHAM

Symmetries of Matter Distributions 179
JAUME CAROT and JOSE da COSTA

Topology and Causal Structure 185
ANDREW CHAMBLIN

Charged Dilaton Black Holes with Unusual Asymptotics 189
KEVIN C. K. CHAN

General Relativity Calculations in Maple 195
KENNETH C-K CHU, CESAR E. FARELL, GREG J. FEE and R. G. MCLENAGHAN

Vacuum Solutions of Gauss-Bonnet Theory and Induced Cosmological Models 201
ALAN A. COLEY

Quasilocal Formalism for Dilaton Gravity with Yang-Mills Fields 205
JOLIEN D. E. CREIGHTON and ROBERT B. MANN

Einstein-Yang-Mills Cosmology with Taub-Type Symmetry 211
BAHMAN DARIAN and H. P. KÜNZLE

Using the Schrödinger Equation for Astrophysical and Cosmological Simulations 217
GEORGE DAVIES

Decoherence of the Unruh Detector 221
JEAN-GUY DEMERS

Black Hole Entropy and Renormalization 227
JEAN-GUY DEMERS, RENÉ LAFRANCE and ROBERT C. MYERS

Matching Conditions for the 1+1 de Sitter Kinked Spacetime 233
KENNETH A. DUNN, TINA A. HARRIOTT and J. G. WILLIAMS

COBE Constraints on Kaluza-Klein Cosmologies 237
VALERIO FARAONI, FRED I. COOPERSTOCK and JAMES M. OVERDUIN

Cosmologies with Two-Dimensional Inhomogeneity 245
A. FEINSTEIN, J. IBÁÑEZ and RUTH LAZKOZ

Stiff Matter G_2 Cosmologies 249
STEPHEN W. GOODE and SCOTT BULLOCK

Projective Symmetry in Space-Times 253
GRAHAM S. HALL

Light Repulsion in Higher-Dimensional Relativity 259
DIMITRI KALLIGAS and PAUL S. WESSON

A No-Go Theorem in String Cosmology 263
NEMANJA KALOPER

MapleTensor: A New System for Performing Indicial and Component Tensor Calculations by Computer 269
MASOUD KAVIAN, R. G. MCLENAGHAN and K. O. GEDDES

Low-Energy Scattering of Extreme Black Holes in String Theory 273
RAMZI R. KHURI and ROBERT C. MYERS

Entropy of Eternal Black Holes 279
ERIK A. MARTINEZ

Nonexistence of Petrov Type III Space-Times on which Weyl's Neutrino Equation and Maxwell's Equations Satisfy Huygens' Principle 285
R. G. MCLENAGHAN and FERNANDO D. SASSE

Symmetry Operators for the Weyl Neutrino Equation on Curved Spacetime 291
R. G. MCLENAGHAN and DAVID M. WALKER

Two-Dimensional Dilaton Black Holes 297
GUY MICHAUD and ROBERT C. MYERS

Gravitationally-Induced Three-Flavor Neutrino Oscillations as a Possible Solution to the Solar Neutrino Problem 303
JONAS R. MUREIKA and ROBERT B. MANN

Diffential Invariants and Regularity 307
PETER MUSGRAVE and KAYLL LAKE

GRTensorII: A Computer Algebra System for General Relativity 313
PETER MUSGRAVE, DENIS POLLNEY and KAYLL LAKE

Exact Solution for Two Charged Spherically Symmetric Sources in Equilibrium 319
G. P. PERRY and FRED I. COOPERSTOCK

The Equation of Motion in Higher-Dimensional Gravity 325
JAIME PONCE de LEON and PAUL S. WESSON

The Nature of Cosmic Time 331
DAVID S. SALOPEK

Numerical Quantum Gravity by Dynamical Triangulation 339
KRISTIN SCHLEICH and DONALD WITT

Conformal Motions in Spherical and Plane Symmetric Spacetimes 345
BRIAN O. J. TUPPER

Dissipative Cosmology 351
ROBERT van den HOOGEN

Counting Kinks in 1+1 Dimensions 357
J. G. WILLIAMS and PETER D. ZVENGROWSKI

Toward a Definition of Chaos for General Relativity 361
DONALD WITT and KRISTIN SCHLEICH

Causal Precedence Formalism 367
ERIC WOOLGAR

Appendix: List of Participants 371

Preface

The Canadian Conferences on General Relativity and Relativistic Astrophysics (CCGRRA) are held every other year at various locations in Canada. They provide an opportunity for researchers in areas involving gravitational physics to discuss their work. In our large country with widely scattered population centres and universities, this enables people who work in closely related fields to compare their findings; and it enables cross-pollination among those in different disciplines. Hence these meetings have featured talks from a mathematical physics perspective, at one end of the spectrum, to reports on the latest discoveries in observational astronomy, at the other.

The Sixth CCGRRA, held at the University of New Brunswick in Fredericton, New Brunswick on May 25-27, 1995, continued this tradition. The conference was attended by about 80 participants, from Canada and the United States, as well as from Latin America and Europe.

The conference was centered around eleven plenary talks, and was rounded out by parallel sessions consisting of shorter contributed talks and a poster session. All the participants who spoke or presented posters were invited to submit their contribution to the Editors, who arranged for the refereeing of the submissions. The papers in this volume are the end product of this procedure.

We would like to thank first of all the members of the National Organizing Committee for their guidance and support. The National Organizing Committee consisted at the time of C. C. Dyer, J. Gegenberg, D. Hobill, G. Kunstatter, and R. G. McLenaghan. We would also like to thank the rest of the members of the Local Organizing Committee for their monumental efforts in organizing the conference. These are F. Söler, B. O. J. Tupper and H. Zaidi. We would also like to thank the students Chris Dunnett and Carla Harper and the secretaries Eleanor Perrin, Janice Phillips and Linda Guthrie in the Mathematics and Statistics Department at UNB for their assistance. We extend special thanks to Jon Thompson, the Chair of the Mathematics and Statistics Department and Israel Unger, the Dean of Science at UNB for their support. Finally we are grateful to the following organizations for their financial support:

- Canadian Institute for Theoretical Astrophysics;
- The Fields Institute for Research in Mathematical Sciences;
- Natural Sciences and Engineering Research Council of Canada;
- Faculté des Études Supérieures et de la Recherche de l'Université de Moncton;
- University of New Brunswick.

S. P. Braham, J. D. Gegenberg, R. J. McKellar
University of New Brunswick
May, 1995

Sixth Canadian Conference on General Relativity and Relativistic Astrophysics - Thursday, May 25

9:00	**Welcome (Ch3) – Israel Unger, Dean of Science, UNB**			
9:10	**Announcements**			
9:15	Werner Israel – Some Recent Advances in Black Hole Theory			
10:15	**Coffee Break**			
10:30	Michael J. West – Black Holes, Galaxy Formation, and the Large-Scale Structure of the Universe			
	Session TA1 (Ch3)		**Session TB1 (S107)**	
11:30	S. Morsink	The Inner Singularity of the Generic Black Hole	M. Kavian	A New System for Performing Indicial and Component Tensor Calculations by Computer
11:50	A. Bonanno	Mass inflation in a rotating charged hole	G. Fee	A new algorithm for computing the scalar curvature
12:10	J. Chan	Mass Inflation in Lower Dimensions	R. McLenaghan	General Relativity Calculations in Maple
12:30	**Lunch**			
2:00	Charles G. Torre – Spinors, Jets, and the Einstein Equations			
3:00	Jorge A. Pullin – Colliding Black Holes: Surprising Success of Linearized Theory			
4:00	**Coffee Break**			
	Session TA2 (Ch3)		**Session TB2 (S107)**	
4:15	R. Khuri	Low-Energy Scattering of Extreme Black Holes in String Theory	J. Mureika	Gravitationally-Induced 3-Flavor Neutrino Oscillations as a Possible Explanation of the Solar Neutrino Problem?
4:35	E. Martinez	Entropy for eternal black holes	T. Harriott	Matching Conditions for the 1+1 de Sitter Kinked Spacetime
4:55	R. Laflamme	Instability of Kaluza Klein Black Holes	P. Musgrave	Differential Invariants and Regularity
5:15	R. Lafrance	Black Hole Entropy and Renormalization	J. Ibáñez Medrano	Cosmologies with Two-Dimensional Inhomogeneity
5:35	G. Michaud	Two-dimensional dilaton black holes		

All invited talks will be in Ch3, Toole Hall Auditorium

Sixth Canadian Conference on General Relativity and Relativistic Astrophysics - Friday, May 26

8:55	**Announcements**			
9:00	Robert M. Wald – The Laws of Black Hole Mechanics in an Arbitrary Theory of Gravity			
10:00	**Coffee Break**			
10:15	Gabor Kunstatter – Charged Black Holes in Generic 1+1 Dimensional Dilaton Gravity			
	Session FA1 (Ch3)		**Session FB1 (S107)**	
11:20	D. Louis Martinez	On the Noether Identities for a Class of Systems with Singular Lagrangians	J. Overduin	COBE Constraints on Inflationary Kaluza-Klein Models
11:40	S. Braham	Quantum Black Holes and Hypertime	R. van den Hoogen	Causal Viscous Fluid Cosmology
12:00	K. Schleich	Computing the Wavefunction of the Universe by Dynamical Triangulation	C. Gauthier	L'âge de l'Univers dans un hyperespace-temps multiconnexe
12:20	J. Gegenberg	Edge Currents and Entropy of the 1+1 Dimensional Black Hole	G. Davies	Using the Schrodinger Equation for Astrophysical and Cosmological Simulations
12:40	**Lunch**			
2:00	Alan A. Coley – Self-Similar Cosmological Models			
3:00	Jim Isenberg – Solving the Einstein Constraint Equations			
4:00	**Coffee Break**			
	Session FA2 (Ch3)		**Session FB2 (S107)**	
4:15	D. Hobill	When Black Holes Collide: Providing a Complete Picture	B. Tupper	Conformal Motions in Spherical and Plane Symmetric Spacetimes
4:35	E. Poisson	Semiclassical enforcement of strong cosmic censorship?	G. Hall	Projective Symmetry in Space-Times
4:55			J. Carot	Symmetries of Matter Distributions
6:30	**Bus to Banquet**			
7:00	**Banquet at Mactaquac Inn**			

All invited talks will be in Ch3, Toole Hall Auditorium

Sixth Canadian Conference on General Relativity and Relativistic Astrophysics - Saturday, May 27

8:55	**Announcements**			
9:00	Clifford P. Burgess – Four Dimensional Black Holes and Duality in Superstring Theory			
10:00	**Coffee Break**			
10:20	John L. Friedman – Field Theory in Spacetimes that are not Globally Hyperbolic			
	Session SA1 (Ch3)		**Session SB1 (S107)**	
11:20	N. Kaloper	A No-Go Theorem for Stringy Cosmology	A. Chamblin	Topology and Causal Structure
11:40	J.-G. Demers	Decoherence and the Unruh detector	R. Vera Mège	A Unified Relativistic Physics from a Standing Wave Particle Model
12:00	A. Steif	Critical behavior in 2+1 dimensional gravitational collapse		
12:20	**Lunch**			
1:10	Eric Poisson – Gravitational Waves from Coalescing Binaries			
	Session SA2 (Ch3)		**Session SB2 (S107)**	
2:10	E. Woolgar	Causal Orders on C^0-Lorentzian Manifolds	R. Nezami	Non-Abelian Stokes Theorem, Magnetic Monopoles and Knot Theory
2:30	D. Witt	A Definition of Chaos for General Relativity	G.P. Perry	New exact solution for two charged spherically symmetric sources in equilibrium
2:50	H. Qayum	Approach to Quantization in 2+1 Dimensional Charged Spacetimes	R. Gautreau	Black Holes – Birth, Death, Metamorphosis
3:10	B. Darian	Locally Rotationally Symmetric Einstein-Yang-Mills Cosmology		
3:30	J. Creighton	Quasilocal Formalism for Dilaton Gravity with Yang-Mills Fields		

All invited talks will be in Ch3, Toole Hall Auditorium

Fields Institute Communications
Volume **15**, 1997

Four Dimensional Black Holes and Duality in Superstring Theory

Clifford P. Burgess
Physics Department
McGill University
3600 University St.
Montréal, Québec, Canada H3A 2T8
cliff@hep.physics.mcgill.ca

Robert C. Myers
Physics Department
McGill University
3600 University St.
Montréal, Québec, Canada H3A 2T8
rcm@hep.physics.mcgill.ca

Fernando Quevedo
Theory Division, CERN,
CH-1211, Genève 23, Switzerland.

Abstract. Some recent results on the applications of duality (and related) transformations to general four-dimensional, spherically symmetric, asymptotically flat and time-independent string configurations are summarized. Two classes of results have been obtained. First, these transformations are used to generate the general such solution to the lowest-order field equations in the α' expansion. Second, the action and implications of duality (based on time-translation) on the general configuration is determined. It is found to interchange two pairs of the six parameters which label these configurations, namely: (1) the mass with the dilaton charge, and (2) the axion charge with the Taub-NUT parameter. For the special case of the Schwarzschild black hole this implies the relation $M \to -k/\alpha' M$, where k is a known, positive, dimensionless number. It is argued that, in some circumstances, dual theories need not be equivalent in the simplest sense.

1991 *Mathematics Subject Classification.* Primary 83E30; Secondary 83C57.

Invited talk presented to the Sixth Canadian Conference on General Relativity and Relativistic Astrophysics, Fredericton, NB, May 1995.

C. B. would like to thank the organizers for their kind invitation to speak at the conference. Our funds have been provided by N.S.E.R.C. of Canada, les Fonds F.C.A.R. du Québec and the Swiss National Foundation.

© 1997 American Mathematical Society

1 Introduction

This article is meant to describe some preliminary applications of duality to unravelling the implications of string theory for black hole physics in four dimensions (Burgess et al. [1995a] and [1995b]). Before summarizing the points to be outlined, some motivation is in order as to why duality transformations, and their applications to black hole physics, are worth thinking about.

Duality transformations (Giveon et al. [1994] gives an extensive review) are a particular type of change of variable which have come to play increasingly important roles in extracting the physical content of string theory, and of some ordinary quantum field theories. Field theories which are related by duality transformations are thought to be completely equivalent, even though they may appear to be very different. Since it is sometimes true that intractible strongly-coupled theories turn out to be dual to calculable weakly-coupled ones, the physical equivalence of the dual theories can be exploited to infer the behaviour of otherwise unsolvable systems.

The development of this new tool may open new approaches to studying old problems of compelling physical interest which have hitherto been too difficult to crack. In particular, a current hope is that they may be useful for extracting the predictions of string theory for the dynamics of spacetimes having very strong, or even singular, curvatures, such as are believed to arise in the final state of runaway gravitational collapse, or in the earliest moments of the universe. The predictions of string theory are of particular interest for this long-standing puzzle, since string theory is the only known theory which gives sensible results for the simpler, but related, problem of the scattering of particles in flat space at energies at and above the Planck-scale.

The good news (so far) is that there are preliminary indications that string theory has qualitatively new features which may figure importantly in our final understanding of gravitational collapse. The main new feature to emerge so far is the realization that string theories appear to be quite forgiving in their notion of what constitutes a physically unacceptable singularity. What might be a malignantly singular field configuration from the point of view of ordinary point-particle field theories, can be completely benign as a background for string propagation. This understanding has emerged as more solutions have been constructed (Tseytlin [1994], for a recent review) to the full string equations, including some singular ones. It is also indicated by the existence of duality transformations which relate spacetimes with singularities to duals which are equivalent, and yet are absolutely nonsingular (including even Minkowski space). Some of the dual spacetimes which have been related in this way even include black hole spacetimes (Giveon [1991], Witten [1991] and Dijkgraaf et al. [1992]), although admittedly only in two spacetime dimensions.

All of this motivates further study of black-hole-type field configurations within the context of string theory. This article reports on the results of two lines of inquiry in this direction. These two lines may be summarized as follows:

1. The first investigation (Burgess et al. [1995a]) uses duality transformations, and some of their extensions, to generate new classical solutions to the string equations of motion, starting from simple initial solutions. In particular, they are used to systematically construct all possible spherically-symmetric, time-independent and asymptotically-flat solutions in four dimensions. There are two new features to these results. (*i*) First, because using duality to generate new solutions is a purely algebraic procedure, it

is much simpler than a direct attempt to solve the relevant set of nonlinear, coupled, partial-differential equations. This simplicity has permitted the inclusion of more nontrivial background fields than previously had been tractable. (*ii*) Second, keeping in mind the broad-mindedness of string theory towards singularities, *all* solutions are presented, including some that are singular. Singular solutions were often omitted in previous constructions.

2. The second application (Burgess et al. [1995b]) is to critically analyze the situation under which dual configurations can be expected to be physically equivalent. In particular, an important class of duality transformations — including most of those applied to black holes — are argued to *not* relate physically-equivalent spacetimes in the usual sense. The same considerations also lead to restrictions on the boundary conditions which must be satisfied by some fields in order for the duality transformations to take their usual form.

These two developments are presented in the remainder of this article in the following way. Section 2 sets the stage by briefly presenting a reminder of the connection between string field configurations, and two-dimensional conformal field theories. This is followed by the two applications listed above. First, Section 3 through Section 5 present the use of duality-related transformations for constructing solutions to the low-energy field equations. Section 3 summarizes the transformations themselves, while Section 4 gives the simplest — i.e., dilaton-metric — solutions. Section 5 then applies the transformations of Section 3 to the field configurations of Section 4, thereby generating solutions also incorporating nonzero axion and gauge fields.

The second line of argument is the topic of Section 6 through Section 8. Section 6 states the form of the duality transformations in a manner which is sufficiently general for the desired applications. Their implications for the general spherically-symmetric, time-independent and asymptotically-flat solutions, including in particular the black hole solutions, are the topic of Section 7. The result, as applied to black hole configurations, makes it difficult to see how the dual solutions can be physically equivalent. Section 8 is dedicated to the resolution of this puzzle.

The conclusions which follow from these sections are finally summarized in Section 9.

2 Strings and 2D Field Theories

This section is meant to review the connection between string field configurations, and two-dimensional conformal field theories, since this underlies all of what follows. This connection allows the application of results for two-dimensional field theories to draw conclusions about solutions to the string field equations.

2.1 String field configurations. A classical solution to the string theory equations of motion is equivalent to a conformally-invariant, two-dimensional field theory. The connection is simplest to see in the case that the two-dimensional field theory is a sigma model, whose two-dimensional scalar fields, $x^\mu(\sigma)$, can be interpreted as describing the worldsheet of a string propagating within a 'target' spacetime. For instance, suppose the sigma model action has the form

$$\begin{aligned} S[x^\mu] &= -\frac{1}{2\pi\alpha'}\int d^2\sigma\,\sqrt{\gamma}\left[G_{\mu\nu}(x)\,\gamma^{\alpha\beta}+B_{\mu\nu}(x)\,\epsilon^{\alpha\beta}\right]\partial_\alpha x^\mu\partial_\beta x^\nu \\ &\quad -\frac{1}{8\pi}\int d^2\sigma\,\sqrt{\gamma}\,\phi(x)\mathcal{R}, \end{aligned} \tag{2.1}$$

where $\gamma_{\alpha\beta}$ and $\epsilon^{\alpha\beta}$ are the two-dimensional metric and antisymmetric Levi-Civita symbol and $\mathcal{R}$ is the curvature scalar for the metric $\gamma_{\alpha\beta}$.

The coupling functions, $G_{\mu\nu}(x)$, $B_{\mu\nu}(x)$ and $\phi(x)$, are interpreted in string theory as being background values for three fields — the metric, the 'axion', and the 'dilaton' — which represent three of the modes of the string. These particular modes would have been massless if the string were propagating through Minkowski space. There are typically other such nearly 'massless' modes as well in a string theory, and we shall consider later a spin-one gauge field, $A_\mu(x)$, in addition to the above three.

Now comes the main point. There are two ways in which one might imagine defining the equations of motion for these fields, given our presently limited understanding of string theory itself. The direct way is to infer the interactions of the various string modes by computing their tree-level scattering on simple spacetimes, such as for Minkowski space. One finds an action which reproduces these scattering amplitudes, and then computes the equation of motion using this action.

The alternative way to determine the equations of motion for $G_{\mu\nu}$, $B_{\mu\nu}$, ϕ and A_μ is to compute the conditions which these quantities must satisfy in order for the corresponding two-dimensional theory to be conformally invariant at the quantum level. It is an amazing fact that these two methods produce results which agree with one another — up to the ubiquitous freedom to perform field redefinitions — in all cases for which they have been compared.

The resulting field equations can be explicitly written as an expansion in powers of derivatives of the background fields times the dimensionful constant α'. In four dimensions the action which reproduces the leading terms in these equations for the fields of interest is (Green et al. [1987]):

$$\mathcal{L} = \frac{e^\phi}{8\pi\alpha'}\,\sqrt{-G}\,\left[R(G)+(\nabla\phi)^2-\frac{1}{12}\,H^{\mu\nu\lambda}H_{\mu\nu\lambda}-\frac{1}{8}\,F^{\mu\nu}F_{\mu\nu}\right]+\cdots, \tag{2.2}$$

where $H_{\mu\nu\lambda}=\partial_\mu B_{\nu\lambda}-\frac{1}{4}\,A_\mu F_{\nu\lambda}+$ (cyclic permutations)[1] and $F_{\mu\nu}=\partial_\mu A_\nu-\partial_\nu A_\mu$ are, respectively, the field strengths for the axion and electromagnetic fields. $R(G)$ is the Ricci scalar for the metric, $G_{\mu\nu}$, and the square root involves the quantity: $G=\det G_{\mu\nu}$. This metric is often called the 'sigma-model' metric to distinguish it from the one for which a field redefinition has been performed to put the Einstein term into standard form. The ellipses in Equation (2.2) represent terms which involve other massless fields and others involving more derivatives that arise at higher orders in α'.

Some of the higher-order corrections to these equations in the derivative expansion have also been computed. For later purposes we quote the quantity which is responsible for the higher-order corrections to configurations involving only the

[1] By including the gauge-field-dependent Chern-Simons term, $\frac{1}{4}\,AF$, in the definition of H we restrict our discussion of gauge fields to the heterotic string. By contrast, the discussion for $A_\mu=0$ applies equally well to bosonic and superstrings.

metric and the dilaton, which turns out to be (Green et al. [1987]):

$$\delta\mathcal{L} = \frac{\lambda e^{\phi}}{2} \sqrt{-G}\, R_{\mu\nu\rho\sigma} R^{\mu\nu\rho\sigma} \ . \tag{2.3}$$

Here λ is $\frac{1}{2}$ for bosonic strings, $\frac{1}{4}$ for heterotic strings, and 0 for supersymmetric strings.

This establishes a direct connection between background string configurations and two-dimensional conformal field theories. The spectrum of string states corresponds to a class of operators in these field theories which represent the conformal group in a particular way. The scattering of these string states can be computed by evaluating appropriate correlation functions involving these operators in the conformal field theory.

Given this connection it is immediate how to interpret the equivalence between any two conformal field theories in terms of string states. If two conformal theories can be shown to have precisely the same content, then they must describe identical string scattering about identical background string configurations. Duality transformations are interesting because they imply precisely this kind of equivalence between two-dimensional field theories.

3 Classical Transformations

There is a broad class of transformations which are guaranteed to map conformal field theories into other conformal field theories, even though the theories which are related in this way need not be physically equivalent. That is to say, this broader class of transformations are guaranteed to take classical string vacua into other classical vacua, but the full Hilbert space of string modes constructed about these vacua can be different. Although these transformations cannot therefore be considered to be *bona-fide* string symmetries, they are nevertheless very useful for generating new classical solutions from known ones.

There are two transformations of this type which are used in what follows. We outline both of these in the following two subsections.

3.1 $SL(2,\mathbb{R})$ **transformations.** The first class of such transformations is a group of $SL(2,\mathbb{R})$ transformations (De Roo [1985], Shapere et al. [1991] and Sen [1993]) which includes the classical string S-duality transformations (Witten [1985], Burgess et al. [1986] and Li et al. [1986] and [1987]) of the low-energy effective theory. To formulate these transformations it is useful to use the scalar variable, $a(x)$, which is dual to the three-index axion field, $H_{\mu\nu\lambda}$:

$$H_{\mu\nu\rho} = -e^{-2\phi} \epsilon_{\mu\nu\rho\kappa} \nabla^{\kappa} a. \tag{3.1}$$

Here all indices are raised and lowered with the Einstein metric, $g_{\mu\nu}$, which (in four spacetime dimensions) is related to the sigma-model metric, $G_{\mu\nu}$ by: $g_{\mu\nu} = e^{\phi}\, G_{\mu\nu}$, and the Levi-Civita tensor is also constructed using $g_{\mu\nu}$. Many properties of the theory take a simple form if the field, a, is combined with the dilaton, ϕ, into the complex combination, $S = a + i\, e^{\phi}$. In terms of this variable the $SL(2,\mathbb{R})$ transformation becomes

$$\begin{aligned} S &\longrightarrow \frac{\mathrm{a}\, S + \mathrm{b}}{\mathrm{c}\, S + \mathrm{d}} \\ (F_+)_{\mu\nu} &\longrightarrow (\mathrm{c}\, S + \mathrm{d})\ (F_+)_{\mu\nu} \\ (F_-)_{\mu\nu} &\longrightarrow (\mathrm{c}\, S^* + \mathrm{d})\ (F_-)_{\mu\nu} \, . \end{aligned} \tag{3.2}$$

where S^* is the complex conjugate of S, and $(F_\pm)_{\mu\nu} \equiv F_{\mu\nu} \pm \frac{i}{2}\, \epsilon_{\mu\nu\rho\kappa}\, F^{\rho\kappa}$. Once again it is the Einstein metric which is involved in these definitions. The quantities a, b, c and d are real numbers which must satisfy ad − bc = 1. If S, $g_{\mu\nu}$ and $F_{\mu\nu}$ all satisfy the string equations of motion, then so must the transformed variables as defined by Equation (3.2).

3.2 $O(1,1)$ transformations. There is a second transformation which can be used to generate new classical string solutions from old ones (Cecotti et al. [1988], Gasperini et al. [1991], Meissner and Veneziano [1991], Hassan and Sen [1992] and Sen [1992]). In the simplest case of a field configuration that is independent of a single coordinate, s, there is an $O(1,1)$ group of such transformations. The action of these transformations is most easily written when the background fields are written as the following 9×9 matrix

$$\mathcal{M} = \begin{pmatrix} \mathcal{K}_-^T G^{-1} \mathcal{K}_- & \mathcal{K}_-^T G^{-1} \mathcal{K}_+ & -\mathcal{K}_-^T G^{-1} A \\ \mathcal{K}_+^T G^{-1} \mathcal{K}_- & \mathcal{K}_+^T G^{-1} \mathcal{K}_+ & -\mathcal{K}_+^T G^{-1} A \\ -A^T G^{-1} \mathcal{K}_- & -A^T G^{-1} \mathcal{K}_+ & A^T G^{-1} A \end{pmatrix} \tag{3.3}$$

where

$$(\mathcal{K}_\pm)_{\mu\nu} = -B_{\mu\nu} - \mathcal{G}_{\mu\nu} \pm \eta_{\mu\nu}, \tag{3.4}$$

and $\eta_{\mu\nu}$ is the flat Minkowski metric in four dimensions. In these expressions the quantity $\mathcal{G}_{\mu\nu}$ is defined by: $\mathcal{G}_{\mu\nu} = G_{\mu\nu} + \frac{1}{4}\, A_\mu A_\nu$. For a detailed statement of the conventions used, see Burgess et al. [1995a].

With these variables the $O(1,1)$ transformations can be expressed in matrix form, $\mathcal{M} \to \Omega \mathcal{M} \Omega^T$, where the transformation matrix is given by

$$\Omega = \begin{pmatrix} I_7 & 0 & 0 \\ 0 & x & \sqrt{x^2-1} \\ 0 & \sqrt{x^2-1} & x \end{pmatrix}. \tag{3.5}$$

Here I_7 is the 7×7 unit matrix, and x is a parameter satisfying $x^2 \geq 1$. The action of the $O(1,1)$ transformations on the dilaton is given by

$$e^\phi \to \left(\frac{\det G}{\det G'} \right)^{\frac{1}{2}} e^\phi \quad , \tag{3.6}$$

where $G'_{\mu\nu}$ denotes the transformed sigma-model metric.

4 Dilaton-Metric Configurations

The goal now is to use these transformations to generate the most general spherically-symmetric, asymptotically-flat and static solutions to the string equations. This will be done by applying the transformations to a particularly simple class of solutions. The first step — identifying this initial simple class of solutions — is the topic of the present section.

4.1 The lowest-order solution. The starting point is the most general four-dimensional, asymptotically flat, static and spherically-symmetric configuration which solves the string equations to leading order in α', and which involves only the metric and dilaton fields. We therefore directly solve the equations:

$$\begin{aligned} R_{\mu\nu}(g) &= \frac{1}{2} \nabla_\mu \phi \nabla_\nu \phi \\ \nabla^2 \phi &= 0, \end{aligned} \tag{4.1}$$

using the *ansatz*

$$\begin{aligned} g_{\mu\nu} dx^\mu dx^\nu &= -f(r)dt^2 + g(r)dr^2 + h^2(r)(d\theta^2 + \sin^2\theta\, d\varphi^2), \\ \phi &= \phi(r), \end{aligned} \tag{4.2}$$

The result is a family of field configurations whose study actually predates the earliest advent of string theory itself (Buchdahl [1959] and Janis et al. [1969]):

$$\begin{aligned} f = \frac{1}{g} &= \Lambda^\delta(r) \\ h^2 &= r^2 \Lambda^{1-\delta}(r) \\ e^\phi &= e^{\phi_0}\, \Lambda^\gamma(r), \end{aligned} \tag{4.3}$$

where $\Lambda(r)$ is a shorthand for the function $1 - (\ell/r)$. ℓ is the only dimensionful parameter in the solutions, and so it simply sets their overall scale. We assume both ℓ and r to be positive in what follows. The two dimensionless quantities, δ and γ, are the arbitrary parameters which label the solutions, subject to the condition $\delta^2 + \gamma^2 = 1$.

ϕ_0 is the asymptotic value which is obtained by the dilaton field as $r \to \infty$. It is convenient in what follows to shift ϕ by a constant to ensure that it tends to zero for large r, and so to completely remove ϕ_0 from the solutions. This is always possible in string theory, with the general result — for classical string solutions — that ϕ_0 only appears as an overall factor in the low-energy lagrangian of Equations (2.2) and (2.3). It can therefore be absorbed into the definition of Newton's constant, which we take (in four dimensions) to be $G_N = \frac{1}{2}\, e^{-\phi_0}\, \alpha'$.

The choice $(\delta, \gamma) = (1, 0)$ yields the Schwarzschild metric with constant dilaton field, and ℓ is in this case related to Newton's constant, G_N, and the black hole mass, M, by $\ell = 2G_N M$. So long as $\gamma\ell \neq 0$, however, the metric of Equation (4.3) has a real curvature singularity at $r = \ell$.

4.2 The conserved charges. The two independent parameters which label this family of solutions have a useful interpretation in terms of the asymptotic behaviour, as $r \to \infty$, of the fields involved. In general, for large r, the solutions take the form:

$$\begin{aligned} f(r) &= 1 - \frac{\mathcal{A}}{r} + \cdots ; \\ g(r) &= 1 + \frac{\mathcal{B}}{r} + \cdots ; \\ h(r) &= r^2 \left[1 + O\left(\frac{1}{r}\right) \right], \\ e^{\phi(r)} &= 1 - \frac{Q_D}{r} + \cdots , \end{aligned} \tag{4.4}$$

with constants, $\mathcal{A} = \mathcal{B} = \delta\ell$ and $Q_D = \gamma\ell$.

These constants have a physical interpretation as specifying the corresponding conserved charges which are carried by the solutions. That is, the conserved (ADM (Arnowitt et al. [1962])) mass, M, of the solution is related to the constant $\mathcal{A}$ by $2G_N\, M = \mathcal{A}$. The constant, Q_D, similarly defines a *dilaton charge* for the solution.

The utility of labelling the solutions by their values for M and Q_D is that these quantities are equally well defined for the complete solution to the string equations. Given these asymptotic expressions, the complete equations may be solved order by order in α', giving a unique solution for *any* choice of M and Q_D. The same is not true for the parameters δ and γ, which are defined in terms of the specific form of the solution to these equations only at lowest order in α'. For instance, if we focus on the black hole solution — which we take to be the solution for which the potential singularity at $r = \ell$ is only a coordinate artifact — and work to next order in α' by including Equation (2.3) into the low-energy string action, then the result is still characterized by the quantities M and Q_D, but with (Callan et al. [1989]):

$$\begin{aligned} 2G_N M = \mathcal{A} = \mathcal{B} &= \ell + \frac{11\lambda\alpha'}{6\ell} + O(\alpha'^2), \\ Q_D &= -\frac{2\lambda\alpha'}{\ell} + O(\alpha'^2). \end{aligned} \tag{4.5}$$

Recall that the constant λ appearing here is $\frac{1}{2}$ in the bosonic string, $\frac{1}{4}$ in the heterotic string, and 0 in the superstring.

5 More General Classical Solutions

It is now possible to compute more general classical solutions, simply by repeatedly applying the $SL(2,\mathbb{R})$ and $O(1,1)$ transformations of the previous sections to the above solutions. Before doing so explicitly, it is first worth characterizing the possible parameters which can describe these solutions by first analyzing the asymptotic behaviour which is permitted for the fields at large r.

5.1 The conserved charges. Once the metric and dilaton fields are supplemented by nonzero axion and gauge fields, more complicated solutions become possible. We take the following spherically-symmetric and time-independent *ansätze* for the axion and gauge fields:

$$a = a(r), \qquad F_{tr} = E(r), \qquad F_{\theta\varphi} = B(r)\sin\theta. \tag{5.1}$$

and we generalize, for later convenience, our metric *ansatz* to include stationary, but not static, metrics:

$$g_{\mu\nu}dx^\mu dx^\nu = -f(r)(dt + 2N\cos\theta\, d\phi)^2 + g(r)dr^2 + h^2(r)(d\theta^2 + \sin^2\theta\, d\varphi^2). \tag{5.2}$$

The parameter N here is called the *NUT* parameter, due to the similarity of Equation (5.2) with the Taub-NUT metric (Taub [1951], Newman et al. [1963] and Misner [1967]).

The corresponding new conserved charges can be inferred by examining the large-r behaviour that is implied by the field equations for these fields. This is given, for the dilaton and metric, by Equations (4.4), and for the axion and gauge fields by:

$$\begin{aligned} a(r) &= -\frac{Q_A}{r} + \cdots; \\ A_t &= \frac{Q_E}{r} + \cdots, \\ A_\varphi &= -Q_M\cos\theta + \cdots. \end{aligned} \tag{5.3}$$

The constants Q_A, Q_E and Q_M are the solution's axion, electric and magnetic charges, respectively.

5.2 Dilaton-axion-metric solutions. The simplest generalization is the inclusion of a nonzero axion field in addition to the original dilaton and metric configurations. These may be generated by applying the $SL(2,\mathbb{R})$ transformations, Equation (3.2), to the solution of Equation (4.3), being careful to preserve the boundary conditions for ϕ and to ensure that $a \to 0$ at large r. (There is no loss of generality in choosing this boundary condition for the axion field, since the definition, Equation (3.1), only defines $a(r)$ up to an additive constant.)

Performing the $SL(2,\mathbb{R})$ transformation, we obtain the same metric configuration as before, but the new dilaton and axion fields, $\hat{\phi}$ and $\hat{a}$:

$$e^{\hat{\phi}} = \left(1+\omega^2\right)\frac{\Lambda^{\gamma}(r)}{\omega^2\Lambda^{2\gamma}(r)+1}, \qquad \hat{a} = \frac{\omega\left[\Lambda^{2\gamma}(r)-1\right]}{\omega^2\Lambda^{2\gamma}(r)+1}, \tag{5.4}$$

ω is the new real parameter of the solution, while δ and γ are the labels of the original dilaton-metric configuration. The starting dilaton-metric solution is re-obtained as the special case $\omega = 0$. These parameters are related to the three conserved charges of the dilaton-axion-metric system by:

$$2G_N M = \delta\ell, \qquad Q_D = \frac{1-\omega^2}{1+\omega^2}\,\gamma\,\ell \qquad \text{and} \qquad Q_A = \frac{2\omega\,\gamma\,\ell}{1+\omega^2}. \tag{5.5}$$

All of these solutions have real singularities at $r = \ell$, except for the special case $\gamma\ell = 0$. An interesting special limit of these solutions is the case $\omega = \pm 1$, for which the dilaton charge, Q_D, vanishes. Notice that even for this choice, however, the dilaton field is nonvanishing due to the nontrivial axion configuration.

5.3 Including a gauge potential. It is conceptually no more difficult, although algebraically more tedious, to generate the general dilaton–axion–metric–gauge-potential configuration. This is obtained by repeatedly performing $SL(2,\mathbb{R})$ and $O(1,1)$ transformations to the previous solutions. Three new parameters, x, ϵ and ρ, enter the solution in this way before these transformations start just regenerating previously-obtained configurations. These three parameters are related to the electric and magnetic charges, Q_E and Q_M, of the background electromagnetic fields, and to the NUT parameter, N, of the metric (which is defined by Equation (5.2)).

The result of this process is a fairly complicated field configuration, whose explicit form is not particularly illuminating and so is not repeated here. Detailed expressions are given in Burgess et al. [1995a]. Instead, we turn to the action of duality on these solutions, and on their extensions to higher order in the α' expansion.

6 Duality Transformations

The remainder of this article is devoted to exploring the implications of duality transformations for the black-hole, and singular, solutions just constructed. Some of the results we obtain also apply to the exact solutions to the string equations.

6.1 The transformation rules. The first step is to define what is meant by duality transformations. A reasonably general algorithm has emerged with which it is always possible to systematically generate dual field theories from a given one (Roček and Verlinde [1992] and Alvarez et al. [1994] and [1995b]), generalizing an earlier construction which had been developed for the earliest known string duality (Buscher [1987] and [1988]). The algorithm applies to any field theory which admits a continuous symmetry.

It is simple to state the result for the case of the sigma model, Equation (2.1), when the symmetry corresponds to the independence of one of the coordinate directions, of the fields $G_{\mu\nu}$, $B_{\mu\nu}$, ϕ and A_μ. Denoting this direction by s, then the sigma model which is dual to the original one involves the fields $\widetilde{G}_{\mu\nu}$, $\widetilde{B}_{\mu\nu}$, $\tilde{\phi}$ and $\widetilde{A}_\mu$, where (Buscher [1987] and [1988], Giveon et al. [1989], Shapere and Wilczek [1989] and Alvarez et al. [1995a]):

$$\begin{aligned}
\widetilde{\mathcal{G}}_{ss} &= 1/\mathcal{G}_{ss}, \qquad \widetilde{\mathcal{G}}_{s\mu} = -B_{s\mu}/\mathcal{G}_{ss}, \\
\widetilde{\mathcal{G}}_{\mu\nu} &= \mathcal{G}_{\mu\nu} - \left[\frac{\mathcal{G}_{s\mu}\mathcal{G}_{s\nu} + B_{s\mu}B_{s\nu}}{\mathcal{G}_{ss}}\right] \\
\widetilde{B}_{s\mu} &= -\mathcal{G}_{s\mu}/\mathcal{G}_{ss}, \qquad \widetilde{B}_{\mu\nu} = B_{\mu\nu} + \frac{\mathcal{G}_{s\mu}B_{s\nu} - \mathcal{G}_{s\nu}B_{s\mu}}{\mathcal{G}_{ss}}, \\
\widetilde{A}_s &= -\frac{A_s}{\mathcal{G}_{ss}} \\
\widetilde{A}_\mu &= A_\mu - A_s\,\frac{\mathcal{G}_{s\mu} - B_{s\mu}}{\mathcal{G}_{ss}} \\
e^{\tilde{\phi}} &= e^{\phi}\left(\frac{\det G}{\det \widetilde{G}}\right)^{1/2}.
\end{aligned} \tag{6.1}$$

As before, the quantity $\mathcal{G}_{\mu\nu}$ is defined by: $\mathcal{G}_{\mu\nu} = G_{\mu\nu} + \frac{1}{4}A_\mu A_\nu$, and the index μ runs over all values except for $\mu = s$. (Alvarez et al. [1994] presents the duality transformations in a more manifestly covariant way.) These transformations in general can acquire higher-derivative corrections in powers of α' as well.

6.2 An example: the torus. Perhaps the simplest example brings out many of the main features of these duality transformations[2]. The simplest example consists of a toroidal spacetime, for which the metric is flat and all other background fields are taken to be trivial. If we base the duality on the symmetry of translations along one of the compact coordinate directions of the torus — call it s, say — then the general expression of Equation (6.1) simply reduces to the replacement:

$$G_{ss} \to \frac{1}{G_{ss}}, \qquad \phi_0 \to \phi_0 + \log\left|G_{ss}\right|. \tag{6.2}$$

Here both G_{ss} and ϕ_0 are constants. The dual configuration is once more a torus, but Equation (6.2) states that if the circumference in the s direction is initially R, then for the dual theory it becomes $\widetilde{R} \propto \alpha'^2/R$.

The beauty of the toroidal example is that for this background the complete spectrum of string fluctuations is known, and the correspondence of states in the dual theories can be explicitly followed. States turn out to be labelled by two integers, m and n, in addition to other quantum numbers, where m labels the quantized momentum in the s direction and n gives the number of times which the string winds around this direction. It turns out that duality acts to interchange m and n, while leaving all of the other quantum numbers fixed. The important role which is played by the winding modes, for which $n \neq 0$, emphasizes the intrinsically 'stringy' nature of the duality symmetry.

When the direction associated with the symmetry coordinate is compact, such as for the translations of a torus just considered, then the sigma models having the dual field configurations can be shown to be completely equivalent, and so describe

[2]See Giveon et al. [1994] for references to the extensive literature on toroidal duality.

exactly the same string physics (Roček and Verlinde [1992]). Equations (6.1) can therefore be considered as full quantum symmetries of string theory.

The relation of the dual theories is less clear when the relevant symmetry coordinate is not compact. We argue in what follows that the equivalence need *not* be true in the noncompact case.

7 Applications of Duality

We now explore the implications of duality for general spherically-symmetric, time-independent and asymptotically-flat field configurations.

7.1 The dilaton-metric configuration. To get an idea for what is implied by a duality transformation, we first apply one to the dilaton-metric configuration considered in Section 4.1, above. We base the duality transformation on the symmetry of time translation of the original solution. This solution, given by Equation (4.3), is characterized by the two constants, δ and γ, together with the dimensional quantity ℓ. Applying the duality transformation, Equation (6.1), to this configuration leads to another solution of the form as Equation (4.3), but with the constants δ and γ interchanged:

$$\delta \longleftrightarrow \gamma. \tag{7.1}$$

Given the relation between these parameters and the conserved charges we see that Equation (7.1) implies:

$$2G_N M \longleftrightarrow Q_D. \tag{7.2}$$

As will now be discussed, this last way of writing the duality transformation law applies equally well to the solutions at higher-orders in α'.

7.2 Application to general configurations. Since part of the promise of studying duality transformations lies in their potential for leading to exact information concerning the theory involved, it is of enormous interest to understand how duality acts on the exact string solutions, rather than simply having its action on the approximate solutions to low orders in the derivative expansion. The obvious difficulty lies in determining this action when explicit expressions for the solutions, and the transformation laws, are themselves not known beyond the leading order in α'.

It is nonetheless possible to draw general conclusions for the case of time-independent, spherically-symmetric, asymptotically flat field configurations, since these are in principle completely characterized by their values for the conserved charges M, N, Q_D, Q_A, Q_E, and Q_M. Furthermore, these charges are themselves completely determined by the behaviour of the fields in the asymptotic, large-r, regime for which all fields are very slowly-varying. As a result, it is possible to determine the action of duality on the conserved charges of a solution using just the leading expressions in the derivative expansion.

By applying the transformation law, Equation (6.1), to the asymptotic forms for the background fields, Equations (4.4) and (5.3), the action of duality on the space of four-dimensional, asymptotically-flat, time-independent and spherically-symmetric string solutions may be determined in general. We find that the solution which is labelled by the charges M, N, Q_D, Q_A, Q_E, and Q_M becomes mapped to the solution whose charges are $\widetilde{M}$, $\widetilde{N}$, $\widetilde{Q}_D$, $\widetilde{Q}_A$, $\widetilde{Q}_E$, and $\widetilde{Q}_M$, where:

$$2G_N \widetilde{M} = Q_D, \qquad \widetilde{Q}_D = 2G_N M, \qquad \widetilde{Q}_A = 2N, \qquad 2\widetilde{N} = Q_A, \tag{7.3}$$

and

$$\widetilde{Q}_E = Q_E, \qquad \widetilde{Q}_M = Q_M. \tag{7.4}$$

That is, duality simply interchanges the mass with the dilaton charge, as well as interchanging the axion charge with the NUT parameter, while leaving the electric and magnetic charges untouched.

7.3 The black hole special case. This transformation law has some odd consequences, as may be seen by focussing on the black-hole solutions. At lowest order in α' these are characterized by the parameters $\delta = 1$ and $\gamma = 0$, and so their image under duality must have $\tilde{\delta} = 0$ and $\tilde{\gamma} = 1$. If we use the expressions for M and Q_D, as given to $O(\alpha')$ in Equations (4.5), then we find for the dual configuration:

$$\begin{aligned} 2G_N\widetilde{M} &= -\frac{2\lambda\alpha'}{\ell} + O(\alpha'^2), \\ \widetilde{Q}_D &= \ell + \frac{11\lambda\alpha'}{6\ell} + O(\alpha'^2). \end{aligned} \tag{7.5}$$

This transformation rule contains the seeds of a puzzle. Notice, in this regard, the mapping from positive to negative mass:

$$\widetilde{M} = -\frac{k}{\alpha' M} + O\left(\frac{1}{\alpha'^2 M^3}\right), \tag{7.6}$$

where $k = \lambda\alpha'^2/2G_N{}^2 = 2\lambda e^{2\phi_0}$ is a dimensionless constant. (In the special case of the superstring, for which $\lambda = 0$, it is necessary to work to still higher order in α' to infer the sign of $\widetilde{M}$, again giving the result (Burgess et al. [1995b]) that M and $\widetilde{M}$ have opposite signs.) The puzzle is to understand how the physical equivalence of dual solutions can be reconciled with a change of sign of the solution's mass. The next section is devoted to the resolution of this puzzle.

8 Equivalence Under Duality

Intuitively, it seems absurd that a background field configuration having negative mass could be physically equivalent to one for which the mass is positive. The goal of this section is to pin down this intuition, in order to better understand the circumstances under which dual spacetimes can be equivalent.

A more precise reason for doubting the equivalence of spacetimes having opposite masses would be the expectation that such spacetimes should gravitationally scatter incident test particles differently. After all, based on experience with the Newtonian force law for universal gravitation, a positive mass should exert an attractive gravitational force on a distant, slowly-moving particle, while the force due to a negative mass should be repulsive in this limit. One might worry, however, whether the presence of other massless background fields — such as the dilaton or axion, etc. — may invalidate any intuition which is based on the Newtonian limit of pure gravity.

8.1 Test-particle scattering. It is straightforward to test these ideas, by computing the scattering of test particles which remain at large r, and so which experience only slowly varying fields. This calculation may be performed explicitly in string theory by choosing massless string states as test particles. The result is particularly simple in the limit for which the wavelength of the incident state is both longer than the Planck length, and shorter than the typical distances over which the background fields vary. This is the regime of geometric optics, for which simple arguments show that massless string states simply follow the null geodesics of the metric, regardless of the presence of axion and dilaton fields (Burgess et al. [1995b]).

For example, in this limit the scattering angle, $\Delta\varphi$, of a photon which remains at large r throughout its interaction with the background field is given (for vanishing NUT parameter) by (Burgess et al. [1995b])

$$\Delta\varphi \simeq \frac{4G_N M}{r_0}, \tag{8.1}$$

where r_0 is the smallest radial coordinate that is attained by the photon. As advertised, this result depends only on the mass M of the source, and not on its other charges. The result for the dual spacetime is given by the same expression, with $M \to \widetilde{M}$[3].

8.2 A better argument. Convincing as the previous calculation may seem, it leaves room for doubt concerning the inequivalence of the dual field configurations we have considered. This is because it leaves open a loophole, whose existence becomes clear once the well-understood example of toroidal duality is reconsidered. As was stated in Section 6.2, duality acts to interchange the labels m and n which respectively characterize the momentum and winding number along the symmetry direction. It is the scattering of 'winding modes' ($n \neq 0$) in the dual theory which is equal to the scattering of 'momentum modes' ($m \neq 0$) in the original theory. The problem with the scattering calculation of Section 8.1 arises because it is not clear whether the photon in the original theory is mapped to the same photon in the dual theory. If not there is no reason for a result like Equation (8.1) to be the same for both the original spacetime and its dual.

In fact, since the symmetry on which duality is based for the configurations considered here is time translation, it is not completely clear precisely how the winding and momentum modes should be defined. As a result, an alternative argument in favour of the inequivalence of the dual solutions is now presented which does not rely on being able to trace the detailed action of duality on the various string states. This is done by identifying a one-parameter family of equivalent background field configurations, and showing that these dualize to a one-parameter family of inequivalent configurations. We are led to the conclusion that Equations (6.1) need not be a string symmetry, at least when the coordinate 's' labels a noncompact direction.

The key point on which this argument relies is the observation that the transformations of Equations (6.1) depend in detail upon the boundary condition which is satisfied at large r by the gauge potential in the time direction, A_t. In particular, the duality transformations obtained in Section 7.2 for the conserved charges of the spherically-symmetric solutions are only correct if it is assumed that A_t falls off like $1/r$. This is required if the quantity $\mathcal{G}_{tt} = G_{tt} + \frac{1}{4}\, A_t A_t$ is to have the same asymptotic form as does G_{tt} — a result which is required in deriving Equations (7.3) and (7.4).

---

[3]This assumes only that the dual NUT parameter also vanishes, which implies the vanishing of the axion charge, Q_A, of the original solution. This class of configurations is amply big for the present purposes.

Suppose, then, we instead assume the following asymptotic behaviour for A_t:

$$A_t = 2v + \frac{Q_E}{r} + \cdots , \tag{8.2}$$

where v is a constant. This form does not affect the asymptotic behavior of the electric and magnetic fields, and so does not change the values of the electric and magnetic charges. In order to not change the asymptotic form for the field strength, $H = dB - \frac{1}{4} A\, dA$, it is also necessary to alter the asymptotic behavior of the Kalb-Ramond field, B, which we assume to have done in what follows.

When performing the duality transformation, Equation (6.1), using the asymptotic form of Equation (8.2) it is necessary to rescale the time coordinate to preserve the limit $G_{tt} \to -1$ as $r \to \infty$, and to shift the dilaton by a constant in order to recover its previous limiting value at infinity. With these choices the action of duality on the asymptotic charges is no longer given by Equations (7.3) and (7.4), but rather by (Burgess et al. [1995b]):

$$\begin{aligned}
2\widetilde{G_N}\widetilde{M} &= \frac{1}{1-v^2}\left[Q_D - 2v^2 G_N M - vQ_E\right] \\
\widetilde{Q}_D &= \frac{1}{1-v^2}\left[2G_N M - v^2 Q_D + vQ_E\right] \\
2\widetilde{N} &= \frac{1}{1-v^2}\left[Q_A - 2v^2 N\right] \\
\widetilde{Q}_A &= \frac{1}{1-v^2}\left[(1-2v^2)2N + v^2 Q_A\right] \\
\widetilde{Q}_E &= \frac{1}{1-v^2}\left[(1+v^2)Q_E + 2v(2G_N M - Q_D)\right] \\
\widetilde{Q}_M &= Q_M + \frac{v}{1-v^2}\left[2N - Q_A\right] .
\end{aligned} \tag{8.3}$$

It is assumed in these equations that $v^2 < 1$. Notice that the previous results are obtained as $v \to 0$. The 'tilde' is written over Newton's constant in the first line to emphasize that this constant is modified because of the shift which was required of the dilaton field in order to preserve its boundary conditions at large r.

We are now in a position to argue against the necessity of physical equivalence between dual solutions when the duality which relates the solutions is based on a noncompact symmetry. The first step is the observation that the asymptotic value, v, which is taken by A_t drops out of all physical quantities because it can be changed by performing a gauge transformation. The one-parameter family of field configurations which differ only in their asymptotic value for A_t are therefore all physically equivalent to one another. In particular, all would agree on their prediction for the scattering of an incident massless string mode.

Now, if duality based on Equations (6.1) were always to generate physically equivalent string configurations, it would follow that the one-parameter family of dual configurations which are obtained from the original gauge-equivalent family must also all be physically equivalent. But Equations (8.3) show that the dual configurations have conserved charges which depend explicitly on the value of the parameter, v. They therefore differ in their physical predictions, such as for the scattering of massless string modes, and so cannot be physically equivalent. We conclude that for the class of configurations considered here, Equations (6.1) are *not* symmetries of string theory.

Notice that the compactness of the symmetry plays a role in this argument. If the symmetry direction were compact, the above argument would fail since it would no longer be possible to shift A_t by an arbitrary constant value simply by performing a gauge transformation. Different constant values for A_t would be physically distinct since they could be characterized by different values for a gauge-invariant quantity: the Wilson lines around the compact direction.

8.3 Flat space revisited. The above considerations can be further focussed by reconsidering the case of flat space and toroidal compactifications. We can avoid the complication of understanding the meaning of winding and momentum modes in the time direction by considering instead duality based on translations in one of the noncompact spatial directions. It is instructive, in this case, to consider this direction as the limit of a compact, circular, direction as its circumference tends to infinity.

For any finite circumference, duality indeed represents an exact string symmetry, with all momentum modes dualizing to winding modes in the symmetry direction, and *vice versa*. For infinite circumference the momentum states degenerate into a continuum of permitted eigenvalues, while the masses of all of the winding modes tend to infinity. As required by duality, precisely the opposite happens as the circumference tends to zero: winding states form a continuum and momentum states become infinitely massive. So long as both winding- and momentum- modes are both kept in the infinite- (and zero-) circumference limit, duality continues to relate equivalent theories.

But string theory, as it is normally defined on (noncompact) flat Minkowski space, is the path integral over all possible string embeddings in this space. This corresponds to what is obtained from a compact space in the infinite-circumference limit *provided* that the winding modes are dropped in this limit. Once the winding modes are omitted, the equivalence of dual configurations is no longer guaranteed.

This picture is similar to that obtained by Alvarez et al. [1994], who analyze more systematically the equivalence of solutions related by duality based on a noncompact symmetry. They find equivalence, but only if one of the dual pair of solutions is quantized as a vortex gas. In this unconventional string theory, there are no local vertex operators carrying momentum in the symmetry direction, but instead one constructs nonlocal 'vortex' operators carrying 'winding-number' for this direction.

This line of argument indicates that a string theory is not uniquely specified by listing its configuration of background fields, since additional choices — such as whether or not to include nonlocal vortex operators — must also be made. Although duality can always produce an equivalent string theory from any given one, the two string theories generically will *not* both be interpretable as the sum over string embeddings in the given background field configurations. In comparing the dual theories, one must keep in mind this distinction. For example, for two-dimensional black holes, the propagation and interactions of tachyons near the singularity are equivalent to the propagation and interactions near the horizon of some dual vortex states, rather than of tachyons.

9 Conclusions

The purpose here has been to present some results of a preliminary investigation (Burgess et al. [1995a] and [1995b]) into the consequences of duality, and related, transformations on our understanding of string propagation through black-hole-like spacetimes. This has been accomplished by outlining two qualitatively different types of results.

The first class of result simply uses these transformations to generate new solutions to the string equations. This technique has been used to generate the most general static, spherically-symmetric and asymptotically-flat string configuration which solves the string equations to lowest order in α'. Solutions of this form are now known which involve all of the metric, dilaton, axion and gauge fields — a broader class than had hitherto been constructed. They are labelled by a collection of five conserved quantities: the mass M, and the dilaton, axion, electric and magnetic charges: Q_D, Q_A, Q_E and Q_M. If the static condition is relaxed to include some stationary metrics, then a sixth charge, the NUT parameter, N, is also required.

Using duality, and related, transformations to generate these solutions has the practical advantage of requiring the use of only algebraic techniques, instead of having to solve several coupled, nonlinear, partial-differential equations. This makes its application relatively straightforward to apply to more complicated, less symmetric, field configurations.

Most of the solutions which were generated in this way have real curvature singularities at fixed, nonzero r, and these singularities are in some cases naked. We regard it to be premature to discard these solutions until it is better understood precisely what configurations string theory considers to be singular.

The second type of investigation reported here concerns the action and implications of duality on these low-energy solutions in particular, and on static, spherically-symmetric and asymptotically-flat string configurations in general.

The result can be presented in a very general form. By investigating the action of duality on the asymptotic form of a general field configuration for large r, it becomes clear that duality simply interchanges the configuration's mass and dilaton charge, as well as interchanging the axion charge with the NUT parameter. The electric and magnetic charges do not change. These results are very robust — applying equally well once higher orders in α' are included — depending as they do only on the large-r behaviour of the solution.

Somewhat surprisingly we find that the dual field configuration depends in an important way on the assumed asymptotic form for the electrostatic potential, A_t. The conclusions of the previous paragraph assumed A_t to be chosen to vanish asymptotically.

For the special case of the four-dimensional 'Schwarzschild-like' solutions — i.e., those that are nonsingular at the Schwarzschild radius — a black hole of mass M is mapped by duality onto a singular configuration whose mass, $\widetilde{M}$, is given by $\widetilde{M} = -k/(\alpha' M)$, where $k = \lambda \alpha'^2/2G_N{}^2 = 2\lambda e^{2\phi_0}$. λ here is $\frac{1}{2}$ in the bosonic, and $\frac{1}{4}$ in the heterotic string. For the superstring, it happens that $\lambda = k = 0$, and so $O(\alpha'^3)$ corrections are required in order to determine the dual mass. The result in this case turns out to be: $\widetilde{M} = -\,k'/(\alpha'^3 M^5)$, where $k' = 3\zeta(3)\, e^{6\phi_0}$.

We have also addressed the potential inequivalence of the two dual solutions in the case where the symmetry on which duality is based is not compact. An argument was presented which led to one of the following two options: (*i*) dual solutions, defined by Equations (6.1), can be physically inequivalent; or (*ii*) constant shifts of the electrostatic potential, A_t, can change the physical content of the theory. Arguments in favour of the first option were presented, using the well-understood example of flat space.

The potential implications of duality are just beginning to be explored, both for black holes, and for other physical systems. May the rewards be both rich and varied!

References

Alvarez, E., Alvarez-Gaumé, L. and Bakas, I. [1995a] *T Duality and Spacetime Supersymmetry*, Nucl. Phys. B **457**, 3–26.

Alvarez, E., Alvarez-Gaumé, L., Barbón, J. L. F. and Lozano, Y. [1994] *Some Global Aspects of Duality in String Theory*, Nucl. Phys. B **415**, 71–100.

Alvarez, E., Alvarez-Gaumé, L. and Lozano, Y. [1995b] *An Introduction to T Duality in String Theory*, Proceedings of the Trieste Spring School, 1994 (Dijkgraaf, R., Narain, K. S. and Randjbar-Daemi, S., eds.), North-Holland, Amsterdam, pp. 1-20.

Arnowitt, R., Deser, S. and Misner, C. W. [1962] *The dynamics of general relativity*, Gravitation: An Introduction to Current Research (Witten, L., ed.), Wiley, New York, pp. 227–265.

Buchdahl, H. [1959] *Reciprocal Static Metrics and Scalar Fields in the General Theory of Relativity*, Phys. Rev. **115**, 1325–1328.

Burgess, C. P., Font, A. and Quevedo, F. [1986] *Low Energy Effective Action for the Superstring*, Nucl. Phys. B **272**, 661–676.

Burgess, C. P., Myers, R. C. and Quevedo, F. [1995a] *On Spherically Symmetrical String Solutions in 4 Dimensions*, Nucl. Phys. B **442**, 75–96.

______ [1995b] *Duality and 4-Dimensional Black Holes*, Nucl. Phys. B **442**, 97–125.

Buscher, T. [1987] *A Symmetry of the String Background Field Equations*, Phys. Lett. B **194**, 59–62.

______ [1988] *Path Integral Derivation of Quantum Duality in Nonlinear Sigma Models*, Phys. Lett. B **201**, 466–469.

Callan, C. G., Myers, R. C. and Perry, M. J. [1989] *Black Holes in String Theory*, Nucl. Phys. B **311**, 673–689.

Cecotti, S., Ferrara, S. and Girardello, L. [1988] *Hidden Noncompact Symmetries in String Theory*, Nucl. Phys. B **308**, 436–447.

De Roo, M. [1985] *Matter Coupling in N=4 Supergravity*, Nucl. Phys. B **255**, 515–532.

Dijkgraaf, R., Verlinde, E. and Verlinde, H. [1992] *String Propagation in a Black-Hole Geometry*, Nucl. Phys. B **371**, 269–314.

Gasperini, M., Maharana, J. and Veneziano, G. [1991] *From Trivial to Nontrivial Conformal String Backgrounds via O(d,d) Transformations*, Phys. Lett. B **272**, 277–284.

Giveon, A. [1991] *Target Space Duality and Stringy Black Holes*, Mod. Phys. Lett. A **6**, 2843–2854.

Giveon, A., Porrati, M. and Rabinovici, E. [1994] *Target-Space Duality in String Theory*, Phys. Rept. **244**, 77–202.

Giveon, A., Rabinovici, E. and Veneziano, G. [1989] *Duality in String Background Space*, Nucl. Phys. B **322**, 167–178.

Green, M., Schwarz, J. and Witten, E. [1987] *Superstring Theory I*, Cambridge University Press, Cambridge.

Hassan, S. F. and Sen, A. [1992] *Twisting Classical Solutions in Heterotic String Theory*, Nucl. Phys. B **375**,103–118.

Janis, A. I., Robinson, D. C. and Winicour, J. [1969] *Comments on Einstein Scalar Solutions*, Phys. Rev. **186**, 1729–1731.

Li, S. P., Peschanski, R. and Savoy, C. [1986] *Global Symmetries and Flat Potentials in the Compactified Superstring*, Phys. Lett. B **178**, 193–199.

______ [1987] *Generalized No-Scale Models and Classical Symmetries of Superstrings*, Phys. Lett. B **194**, 226–234.

Meissner, K. and Veneziano, G. [1991] *Symmetries of Cosmological Superstring Vacua*, Phys. Lett. B **267**, 33–36.

Misner, C. W. [1967] *Taub-NUT space as a counterexample to almost anything*, Relativity Theory and Astrophysics I, Relativity and Cosmology, Lectures in Applied Mathematics, Vol. 8 (Ehlers, J., ed.), American Mathematical Society, Providence, RI, pp. 160.

Newman, E., Tamburino, L. and Unti, T. [1963] *Empty Space Generalization of the Schwarzschild Metric*, J. Math. Phys. **4**, 915–923.

Roček, M. and Verlinde, E. [1992] *Duality, Quotients, and Currents*, Nucl. Phys. B **373**, 630–646.

Sen, A. [1992] *Twisted Black P-Brane Solutions in String Theory*, Phys. Lett. B **274**, 34–40.

——— [1993] *Electric-Magnetic Duality in String Theory*, Nucl. Phys. B **404**, 109–126.

Shapere, A., Trivedi, S. and Wilczek, F. [1991] *Dual Dilaton Dyons*, Mod. Phys. Lett. A **6**, 2677–2686.

Shapere, A. and Wilczek, F. [1989] *Self-Dual Models with Theta Terms*, Nucl. Phys. B **320**, 669–684.

Taub, A. H. [1951] *Empty Spacetimes Admitting a Three-Parameter Group of Motions*, Ann. Math. **53**, 472–490.

Tseytlin, A. A. [1994] *Proceedings of the 2nd Journée Cosmologie, Observatoire de Paris, June 2-4, 1994*, World Scientific, Singapore, to appear.

Witten, E. [1985] *Dimensional Reduction of Superstring Models*, Phys. Lett. B **155**, 151–163.

——— [1991] *On String Theory and Black Holes*, Phys. Rev. D **44**, 314–324.

Fields Institute Communications
Volume **15**, 1997

Self-Similarity and Cosmology

Alan A. Coley
Department of Mathematics, Statistics and Computing Science
Dalhousie University
Halifax, Nova Scotia, Canada B3H 3J5
aac@cs.dal.ca

Abstract. Self-similarity in general relativity is discussed and self-similar solutions of Einstein's field equations are reviewed. In particular, spacetimes admitting a homothetic vector (self-similarity of the "first" kind) which arise as exact solutions of Einstein's field equations (especially in the cosmological context) are considered, and particular attention is focussed on the relationship of these exact self-similar solutions with the asymptotic states of more general (cosmological) models. Applications of self-similar models in both the cosmological and astrophysical context are reviewed in the special case of spherical symmetry. Finally, the more general case of self-similarity of the "second" kind is discussed.

1 Introduction

Self-similar solutions in Newtonian hydrodynamics occur when the corresponding physical quantities depend on functions depending on $x/l(t)$, where x and t are independent (space and time) variables and l is a time dependent scale. Self-similarity refers to the fact that the spatial distribution of the characteristics of motion remains similar to itself at all times during the motion (Barenblatt and Zeldovich [1972]). Self-similar solutions represent solutions of degenerate problems in which all dimensional constant parameters entering the initial and boundary conditions vanish or become infinite (Barenblatt and Zeldovich [1972]).

Self-similar solutions were originally of interest since the governing equations of the problem simplify and often systems of partial differential equations (PDEs) reduce to ordinary differential equations. Indeed, self-similarity in the broadest (Lie) sense refers to an invariance which simply allows the reduction of a system of PDEs. It was later realised (Barenblatt and Zeldovich [1972]) that self-similar solutions describe the "intermediate-asymptotic" behaviour of solutions in the region

---

1991 *Mathematics Subject Classification.* Primary 83C20; Secondary 83F05.
I would like to thank Bernard Carr, Conrad Hewitt and John Wainwright for helpful comments. I would also like to thank the organizers for inviting me to participate in this meeting and I would like to acknowledge the Natural Sciences and Engineering Research Council of Canada for financial support.

© 1997 American Mathematical Society

in which a solution no longer depends on the details of the initial and/or boundary conditions (but in which the system may still be far from being in a state of equilibrium).

Cases in which the form of the self-similar asymptotics can be obtained from dimensional considerations are referred to as self-similar solutions of the first kind (Barenblatt and Zeldovich [1972]), and examples of such solutions appear in the study of a strong (nuclear) explosion (Sedov [1967] and Taylor [1950]) and a thermal wave (Zeldovich and Kompaneets [1950], Barenblatt [1952] and Zeldovich and Raizer [1963]).

Similarity solutions are also of interest within general relativity. For example, an explosion in a homogeneous background produces fluctuations which may be very complicated initially, but they tend to be described more and more closely by a spherically symmetric similarity solution as time evolves (Sedov [1967]), and this applies even if the explosion occurs in an expanding cosmological background (Schwartz et al. [1975] and Ikeuchi et al. [1983]). In addition, the evolution of voids is also described by similarity solutions at late times (Bertschinger [1985]).

In cosmology, the expansion of the universe from the big bang (and the collapse of a star to a singularity) might (both) exhibit self-similarity in some form since it might be expected that the initial conditions 'are forgotten' in some sense. Indeed, the possibility that (at least in some circumstances)† fluctuations naturally evolve via the Einstein Field Equations (EFEs) to a self-similar form (in which, e.g., the non-linear perturbations grow at the same rate as the universe) in the non-linear regime from initially more complex initial conditions (what Carr [1995] terms the "similarity hypothesis") has been studied by several authors previously (and a review will be given below).

Indeed, it is one of the aims here to discuss under what circumstance the "similarity hypothesis" holds in various classes of cosmological models.

We shall be particularly concerned with the case in which the source of the gravitational field is that of a perfect fluid; i.e., the energy-momentum tensor is given by

$$T_{ab} = (\mu + p)u_a u_b + p g_{ab}, \tag{1}$$

where u^a is the normalized fluid 4-velocity and μ and p are, respectively, the density and pressure. In natural units $c = 8\pi G = 1$, the EFEs of general relativity then read

$$G_{ab} = T_{ab}, \tag{2}$$

where $\boldsymbol{G}$ is the Einstein tensor defined by $G_{ab} = R_{ab} - \frac{1}{2}R g_{ab}$. Unless otherwise stated, a linear barotropic equation of state of the form

$$p = (\gamma - 1)\mu, \tag{3}$$

will be assumed, where the constant γ obeys $1 \leq \gamma \leq 2$ for ordinary matter. [The case $0 \leq \gamma < 2/3$ is of interest in studying models that undergo inflation.]

Cahill and Taub [1971] were the first to study similarity solutions in general relativity. They did so in the cosmological context and under the assumption of a

†This is certainly not the case in all circumstances. For example, the evolution of spherically symmetric perturbations in a dust universe is described by the Tolman solutions which are not, in general, self-similar. Carr [1995] has commented, however, that the addition of pressure (adding to the non-linearity of the problem) might be necessary to force self-similarity. The situation in spatially homogeneous models is reviewed in the text.

spherically symmetric distribution of a self-gravitating perfect fluid. They assumed that the solution was such that the dependent variables are essentially functions of a single independent variable constructed as a dimensionless combination of the independent variables (r and t, and the solution is invariant under the transformation $\bar{t} = at$, $\bar{r} = ar$; a constant) and that the model contains no other dimensional constants.

Cahill and Taub [1971] showed that the existence of a similarity (of the first kind) in this situation could be invariantly formulated in terms of the existence of a homothetic vector. A proper homothetic vector (HV) is a vector field $\boldsymbol{\xi}$ which satisfies (after a constant rescaling)

$$\mathcal{L}_\xi g_{\mu v} = 2 g_{\mu v}, \tag{4}$$

where $g_{\mu v}$ is the metric and $\mathcal{L}$ denotes Lie differentiation along $\boldsymbol{\xi}$. A homothetic motion or homothety captures the geometric notion of "invariance under scale transformations".

From (4) it follows that

$$\mathcal{L}_\xi R^a{}_{bcd} = 0, \tag{5}$$

and hence

$$\mathcal{L}_\xi R_{ab} = 0; \quad \mathcal{L}_\xi G_{ab} = 0. \tag{6a,b}$$

From dimensional considerations, in the case of self-similarity of the first kind and if the source of the gravitational field is a perfect fluid [Equation (1)], the physical quantities transform according to

$$\mathcal{L}_\xi u^a = -u^a, \tag{7}$$

and

$$\mathcal{L}_\xi \mu = -2\mu, \quad \mathcal{L}_\xi p = -2p. \tag{8}$$

From these equations it follows that

$$\mathcal{L}_\xi T_{ab} = 0, \tag{9}$$

which is consistent with the EFEs (2).

Indeed, in the case of a perfect fluid it follows that (Cahill and Taub [1971] and Eardley [1974]) Equations (7) and (8) result from Equations (4) [through Equations (2), (6) and (9)] so that the physical quantities transform appropriately. Hence in this case "geometric" self-similarity and "physical" self-similarity coincide. However, this need not be the case (see below). Unfortunately, terminology has been introduced (Eardley [1974]) which equates self-similarity with the existence of a homothety (e.g., the homothetic group is referred to as the similarity group) and this is misleading. Perfect fluid spacetimes admitting a HV within general relativity have been comprehensively studied (Eardley [1974]).

If the source is not a perfect fluid, then it does not in general follow that the spacetime symmetries are "inherited" by the matter (through the stress-energy tensor via the EFEs; Coley and Tupper [1989]). Consequently a homothety is not a self-similarity; rather it is a purely geometric property of a spacetime, and it is only through the EFEs that properties of the matter (including similarity) can be associated. The question of whether the matter field exhibits the same symmetries as the geometry within general relativity is called the symmetry "inheritance" problem, and the question (for homotheties) for a general (imperfect) fluid source was completely answered in Coley and Tupper [1989]. (The answer is that inheritance occurs if and only if $\mathcal{L}_\xi u^a = -u^a$: — note the connection with the invariant definition of self-similarity of the second kind below.)

However, let us restate that for perfect fluids (within general relativity) the symmetry is inherited (Coley and Tupper [1989]) and self-similarity follows from the existence of a HV, and in this context (and for the rest of the article) the terminology of self-similarity for these models is appropriate.

On the other hand, if the matter fields exhibit self-similarity, then through the EFEs this places restrictions on the geometry. For example, if self-similarity results from dimensional considerations (i.e., self-similarity of the first kind – for self-similarity of the second kind see Section 5) then $\mathcal{L}_\xi T_{ab} = 0$ [e.g., see Equations (7), (8)], whence from the EFEs

$$\mathcal{L}_\xi G_{ab} = 0; \tag{10}$$

that is, ξ is a so-called matter collineation (Kramer et al. [1980] and Carot et al. [1994b]). Although a HV [Equation (4)] satisfies Equation (10) [see Equation (6b)], a matter collineation is not necessarily a HV. Indeed, Equation (10) does not imply Equation (6a) (Ricci collineation) which in turn does not imply Equation (5) (curvature collineation); i.e., a curvature or even a Ricci collineation need not be a HV. [For example, an affine vector field or affine collineation defined by

$$\mathcal{L}_\xi g_{ab} = \bar{h}_{ab}; \ \bar{h}_{ab} = \bar{h}_{ba} \quad \text{and} \quad \bar{h}_{ab;c} = 0, \tag{11}$$

satisfies Equations (5) (and (6a)). The maximum dimension of the Lie algebra of affine vector fields is 20, and this occurs if the spacetime is flat, else the maximum dimension is 10 (non-flat case). An affine collineation is called proper if it is not a HV (i.e., $\bar{h}_{ab} = cg_{ab}, c \in \mathbb{R}$) or a Killing vector ($\bar{h}_{ab} = 0$; $c = 0$).] The general problem of determining the constraints on the form of the metric (or $\mathcal{L}_\xi g_{ab}$) from an equation like (10) has been termed the "inverse" symmetry inheritance problem, and the study of matter collineations in particular (in which $\mathcal{L}_\xi C^a{}_{bcd} \neq 0$, else ξ is necessarily a HV) was undertaken recently by Carot et al. [1994b].

2 Spacetimes Admitting a Homothetic Vector

Let us review spacetimes admitting a HV that satisfy the EFEs of general relativity. A HV ξ satisfies

$$\mathcal{L}_\xi g_{ab} = 2cg_{ab} = 2\xi_{(a;b)}, \tag{4$'$}$$

where c is a constant. If $c = 0$, then $\boldsymbol{\xi}$ is a Killing vector field (KV), else the HV is said to be non-trivial [and the constant can be normalized; i.e., $c = 1$ –see Equation (4)].

The differential geometric properties of HV were studied by Yano [1955]. The totality of HV on a spacetime form a Lie algebra H_n (of dimension n) which contains (if H_n is non-trivial) an $(n-1)$ dimensional (sub)algebra of KV, G_{n-1}. If a given spacetime is not conformal to an exceptional 'plane wave spacetime', it follows that if the orbits of H_n are r-dimensional, then the orbits of G_{n-1} are $(r-1)$-dimensional (Eardley [1974]), and if a spacetime is not conformally flat, the spacetime is conformally related to a spacetime for which the Lie algebra H_n is the Lie algebra of KV (Defrise-Carter [1975]).

Vacuum spacetimes admitting a HV were studied by McIntosh [1975], in which it was shown that a non-flat vacuum spacetime can only admit a non-trivial HV if that HV is non-null and is not hypersurface orthogonal.

Hereafter, we shall assume that the energy-momentum tensor is that of a perfect fluid. McIntosh showed that a perfect fluid spacetime cannot admit a non-trivial HV which is orthogonal to the fluid 4-velocity (unless $p = \mu$). If a spacetime admits a non-trivial HV and there is an equation of state of the form $p = p(\mu)$, then necessarily $p = (\gamma - 1)\mu$ (Cahill and Taub [1971] and Wainwright [1985]); i.e., Equation (3) results from Equation (8). We note that homothetic initial data is preserved by the EFEs (Eardley [1974]). In the case of radiation (i.e., $p = \frac{1}{3}\mu$, so that $T = 0$) the existence of a HV implies the existence of a conserved quality. In general, if we define the current $P^a = T^{ab}\xi_b$, then from energy-momentum conversation $P^a{}_{;a} = T^{ab}g_{ab} \equiv T$. For radiation, $P^a = \frac{\mu}{3}(4u^a[u_b\xi^b] + \xi^a)$, and $P^a{}_{;a} = 0$.

In all of the exact solutions that follow a perfect fluid source is assumed, and, unless specifically stated, $p = (\gamma - 1)\mu$.

Exact solutions

In addition to flat Minkowski spacetime, the following exact solutions admitting a HV are known (cf. Kramer et al. [1980]).

FRW: All ($k = 0, \pm 1$) FRW models admit a timelike HV in the special case of stiff matter ($\gamma = 2$). Otherwise, only the $k = 0$ models admit a HV, and this occurs for all such models in which $p = (\gamma - 1)\mu$ and hence the scale function has power-law dependence on time (Maartens and Maharaj [1986]).

Spatially homogeneous: Many people have studied self-similar spatially homogeneous models*, including: Bogoyavlenski [1985]‡, Bogoyavlenski and Novikov [1973]‡, Collins [1971]‡, Collins [1973]‡, Collins and Ellis [1979]‡ (tilted Bianchi V), Collins and Stewart [1971]‡, Doreskovich et al. [1973]‡, Eardley [1974], Hewitt and Wainwright [1992]‡ (tilted Bianchi V), Hewitt and Wainwright [1993], Hsu and Wainwright [1986], Jacobs [1968], Jantzen and Rosquist [1986]‡; Koutras [1992], Rosquist and Jantzen [1985] and [1988] (tilted perfect fluid), Siklos [1981], Uggla et al. [1995] and Wainwright and Hsu [1989]‡; see also Kramer et al. [1980].

In Hsu and Wainwright [1986], all transitivity self-similar orthogonal spatially homogeneous perfect fluid and vacuum solutions were summarized. Kantowski-Sachs models were studied in Collins [1977]. Exact self-similar solutions for imperfect fluid sources were given in Burd and Coley [1994]‡ and Coley and van den Hoogen [1994a] and [1995]‡.

Spherically symmetric: Self-similar spherically symmetric models have also been studied by many authors.

Cahill and Taub [1971] (perfect fluid), Carr and Hawking [1974] (black holes; $p = 0$, $p = \frac{1}{3}\mu$), Lin et al. [1976] (black holes; $p = \mu$), Miller [1976] (negative

*Although the exact spatially homogeneous solutions were first displayed in the cited references, it was not until later (post 1985) that many of them were recognized by Wainwright and by Jantzen and Rosquist as being self-similar.

‡In the designated references not only were exact self-similar solutions given but also a qualitative analysis of (a class of) models was presented with emphasis on the significance of the self-similar solutions.

mass retarded cores); Bicknell and Henriksen [1978a] and [1978b] (black holes; $p = (\gamma - 1)\mu$), Henriksen and Wesson [1978a] and [1978b] (static), Wesson [1978], [1981] and [1982] (hierarchical models) and Henriksen et al. [1983] (cosmological constant).

Collins [1977] (Kantowski-Sachs, $p = (\gamma - 1)\mu$ $(1 \leq \gamma \leq 2)$; all solutions, including self-similar ones), Ponce de Leon [1988] (Kantowski-Sachs, $p \neq (\gamma - 1)\mu$ $(p < 0) \ldots$), Wesson [1989] ($\ldots$ interpreted as mixture of false vacuum $(p = -\mu)$ and dust $(p = 0)$) and Ponce de Leon [1990] (viscous Kantowski-Sachs).

Carr and Yahil [1990] (perturbations of Friedmann), Carr and Koutras [1992] (perturbations of Kantowski-Sachs) and Alexander et al. [1989] (perturbations of de Sitter).

Bogoyavlenski [1985] (qualitative analysis) and Ori and Piran [1990] (relativistic gravitational collapse; numerical).

We note that the self-similar Tolman-Bondi solutions are the most general self-similar, static spherically symmetric models (Wainwright [1985]). In the above we have concentrated mainly on solutions of cosmological interest (see also Carr [1995]). Self-similar spherically symmetric solutions that have arisen in the study of astrophysical phenomena are noted later (see Section 4).

G_2 spacetimes: In these solutions there are two commuting spacelike KV acting orthogonally transitively, which together with the HV form a 3-dimensional homothety group, H_3. If the 2 KV are hypersurface orthogonal, the G_2 model will be called 'diagonal'. There are three cases depending upon the nature of the orbits of the H_3.

1. orbits of H_3 spacelike: Eardley [1974], Luminet [1978], Chao [1981] and Carot et al. [1994a] (diag.).

2. orbits of H_3 timelike: Hewitt and Wainwright [1990]‡, Hewitt et al. [1988]‡ (diag.), Hewitt et al. [1991]‡ (diag.), Uggla [1992] (diag.) and Carot et al. [1994a] (diag.).

3. orbits of H_3 null: Carot et al. [1994a] (diag.).

See also Wainwright and Yaremovich [1975]. The case in which one of the commuting KV is timelike (i.e., stationary axisymmetric spacetimes) is of astrophysical interest; see Kramer et al. [1980].

Other models:

Plane symmetric (Taub [1972]).

Weyl spacetimes (Godfrey [1972]).

Diagonal hypersurface homogeneous (Uggla et al. [1995]).

Some specific examples of physical importance:

- Minkowski space (M):

$$ds^2 = -dt^2 + dx^2 + dy^2 + dz^2 \tag{12a}$$

$$\boldsymbol{\xi} = t\frac{partial}{\partial t} + x\frac{\partial}{\partial x} + y\frac{\partial}{\partial y} + z\frac{\partial}{\partial z}. \tag{12b}$$

- $k = 0$, FRW (F_0):

$$ds^2 = -dt^2 + t^{4/3\gamma}(dx^2 + dy^2 + dz^2) \tag{13a}$$

$$\boldsymbol{\xi} = t\frac{\partial}{\partial t} + \left(1 - \frac{2}{3\gamma}\right)\left(x\frac{\partial}{\partial x} + y\frac{\partial}{\partial y} + z\frac{\partial}{\partial z}\right) \tag{13b}$$

where

$$p = (\gamma - 1)\mu; \;\; \mu = \frac{4}{3\gamma^2}t^{-2} \;\; \left(\boldsymbol{u} = \frac{\partial}{\partial t}\right). \tag{13c}$$

- Kasner (vacuum) (K):

$$ds^2 = -dt^2 + t^{2p_1}dx^2 + t^{2p_2}dy^2 + t^{2p_3}dz^2 \tag{14a}$$

with

$$\Sigma p_i = \Sigma p_i^2 = 1 \tag{14b}$$

$$\boldsymbol{\xi} = t\frac{\partial}{\partial t} + (1 - p_1)x\frac{\partial}{\partial x} + (1 - p_2)y\frac{\partial}{\partial y} + (1 - p_3)z\frac{\partial}{\partial z}. \tag{14c}$$

All transitively self-similar orthogonal spatially homogeneous perfect fluid solutions (with $\frac{2}{3} < \gamma \leq 2$) and spatially homogeneous vacuum solutions are summarized in Hsu and Wainwright [1986].

3 Self-Similar Models as Asymptotic States of More General Models

There are two very important reasons for studying self-similar solutions of the EFEs. First, the assumption of self-similarity reduces the mathematical complexity of the governing DEs, often leading in problems of physical interest to the simplification of PDEs to ODES, which makes such solutions relatively easier to find. Second, self-similar models are often related to the asymptotic states of more general models (Hsu and Wainwright [1986]). In the cosmological context the governing equations often reduce to a dynamical system which facilitates the qualitative analysis of the models. Indeed, it is true that self-similar models do indeed play an important role in the asymptotic properties of spatially homogeneous models, spherically symmetric models, G_2 cosmologies and silent universe models. [The ODEs governing silent universe models reduce to a dynamical system for which the singular points do indeed represent self-similar models (Bruni et al. [1995]).]

Spatially homogeneous models

Spatially homogeneous models have attracted the most attention since the governing equations reduce to a relatively simple finite-dimensional dynamical system. Planar systems were initially analyzed by Collins [1971] and [1974] and a comprehensive study of general Bianchi models was first made by Bogoyavlenski and Novikov [1973] (see also Bogoyavlenski [1985]) and more recently (using automorphism variables and Hamiltonian techniques) by Jantzen and Rosquist (Jantzen [1984], Rosquist [1984], Rosquist and Jantzen [1988] and Rosquist et al. [1990]). Perhaps the most illuminating approach has been by Wainwright and collaborators

(Hsu and Wainwright [1986], Wainwright and Hsu [1989] and Hewitt and Wainwright [1993]), in which (the more physically or geometrically natural) expansion-normalized (and thus dimensionless) configuration variables are used so that physically admissible states lie within a bounded region, the dynamical system remains analytic both on the physical region and its boundaries, and typically the asymptotic states lie in the boundary represented by exact physical solutions (rather than singular behaviours).

Wainwright utilizes the orthonormal frame method (Ellis and MacCallum [1969]) and introduces expansion-normalized (commutation function) variables and a new "dimensionless" time variable to study spatially homogeneous perfect fluid models satisfying $p = (\gamma - 1)\mu$. The equations governing the models form an N-dimensional system of coupled autonomous ODEs. The system of ODEs admits a symmetry (when written in expansion-normalized variables), which allows one equation to decouple (the time evolution equation for the expansion – the Raychaudhuri equation). The reduced $N-1$-dimensional dynamical system is then studied. Consequently, at all of the singular points of the reduced system we have $\dot{\theta} = \text{const} \times \theta^2$, and hence all singular points correspond to self-similar cosmological models (Hsu and Wainwright [1986]). It is in this way that the self-similar models play an important role in describing the dynamics of the Bianchi models asymptotically.

In particular, for orthogonal Bianchi models of class A the resulting reduced state space is five-dimensional (Wainwright and Hsu [1989]). Orthogonal Bianchi cosmologies of class B were studied in Hewitt and Wainwright [1993], and these models are governed by a five-dimensional system of analytic ODEs with constraints.

In further work imperfect fluid Bianchi models have been studied under the assumption that all physical quantities satisfy "dimensionless equations of state" thereby ensuring that the singular points of the resulting reduced dynamical system are represented by exact self-similar solutions (Coley and van den Hoogen [1994a] and [1994b]). Models satisfying the linear Eckart theory of irreversible thermodynamics were studied in Burd and Coley [1993] and Coley and van den Hoogen [1994a], models satisfying the truncated causal theory of Israel-Stewart were studied in Coley and van den Hoogen [1995], and models satisfying the full (i.e., non-truncated) relativistic Israel-Stewart theory are presently under study (Coley et al. [1995]).

In addition, spatially homogeneous models (and, in fact, G_2 models – see below) with a homogeneous scalar field and an exponential potential (which includes the case of no potential), so that the source is of the form of a perfect fluid, have also been studied (van den Hoogen [1995], Ibañez et al. [1995] and Wainwright [1995]).

It is interesting to ask under what circumstances the singular points correspond to (and hence the asymptotic properties of Bianchi models are represented by) self-similar models. This depends critically on the equations of state employed. For perfect fluid models for which p/μ is asymptotically constant it is plausible that self-similarity of the asymptotic limits is preserved (Wainwright and Hsu [1989]) and this was in fact proven for a class of two-fluid models (Coley and Wainwright [1992]). However, this property is not robust, and self-similarity is broken for perfect fluid models with more complicated equations of state, for a source consisting of a homogeneous scalar field with a non-exponential potential if the strong energy condition is violated (e.g., if there exists a cosmological constant), or for imperfect fluid models that do not "have dimensionless equations of state" (Coley and van den Hoogen [1994b]).

The asymptotic states of perfect fluid Bianchi models will be summarized, and the role of exact self-similar models will be described, in these notes after the examples below.

Spherically symmetric models

In the spherically symmetric case under the assumption of self-similarity the EFEs reduce to a system of ODEs. In the particular case in which the metric is written in conformally static form these equations reduce to a dynamical system, which has been intensively studied by Bogoyavlenski [1985]. These models are of particular physical interest, for example, in the study of explosions in an (expanding) homogeneous background, cosmological models containing black holes and in the study of the evolution of voids (Bogoyavlenski [1985] and Carr [1995]).

Preliminary calculations indicate that all self-similar, perfect fluid (with linear equation of state, $p = (\gamma - 1)\mu$) spherically symmetric solutions (that are not static and are sufficiently well-behaved) are asymptotic to either the flat FRW model or a Kantowski-Sachs exact similarity solution (Carr and Coley [1995]). In further analysis the stability of these (two) self-similar models within the class of perfect fluid spherically symmetric models will be determined, and the possibility that spherically symmetric solutions may evolve to a self-similar form will be investigated.

G_2 models

Inhomogeneous perfect fluid (with $p = (\gamma - 1)\mu$) cosmological models admitting two commuting spacelike KV acting orthogonally transitively – the so-called G_2 cosmologies — have been studied by Hewitt and Wainwright [1990] with a view to describing their asymptotic behaviour near to the big bang and at late times.

In particular, Hewitt and Wainwright [1990] showed that the EFEs can be written as an autonomous system of first-order, quasi-linear (formally hyperbolic) PDEs (without constraints) in terms of (two independent) dimensionless variables. (Here the state space is an infinite dimensional function space.) By defining the dynamical equilibrium states (DES) by an appropriate (invariantly) defined time derivative of the state vector being zero, Hewitt and Wainwright [1990] proved that the DES correspond to cosmological models that are self-similar (but not necessarily spatially homogeneous). For the DES the EFEs reduce to a system of autonomous ODEs (with spatial dependence) that can be studied qualitatively by normal techniques (and the spatially homogeneous subcases have been studied previously - see above).

Hewitt and Wainwright [1990] conjecture that the DES may describe the asymptotic (or intermediate) dynamical behaviour of the orthogonally transitive G_2 models. In particular, they showed that in the subclass of separable diagonal G_2 cosmologies (in which the 2 KV are hypersurface orthogonal), the models do indeed asymptote toward the DES; i.e., the models in this special subclass are asymptotically self-similar. The qualitative properties of these diagonal self-similar G_2 cosmologies (in which the orbits of the 3-dimensional homothetic group are timelike; i.e., the velocity vector is tangent to the group orbits), whose time evolution is completely determined and whose spatial structure is governed by a 2-dimensional plane autonomous system of ODEs, have been studied by Hewitt et al. [1988] and [1991].

Example (Bianchi models)

The expansion scalar θ, shear scalar σ ($\sigma^2 = \frac{1}{2}\sigma_{ab}\sigma^{ab}$), and Ricci 3-curvature (orthogonal to u^a) ${}^3R$, are defined in Ellis [1971].
Basic equations:

$$\text{Raychaudhuri Equation}: \dot{\theta} = -\frac{1}{3}\theta^2 - 2\sigma^2 - \frac{1}{3}(\mu + 3p). \tag{15}$$

$$\text{Conservation Equation}: \dot{\mu} = -(\mu + p)\theta. \tag{16}$$

$$\text{Generalized Friedmann Equation}: \theta^2 = -\frac{1}{3}\,{}^3R + 3\sigma^2 + 3\mu. \tag{17}$$

Ricci identity: for $\dot{\sigma}$ (Ellis [1971]).

We define the expansion-normalized variables ($\theta > 0$):

$$\Omega = \frac{3\mu}{\theta^2}, \qquad \text{density parameter:} \tag{18}$$

$$\beta = 2\sqrt{3}\frac{\sigma}{\theta}, \qquad \text{normalized rate of shear:} \tag{19}$$

$$\frac{d\tau}{dt} = \frac{1}{3}\theta, \qquad [\tau \to -\infty \text{ as } t \to 0^+; \text{a dash (dot) denotes differentiation with respect to } \tau\ (t)]. \tag{20}$$

The equation of state is (Equation (4)):

$$p = (\gamma - 1)\mu \qquad (1 \leq \gamma \leq 2 \text{ for normal matter}^*)$$

and an energy condition is:

$$\mu \geq 0.$$

A. FRW: The metric is given by

$$ds^2 = -dt^2 + R^2(t)\left[\frac{dr^2}{1 - kr^2} + r^2 d\Omega^2\right], \tag{21}$$

where k is the curvature constant. Here $\sigma = 0$ ($\theta = 3\dot{R}/R$), and the equations reduce to the single ODE:

$$\Omega' = (2 - \gamma)\Omega(1 - \Omega), \tag{22}$$

where $\Omega \geq 0$ and $1 - \Omega = -\frac{3}{2}\,{}^3R\theta^{-2}$, so that $\Omega < 1, = 1, > 1$ according to whether $k < 0, = 0, > 0$ (models open, flat or closed), respectively.

- Singular points ($\gamma \neq 2$):
  - $\Omega = 1$ (F_0); past attractor (flat FRW model describes the dynamics near the big bang).
  - $\Omega = 0$ (M); future attractor (open models approach (Milne form) of Minkowski space at later times).

---

*Models with $0 \leq \gamma < \frac{2}{3}$ are of interest in connection with inflationary models of the universe (see, for example, Wald [1983]).

[For closed models $\theta = 0$ at the point of maximum expansion, so the models are only valid up to this point. The models recollapse and asymptotically approach the flat FRW model F_0 to the future].

B. Bianchi V: The metric is given by

$$ds^2 = -dt^2 + A^2(t)dx^2 + e^{2x}(B^2(t)dy^2 + A^4(t)B^{-2}(t)dz^2), \tag{23}$$

where $\theta = 3\frac{\dot{A}}{A}$, $\sigma^2 = \left(\frac{\dot{A}}{A} - \frac{\dot{B}}{B}\right)$, and ${}^3R = -\frac{6}{A^2}$. The Ricci identity leads to

$$\dot{\sigma} = -\theta\sigma. \tag{24}$$

The EFEs then reduce to the plane autonomous system:

$$\begin{aligned} \Omega' &= (2-3\gamma)\Omega(1-\Omega) + \beta^2\Omega \\ \beta' &= \frac{\beta}{2}[(3\gamma-2)\Omega + \beta^2 - 4], \end{aligned} \tag{25}$$

where

$$\beta^2 + 4\Omega - 4 = -\frac{1}{A^2\theta^2} \leq 0$$

(from the Generalized Friedmann Equation) and

$$\Omega \geq 0,$$

so that phase space is compact.

- invariant sets:
  $\beta = 0$ (FRW, dealt with above).

  $\beta^2 + 4\Omega - 4 = 0$ (Bianchi I, boundary) (in this case we have a single ODE and the models evolve from K to F_0).

In addition, the equations above are symmetric about the Ω-axis.

- singular points (Ω, β) ($\gamma \neq 2$; see Figure 3a in Collins [1971]):
  - $(0,0)$ (M) : future attractor

    $(1,0)$ (F_0) : saddle (past attractor for open FRW models, intermediate asymptote otherwise).

  - $(0,2)$ (K_+) : , $(0,-2)(K_-)$: } past attractor (initial cigar singularity is of Kasner-type).

In the special case $\gamma = 2$, the boundary $\beta^2 + 4\Omega - 4 = 0$ becomes a line of repelling singular points. The corresponding Bianchi I stiff matter models evolve from the general self-similar (stiff matter) Bianchi I solution of Jacobs [1968] to the Milne (flat) universe (see Collins [1971]).

C. Bianchi models (Remarks):

The determination of the asymptotic behaviour of more general Bianchi models is much more difficult. Let us summarize the asymptotic states of Bianchi models [assuming again $p = (\gamma - 1)\mu$ $(1 \leq \gamma \leq 2)$] and describe the primary role of exact self-similar models. These remarks summarize the work of Wainwright and Hsu [1989] and Hewitt and Wainwright [1993], who studied the asymptotic states of orthogonal spatially homogeneous (OSH) models in terms of attractors of the associated dynamical system for class A and class B models, respectively.

- A large class of OSH models (including all class B models) are asymptotically self-similar at the initial singularity and are approximated by exact self-similar power law models (either perfect fluid or vacuum). [E.g., self-similar Kasner vacuum models or self-similar, locally rotationally symmetric (of class III) perfect fluid models of Collins and Stewart [1971] (Bianchi type II– see also Collins [1971] and Doreshkevich et al. [1973])].

However, this behaviour is not generic; general orthogonal models of Bianchi type IX (and VIII) have an oscillatory behaviour with chaotic-like characteristics (and the matter density becomes dynamically negligible) as one follows the evolution into the past towards the initial singularity. Ma and Wainwright [1994] show that the orbits of the associated cosmological dynamical system are negatively asymptotic to a lower ($= 2$) dimensional attractor which is the union of three ellipsoids (in $\mathbb{R}^5$) consisting of the Kasner ring joined by Taub separatrices, where the orbits spend most of the time near the Kasner vacuum equilibrium points. Clearly the self-similar Kasner models play a primary role in the asymptotic behaviour of these models.

- Exact self-similar power law models can also approximate general Bianchi models at intermediate stages of their evolution. [E.g., radiation Bianchi VII$_h$ models (Doreshkevich et al. [1973])]. Of special interest are those models that can be approximated by an isotropic model (e.g., those models whose orbits spend a period of time near to a flat FRW equilibrium point) at an intermediate stage of their evolution.

- Self-similar solutions can describe the behaviour of Bianchi models at late times (i.e., as $t \to \infty$). [E.g., self-similar flat space and self-similar homogeneous vacuum plane waves (see, for example, Collins [1971] and Wainwright [1985])].

All models expand indefinitely except for the Bianchi IX models. The question of which Bianchi models can possibly isotropize was dealt with by Collins and Hawking [1973].

All vacuum models of Bianchi (B) types IV, V, VI$_h$ and (especially) VII$_h$ are asymptotic to plane wave states to the future (type V models tend to the Milne form of flat spacetime –Hewitt and Wainwright [1993]) and typically (and it has been conjectured generically - Hewitt and Wainwright [1993]) non-vacuum models are asymptotic in the future to either plane wave vacuum solutions (see also Doroshkevich et al. [1973] and Siklos [1981]) or non-vacuum Collins type VI$_h$ solutions (Collins [1971]).

Bianchi (A) models of types VII$_o$ (non-vacuum) and VIII expand indefinitely but are found to have oscillatory (but non-chaotic) behaviour in the Weyl curvature (Wainwright [1995]). Bianchi type IX models obey the closed universe recollapse conjecture (Lin and Wald [1989]); the description of these models in terms of expansion-normalized variables is only valid up to the point of maximum expansion (where $\theta = 0$).

4 Spherically Symmetric Similarity Solutions and their Applications

In the case of spherical symmetry, similarity solutions are especially important. In particular, the governing PDEs reduce to ODEs in terms of the similarity variable.

Cosmology

Cahill and Taub [1971] were the first to study such similarity solutions in the cosmological context. In particular, they examined the nature of similarity solutions of the EFEs fitted across the shock front; further study of the solutions at the sonic point has been done by Bogoyavlenski [1985] and Carr and collaborators (see Carr [1995]). This work is partially motivated by the similarity hypothesis (Carr [1995]) which asserts that cosmological solutions may naturally evolve to self-similar forms even if they start out more complicated.

There are two classes of solutions of special cosmological importance.

A. The class of models asymptotic to the exact $k = 0$ FRW model (which is included in the class) (Carr and Yahil [1990]).

The first cosmological solutions studied were those containing black holes. Carr and Hawking [1974] showed that such solutions exist (for radiation and dust – see also Bicknell and Henriksen [1978]) but only if the universe is asymptotically (rather than exactly) FRW; i.e., black holes formed through purely local processes in the early universe cannot grow as fast as the particle horizon. In fact, *all* solutions which can be attached to an exact FRW model via a sound wave (i.e., not just the black hole solutions) are non-physical (as one goes inward from the sound wave they either enter a negative mass regime or reach another sonic point at which the pressure diverges; Carr and Yahil [1990]).

As well as black hole solutions there are perturbations of FRW models; i.e., models which represent density fluctuations (growing at the same rate as the universe) which are asymptotic to a $k = 0$ FRW universe (Carr and Yahil [1990]). For each value of γ ($p = (\gamma - 1)\mu$) there exists a 2-parameter family of solutions, which contain a 1-parameter subfamily of solutions (on a restricted part of phase-space) that are regular at the sonic point. There exists one supersonic solution (asymptotically FRW) attached to each subsonic solution which is regular at the origin (transonic solutions).

There exist both *underdense* and *overdense* (relative to the exact FRW) solutions. The underdense solutions might be relevant in describing large-scale structure; for example, such similarity solutions might describe large scale voids (60 Mpc) or might be of relevance to the formation of bubbles at a cosmological phase-transition (see also Carr and Whinnett [1995]). The significance of the overdense solutions is less clear; perhaps they are relevant in the formation of highly overdense blobs at a cosmic phase transition. Unlike the underdense solutions, acceptable solutions (regular at the sonic point) only occur in successive narrow bands (of very large relative overdensity) in the overdense case. Finally, for each value of γ a static similarity solution can be uniquely matched to an asymptotic FRW model via an expanding sound wave.

B. The class of models asymptotic to a self-similar member of the Kantowski-Sachs family (Carr and Koutras [1992]).

The class of spherically symmetric, spatially homogeneous solutions were first studied by Kantowski and Sachs [1966] in the case $p = 0$ and Collins [1977] for $p = (\gamma - 1)\mu$. Not all such solutions are self-similar; there exists a unique self-similar exact solution for each value of γ.

The physical significance of such models is not very clear; perhaps they are related to the formation of bubbles (at a cosmological phase transition) in the early universe (Wesson [1989]). For each γ, there exists a 2-parameter family of solutions asymptotic to a Kantowski-Sachs solution, but there exist only isolated regular solutions and any such solution which hits the sonic surface is unlikely to be regular there (Carr and Koutras [1992]).

The positive pressure solutions ($1 < \gamma < 2$) are unphysical in that the signature is non-Lorentzian and $m < 0$ (although $\mu > 0$)! The negative pressure models ($\gamma < 1; p < 0$ perhaps due to particle production at a cosmological phase transition) have a Lorenzian signature, $m > 0$ (if $\gamma < \frac{2}{3}$, whereby the strong energy condition is violated) and $\mu > 0$, and are consequently more interesting (Carr and Koutras [1992]); perhaps these models describe the growth of bubbles with $p > 0$ in a background of negative pressure. Generalized Kantowski-Sachs similarity solutions (with $p < 0$ and $p \neq (\gamma - 1)\mu$) with a mixture of false vacuum and dust have been studied by Ponce de Leon [1988] and Wesson [1989].

Spherically symmetric self-similar cosmological models are discussed further in the excellent review by Carr [1995]. The stability of such solutions has been further studied in Carr and Coley [1995].

Gravitational Collapse and Naked Singularities

The cosmic censorship hypothesis asserts, in very general terms, that singularities which develop from regular initial conditions have no causal influence on spacetime (Penrose [1969] and Israel [1984]). Spherically symmetric homothetic spacetimes contain an interesting class of naked singularities, whose "nakedness" is a consequence of their self-similarity (Henriksen and Patel [1991]). Until recently possible counterexamples to the cosmic censorship hypothesis have been restricted to spherically symmetric spacetimes, and typically involve shell crossing, shell focussing or self-similarity (Lake [1992] and Eardley and Smarr [1979]). Indeed, despite the paucity of exact self-similar solutions, they form the bulk of examples in the literature on shell-focussing singularities (Eardley et al. [1986], Zannias [1991] and Lake [1992]). In particular, naked singularities in spherically symmetric, perfect fluid, self-similar collapse have been studied by Ori and Piran [1987] and [1990], Waugh and Lake [1988] and [1989], Lake and Zannias [1990], and Henriksen and Patel [1991] (in addition, see Reference 2 in Lake [1992]). The spherically symmetric collapse of a self-similar massless scalar field has also been studied by Brady [1995].

Recently, studies have shown that naked singularities may also arise in non-spherical collapse, both theoretically (Barrabas et al. [1991]) and numerically (Shapiro and Teukolsky [1992]). In the latter case the axisymmetric collapse of a prolate cluster (of collisionless matter) to a singular "spindle state" was studied.

Critical Phenomena

Recently, in the study of the gravitational collapse of a spherically symmetric massless (minimally coupled) scalar field in general relativity, the existence of critical phenomena has been discovered (Choptuik [1993]). Depending upon initial data parametrized smoothly by a parameter p, the final outcome is a spacetime that contains a black hole ($p > p^*$) or a regular spacetime that does not ($p < p^*$) separated by a critical value p^*. In numerical work the behaviour of the models in the near critical regime ($|p - p^*| < \delta$) close to the threshold of black hole formation was found to be characterized by a number of non-linear phenomena (analogous to critical phenomena observed in statistical systems). These include (Choptuik [1993] and [1994]); (i) exponential sensitivity to initial conditions (as $p \to p^*$), (ii) a type of discrete scale invariance referred to as 'scale-periodicity', and (iii) a universal power-law dependence (with critical exponent) of the black hole mass on $p - p^* (> 0)$. Choptuik [1994] remarked that the scale-periodicity is perhaps the most intriguing feature of the critical behaviour, and noted that the universality (e.g., the independence of the critical exponent on the details of the initial data) suggests that near critical solutions may be described by a single universal solution.

A remarkably similar picture has emerged from the numerical analysis of spherically symmetric scalar field collapse for both a non-minimally coupled scalar field and a self-interacting scalar field (Choptuik [1994]), and in the case of vacuum, axisymmetric gravitational wave collapse (Abrahams and Evans [1993]).

The strong field 'precisely critical' behaviour is described by a unique, critical dynamics which is invariant under a certain discrete scaling. However, obtaining analytical results with a discrete self-similarity is very difficult. Therefore, attempts have been made to explain the phenomena theoretically by studying exact continuous self-similar collapse, and spherically symmetric homothetic spacetimes have been investigated in both the case of a massless scalar field (Brady [1995]) and a perfect fluid (Evans and Coleman [1994]). In addition, there is evidence for a unique self-similar critical solution in that some work has suggested that continuous self-similarity might be a good approximation in the near critical regime (Evans and Coleman [1994]).

In particular, Evans and Coleman [1994] have studied radiation perfect fluid ($p = \mu/3$) spherically symmetric collapse under the assumption of continuous self-similarity. Analogous nonlinear behaviour was again found numerically, and an exact self-similar critical solution was obtained. Perhaps most importantly, however, it was found that as the critical point is asymptotically approached the evolution (of the fluid and gravitational field) develops a self-similar region near the centre of collapse (given by the exact critical solution). Only a precisely critical model (in which the rate of collapse may balance the rate of radiative energy) asymptotically approaches the exact self-similar solution; near but not precisely critical spacetimes develop a self-similar local region, but eventually the self-similarity of the collapse is broken (this is referred to as 'intermediate asymptotics'-Barenblatt and Zeldovich [1972]). Since the self-similar fluid solution is not asymptotically flat, the 'local' self-similar region needs to be matched to an asymptotically flat region in a more sophisticated model. Nonetheless, the self-similar solution is anticipated to represent the asymptotic behaviour within the central region of the collapse.

Finally, in the actual numerical simulations performed no shocks have been observed to have formed. Therefore, the unique critical solutions may be connected with exact self-similar spherically symmetric solutions which are regular at the origin and regular at the sonic point; these models have been studied extensively (see, for example - Bogoyavlenski [1985], Ori and Piran [1990], Carr [1995] and references within) and were discussed earlier in this section.

5 Self-Similarity of the Second Kind

The existence of self-similar solutions of the first kind is related to the conservation laws and to the invariance of the problem with respect to the group of similarity transformations of quantities with independent dimensions, in which case a certain regularity of the limiting process in passing from the original non-self-similar regime to the self-similar regime is implicitly assumed. However, in general such a passage to this limit need not be regular, whence the expressions for the self-similar variables are not determined from dimensional analysis of the problem alone. Solutions are then called *self-similar solutions of the second kind.* Characteristic of these solutions is that they *contain dimensional constants* that are not determined from the conservation laws (but can be found by matching the self-similar solutions with the non-self-similar solutions whose asymptotes they represent) (Barenblatt and Zeldovich [1972]).

Self-similarity in the broadest sense refers to the general situation in which a system is not restricted to be strictly invariant under the relevant group action, but merely to be appropriately rescaled. The basic condition characterizing a manifold vector field $\boldsymbol{\xi}$ as a self-similar generator is that there exist constants d_i such that for each independent physical field Φ^i,

$$\mathcal{L}_\xi \Phi^i = d_i \Phi^i, \tag{26}$$

where for each i, d_i is the scalar product of the dimensionality covector $\vec{d}_i$ of Φ^i (with respect to the rescaling algebra) and some rescaling algebra vector $\vec{a}$ (Carter and Henriksen [1991]).

In the Newtonian case the physical fields consist of μ, p and ϕ (the Newtonian gravitational potential), and $\boldsymbol{\xi}$ is the generator of a self-similarity with respect to a 3-dimensional rescaling algebra vector (time, length and mass). [Here μ, p and ϕ are scalar fields and $\mathcal{L}_\xi$ thus denotes their directional derivative.] The ordinary fluid mechanical equations (i.e., the continuity equation and the Eulerian dynamical evolution equations) are preserved by the action of the 3-parameter rescaling group. Unfortunately, in general the self-similarity will not survive when additional laws governing p (equation of state) and ϕ (Poisson's equation) complete the system of equations (Carter and Henriksen [1991]); in order for the system to remain invariant further restrictions are imposed. The effective invariance of the Poisson equation leads to restricting attention to a 2-dimensional subalgebra. Further restrictions then arise from the imposition of an equation of state; for a barotropic equation of state $p = p(\mu)$, except in the special case of dust ($p = 0$ affords no additional restrictions), the special polytropic case $p = p_0\mu^{\overline{\gamma}}$ (where $\overline{\gamma}$ is the polytropic index) effects reduction to a 1-parameter rescaling subalgebra, resulting in a specific 'self-similar' index. It is through such an index that a formal basis can be provided for classifying self-similarity as first class, as the more general second class, or the less-often studied special zeroth case (cf. Barenblatt and Zeldovich [1972]).

It is through comparisons with the situation in Newtonian fluid mechanics that these ideas have been adapted to the relativistic context (Carter and Henriksen

[1989]). Clearly, the characterizing equations must be generalized to their tensorial counterparts of the covariant form

$$\mathcal{L}_\xi \Phi^i_A = d_i \Phi^i_A, \tag{27}$$

where the fields Φ_A can be scalar (e.g., μ), vectorial (e.g., u_a) or tensorial (e.g., g_{ab}). In particular, in general relativity the gravitational potential ϕ is replaced by the metric tensor g_{ab}, and an appropriate definition of "geometrical" self-similarity is necessary. In the seminal work by Cahill and Taub [1971] the simplest generalization was effected whereby the metric itself satisfies an equation of the form (27), namely $\boldsymbol{\xi}$ is a HV, this evidently corresponding to Zeldovich's similarity of the first kind.

However, in relativity it is not the energy-momentum tensor itself that must satisfy (27), but each of the physical fields making up the energy-momentum tensor must separately satisfy an equation of the form (27). In the case of a fluid characterized by the timelike congruence u_a the energy-momentum tensor can be uniquely decomposed with respect to u_a (Ellis [1971]), and each of these uniquely defined components (each of which has a physical interpretation in terms of the energy, pressure, heat flow and anisotropic stress as measured by an observer comoving with the fluid) must separately satisfy an equation of the form (27). In the same way, if the metric can be uniquely, physically, and covariantly decomposed then the homothetic condition can be replaced by the conditions that each uniquely defined component must satisfy (27), maintaining self-similarity. For example, in the case of a fluid, the metric can be decomposed uniquely in terms of u_a, through the projection tensor

$$h_{ab} = g_{ab} + u_a u_b, \tag{28}$$

into parts h_{ab} and (minus) $u_a u_b$. The projection tensor represents the projection of the metric into the 3-spaces orthogonal to u^a (i.e., into the rest frame of the comoving observers), and if u_a is irrotational these 3-spaces are surface forming, the decomposition is global, and h_{ab} represents the intrinsic metric of these 3-spaces. [h_{ab} is the first fundamental form of the hypersurfaces orthogonal to u^a and can be regarded as the relativistic counterpart of the Newtonian metric tensor, when the flow independent u_a is defined as the relativistic counterpart of the preferred (irrotational) Newtonian time covector $-t_{,u}$ (Carter and Henriksen [1989]).]

Kinematic self-similarity

It is arguments similar to these, and, more importantly, a detailed comparison with self-similarity in a continuous Newtonian medium, that has led Carter and Henriksen [1989] to the covariant notion of *kinematic self-similarity* in the context of relativistic fluid mechanics. A kinematic self-similar vector $\boldsymbol{\xi}$ satisfies the condition

$$\mathcal{L}_\xi u_a = \alpha u_a, \tag{29}$$

where α is a constant (i.e., $\boldsymbol{\xi}$ is a continuous kinematic self-similar generator with respect to the flow u_a). Furthermore,

$$\mathcal{L}_\xi h_{ab} = 2h_{ab} \tag{30}$$

($\boldsymbol{\xi}$ has been normalized so that the constant in (30) has been set to unity). Evidently, in the case $\alpha = 1$ it follows that $\boldsymbol{\xi}$ is a HV (Cahill and Taub [1971]), corresponding to self-similarity of the first kind (Barrenblatt and Zeldovich [1972]). Carter and Henriksen [1989] then argue that the case $\alpha \neq 1$ ($\alpha = 0$) is the natural relativistic counterpart of self-similarity of the more general second kind (zeroth kind).

The parameter α represents the constant relative proportionality factor governing the rates of dilation of spatial length scale and amplification of time scale. Evidently, when $\alpha \neq 1$ (i.e., $\boldsymbol{\xi}$ is not a HV), the relative rescaling of space and time (under $\boldsymbol{\xi}$) are not the same (and in the zeroth case there is a space dilation without any time amplification). The self-similarity index (referred to above) is automatically the same as α (defined in (29)) under the usual normalization ($u^a t_{,a} = 1$ in the Newtonian case).

Remarks

- All of the fields must satisfy equations of the form (27). However, not all of the d_i need be independent. In practice it is convenient to assume independence and then determine the constraints on the d_i (in addition to the constraints on the form of the solutions) that arise from the imposition of the equations of state and the EFEs. In the case of a perfect fluid we have the following:

$$\text{“kinematic” self-similarity}: \quad \mathcal{L}_\xi u_a = \alpha u_a, \tag{29}$$

$$\text{“geometric” self-similarity}: \quad \mathcal{L}_\xi h_{ab} = 2h_{ab}, \tag{30}$$

$$\text{“physical” self-similarity}: \quad \mathcal{L}_\xi \mu = a\mu, \quad \mathcal{L}_\xi p = bp, \tag{31}$$

  where a and b are constants.

- Additional constraints arise from the equation of state. The least restrictive case arises from the exceptional pressure-free case. Here $p = 0$, whence

$$\mathcal{L}_\xi T_{ab} = (2\alpha + a)T_{ab}, \tag{32}$$

  and no further restrictions apply. Otherwise, if, for example, $p = p(\mu)$, then necessarily $p = p_0\mu^{\overline{\gamma}}$, and immediately we see from equations (31) that $b = a\overline{\gamma}$.

- There will be further constraints arising from the compatibility of the EFEs and Equations (29)–(31) (“integrability” conditions). It is curious to note that unlike the inheritance problem, in which the conditions under which the matter fields inherit a geometrical symmetry are sought (Coley and Tupper [1989] and [1990]), and the inverse inheritance problem, in which the conditions on the geometry arising from a symmetry on the matter fields are sought (Carot et al. [1994b]), in the case of self-similarity both the geometry and matter fields obey symmetry properties (Equations (29)–(31)) and the problem becomes a question of the compatibility of these properties and the EFEs.

- However, there are some relationships between these various problems. First, we notice the similarity between Equation (29) and the inheritence condition (Coley and Tupper [1989] and [1990]), and also note that in the case of kinematic self-similarity the matter necessarily ‘inherits’ the symmetry [see, for example, Equation (31)]. Second, in the case of dust, for example, the energy-momentum tensor satisfies the condition (32); this condition generalizes the definition of a matter symmetry or matter collineation. However, in the case of self-similar dust (in which the energy-momentum tensor is degenerate and of rank 1), in which μ satisfies Equation (31a) and u_a satisfies Equation (29) (and $\boldsymbol{\xi}$ is a HV if and only if $\alpha = 1$), whence T_{ab} satisfies (32), it immediately follows that

$$\mathcal{L}_\xi \phi_a = 0, \tag{33}$$

where ϕ_a is the 1-form defined by $\phi_a = \mu^{-\alpha/a} u_a$, and the results of Carot et al. [1994b] can be applied.

Examples

Let us consider kinematic self-similarity in the case of spherical symmetry. In the *spherically symmetric* case Carter and Henriksen [1989] have shown that there exist comoving coordinates in which the self-similar generator is given by

$$\xi^a \frac{\partial}{\partial x^a} = (\alpha t + \beta)\frac{\partial}{\partial t} + r\frac{\partial}{\partial r}, \tag{34}$$

and the metric is given by

$$ds^2 = -e^{2\phi} dt^2 + e^{2\psi} dr^2 + r^2 S^2 d\Omega^2, \tag{35}$$

where $d\Omega^2 \equiv d\theta_1^2 + sin^2\theta_1 d\theta_2^2$, and ϕ, ψ and S depend only on the self-similarity coordinate ξ. The metric (35) is consequently manifestly of the same form as in Cahill and Taub [1971], and the resulting governing differential equations do indeed reduce to a system of ODEs.

In the case of self-similarity of the first kind, i.e., the homothetic case, $\alpha = 1$ (and β can be rescaled to zero) and $\xi = r/t$, as usual (Cahill and Taub [1971]).

In the lesser-studied zeroth case, $\alpha = 0$ (and β can be rescaled to unity) and $\xi = re^{-t}$. An example of this case is provided by the solution of Henriksen et al. [1983] in which a dimensional constant (and hence a fundamental scale) is introduced via the cosmological constant (see also Carter and Henriksen [1989]).

In the general case $\alpha \neq 0, 1$ (β rescaled to zero), corresponding to *self-similarity of the second kind*, the self-similarity coordinate is given by

$$\xi = r(\alpha t)^{-1/\alpha}. \tag{36}$$

An important example of the second-kind self-similarities is provided by a class of zero-pressure perfect fluid models (i.e., dust models in which u^a is geodesic; i.e., $\phi' = 0$ in Equation (35)) first studied by Lynden-Bell and Lemos [1988] and described in detail by Henriksen [1989] and Carter and Henriksen [1989]. This class of models has recently been generalized to the non-zero-pressure case in which the pressure contains a physical dimensional constant (Benoit and Coley [1996]).

Partial homothety

Similarity, or invariance under scale transformations, implies the purely multiplicative rescaling of certain physical quantities, whereby it is the relative (and not the absolute) values of the quantities that are relevant in the physical system under study. We note that in all cases of self-similarity there exist coordinates in which all dimensionless quantities in the theory can be put into a form in which they are functions of dimensionless independent variables. In the case of spherical symmetry all dimensionless quantities are functions of a single independent variable ξ, which is a dimensionless combination of space coordinates and time. In the absence of any physical dimensional scale this reduces (through the use of c) to $\xi = r/t$ (similarity of the "first" kind). However, in the more general case ξ can be a more complicated expression.

In an alternative, and possibly more general, approach Ponce de Leon [1993] has studied spherically symmetric fluid models admitting a so-called *partial homothety* (or intrinsic symmetry, Collins [1979]) $\boldsymbol{\zeta}$, which is simply defined by

$$\mathcal{L}_\zeta h_{ab} = 2h_{ab}. \tag{37}$$

There exist coordinates in which the partial homothety is of the form

$$\zeta^a = (F/F_t, r, 0, 0)$$

and the similarity variable is given by

$$\zeta = r/F(t), \tag{38}$$

where $F(t)$ is an arbitrary function. Ponce de Leon [1993] showed that partially homothetic spherical distributions are manifestly self-similar (i.e., the metric is given by (35) with the similarity variable ζ given by (38)) and that, necessarily,

$$\mathcal{L}_\zeta u^a = -\psi u^a, \tag{39}$$

where

$$\psi \equiv -u_a \mathcal{L}_\zeta u^a = 1 - \frac{F F_{tt}}{F_t^2}. \tag{40}$$

Evidentally, partial homotheties contain kinematic self-similarities as a special (restricted) case (compare Equations (29) and (30) with (37) and (39)) when the function $F(t)$ is of a restricted form (compare, for example, Equation (36) with (38)) and hence ψ in (40) is constant. In particular, $\boldsymbol{\zeta}$ is a (full) HV when $F = F_0 t$ (F_0 constant) and hence $\psi = 1$; when $F \neq F_0 t$, $\boldsymbol{\zeta}$ is a generalized self-similarity. Examples of spherically symmetric self-similar fluid models with ψ not constant have been given in Ponce de Leon [1988] and [1991]; in particular, in the family of homogeneous Kantowski-Sachs models found by Ponce de Leon [1988] an intrinsic time-scale is introduced due to the fact that the models recollapse.

Ponce de Leon [1993] has shown the existence of a relationship between partial homotheties and self-similarity only in the special case of spherical symmetry in which the matter is represented by a fluid with viscosity (satisfying the first order equations of state of Eckart [1940]) but with no heat conduction. Potential problems in this approach include the fact that in the shear-free case self-similar distributions are not necessarily partially homothetic (Ponce de Leon [1993]) and the fact that the matter quantities occuring in the energy-momentum tensor do not inherit the self-similar symmetry associated with the geometry (Ponce de Leon [1988] and [1991]).

References

Abrahams, A. M. and Evans, C. R. [1993] *Critical behavior and scaling in vacuum axisymmetrical gravitational collapse,* Phys. Rev. Lett. **70**, 2980–2983.

Alexander, D., Green, R. M. and Emslie, A. G. [1989] *Spatially inhomogeneous spacetimes with a self-symmetry of the 2nd kind,* MNRAS **237**, 93–107.

Barenblatt, G. I. and Zeldovich, Ya. B. [1972] *Self-similar solutions as intermediate asymptotics,* Ann. Rev. Fluid Mech. **4**, 285–312.

Barrabes, C., Israel, W. and Letelier, P. S. [1991] *Analytic models of nonspherical collapse, cosmic censorship and the hoop conjecture,* Phys. Lett. A **160**, 41–44.

Benoit, P. M. and Coley, A. A. [1996] *Spherically symmetric spacetimes and kinematic self-similarity,* preprint.

Bertschinger, E. [1985] *Cosmological self-similar shock waves and galaxy formation,* Ap. J. **268**, 17–29.

Bicknell, G. V. and Henriksen, R. N. [1978a] *Self-similar growth of primordial black holes. I. Stiff equations of state,* Ap. J. **219**, 1043–1057.

——— [1978b] *Self-similar growth of primordial black holes. II. General sound speed,* Ap. J. **225**, 237–251.

Bogoyavlenski, O. I. [1985] *Methods in the Qualitative Theory of Dynamical Systems in Astrophysics and Gas Dynamics*, Springer-Verlag, New York.

Bogoyavlenski, O. I. and Novikov, S. P. [1973] *Singularities of the cosmological model of the Bianchi IX type according to the qualitative theory of differential equations,* Sov. Phys.-JETP **37**, 747–755.

Brady, P. R. [1995] *Self-similar scalar field collapse: Naked singularities and critical behavior,* Phys. Rev. D **51**, 4168–4176.

Bruni, M., Matarrase, S. and Pantano, O. [1995] *Dynamics of silent universe* Ap. J. **445**, 958–977.

Burd, A. B. and Coley, A. A. [1994] *Viscous fluid cosmology* Class. Quantum Grav. **11**, 83–105.

Cahill, A. H. and Taub, M. E. [1971] *Spherically symmetric similarity solutions of the Einstein field equations for a perfect fluid,* Comm. Math. Phys. **21**, 1–40.

Carot, J., Mas, L. and Sintes, A. M. [1994a] *Space-times admitting a 3-parameter similarity group,* J. Math. Phys. **35**, 3560–3570.

Carot, J., da Costa, J. and Vaz, E. G. L. R. [1994b] *Matter collineations - The inverse symmetry inheritance problem,* J. Math. Phys. **35**, 4832–4838.

Carr, B. J. [1995] *Spherically symmetric similarity cosmology,* preprint.

Carr, B. J. and Coley, A. A. [1995] *Self-similar spherically symmetric cosmological models and their properties,* preprint.

Carr, B. J. and Hawking, S. W. [1974] *Black holes in the early universe,* MNRAS **168**, 399–415.

Carr, B. J. and Koutras, A. [1992] *Self-similar perturbations of a Kantowski-Sachs model,* Ap. J. **405**, 34–50.

Carr, B. J. and Whinnett, A. [1995] *Self-similar voids,* preprint.

Carr, B. J. and Yahil, A. [1990] *Self-similar perturbations of a Friedmann universe,* Ap. J. **360**, 330–342.

Carter, B. and Henriksen, R. N. [1989] *A covariant characterization of kinematic self-similarity,* Ann. Physique Supp. **14**, 47–53.

——— [1991] *A systematic approach to self-similarity in Newtonian space-time,* J. Math. Phys. **32**, 2580–2597.

Chao, W. Z. [1981] *Self-similar cosmological models,* Gen. Rel. Grav. **13**, 625–647.

Choptuik, M. W. [1993] *Universality and scaling in gravitational collapse of a massless scalar field,* Phys. Rev. Lett. **70**, 9–12.

——— [1994], Deterministic Chaos in General Relativity (Hobill, D. et al., eds.), Plenum, New York.

Coley, A. A. and van den Hoogen, R. J. [1994a] *Qualitative analysis of diagonal Bianchi type-V imperfect fluid cosmological models,* J. Math. Phys. **35**, 4117–4144.

——— [1994b] *Self-similar cosmology*, Deterministic Chaos in General Relativity (Hobill, D. et al., eds.), Plenum, New York.

——— [1995] *Qualitative analysis of viscous fluid cosmological models satisfying the Israel-Stewart theory of irreversible thermodynamics,* Class. Quantum Grav. **12**, 2335–2354.

Coley, A. A., van den Hoogen, R. J. and Maartens, R. [1995] *Qualitative viscous cosmology,* Phys. Rev. D, to appear.

Coley, A. A. and Tupper, B. O. J. [1983] *Zero-curvature Friedman-Robertson-Walker models as exact viscous magnohydrodynamic cosmologies,* Ap. J. **271**, 1–8.

——— [1989] *Special conformal Killing vector space-times and symmetry inheritance,* J. Math. Phys. **30**, 2616–2625.

——— [1990] *Spacetimes admitting inheriting conformal Killing vector-fields,* Class. Quantum Grav. **7**, 1961–1981.

Coley, A. A. and Wainwright, J. [1992] *Qualitative-analysis of 2-fluid Bianchi cosmologies,* Class. Quantum Grav. **9**, 651–665.

Collins, C. B. [1971] *More qualitative cosmology,* Comm. Math. Phys. **23**, 137–158.

——— [1974] *Tilting at cosmological singularities,* Comm. Math. Phys. **39**, 131–151.

——— [1977] *Global structure of "Kantowski-Sachs" cosmological model,* J. Math. Phys. **18**, 2116–2124.

——— [1979] *Intrinsic symmetries in general relativity,* Gen. Rel. Grav. **10**, 925–929.

Collins, C. B. and Ellis, G. F. R. [1979] *Singularities in Bianchi cosmologies,* Phys. Rep. **56**, 65–105.

Collins, C. B. and Hawking, S. W. [1973] *Why is the universe isotropic?,* Ap. J. **180**, 317–334.

Collins, C. B. and Lang, J. M. [1987] *A class of self-similar spacetimes and a generalization,* Class. Quantum Grav. **4**, 61–78.

Collins, C. B. and Stewart, J. M. [1971] *Qualitative cosmology,* MNRAS **153**, 419–434.

Defrise-Carter, L. [1975] *Conformal groups and conformally equivalent isometry groups,* Comm. Math. Phys. **40**, 273–282.

Doroshkevich, A. G., Lukash, V. N. and Novikov, I. D. [1973] *The isotropization of homogeneous cosmological models,* Sov. Phys.-JETP **37**, 739–746.

Eardley, D. M. [1974] *Self-similar spacetimes: Geometry and dynamics,* Comm. Math. Phys. **37**, 287–309.

Eardley, D. M., Isenberg, J., Marsden, J. and Moncrief, V. [1986] *Homothetic and conformal symmetries of solutions to Einstein's equations,* Comm. Math. Phys. **106**, 137–158.

Eardley, D. M. and Smarr, L. [1979] *Time functions in numerical relativity: Marginally bound dust collapse,* Phys. Rev. D **19**, 2239–2259.

Eckart, C. [1940] *The thermodynamics of irreversible processes, III: Relativistic theory of the simple fluid,* Phys. Rev. **58**, 919–924.

Ellis, G. F. R. [1971] *Relativistic Cosmology,* General Relativity and Cosmology, XLVII Corso, Varenna, Italy (1969) (Sachs, R., ed.), Academic, New York.

Ellis, G. F. R. and MacCallum, M. A. H. [1969] *A class of homogeneous cosmological models,* Comm. Math. Phys. **12**, 108–141.

Evans, C. R. and Coleman, J. S. [1994] *Critical phenomena and self-similarity in the gravitational collapse of radiation fluid,* Phys. Rev. Lett. **72**, 1782–1785.

Godfrey, B. B. [1972] *Horizons in Weyl metrics exhibitting extra symmetries,* Gen. Rel. Grav. **3**, 3–16.

Hall, G. S., Low, D. J. and Pulham, J. R. [1994] *Affine collineations in general-relativity and their fixed-point structure,* J. Math. Phys. **35**, 5930–5944.

Henriksen, R. N. [1989] *Spherically symmetric, self gravitating dust is always 'self-similar',* MNRAS **240**, 917–929.

Henriksen, R. N., Emslie, A. G. and Wesson, P. S. [1983] *Space-times with constant vacuum energy density and a conformal Killing vector,* Phys. Rev. D **27**, 1219–1227.

Henriksen, R. N. and Patel, K. [1991] *Null charts and naked singularities in spherically symmetric, homothetic spacetimes,* Gen. Rel. Grav. **23**, 527–581.

Henriksen, R. N. and Wesson, P. S. [1978a] *Self-similar spacetimes I: Three solutions,* Ap. Sp. Sci. **53**, 429–444.

——— [1978b] *Self-similar spacetimes II: Perturbation scheme,* Ap. Sp. Sci. **53**, 445–457.

Hewitt, C. G. and Wainwright, J. [1990] *Orthogonally transitive G_2 cosmologies,* Class. Quantum Grav. **7**, 2295–2316.

——— [1992] *Dynamic-systems approach to tilted Bianchi cosmologies - Irrotational models of type V,* Phys. Rev. D **46**, 4242–4252.

——— [1993] *A dynamic-systems approach to Bianchi cosmologies - Orthogonal models of class B,* Class. Quantum Grav. **10**, 99–124.

Hewitt, C. G., Wainwright, J. and Glaum, M. [1991] *Qualitative analysis of a class of inhomogeneous self-similar cosmological models: II,* Class. Quantum Grav. **8**, 1505–1518.

Hewitt, C. G., Wainwright, J. and Goode, S. W. [1988] *Qualitative analysis of a class of inhomogeneous self-similar cosmological models,* Class. Quantum Grav. **5**, 1313–1327.

Hsu, L. and Wainwright, J. [1986] *Self-similar spatially homogeneous cosmologies: Orthogonal perfect fluid and vacuum solutions,* Class. Quantum Grav. **3**, 1105–1124.

Ibañez, J., van den Hoogen, R. J. and Coley, A. A. [1995] *Isotropization of scalar field Bianchi models with an exponential potential,* Phys. Rev. D **51**, 928–930.

Ikeuchi, S., Tomisaka, K. and Ostriker, J. P. [1983] *The structure and expansion law of a shock wave in an expanding universe,* Ap. J. **265**, 583–596.

Israel, W. [1984] *Does a cosmic censor exist?,* Found. Phys. **14**, 1049–1059.

Jacobs, K. C. [1968] *Spatially homogeneous and Euclidean cosmological models with shear,* Ap. J. **153**, 661–678.

Jantzen, R. T. [1984] *Cosmology of the Early Universe* (Ruffini, R. and Fang, L., eds.), World Scientific, Singapore.

Jantzen, R. T. and Rosquist, K. [1986] *Exact power law metrics in cosmology,* Class. Quantum Grav. **3**, 281–309.

Kantowski, R. and Sachs, R. [1966] *Some spatially homogeneous anisotropic relativistic cosmological models,* J. Math. Phys. **7**, 443–446.

Koutras, A. [1992] *Ph.D. Thesis*, Queen Mary and Westfield College.

Kramer, D., Stephani, H., MacCallum, M. and Herlt, E. [1980] *Exact Solutions of Einstein's Field Equations*, Cambridge University Press, Cambridge.

Lake, K. [1992] *Precursor singularities in spherical gravitational collapse,* Phys. Rev. Lett. **68**, 3129–3132.

Lake, K. and Zannias, T. [1990] *Naked singularities in self-similar gravitational collapse,* Phys. Rev. D **41**, 3866–3889.

Lin, D. N. C., Carr, B. J. and Fall, S. D. M. [1978] *The growth of primordial black holes in a universe with a stiff equation of state,* MNRAS **177**, 51–64.

Lin, X. and Wald, R. M. [1989] *Proof of the closed-universe-recollapse conjecture for the diagonal Bianchi Type-IX cosmologies,* Phys. Rev. D **40**, 3280–3286.

Luminet, J. P. [1978] *Spatially homothetic cosmological models,* Gen. Rel. Grav. **9**, 673–685.

Lynden-Bell, D. and Lemos, J. P. S. [1988] *On Penston's self-similar solution for cold collapse,* MNRAS **233**, 197–208.

Ma, P. K.-H. and Wainwright, J. [1994] *Deterministic Chaos in General Relativity* (Hobill, D. et al., eds.), Plenum, New York.

Maartens, R. and Maharaj, S. D. [1986] *Conformal Killing vectors in Robertson-Walker spacetimes,* Class. Quantum Grav. **3**, 1005–1011.

McIntosh, C. B. G. [1975] *Homothetic motions in general relativity,* Gen. Rel. Grav. **7**, 199–213.

Miller, B. D. [1976] *Negative-mass lagging cores of the Big Bang,* Ap. J. **208**, 275–285.

Ori, A. and Piran, T. [1987] *Naked singularities in self-similar spherical gravitational collapse,* Phys. Rev. Lett. **59**, 2137–2140.

______ [1990] *Naked singularities and other features of self-similar general-relativistic gravitational collapse* Phys. Rev. D **42**, 1068–1090.

Penrose, R. [1969] *Gravitational collapse: The role of general relativity,* Nuovo Cim. **1**, 252–276.

Ponce de Leon, J. [1988] *Solutions of Einstein's equations relevant to the description of "bubbles" in the early universe,* J. Math. Phys. **29**, 2479–2485.

______ [1990] *Inflation in a spatially closed anisotropic universe,* J. Math. Phys. **31**, 371–377.

______ [1991] *Self-similar symmetry in Tolman models,* MNRAS **250**, 69–74.

______ [1993] *Self-similar symmetry in general relativity,* Gen. Rel. Grav. **25**, 865–880.

Rosquist, K. [1984] *Regularised field equations for Bianchi type VI spatially homogeneous cosmology,* Class. Quantum Grav. **1**, 81–93.

Rosquist, K. and Jantzen, R. T. [1988] *Unified regularization of Bianchi cosmology,* Phys. Rep. **166**, 89–124.

Rosquist, K., Uggla, C. and Jantzen, R. T. [1990] *Extended dynamics and symmetries in perfect fluid cosmologies,* Class. Quantum Grav. **7**, 625–637.

Schwartz, J., Ostriker, J. P. and Yahil, A. [1975] *Explosive events in the early universe,* Ap. J. **202**, 1–6.

Sedov, L. I. [1967] *Similarity and Dimensional Methods in Mechanics*, Academic Press, New York.

Shapiro, S. L. and Teukolsky, S. A. [1992] *Gravitational collapse of rotating spheroids and the formation of naked singularities,* Phys. Rev. D **45**, 2006–2012.

Siklos, S. T. C. [1981] *Some Einstein spaces and their global properties,* J. Phys. A **14**, 395–409.

Taub, A. H. [1972] *General Relativity, papers in honour of J. L. Synge* (O'Raifeartaigh, L., ed.), Oxford University Press, London.

Taylor, G. I. [1950] *The formation of a blast wave by a very intense explosion. II. The atomic explosion of 1945,* Proc. Roy. Soc. London **A201**, 175–186.

Uggla, C. [1992] *Inhomogeneous self-similar cosmological models,* Class. Quantum Grav. **9**, 2287–2295.

Uggla, C., Jantzen, R. T. and Rosquist, K. [1995] *Exact hypersurface-homogeneous solutions in cosmology and astrophysics,* Phys. Rev. D **51**, 5522–5557.

van den Hoogen, R. J. [1995] *Ph.D. Thesis*, Dalhousie University.

Wainwright, J. [1985] *Galaxies, Axisymmetric Systems and Relativity* (MacCallum, M., ed.), Cambridge University Press, Cambridge.

——— [1995] private communications.

Wainwright, J. and Hsu, L. [1989] *A dynamical-systems approach to Bianchi cosmologies - Orthogonal models of class-A,* Class. Quantum Grav. **6**, 1409–1431.

Wainwright, J., Ince, W. and Marshman, B. [1979] *Spatially homogeneous and inhomogeneous cosmologies with equation of state $p = \mu$*, Gen. Rel. Grav. **10**, 259–271.

Wainwright, J. and Yaremovicz, P. E. A. [1975] *Killing vector fields and the Einstein-Maxwell field equations with perfect fluid source,* Gen. Rel. Grav. **7**, 345–359.

Wald, R. M. [1983] *Asymptotic behavior of homogeneous cosmological models in the presence of positive cosmological constant,* Phys. Rev. D **28**, 2118–2120.

Waugh, B. and Lake, K. [1988] *Strengths of shell-focusing singularities in marginally bound collapsing self-similar Tolman spacetimes,* Phys. Rev. D **38**, 1315–1316.

——— [1989] *Shell-focusing singularities in spherically symmetric self-similar spacetimes,* Phys. Rev. D **40**, 2137–2139.

Wesson, P. S. [1979] *Observable relations in an inhomogeneous self-similar cosmology,* Ap. J. **228**, 647–663.

——— [1981] *Improved standard cosmology,* Phys. Rev. D **23**, 2137–2141.

——— [1982] *Improved standard cosmology: Comparison with observation,* Ap. J. **259**, 20–24.

——— [1989] *A class of solutions in general relativity of interest for cosmology and astrophysics,* Ap. J. **336**, 58–60.

Yano, K. [1955] *The Theory of Lie Derivatives*, North-Holland, Amsterdam.

Zannias, T. [1991] *Initial value constraints and generation of isometries and homothetic motions,* Phys. Rev. D **44**, 2397–2408.

Zeldovich, Ya. B. and Kompaneets, A. S. [1950] *Collection Dedicated to Joffe* (Lukirsky, P. I., ed.), Izd. Akad. Nauk SSSR, Moscow **61**.

Zeldovich, Ya. B. and Raizer, Yu. P. [1963] *Physics of Shock Waves and High Temperature Phenomena*, Academic, New York.

Fields Institute Communications
Volume **15**, 1997

Field Theory on Spacetimes that are not Globally Hyperbolic

John L. Friedman
Department of Physics
University of Wisconsin
Milwaukee, Wisconsin USA 53201
friedman@thales.phys.uwm.edu

Abstract. The classical initial value problem makes sense for a class of spacetimes broader than the globally hyperbolic geometries for which existence and uniqueness theorems are traditionally proved. These include some spacetimes with closed timelike curves (CTCs) as well as Lorentzian universes-from-nothing, spacetimes with a single spacelike boundary that nevertheless have a smooth Lorentzian metric. A set of simple spacetimes that are not globally hyperbolic are reviewed. The initial value problem for classical fields is discussed for the sample spacetimes and results and conjectures are presented for more general spacetimes.

Constructing a quantum field theory is more problematic. In the CTC spacetimes, a path integral approach to scattering leads to nonunitary evolution of interacting fields, although alternative evolutions that evade the problem have been proposed. In the time-nonorientable spacetimes, one can define a family of local algebras on an atlas of globally hyperbolic subspacetimes, with overlap conditions on the intersections of neighborhoods. But the extension of a generic positive linear functional from a single algebra to the collection of all local algebras violates positivity: One cannot find a collection of quantum states satisfying the physically appropriate overlap conditions. It is possible to overcome the difficulty by artificially restricting the size of neighborhoods in a way that has no classical counterpart. The resulting "theory", however, allows less information than quantum field theory in a globally hyperbolic spacetime, because there are always sets of points in the spacetime for which no correlation function is defined.

1 Expectation of Spacetimes that are not Globally Hyperbolic

If the notion of spacetime — of manifold and Lorentz metric — retains its meaning at scales where quantum fluctuations of the metric are of order unity, closed timelike curves (CTCs) seem likely to pervade a microscopic quantum geometry. In a sum over all Lorentzian histories, almost all histories will have microscopic

1991 *Mathematics Subject Classification.* Primary 83C05; Secondary 58G16.
This work was supported in part by NSF Grant No. PHY 95-07740

© 1997 American Mathematical Society

CTC's, and a similar sea of microscopic topological structures would seem to lead to a loss of space and time orientability. One apparently avoids microscopic causal pathology only by adopting a theory in which causal structure is fundamental or one in which the metric is not meaningful or not Lorentzian at small scales.

Black-hole evaporation provides an additional motivation for considering CTC's, if one favors a scenario in which the collapsing matter re-expands to form a child universe. (The older term "child universe" is used here instead of "baby universe" to refer to a Lorentzian spacetime that retains information from its parent.) At least in the semiclassical theory, once a black-hole forms, information is lost to external observers. If one does not know the initial state, arbitrarily accurate measurements in the exterior spacetime cannot recover the state within a horizon. But if collapsing matter forms a child universe, information may not be lost in the full spacetime. Although the usual picture of a single S^3 bifurcating to two S^3's allows no Lorentzian metric, one can construct child universes with topologies that do allow Lorentzian metrics; one such topology is obtained by removing 3 balls from $\mathbb{C}P^2$. With a nondegenerate Lorentzian metric, the price of this topology change is the existence of closed timelike curves (Geroch [1967]).

A quite different class of spacetimes that are not globally hyperbolic are the *Lorentzian universes-from-nothing.* In the Hartle-Hawking proposal one works in a Euclidean framework and obtains a wave function of the universe by summing Euclidean geometries with a single boundary. In a Lorentzian path-integral approach to quantum gravity, one can, as in the Euclidean case, imagine constructing a wave function of the universe as a sum over all Lorentzian 4-geometries with a single spacelike boundary. There is a large class of such Lorentzian spacetimes, compact on one side of their spacelike boundary, and which nevertheless have a smooth Lorentzian metric. They provide the only examples of topology change in which one has a smooth, nondegenerate Lorentzian metric without closed timelike curves. Instead, the spacetimes are time nonorientable.

2 Sample Spacetimes

It is useful in thinking about field theory without global hyperbolicity to have a collection of sample spacetimes on which the initial value problem for classical fields is easily understood. Several of these are are two-dimensional and flat, obtained from pieces of Minkowski space by identifications. A list of spacetimes is presented in this section; then, in Section 3, a brief discussion is given of the initial value problem on each, with reference to the recent literature.

2.1 Sample Lorentzian universes-from-nothing The simplest four-dimensional Lorentzian universes-from-nothing (Friedman and Higuchi [1995]) have the topology of a finite timelike cylinder, $S^3 \times \mathbb{R}^+$, with diametrically opposite points of its past spherical boundary identified. This is the topology of antipodally identified deSitter space. Its 2-dimensional analog is the Möbius strip, which can be constructed from a finite 2-dimensional timelike cylinder $S^1 \times \mathbb{R}^+$ by identifying diametrically opposite points of its circular past boundary, $\overline{\Sigma}$. The median circle Σ of the Möbius strip results from identifying points of $\overline{\Sigma}$. The orientable double-covering space of the strip is a cylinder $\overline{M}$ of double the timelike length, and M is constructed from $\overline{M}$ by identifying antipodal points. If the covering space has the metric of a flat, timelike cylinder, the Möbius strip M will be time nonorientable with a locally flat metric and a timelike Killing vector (defined globally only up to sign) perpendicular to Σ. If the covering spacetime $\overline{M}$ is given the 2-dimensional deSitter metric, the Möbius strip will acquire the metric of antipodally identified deSitter space.

Comparably simple two-dimensional spacetimes with CTC's can be obtained from strips of two-dimensional Minkowski space lying between two parallel timelike, spacelike or null lines. A static closed universe results from identifying the edges of the strip of Minkowski space between $x = 0$ and $x = d$,

$$(t, x = 0) \equiv (t, x = d). \tag{2.1}$$

One can think of identifying points on right and left walls of a room at the same time. CTC's arise if one identifies the right wall at one time with the left wall at a later time, or if the walls are closing in on each other. In the latter case, the walls will be identified with a time-dependent time-shift, relative to a choice of Minkowski time.

2.2 Spacelike and timelike cylinders Flat cylinders with infinite extent in a timelike direction are obtained by identifying the left and right edges of the strip of Minkowski space between $x = 0$ and $x = d$, after a time translation by τ:

$$(t, x = 0) \equiv (t + \tau, x = d). \tag{2.2}$$

For $\tau > d$, lines joining identified points are CTC's. These spacetimes are everywhere nonchronal: a CTC passes through every point.

Turning these cylinders on their sides, one obtains a family of spacetimes that are infinite in a spacelike direction. That is, exchanging the roles of d and τ, one identifies the top and bottom edges of the strip between $t = 0$ and $t = \tau$ after a spatial translation by d:

$$(t = 0, x) \equiv (t = \tau, x + d). \tag{2.3}$$

Again for $\tau > d$ the spacetimes are everywhere nonchronal.

2.3 Misner space From identifications at the boundary of a flat strip bounded by two parallel null lines, one obtains Misner space (Misner [1967], Hawking and Ellis [1973] and Thorne [1994]), a geometry in which the nonchronal region $\mathcal{A}$ is only a part of spacetime. To the past of $\mathcal{A}$ there are no CTC's. This past part of Misner space can obtained from a 1-dimensional room whose walls are moving toward each other – by identifying left and right walls at the same proper time read by clocks on each wall (see, e.g., Thorne [1994]).

To construct the full space, let $u = t - x$, $v = t + x$, and consider the null strip $Bu_0 < u < u_0$, where $u_0 < 0$, and a boost $\mathcal{B}$ of Minkowski space corresponding to velocity $V > 0$ is given by

$$u \to Bu, \qquad v \to B^{-1}v, \tag{2.4}$$

with

$$B = \sqrt{\frac{1+V}{1-V}}. \tag{2.5}$$

Points at the boundaries of the strip are identified after a boost:

$$(u_0, v) \equiv (Bu_0, B^{-1}v). \tag{2.6}$$

Identified points are spacelike separated for $v < 0$, null separated at $v = 0$ and timelike separated for $v > 0$. Closed timelike curves thus pass though each point of the region $v > 0$.

2.4 Geroch-Horowitz-Politzer spaces The examples so far have been smooth and flat. Another class of simple spacetimes with CTC's and a single conical singularity is obtained by removing two slits from Minkowski space and identifying the four line segments that form the resulting boundary. Consider, for example, two-dimensional Minkowski space with two parallel timelike or spacelike segments L_I and L_{II} of equal length removed, and each side of each segment identified with a side of the other segment after translation by a timelike vector.

The resulting geometries are flat with two conical singularities, corresponding to the removed endpoints of the segments. Similar spacetimes have been discussed by Geroch and Horowitz [1979] and Politzer [1992]. Some identified points are timelike related, and timelike curves in Minkowski space joining identified points of the boundary segments become CTCs in the spacetime $\mathcal{N}$. In each spacetime $\mathcal{N}$ there is a spacelike hypersurface $\mathcal{M}$ to the past of the nonchronal region on which initial data can be set.

The final two classes of spacetimes considered here are in three and four spacetime dimensions, respectively.

2.5 Gott spacetimes In three dimensions at least some of the "string" spacetimes introduced by Gott [1991] have a tractable initial value problem. These are spacetimes that are flat everywhere except on two lines corresponding to conical singularites of the metric. A three-dimensional space with a single such conical singularity can be obtained from flat Minkowski space with polar coordinates $\{t, r, \phi\}$ by removing a wedge $0 < \phi < \alpha$ and identifying corresponding points along the bounding timelike faces:

$$(t, r, \phi = 0) \equiv (t, r, \phi = \alpha). \tag{2.7}$$

To construct a Gott spacetime, one removes two relatively boosted wedges and for each wedge identifies points as indicated by Equation (2.7), in the the rest frame of that wedge.

The simplest example (Boulware [1992]) has two conical singularities of deficit angle π. It can be obtained from Minkowski space by removing two wedges of angle π, leaving a slab between two parallel timelike faces. In a cartesian chart t, x, y, one can take the slab to be the region $0 \le y \le L$. Identifications will produce conical singularities on each face along the lines $x = Vt, y = L$, $x = -Vt, y = 0$. They are the boosted versions of identifications $(t,\ x) \equiv (t,\ -x)$ on each face; in terms of null coordinates $u = t - x$, $v = t + x$, the boosted identifications have the form

$$(u,\ v,\ y = 0) \equiv (B^{-2}v,\ B^2u,\ y = 0) \tag{2.8}$$

$$(u,\ v,\ y = L) \equiv (B^2v,\ B^{-2}u,\ y = L), \tag{2.9}$$

with B given in terms of V by Equation (2.5).

The causal structure of Gott spacetimes with deficit angles smaller than π have been analyzed by Cutler [1992], who shows that a spacelike hypersurface Σ extends to spatial infinity and lies to the past of the nonchronal region. The Cauchy horizon is not compactly generated, and CTC's run to spatial infinity. (A Cauchy horizon is compactly generated if there is a compact set A such that every past directed null generator eventually enters and remains in A.) These characteristics hold for the particular Gott spacetime introduced here, but an additional key feature is that its covering space is three-dimensional Minkowski space (with images of the string singularities removed)(Boulware [1992] and Carinhas [1993] and [1995]).

2.6 Wormhole (time-tunnel) spacetimes The low dimensionality of the previous examples and the conical singularities of some of them add pathologies that need not be present in four dimensions. Among geometries with CTC's, candidates for the least pathological spacetimes are smooth and asymptotically flat, with compact nonchronal regions. The lower dimensional examples lack at least one of these features, and the simplest examples of these least-pathological geometries are still the wormhole spacetimes (Morris and Thorne [1988], Morris et al. [1988], Friedman et al. [1990] and Friedman and Morris [1991] and [1995]). The manifold has topology $\mathcal{N} = M \times \mathbb{R}$, where M is a plane with a handle (wormhole) attached: $M = \mathbb{R}^3 \# (S^2 \times S^1)$. Equivalently, $\mathcal{N}$ is obtained from $\mathbb{R}^4$ by removing two solid rods, copies of $B^3 \times \mathbb{R}$, and identifying their cylindrical ($S^3 \times \mathbb{R}$) boundaries.

Two simple examples are obtained from flat space by removing identical solid cylinders. Call spacetime (i) the result of removing identical cylinders whose axes are parallel timelike lines and identifying points of the cylindrical boundaries by a time translation. The spacetime is static, with timelike Killing vector t^α, and the time-translation is defined by means of the family of surfaces orthogonal to t^α. Here the CTC's persist forever.

To construct spacetime (ii) one removes two cylinders that are related to one another by a boost by a velocity V as well as by a spatial translation, and the boost + translation serves to identify the boundaries of the cylinders. The wormhole mouths are moving relative to one another, in a geometry that is static outside disjoint neighborhoods of the wormhole mouth. There are then future and past Cauchy horizons, enclosing a compact nonchronal region.

In each case, one can smooth the geometry near the wormhole mouths – that is, in two disjoint neighborhoods surrounding the cylinders. By making the two smoothed regions outside the two cylinders isometric and locally static, one can again relate the cylinders to one another by (i) a time-translation or (ii) a boost relative to a flat exterior outside the two smoothed neighborhoods.

3 The Classical Cauchy Problem

3.1 Sample Lorentzian universes-from-nothing In the Lorentzian universes-from-nothing of Section 2.1, there is a time-nonorientable, spacelike hypersurface that serves as an *initial value surface* for the spacetime: freely specified initial data for wave equations have a unique global time-evolutions. So defined, an initial value surface is a natural generalization of a Cauchy surface to spacetimes that are not globally hyperbolic.

On the Mobiüs strip with either a flat metric or with a deSitter metric, the initial value surface Σ is the median circle. Data for, e.g., the scalar wave equation is given as it is for a Cauchy surface, by specifying the scalar field ϕ and its gradient $\nabla_\alpha \phi$ on Σ. (This way of speaking avoids mention of a normal to Σ, but one can alternatively choose a smooth direction field $\pm n^\alpha$ normal to Σ and specify $\pm n^\alpha \nabla_\alpha \phi$.) The Cauchy problem is well defined for data on Σ because the covering spacetime $\overline{M}, \overline{g}$ is globally hyperbolic – in the Möbius strip examples, $\overline{M}, \overline{g}$ is a cylinder, infinite in a timelike direction, with either a flat metric or a deSitter metric. If $\overline{x}, \overline{x}'$ are the two images in $\overline{M}$ of a point $x \in M$, the isometry $A : \overline{x} \to \overline{x}'$ is called the antipodal map. The lift to $\overline{\Sigma}$ of data on Σ is an initial data set that is antipodally symmetric (invariant under A). Because $\overline{M}, \overline{g}$ is globally hyperbolic, the data has a unique time evolution, $\overline{\phi}$, and because $\overline{M}, \overline{g}$ is antipodally symmetric, $\overline{\phi}$ is antipodally symmetric. The field $\overline{\phi}$ on $\overline{M}$ is therefore be the lift of a unique solution ϕ to the scalar wave equation on M.

The Möbius strip with a flat metric has CTC's if one extends the strip to a timelike thickness greater than its circumference. (These are smooth timelike curves $c(\lambda)$ that intersect the same point twice: $c(\lambda_0) = c(\lambda_1)$. The segment near λ_0, is, in general, transverse to the segment near λ_1, and the tangent vectors at the point of intersection have opposite time-orientation.) With a deSitter metric, on the other hand, there are no CTC's, and the only causal anomaly is the time non-orientability. Although neither spacetime is globally hyperbolic, the Cauchy problem is, as we have seen, well defined for both.

For the other lower dimensional spacetimes with CTC's, however, there are difficulties in the Cauchy problem. Some are related to the grandfather paradox, others to conical singularities or to a Cauchy horizon that is not compactly generated.

3.2 Spacelike and timelike cylinders For the cylinder with infinite spatial extent, the only candidates for an initial value surface are spacelike lines extending to spatial infinity. Data placed on such a surface can be uniquely evolved in the covering space, but the resulting solution corresponds to a solution on the original spacetime only if it is suitably periodic: almost no data allows a solution. This is essentially the grandfather paradox – locally one can construct a unique solution, but when extended to a region that loops around the cylinder and back to Σ, the locally evolved solution returns to Σ with a value that is inconsistent with its initial data.

The cylinder with infinite timelike extent and a flat metric that makes it everywhere nonchronal has no candidate for an initial value surface – no complete spacelike surface transverse to all timelike curves. One cannot ask whether the initial value problem is well defined, but one can ask the less ambitious question of whether it is *benign* (Yurtsever [1994]): For each point x, is there a finite spacelike surface S containing x for which arbitrary smooth data on S can be extended to a solution on the spacetime? For the massive wave equation, the cylinder spacetimes seem not even to be benign: A single massive particle leaving any spacelike surface S can be aimed to return to S at a point different from the point of departure.

Problem *Prove (or disprove) the conjecture that the flat cylinders, given by the identifications (2.2) or (2.3) are not benign for the massive scalar wave equation.*

3.3 Misner space The half of Misner space to the past of its Cauchy horizon is globally hyperbolic. If P and P' are points of the null edges of the strip of Minkowski space that are identified to form Misner space, a Cauchy surface Σ is obtained from the segment of the boost hyperbola that joins P to P' (i.e., when extended beyond the strip, the hyperbola contains the images of P under successive boosts $\mathcal{B}$). Data on Σ does not give a smooth solution on Misner space, however. A future directed null geodesic from P loops about the space an infinite number of times before reaching the Cauchy horizon, and with each loop it is boosted: any continuously defined affine parameter decreases by the boost factor B^{-1} with each loop[1]. Thus the null geodesic reaches the Cauchy horizon and terminates in finite affine parameter length; its total length has bound proportional to the convergent sum $B^{-1} + B^{-2} + \cdots$. Solutions to wave equations diverge at the Cauchy horizon for a similar reason. A left-moving solution effectively loops around the strip an infinite number of times before reaching the Cauchy horizon, and with each loop it is blue-shifted by B. The flux of the solution and its gradient diverge on the Cauchy horizon.

3.4 Geroch-Horowitz-Politzer spaces For each spacetime $\mathcal{N}$ there is a spacelike hypersurface $\mathcal{M}$ to the past of the nonchronal region on which initial data can be set, and is not difficult to find an L_2 solution (Friedman and Morris [1995]). The result for the massless scalar wave equation is this:

Proposition *For any initial data $\phi, \dot{\phi}$ in $L_2(\Sigma) \times L_2(\Sigma)$ there is a unique solution in $L_2^{\rm loc}$ to the massless scalar wave equation on spacetimes of the form described above. In a neighborhood of spatial infinity the solution agrees with the Minkowski space solution for the same initial data.*

The solution is constructed as a sum of right-moving and left-moving parts, which can be evolved independently on strips bounded by null rays that begin at each of the endpoints of the two removed slits. Smooth data, however, leads to solutions that are discontinous across the strips; the Cauchy problem does not have generic solutions in the space of finite energy fields.

3.5 Gott spacetimes For the Gott spacetimes given by the identifications (2.8) and (2.9), Cutler's partial Cauchy surface can be taken to be any of the hyperbolas $t = -\sqrt{a^2 + x^2}$. Carinhas [1993, 1995] has shown for the massless scalar wave equation that data on Σ satisfying suitable asymptotic conditions leads to solutions on these spacetimes. There are a number of related problems that could be done here, some more straightforward than others:

Related problems *Show that the solution Carinhas obtains is smooth for smooth initial data. Show that it is unique if and only if one requires zero data on the part of past null infinity that lies to the future of Σ. Extend the existence theorem to the massive scalar wave equation, and show that the solution is unique (because massive fields have zero data at null infinity). Extend these results to the broader class of Gott spacetimes having conical singularities with deficit angles less than π.*

3.6 Wormhole (time-tunnel) spacetimes For time-tunnels of the form (ii), with compact nonchronal region, Morris, Thorne and Yurtsever [1988] have given a persuasive argument, based on the geometric optics approximation, that the Cauchy problem is well defined. As in the Misner-space example, there is an incomplete null geodesic that loops through the wormhole and approaches a closed null geodesic on the Cauchy horizon; and the null geodesic is boosted with each successive loop. The wormhole's geometry, however, can make a congruence of rays diverge as it loops about the wormhole, and if the divergence dominates the blue shift, then in

[1]That is, trajectories of a (locally-defined) timelike Killing vector cross the null geodesic at a sequence of points. The Killing vector can be used to compare the affine parameter at successive crossing points by time-translating a segment of the geodesic to successively later segments. Compared in this way, the affine parameter of a given segment will will be less than that of the next segment by the blueshift factor B^{-1}.

this geometric optics approximation, the energy of the field will remain finite at the Cauchy horizon.

The force of the argument is that for smooth fields, one can construct a solution in a similar way by summing a multiple-scattering series. The field incident on each mouth of a wormhole is partly transmitted and partly reflected; the successive scatterings lead to a series that converges most slowly (or diverges most rapidly) as the wavelength decreases. Thus if one finds convergence in the geometric optics approximation, one expects the multiple scattering series itself to converge.

The only formal existence proof available for wormhole spacetimes, however, is for smooth, static wormholes, case (i) (Friedman and Morris [1991] and [1995]). The spacetimes have no spacelike initial value surfaces, but for massless wave equations, data can be posed at $\mathcal{I}^-$.

Proposition 1 *For almost all such spacetimes, the following existence theorem holds. Let f be initial data on $\mathcal{I}^-$ for which f and all its derivatives are in $L_2(\mathcal{I}^-)$. Then there exists a solution Φ to the massless scalar wave equation which is smooth and asymptotically regular at null and spatial infinity and which has f as initial data.*

Because the spacetime is static, one can write the solution as a sum of harmonics, solutions with time dependence $e^{i\omega t}$. The difficulty of the proof is a lack of orthogonality among eigenfunctions for each harmonic; the problem arises from the lack of a global spacelike surface orthogonal to the timelike Killing vector. It is easy to show existence of smooth eigenfunctions of the spatial laplacian $\vec{\nabla}^2$, harder to show convergence of the spectral decomposition.

The extension of the proof to fields of other spins is sketched, but not all details are filled in.

Problem *Complete the proof of the Proposition for spin-1/2 and spin-1 fields.*

3.7 More general spacetimes One can rule out, for the static wormhole spacetimes, a lack of uniqueness corresponding to a field forever trapped *inside* the nonchronal region, a smooth analog of a closed null geodesic. For spacetimes with a compact nonchronal region, $\mathcal{A}$, there is a complementary result – that when solutions to the Cauchy problem exist, they are unique *outside* $\mathcal{A}$ (Friedman and Morris [1995]). Because data is now given on a spacelike surface (instead of $\mathcal{I}^-$), one need no longer restrict consideration to massless wave equations.

Proposition 2 *Let $\mathcal{N}, g_{\alpha\beta}$ be a smooth, asymptotically flat spacetime with regions $\mathcal{F}$ and $\mathcal{P}$ to the future and past of a compact 4-dimensional submanifold $\mathcal{A}$ defined by $\mathcal{F} = \mathcal{N}\backslash J^-(\mathcal{A})$ and $\mathcal{P} = \mathcal{N}\backslash J^+(\mathcal{A})$, where both $\mathcal{F}$ and $\mathcal{P}$ are globally hyperbolic and foliated by complete spacelike 3-manifolds. Suppose for arbitrary smooth data with finite energy and a Cauchy surface $\mathcal{M}_P$ of $\mathcal{P}$ that the scalar wave equation has a solution on $\mathcal{N}$ with finite energy on $\mathcal{F}$, and suppose that for arbitrary smooth data with finite energy and a Cauchy surface $\mathcal{M}_F$ of $\mathcal{F}$ that the scalar wave equation has a solution on $\mathcal{N}$ with finite energy on $\mathcal{P}$. Then, solutions on $\mathcal{N}$ with finite energy in $\mathcal{N}\backslash\mathcal{A}$ are unique in $\mathcal{N}\backslash\mathcal{A}$.*

The key mathematical question remains open:

Problem *Characterize the class of spacetimes on which the generalized Cauchy problem is well defined – for which smooth data on a spacelike hypersurface deter-*

mines unique solutions to free-field wave equations on the spacetime. Is the class independent of the spin and mass of the field?

Because the interacting billiard-ball examples considered by Echeverria et al. [1991] have a multiplicity of solutions for the same initial data, uniqueness in spacetimes with CTCs is likely to hold only for free or weakly interacting fields. Because solutions seem always to exist for the billiard ball examples in the spacetimes they considered, it may be that classical interacting fields have solutions on spacetimes for which solutions to the free field equations exist.

Based on the Morris-Thorne-Yurtsever argument for existence of solutions of wormhole spacetimes and the fact that incomplete null geodesics generically mark a compactly generated chronology horizon (Hawking [1992]), we suggest that a plausible criterion for a well-defined Cauchy problem is, roughly speaking, that a spacetime is stable in the geometric optics approximation. This is made more precise by calling a spacetime *optically stable* if congruences of null geodesics always carry finite energy across finite spacelike hypersurfaces. (Caustics do not lead to infinite integrated energies, but divergent integrated energies can arise from congruences centered on looping null geodesics (Friedman and Morris [1995]).) The criteria may apply to fields with nonzero rest mass, because divergent energies arise from high-frequency behavior, and the proof plausibly ignores rest mass. Examples given in the previous reference show that, without some requirement akin to the optical stability, there can be too many closed geodesics to allow a well-defined initial value problem.

Existence Conjecture *Let $\mathcal{N}, g$ be a smooth, asymptotically flat spacetime for which past and future regions $\mathcal{P} = \mathcal{N} \backslash J^+(\mathcal{A})$ and $\mathcal{F} = \mathcal{N} \backslash J^-(\mathcal{A})$ of a compact 4-dimensional submanifold $\mathcal{A}$ are globally hyperbolic. If $\mathcal{N}, g$ is optically stable, the Cauchy problem for massless wave equations (for scalar, Maxwell, and Weyl fields) is well defined.*

An analogous conjecture relates uniqueness for massless fields to uniqueness in a geometic-optics sense.

Uniqueness Conjecture *Again let $\mathcal{N}, g$ be a smooth, asymptotically flat spacetime for which past and future regions $\mathcal{P} = \mathcal{N} \backslash J^+(\mathcal{A})$ and $\mathcal{F} = \mathcal{N} \backslash J^-(\mathcal{A})$ of a compact 4-dimensional submanifold $\mathcal{A}$ are globally hyperbolic. Let $\Sigma_\pm$ be Cauchy surfaces for $\mathcal{N} J^\pm(\mathcal{A})$. If all but a set of measure zero of null geodesics intersect Σ_+ and Σ_-, then solutions to massless wave equations on $\mathcal{N}$ are unique for initial data on Σ_- (and for initial data on Σ_+).*

4 Quantum Field Theory

A generalization of quantum field theory to nonchronal regions within a path integral framework is discussed by Friedman et al. [1992b] and in greater detail by Hartle [1994]. Alternative algebraic approaches are introduced by Kay [1992] and by Yurtsever [1994]; and Fewster and Higuchi [1996] clarify the meaning of these proposals for a set of simple spacetimes that are not globally hyperbolic. Discussion of the algebraic approach (and its limitations) will be limited here to time-nonorientable spacetimes (Section 4.3 below).

A number of authors, beginning with Hiscock and Konkowski [1982] (two-dimensional Misner space) and Kim and Thorne [1991] (four-dimensional wormhole spacetimes) have found that at chronology horizons, the renormalized stress tensor of a quantum field diverges. Because Thorne [1993] has given a comprehensive review of this work, and a further review is part of Visser's recent book (Visser [1995]), the quantum instability will not be treated here. After the reviews were written, Kay et al. [1995] obtained a general theorem implying that the renormalized stress tensor of a free field is singular at any compactly generated Cauchy horizon.

A divergent stress tensor underlies Hawking's chronology protection conjecture (Hawking [1992]), that (at least) macroscopic CTC's are forbidden. For microscopic CTCs, however, one is not working in a semiclassical framework, and a more natural question is whether one can make sense of a set of interacting quantum fields, including the metric, when the allowed geometries generically have CTCs. To obtain a tractable model problem, one freezes a background spacetime with CTCs and asks whether the scattering of free and interacting quantum fields makes sense on this background.

4.1 Unitarity of free-fields In nonchronal spacetimes where the Cauchy problem is known to be well defined, solutions to the free-field wave equation are not trapped by closed geodesics. That is, for spacetimes on which the scattering matrix exists, formal unitarity is a consequence of a conserved inner product on the 1-particle Hilbert space: conservation of the classical symplectic product implies conservation of probability in the Fock space.

Consider for example a spacetime in which a compact nonchronal region lies between two spacelike surfaces Σ_1 and Σ_2. If ϕ and ψ are solutions to the scalar wave equation,

$$(-\nabla_\alpha \nabla^\alpha + m^2)\phi = 0, \tag{4.1}$$

there is a divergencefree current,

$$\nabla_\alpha \left[\psi \overleftrightarrow{\nabla}^\alpha \phi \right] = 0. \tag{4.2}$$

Then Stokes' Theorem, valid independent of the causal structure between Σ_1 and Σ_2, implies a conserved symplectic product,

$$\omega(\psi, \phi) := \int_{\Sigma_1} \psi \overleftrightarrow{\nabla}^\alpha \phi \, dS_\alpha = \int_{\Sigma_2} \psi \overleftrightarrow{\nabla}^\alpha \phi \, dS_\alpha. \tag{4.3}$$

Note that, as in globally hyperbolic spacetimes, a well-defined S-matrix depends on finite particle production and is representation-dependent. Formal unitarity does not imply that an S-matrix exists; when it does exist, the S-matrix is unitary.

4.2 Failure of unitarity for interacting fields The formal steps in the path-integral reduction of S-matrix elements to products of Feynman propagators make no mention of causal structure and can be carried through in spacetimes with a compact nonchronal region. Perturbative unitarity of interacting field theories, however, rests on the structure of the propagator. As shown in Friedman, Papastamatiou and Simon [1992a], it relies on a series of identities, all of which are satisfied if the Feynman propagator, Δ_F, has a form that reflects the causal structure of a globally hyperbolic spacetime,

$$\Delta_F(x,y) = \begin{cases} D(x,y), & \text{if } x \text{ is to the future of } y \\ D(x,y) = \overline{D}(x,y), & \text{if } x \text{ and } y \text{ are spacelike separated} \\ \overline{D}(x,y), & \text{if } y \text{ is to the future of } x. \end{cases} \tag{4.4}$$

Here $D(x,y)$ is the Wightman function. When there are CTCs, one cannot assign a time-ordering to spacetime points x and y, and this causal form is lost. The difficulty turns out to be fatal: The generalization to curved spacetime of the Cutkosky relations that govern perturbative unitarity are lost, and the scattering of interacting fields is not unitary on spacetimes with closed timelike curves. (Boulware [1992] and Friedman et al. [1992b]. Instructive examples are also given in Politzer [1992]; see also Goldwirth et al. [1993].)

Unless one can overcome this loss of unitarity, one cannot consistently use a Copenhagen interpretation to describe observations both before and after a region of CTCs. A sum-over-histories interpretation, on the other hand, may allow one to make sense of quantum field theory (Friedman et al. [1992b] and Hartle [1994]). One can obtain a consistent assignment of probabilities if one considers only paths that start to the past of *any* region of CTCs and terminate to the future of every such region. Experiments at any intermediate time can, as usual, be described by including measuring instruments in the quantum system.

There is, however, an unexpected loss of causality even *before* any CTCs form. That is, one can set up experiments whose outcomes depend on whether CTCs form to the future of the experiment. In addition, the evolution described is not linear; an initial state Ψ is mapped to a final state $X\Psi$ by a nonunitary X, and to find probabilities for restricted classes of histories one renormalizes the final state, replacing $X\Psi$ by $X\Psi/ < \Psi|X^\dagger X\Psi >^{1/2}$.

Arlen Anderson proposes a simple, if ad hoc, alternative that restores unitarity to an evolution given initially by a nonunitary operator $X : \mathcal{H} \to \mathcal{H}$. One can define a polar decomposition, $X = RU$, where $R = (XX^\dagger)^{-\frac{1}{2}}$, and U is unitary. Anderson [1995] postulates that the evolution is given by the unitary operator

$$U = (XX^\dagger)^{-\frac{1}{2}}X. \tag{4.4}$$

The composition law, however, has a difficulty (Fewster and Wells [1995]): If U_1 and U_2 are the unitary operators associated with X_1 and X_2, it is not the case that $U_{21} = U_1U_2$, where U_{21} is the unitary part of $X_{21} = X_2X_1$. Instead, to obtain a consistent interpretation of the composition, Anderson [1995] introduces a second inner product (,) on $\mathcal{H}$, the pullback by X_1^{-1} of the original inner product < , >:

$$(\Psi,\Phi) :=< X^{-1}\Psi, X^{-1}\Phi > . \tag{4.5}$$

Then $U_{21} = \tilde{U}_2U_1$, where $\tilde{U}_2 = R_1^{-1}\hat{U}_2R_1$, where $\hat{U}_2$ is the unitary part of U_2 with respect to (,).

Several other alternatives have been proposed. Fewster and Wells [1995] introduce a unitary dilation of X by enlarging the state space. The reduction to the physical states, however, leads to a loss of coherence. This is reminiscent of an earlier schematic proposal by Deutsch [1991]. Deutsch replaces a spacetime region with CTC's by a black box and requires the evolution of a density matrix to be ordinary before and after the black-box is encountered. Boundary conditions at the time of encounter and a requirement on entropy fix the evolution. Again coherence is lost – pure states evolve to mixed states. It is not clear how one is to apply Deutsch's proposal to a field evolving on a generic spacetime, but Politzer [1994] and Cassidy [1995] examine the proposal and a variant suggested by Politzer for a model quantum mechanical system of two interacting fermions that travel on fixed paths in a two-dimensional Geroch-Horowitz-Politzer spacetime. The Deutsch and Politzer proposals lead to nonlinear evolution as well as to a loss of coherence.

A recent paper by Hawking [1995] introduces another alternative, computing a superscattering matrix by analytically continuing parameters of a Lorentzian spacetime to obtain a Euclidean geometry. The method leads to a loss of coherence, apparently even in the case of free fields where other methods give well-behaved unitary scattering.

4.3 Quantum field theory on Lorentzian universes-from-nothing Several recent authors have noted that time-nonorientability prevents one from carrying through the standard construction of a Fock space or a Weyl algebra of observables (Gibbons [1986], Sanchez and Whiting [1987] and Kay [1992]). Kay [1992] requires existence of a globally defined *-algebra and imposes what he terms the "F-locality condition", which demands in essence, that the *-algebra satisfy the canonical commutation relations in a neighborhood of any point with respect to one time orientation. Under these conditions he proves that the spacetime must be time orientable. An independent study by Gibbons [1986] concludes that one is forced to use a real (i.e., noncomplex) Hilbert space to describe quantum field theory in some time-nonorientable spacetimes including antipodally identified deSitter space. This also suggests that one cannot construct a globally defined conventional quantum field theory in a time-nonorientable spacetime.

The conclusion is surprising, because an observer in antipodally identified deSitter space, M, cannot classically distinguish the spacetime from deSitter space, $\overline{M}$. The past of a timelike worldline in M (defined by a choice of orientation near the worldline) is isometric to the past of either of the two corresponding worldlines in $\overline{M}$. As noted above, the Cauchy problem is well defined on M for fields with initial data on Σ; and the solutions will be identical to those seen by an observer on $\overline{\Sigma}$ who travels along the corresponding worldline and sees the same data.

One can evade Kay's no-go theorem if, instead of a globally defined *-algebra, one demands only a set of *-algebras, each defined in a local neighborhood (Friedman and Higuchi [1995]). Overlap conditions on the *-algebras then ensure that the local algebraic structure coincides with that on a globally hyperbolic spacetime. Explicitly, for a scalar field Φ, one defines abstract algebras for each choice of orientation $\mathcal{O}$ and $\check{\mathcal{O}}$, of a globally hyperbolic subspacetime $U, g|_U$. The oppositely oriented algebras, $\mathcal{A}$ and $\check{\mathcal{A}}$, are the free complex algebras generated by symbols of the form $\{\hat{\phi}(F), F \in C_0^\infty(U)\}$ and $\{\check{\phi}(F), F \in C_0^\infty(U)\}$, modulo the commutation relations,

$$\left[\hat{\phi}(E), \hat{\phi}(F)\right] = i \int E(x)(G_{adv} - G_{ret})(x,y)F(y)d^4V_x d^4V_y,$$
$$[\check{\phi}(E), \check{\phi}(F)] = i \int E(x)(\check{G}_{adv} - \check{G}_{ret})(x,y)F(y)d^4V_x d^4V_y\,. \qquad (4.6)$$

On each overlap, $U \bigcap U'$, there is a linear or antilinear isomorphism $\mathcal{I}$ between the restrictions of $\mathcal{A}_{(U,\mathcal{O})}$ and $\mathcal{A}_{(U',\mathcal{O}')}$ to ϕ's smeared with functions having support on $U \bigcap U'$. For $F \in C_0^\infty(U \cap U')$

$$\mathcal{I}\hat{\phi}(F) = \hat{\phi}'(F),\ \mathcal{I}i = i,\ \text{if}\ \mathcal{O} = \mathcal{O}',$$
$$\mathcal{I}\hat{\phi}(F) = \check{\phi}'(F),\ \mathcal{I}i = -i,\ \text{if}\ \mathcal{O} \neq \mathcal{O}'\,. \qquad (4.7)$$

One would like to use this structure of local algebras to define quantum states as collections of positive linear functions (plf's) on the local algebras, again with consistency conditions on the intersection of globally hyperbolic neighborhoods. We find, however, that if one considers the algebras of observables on an atlas consisting of *all* globally hyperbolic subspacetimes that inherit their causal structure from the spacetime M, g, then one cannot consistently define states. In particular, if the union of a pair of neighborhoods is time nonorientable, one cannot consistently extend a generic plf to the pair of neighborhoods without violating positivity.

One can define a collection of local states on *smaller* atlases, restricted so that the union of any two neighborhoods is time orientable. The collection of algebras and states is then locally indistinguishable from that on the globally hyperbolic covering space of any Lorentzian universe-from-nothing. In particular, any local state can be consistently extended to a collection of states on all algebras associated with the atlas. But the restriction on the size of neighborhoods amounts to a restriction on the size of regions over which one can define correlations between field operators, and this has unpleasant implications. The specification of a collection of states on the neighborhoods that cover and share an initial value surface does not *uniquely* determine a time-evolution: The extension to a collection of states on the set of all algebras is not unique. In addition, in showing that one can extend a local state to a collection of states, Friedman and Higuchi use an antipodally symmetric state on the covering space, and such a state would not yield a well-defined state on the spacetime if all correlations could be measured. Finally, the families of states and algebras depend on the choice of atlas, and there is no unique maximal atlas.

Finally, a few comments on spinor fields may be useful. It is common in the general-relativity literature to regard two-component spinors (Weyl spinors) as fields built from the fundamental representation of $SL(2,\mathbb{C})$ (Bichteler [1968], Geroch [1968], Penrose and Rindler [1984] and Wald [1984]), and readers may be left with the misimpression that one can define two-component spinors only in time-orientable spacetimes. Weyl spinors, however, are really acted on by a larger group that includes time reversal; they belong to an action of a covering group of that subgroup L_+ of the full Lorentz group comprising the component of the identity and the component of time-reversing, space-preserving Lorentz transformations.

The group L_+ has two double covers, Sin^+ and Sin^-, depending on the sign of $\mathcal{T}^2$, where $\mathcal{T}$ is either of the two elements of the covering group that correspond to the choice T of time-reversal. Only Sin^- acts on the usual two-component spinors associated with Weyl neutrinos in Minkowski space. In an orientable spacetime, the difference between an $SL(2,\mathbb{C})$-spinor structure and a Sin^--spinor structure is unimportant, unless one wishes explicitly to discuss time reversal. In a time-nonorientable spacetime, however, the difference is essential. Because one cannot pick a bundle of time-oriented frames, time-nonorientable spacetimes have no $SL(2,\mathbb{C})$-spinor structure, and two-component spinors rely for their definition on an action of a the covering group Sin^- of L_+. Other authors have considered generalized spinor structures on generic nonorientable spacetimes (Karoubi [1968], Whiston [1975], Dabrowski [1988], Chamblin [1994] and Gibbons [1994]). For these generic spacetimes, the situation is somewhat different, because one must consider actions of the full Lorentz group; and the usual action of parity requires four-component spinors.

A class of time-nonorientable spacetimes, however, does allow sinors (Sin^--structures) (Friedman [1995] and Chamblin and Gibbons [1995]). For the special case of Lorentzian universes-from-nothing, the precise statement is as follows:

Proposition 3 *Let M, g be a Lorentzian universe-from-nothing and let $\overline{M}$ be its orientable double cover. Then the inequivalent sinor structures are in 1-1 correspondence with homomorphisms $h \in H^1(\overline{M})$ that respect the antipodal map A and for which $h[\overline{c^2}] = -1$, for every time reversing curve c in M.*

Here $\overline{c^2}$ is a lift of c^2 to $\overline{M}$. A sinor field is then a cross section of a bundle associated to a sinor structure. Lorentzian universes-from-nothing for which Σ is a 3-torus have both Sin^+ and Sin^--spinor structures, while antipodally identified deSitter space has only a Sin^+-spinor structure and so does not admit global fields of the usual kind of chiral fermions. Even in a time-nonorientable spacetime that allows the usual 2-component spinors, however, one cannot construct a global Lagrangian density that violates time-reversal invariance, and one therefore cannot globally implement the standard model of particle physics.

References

Anderson, A. [1995] *Unitarity restoration in the presence of closed time-like curves,* Phys. Rev. D **51**, 5707–5715.

Bichteler, K. [1968] *Global existence of spin structures for gravitational fields,* J. Math. Phys. **9**, 813–815.

Boulware, D. G. [1992] *Quantum field theory in spaces with closed timelike curves,* Phys. Rev. D **46**, 4421–4441.

Carinhas P. [1993] Doctoral Dissertation, University of Wisconsin-Milwaukee.

——— [1995] *Cauchy problem for Gott spacetime*, Eprint gr-qc/9507020.

Cassidy, M. J. [1995] *Nonlinearity in quantum theory and closed time-like curves,* Phys. Rev. D **52**, 5676–5680.

Chamblin, A. [1994] *On the obstructions to non-Cliffordian pin structures,* Commun. Math. Phys. **164**, 65–85.

Chamblin, A. and Gibbons, G. W. [1995] *A judgment on sinors,* Class. Quantum Grav. **12**, 2243–2248.

Cutler, Curt [1992] *Global structure of Gott's two-string spacetime,* Phys. Rev. D **45**, 487–494.

Dabrowski, L. [1988] *Group Actions on Spinors* Monographs and Textbooks in Physical Science, Bibliopolis, Naples.

Deutsch, D. [1991] *Quantum mechanics near closed timelike lines,* Phys. Rev. D **44**, 3197–3217.

Echeverria, F., Klinkhammer, G. and Thorne, K. S. [1991] *Billiard balls in wormhole spacetimes with closed timelike curves: Classical theory,* Phys. Rev. D **44**, 1077–1099.

Fewster, C. J. and Higuchi, A. [1996] *Quantum field theory on certain non-globally hyperbolic spacetimes,* Class. Quantum Grav. **13**, 51–61.

Fewster, C. J. and Wells, C. G. [1995] *Unitarity of quantum theory and closed timelike curves,* Phys. Rev. D **52**, 5773–5782.

Friedman, J. L. [1995] *2-component spinor fields in time-nonorientable spacetimes,* Class. Quantum Grav. **12**, 2231–2241.

Friedman, J. L. and Higuchi, A. [1995] *Quantum field theory in Lorentzian universes from nothing,* Phys. Rev. D **52**, 5687–5697.

Friedman, J. L. and Morris, M. S. [1991] *The Cauchy problem for the scalar wave equation is well defined on a class of spacetimes with closed timelike curves,* Phys. Rev. Lett. **66**, 401–404.

——— [1995] *Existence and uniqueness theorems for massless fields on a class of spacetimes with closed timelike curves*, Comm. Math. Phys., in press.

Friedman, J. L., Morris, M. S., Echeverria, F., Klinkhammer, G., Thorne, K. S. and Yurtsever, U. [1990] *Cauchy problem in spacetimes with closed timelike curves,* Phys. Rev. D **42**, 1915–1930.

Friedman, J. L., Papastamatiou, N. J. and Simon, J. Z. [1992a] *Unitarity of interacting fields in curved spacetime,* Phys. Rev. D **46**, 4442–4455.

——— [1992b] *Failure of unitarity for interacting fields on spacetimes with closed timelike curves,* Phys. Rev. D **46**, 4456–4469.

Geroch, R. P. [1967] *Topology in general relativity,* J. Math. Phys. **8**, 782–786.

______ [1968] *Spinor structure of space-times in general relativity. I,* J. Math. Phys. **9**, 1729–1744.

Geroch, R., and Horowitz, G. [1979] *Global structure of spacetimes*, General Relativity (Hawking, S. W. and Israel, W., eds.), Cambridge University Press, Cambridge, pp. 212–293.

Gibbons, G. W. [1986] *The elliptic interpretation of black holes and quantum mechanics,* Nucl. Phys. B **271**, 497–508.

______ [1994] *Changes of topology and changes of signature,* Int. J. Mod. Phys. **3**, 61–70.

Goldwirth, Dalia S., Perry, Malcolm J., Piran, Tsvi and Thorne, Kip S. [1994] *Quantum propagator for a nonrelativistic particle in the vicinity of a time machine,* Phys. Rev. D **49**, 3951–3957.

Gott, J. R. [1991] *Closed timelike curves produced by pairs of moving cosmic strings: Exact solutions,* Phys. Rev. Lett. **66**, 1126–1129.

Hartle, J. B. [1994] *Unitarity and causality in generalized quantum mechanics for nonchronal spacetimes,* Phys. Rev. D **49**, 6543–6555.

Hawking, S. W. [1995] *Quantum coherence and closed timelike curves,* Phys. Rev. D **52**, 5681–5686.

______ [1992] *Chronology protection conjecture,* Phys. Rev. D **46**, 603–611.

Hawking, S. W. and Ellis, G. F. R. [1973] *The Large Scale Structure of Space-time*, Cambridge University Press, Cambridge.

Hiscock, W. A. and Konkowski, D. A. [1982] *Quantum vacuum energy in Taub-NUT (Newman-Unti-Tamburino)-type cosmologies,* Phys. Rev. D **26**, 1225–1230.

Karoubi, M. [1968] Ann. Scient. Éc. Norm. Sup. Series 4 **1**, 161.

Kay, B. S. [1992] *The principle of locality and quantum field theory on (non-globally-hyperbolic) curved spacetimes,* Reviews in Mathematical Physics **167**.

Kay, B. S., Radzikowski, M., and Wald, R. M. [1995] *Quantum field theory on spacetimes with compactly generated Cauchy horizons* Comm. Math. Phys., in press.

Kim, Sung-Won and Thorne, Kip S. [1991] *Do vacuum fluctuations prevent the creation of closed timelike curves?,* Phys. Rev. D **43**, 3929–3947.

Misner, C. W. [1967] *Taun-NUT space as a counterexample to almost anything,* Relativity Theory and Astrophysics I. Relativity and Cosmology (Ehlers, J., ed.), Amer. Math. Soc., Providence, pp. 160–169.

Morris, M. S. and Thorne, K. S. [1988] *Wormholes in spacetimes and their use for interstellar travel: A tool for teaching general relativity,* Am. J. Phys. **56**, 395–412.

Morris, M. S., Thorne, K. S. and Yurtsever, U. [1988] *Wormholes, time machines, and the weak energy condition,* Phys. Rev. Lett. **61**, 1446–1449.

Penrose, R. and Rindler, W. [1984] *Spinors and Space-Time. I,* Cambridge University Press, Cambridge.

Politzer, H. David [1992] *Simple quantum systems in spacetimes with closed time-like curves,* Phys. Rev. D **46**, 4470–4476.

______ [1994] *Path integrals, density matrices, and information flow with closed time-like curves,* Phys. Rev. D **49**, 3981–3989.

Sanchez, N. and Whiting, B. F. [1987] *Quantum field theory and the antipodal identification of black holes,* Nucl. Phys. B **283**, 605–623.

Thorne, K. S. [1993] *Closed timelike curves*, General Relativity and Gravitation 1992 (Gleiser, R. J., Kozameh, C. N. and Moreschi, D. M., eds.), Institute of Physics Publishing, Bristol.

______ [1994] *Misner space as a prototype for almost any pathology,* Directions in General Relativity, V. 1 (Hu, B. L., Ryan, M. P. and Vishveshwara, C. V., eds.), Cambridge Univ. Press, Cambridge.

Visser, M. [1995] *Lorentzian Wormholes*, AIP Press, Woodbury, NY.

Wald, R. M. [1984] *General Relativity,* University of Chicago Press, Chicago.

Yurtsever, U. [1994] *Algebraic approach to quantum field theory on non-globally-hyperbolic spacetimes,* Class. Quantum Grav. **11**, 999–1012.

Whiston, G. S. [1975] *Lorentzian characteristic classes,* Gen. Rel. Grav. **6**, 463–475.

Fields Institute Communications
Volume **15**, 1997

Solving the Einstein Constraint Equations

James Isenberg
Department of Mathematics
and
Institute for Theoretical Science
University of Oregon
Eugene, OR USA 97403
jim@newton.uoregon.edu

Abstract. We review what has been learned over the past 25 years about solutions of the Einstein constraint equations on compact manifolds. Our focus is on the study of these equations and their solutions via the conformal method. We first discuss the now complete story concerning constant mean curvature data, and then describe recent progress with non constant mean curvature data.

One of the most useful ways both to find solutions of the Einstein vacuum field equations ($G_{\mu\nu} = 0$) and to study the space of all globally hyperbolic solutions of these equations is via the Cauchy formulation. The idea for constructing a solution via the Cauchy formulation is as follows: One first picks a three-dimensional manifold Σ^3 (e.g., choose $\Sigma^3 = S^3$). Next, one chooses the initial data fields

$$\gamma_{ab} \quad - \quad \text{a Riemannian metric on } \Sigma^3 \tag{1a}$$

and

$$K^{cd} \quad - \quad \text{a symmetric tensor field on } \Sigma^3 \tag{1b}$$

in such a way that the Einstein constraint equations

$$R - K^{ab}K_{ab} + (K^c{}_c)^2 = 0 \tag{2a}$$

$$\nabla_a K^a{}_b - \nabla_b(K^c{}_c) = 0 \tag{2b}$$

are satisfied by this initial data. One then evolves the initial data fields in time according to the Einstein evolution equations

$$\frac{d}{dt}\gamma_{ab} = -2NK_{ab} \tag{3a}$$

$$\frac{d}{dt}K^c{}_d = N(R^c{}_d + K^m{}_m K^c{}_d) - \nabla^c\nabla_d N \tag{3b}$$

1991 *Mathematics Subject Classification.* 83C05.

I thank Jack Gegenberg, Bob McKellar, and the rest of the organizing committee for inviting me to speak at the 1995 Canadian Conference on General Relativity and Relativistic Astrophysics. This talk is largely based on work carried out with V. Moncrief. The work is supported in part by NSF grant 93-08117 at the University of Oregon.

© 1997 American Mathematical Society

with N – the "lapse function" – being chosen freely as a time-dependent scalar function on Σ^3. Finally if one sets

$$M^4 = \Sigma^3 \times \mathbb{R}^1 \tag{4a}$$

and

$$g = -N^2 dt^2 + \gamma_{ab} dx^a dx^b \tag{4b}$$

then (M^4, g) is a globally hyperbolic solution of Einstein's equations.

It was shown 35 years ago by Choquet-Bruhat [1962] that the Cauchy formulation of Einstein's equations is "well-posed" in the sense that *any* choice of the initial data (γ, K) which satisfies the constraint equations can indeed be evolved into a spacetime solution. It has also been shown (Choquet-Bruhat and Geroch [1969]) that for each choice of initial data satisfying (1), there is a "maximal development" spacetime which is unique up to a diffeomorphism and contains all possible spacetime evolutions of that data. Hence the set of maximal development (globally hyperbolic) spacetimes can be identified with (and therefore parametrized by) the set of all solutions of the constraints.

There are a number of questions one can ask about the Cauchy formulation of Einstein's equations and the solutions it produces:

1. How does one find initial data (γ, K) which satisfy the constraints?
2. Does the set of solutions of the constraint equations (and therefore the set of maximal development solutions of Einstein's equations) have an interesting and useful mathematical structure?
3. What is the long-time behavior of the maximum development spacetimes?
4. How should one choose the lapse function N so as to avoid coordinate singularities and so as to most effectively explore the maximal development spacetimes?
5. Can the Cauchy formulation be used as a practical tool for studying the physics of solutions of Einstein's equations?

While all of these questions are being actively studied, we shall focus here on the first one.

The Conformal Method

The most successful method used during the past 25 years or so for producing solutions of the Einstein constraint equations and for collectively parametrizing them is the conformal method. The key idea of this method, which was developed primarily by Lichnerowicz, Choquet-Bruhat, and York (Choquet-Bruhat and York [1980]), is to split the data (γ, K) into two parts: One part is to be chosen freely, while the second part is to be determined by the constraint equations. More explicitly, the free (or "conformal") data consists of

λ_{ab} – a Riemannian metric on Σ^3 (specified up to a conformal factor) (5a)

σ^{cd} – a divergence-free and trace-free symmetric tensor on Σ^3 $(\nabla_c \sigma^{cd} = 0)$ $(\lambda_{cd}\sigma^{cd} = 0)$ (5b)

and τ – a scalar field on Σ^3; (5c)

while the determined data consists of

W^b – a vector field on Σ^3 (6a)

and ϕ – a positive-definite scalar field on Σ^3. (6b)

To construct a solution of the constraints via this method, one first makes a choice of the conformal data (λ, σ, τ), and then one attempts to solve the equations[1]

$$\nabla^2\phi = \frac{1}{8}R\phi - \frac{1}{8}(\sigma^{ab} + LW^{ab})(\sigma_{ab} + LW_{ab})\phi^{-7} + \frac{1}{12}\tau^2\phi^5 \tag{7a}$$

and

$$\nabla_a(LW)^a{}_b = \frac{2}{3}\phi^6\nabla_b\tau \tag{7b}$$

for W and ϕ. (Here ∇ is the covariant derivative compatible with λ, R is its scalar curvature, and L is the conformal Killing operator: $LW_{ab} := \nabla_a W_b + \nabla_b W_a - \frac{2}{3}\lambda_{ab}\nabla_c W^c$). If indeed one can find a set (ϕ, W) which solves equations (7), then the data

$$\gamma_{ab} = \phi^4\lambda_{ab} \tag{8a}$$

and

$$K^{cd} = \phi^{-10}(\sigma^{cd} + LW^{cd}) + \frac{1}{3}\phi^{-4}\lambda^{cd}\tau \tag{8b}$$

will be a solution of the constraint equations (2).

Does this method always work, in the sense that for any choice of conformal data (λ, σ, τ) one can always solve equations (7) for (ϕ, W)? It is easy to see that in fact it does not: Say we choose the three dimensional manifold Σ^3 to be the three dimensional sphere S^3, and say we choose for the conformal data $\sigma^{cd} = 0$, $\tau = 1$, and $\lambda =$ the round sphere metric with $R = 1$. With τ constant, equation (7b) becomes

$$\nabla_a(LW)^a{}_b = 0\,. \tag{9}$$

Since the operator $\nabla \cdot L$ is elliptic, we readily verify that equation (9) has a collection[2] of solutions W, all of which satisfy the condition $LW = 0$. Using this result, we find that the remaining constraint equation (7a) takes the form

$$\nabla^2\phi = \frac{1}{8}\phi + \frac{1}{12}\phi^5\,. \tag{10}$$

Since we seek solutions ϕ which are positive definite, for any such solution the right-hand side of equation (8) is positive definite. But the maximum principle tells us that there is no function ϕ for which $\nabla^2\phi > 0$ on a compact manifold. Hence, for this choice of Σ^3 and for this choice of the conformal data, there is no solution (ϕ, W) to equations (7) and therefore no corresponding solution to the Einstein constraint equations.

While this example shows that the conformal method may fail for certain choices of the conformal data, it doesn't tell us what happens more generally. What we would like to have is a set of criteria which we can use to identify those sets of conformal data which lead to solutions of the constraints, and those which do not. Such a set of criteria is not yet available in general. However, if one restricts one's attention to closed manifolds Σ^3 and to constant mean curvature ("CMC") conformal data – i.e., data with τ constant and therefore $K^c{}_c = \tau$ constant on Σ^3 – then in fact such criteria have been found. We will discuss these below. While much less is known for conformal data on closed Σ^3 with non constant mean curvature, there has been considerable progress during the last few years in finding at least sufficient conditions on (λ, σ, τ) so that equations (7) can be solved for (ϕ, W), and

[1] These equations are obtained by substituting expressions (8) for γ and K into the constraint equations (2).

[2] If λ has no conformal Killing fields, then (9) has a unique solution. However the round sphere metric has a nontrivial set of conformal Killing fields, and all are solutions of (9). They are the only solutions in this case.

therefore a solution of the Einstein constraints can be produced. We will discuss these results here as well.

Constant Mean Curvature Data

Constant mean curvature data is notably simpler to work with than more general (non CMC) data because in the CMC case, one needs to satisfy just one equation rather than four. This is true because, as in the simple S^3 example discussed above, the $\nabla\tau = 0$ condition transforms three of the constraint equations – (7b) – into the easily-handled condition $\nabla_a(LW)^a{}_b = 0$, which implies that $(LW)^a{}_b = 0$. Hence, for CMC conformal data, one needs only to work with the one equation (7a), which now takes the ("Lichnerowicz") form

$$\nabla^2\phi = \frac{1}{8}R\phi - \frac{1}{8}\sigma^2\phi^{-7} + \frac{1}{12}\tau^2\phi^5 \,. \tag{11}$$

and is to be solved for the one function ϕ.

For which choices of CMC conformal data (λ, σ, τ) does there exist a solution ϕ? We can use the maximum principle to easily establish that for certain collections of conformal data, a solution does *not* exist: Recall that (one version of) the maximum principle tells us that there is no smooth function ϕ on a closed manifold such that $\nabla^2\phi \geq 0 \ (\not\equiv 0)$ or $\nabla^2\phi \leq 0 \ (\not\equiv 0)$. So if we choose (λ, σ, τ) with $R \leq 0 \ (\not\equiv 0)$ and with $\tau^2 = 0$, then the Lichnerowicz equation (11) becomes

$$\begin{aligned} \nabla^2\phi &= -\frac{1}{8}|R|\phi - \frac{1}{8}\sigma^2\phi^{-7} \\ &\leq 0 \ (\not\equiv 0), \end{aligned} \tag{12}$$

and there is no solution. Similarly, if we choose (λ, σ, τ) with $R \geq 0 \ (\not\equiv 0)$ and with $\sigma^2 \equiv 0$, then (11) becomes

$$\begin{aligned} \nabla^2\phi &= \frac{1}{8}|R|\phi + \frac{1}{12}\tau^2\phi^5 \\ &\geq 0 \ (\not\equiv 0)\,, \end{aligned} \tag{13}$$

and again there is no solution. The same thing happens as well if we choose (λ, σ, τ) with $R \equiv 0$ and $\sigma^2 \equiv 0$ and $\tau^2 \neq 0$, or with $R \equiv 0$ and $\sigma^2 \not\equiv 0$ and $\tau^2 = 0$.

For most Riemannian metrics on a given closed manifold Σ^3, the scalar curvature has *indefinite* sign, so it appears as if the sets of conformal data just discussed are very special. However, two important results – the Yamabe Theorem and the conformal covariance of the Lichnerowicz equation (11) – can be combined with what we have just shown for these special sets of conformal data to determine much more.

The Yamabe Theorem (Schoen [1984], Lee and Parker [1987]) says that for any Riemannian metric λ on a closed manifold Σ^3, there is a conformal transformation $\lambda_{ab} \to \theta^4\lambda_{ab}$ such that the scalar curvature of $\theta^4\lambda_{ab}$ is a constant on Σ^3. Further, it tells us that for any fixed metric λ_{ab}, *all* conformal transformations which lead to constant scalar curvature produce a scalar curvature of the *same sign.* Thus the set of all Riemannian metrics on Σ^3 can be partitioned into three[3] "Yamabe classes": $\mathcal{Y}^+(\Sigma^3)$ includes those metrics which can be conformally transformed to a metric with constant $R > 0$, $\mathcal{Y}^0(\Sigma^3)$ includes those which can be conformally transformed so $R = 0$, and $\mathcal{Y}^-(\Sigma^3)$ includes the metrics which can be conformally transformed to a metric with constant $R < 0$.

[3]Note that for some closed manifolds Σ^3, $\mathcal{Y}^+(\Sigma^3)$ is empty, and for others, both $\mathcal{Y}^+(\Sigma^3)$ and $\mathcal{Y}^0(\Sigma^3)$ are empty. For all Σ^3, $\mathcal{Y}^-(\Sigma^3)$ is non-empty.

The Yamabe Theorem is important for the study of the Einstein constraint equations because the Lichnerowicz equation is conformally covariant in the following (easily verified) sense: There is a solution ϕ to the Lichnerowicz equation (11) for conformal data (λ, σ, τ) iff for every function $\theta : \Sigma^3 \to \mathbb{R}^+$ there is a solution[4] ψ to the Lichnerowicz equation for conformal data $(\theta^4\lambda, \theta^{-10}\sigma, \tau)$. Thus in trying to determine whether the conformal method works for a given set of data (λ, σ, τ), it is sufficient to make this determination for a conformally related set of conformal data $(\theta^4\lambda, \theta^{-10}\sigma, \tau)$, which may always be chosen so that the scalar curvature is constant.

This fact significantly simplifies the work needed to determine when the conformal method works and when it does not. For example, we can now use the maximum principle to show that if we choose conformal data (λ, σ, τ) with $\lambda \in \mathcal{Y}^-(\Sigma^3)$ – this is a much weaker restriction than the condition we considered earlier: that $R \leq 0$ $(\not\equiv 0)$ – and with $\tau^2 = 0$, then the Lichnerowicz equation has no solution. Similarly we can show that there is no solution for data with $\lambda \in \mathcal{Y}^+(\Sigma^3)$ and $\sigma^2 \equiv 0$, etc.

Ultimately, what we find for constant mean curvature initial data is that to determine if the conformal method will work or not for a given set of conformal data (λ, σ, τ), all we need to know is (a) whether τ^2 is zero or not, (b) whether σ^2 is zero everywhere on Σ^3 or not, and (c) which Yamabe class λ is contained in. Hence there are twelve classes of conformal data to consider: $(\mathcal{Y}^+, \sigma^2 \equiv 0, \tau = 0)$, $(\mathcal{Y}^+, \sigma^2 \not\equiv 0, \tau \neq 0)$, etc. We can conveniently organize these classes into a table, noting that for six of the classes in the table (as we have shown using the maximum principle) the conformal method doesn't work.

	$(\sigma^2 \equiv 0, \tau = 0)$	$(\sigma^2 \equiv 0, \tau \neq 0)$	$(\sigma^2 \not\equiv 0, \tau = 0)$	$(\sigma^2 \not\equiv 0, \tau \neq 0)$
$\mathcal{Y}^+$	No	No		
$\mathcal{Y}^0$		No	No	
$\mathcal{Y}^-$	No		No	

What happens for the other six classes? The work of Choquet-Bruhat, York, O'Murchadha, and Isenberg (Choquet-Bruhat and York [1980], Isenberg [1995]) shows that[5] for all of these other classes of conformal data, the Lichnerowicz equation can be solved, and hence the conformal method works. While the original proof of this for most of these classes was done using the Leray-Schauder method, the only complete proof which we know works for all of the classes is based upon the sub and super solution theorem.

Ignoring the details of function spaces, the sub and super solution theorem, as adapted to the Lichnerowicz equation, says the following: If one can find a function $\phi_+ > 0$ such that

$$\nabla^2\phi_+ \leq \frac{1}{8}R\phi_+ - \frac{1}{8}\sigma^2{\phi_+}^{-7} + \frac{1}{12}\tau^2{\phi_+}^5 \tag{14a}$$

and a function $\phi_- > 0$ such that

$$\nabla^2\phi_- \geq \frac{1}{8}R\phi_- - \frac{1}{8}\sigma^2{\phi_-}^{-7} + \frac{1}{12}\tau^2{\phi_-}^5 \tag{14b}$$

and

$$\phi_- \leq \phi_+ \tag{14c}$$

[4]Indeed, one has $\psi = \phi\theta^{-1}$.

[5]with appropriate specifications of function spaces for λ and σ

then there exists a function ϕ such that

$$\nabla^2\phi = \frac{1}{8}R\phi - \frac{1}{8}\sigma^2\phi^{-7} + \frac{1}{12}\tau^2\phi^5 \tag{15a}$$

and

$$\phi_- \leq \phi \leq \phi_+ \,. \tag{15b}$$

So to prove that the conformal method works for a given class of conformal data, all we need to do is show that for every set of conformal data in that class, there is a super solution $\phi_+ > 0$ satisfying inequality (14a) and a sub solution $\phi_- > 0$ satisfying inequalities (14b) and (14c). Exploiting the conformal covariance of the Lichnerowicz equation, we need only do this for sets of conformal data with $R = +1$, $R = 0$, or $R = -1$.

For three of the classes, finding a sub solution and a super solution is quite easy: indeed, for these three classes, both ϕ_+ and ϕ_- can be chosen to be *constants.* We list each of these classes here, and for each we write down the form that the Lichnerowicz equation takes, and we write out choices for ϕ_+ and ϕ_- which one can make:

- $(\mathcal{Y}^0, \sigma^2 \equiv 0, \tau = 0)$

$$\nabla^2\phi = 0 \tag{16a}$$
$$\phi_+ = 2 \tag{16b}$$
$$\phi_- = 1 \tag{16c}$$

(Here, for any constant C, $\phi = C$ solves (16a).)

- $(\mathcal{Y}^-, \sigma^2 \equiv 0, \tau \neq 0)$

$$\nabla^2\phi = -\frac{1}{8}\phi + \frac{\tau^2}{12}\phi^5 \tag{17a}$$
$$\phi_+ = \sqrt[4]{(\frac{3}{2})(\frac{1}{\tau^2})} \tag{17b}$$
$$\phi_- = \sqrt[4]{(\frac{3}{2})(\frac{1}{\tau^2})} \tag{17c}$$

(Here $\phi = \sqrt[4]{(\frac{3}{2})(\frac{1}{\tau^2})}$ solves (17a).)

- $(\mathcal{Y}^-, \sigma^2 \not\equiv 0, \tau \neq 0)$

$$\nabla^2\phi = -\frac{1}{8}\phi - \frac{1}{8}\sigma^2\phi^{-7} + \frac{\tau^2}{12}\phi^5 \tag{18a}$$
$$\phi_+ = \max\{1, \sqrt[4]{(\frac{3}{2})(\frac{1}{\tau^2})(1 + \max_{\Sigma^3}\sigma^2)}\} \tag{18b}$$
$$\phi_- = \sqrt[4]{(\frac{3}{2})(\frac{1}{\tau^2})} \tag{18c}$$

Notice that we can get away with a constant sub solution if we can find a constant such that the right-hand side of the Lichnerowicz equation is everywhere nonpositive. For the class $\{\mathcal{Y}^-, \sigma^2 \not\equiv 0, \tau = 0\}$, for example, we find that

$$\begin{aligned} -\frac{1}{8}\phi_- - \frac{1}{8}\sigma^2{\phi_-}^{-7} + \frac{1}{12}\tau^2{\phi_-}^5 &= -\frac{1}{8}\sigma^2[(\frac{3}{2})(\frac{1}{\tau^2})]^{-\frac{7}{4}} \\ &\leq 0\,. \end{aligned} \tag{19}$$

Similarly, we can use a constant ϕ_+ as a super solution if, when ϕ_+ is substituted into the right-hand side of the Lichnerowicz equation, we have a nonnegative quantity.

For the remaining three classes, we can find constant super solutions, but we generally can not find constant sub solutions. Let us consider, for example, the class $\{\mathcal{Y}^+, \sigma^2 \not\equiv 0, \tau = 0\}$. The Lichnerowicz equation takes the form

$$\nabla^2 \phi = \frac{1}{8}\phi - \frac{1}{8}\sigma^2 \phi^{-7} . \tag{20}$$

Clearly if we set

$$\phi_+ = \sqrt[8]{\max_{\Sigma^3} \sigma^2} , \tag{21}$$

we find that $\frac{1}{8}\phi_+ - \frac{1}{8}\sigma^2 {\phi_+}^{-7} \geq 0$, and so we have a constant super solution. If conformal data is chosen with σ^2 never zero, then $\phi_- = \sqrt[8]{\min_{\Sigma^3} \sigma^2}$ serves a constant sub solution as well. However if σ^2 has zeroes, then $\phi_- = \sqrt[8]{\min_{\Sigma^3} \sigma^2} = 0$ doesn't work, and indeed there are no positive constants which make the right-hand side of equation (20) nonpositive, so there are no constant sub solutions for such data.

Non constant sub solutions are harder to find than constant ones; generally some analysis is needed. However, we have shown (Isenberg [1987], [1995]) that for any set of conformal data in the class $\{\mathcal{Y}^+, \sigma^2 \not\equiv 0, \tau = 0\}$, there is a non constant sub solution.

The situation for the classes $\{\mathcal{Y}^0, \sigma^2 \not\equiv 0, \tau \neq 0\}$ and $\{\mathcal{Y}^+, \sigma^2 \not\equiv 0, \tau \neq 0\}$ is similar to that for $\{\mathcal{Y}^+, \sigma^2 \not\equiv 0, \tau = 0\}$; we readily find a constant super solution, but in general there is no constant sub solution. In earlier work (Isenberg [1995]), we handled these classes using the "O'Murchadha trick" – we found a scheme for producing conformal transformations of λ which have *non constant* scalar curvature R, with R going negative wherever σ^2 vanishes. Then, one *can* find a constant sub solution as well as a constant super solution. More recently (in the context of studying non constant mean curvature conformal data (Isenberg and Moncrief [1995]), we have found that we can leave R constant, and we can use methods similar to those employed for data in $\{\mathcal{Y}^+, \sigma^2 \not\equiv 0, \tau = 0\}$ to obtain non constant sub solutions. The table is now filled in:

	$(\sigma^2 \equiv 0, \tau = 0)$	$(\sigma^2 \equiv 0, \tau \neq 0)$	$(\sigma^2 \not\equiv 0, \tau = 0)$	$(\sigma^2 \not\equiv 0, \tau \neq 0)$
$\mathcal{Y}^+$	No	No	Yes	Yes
$\mathcal{Y}^0$	Yes	No	No	Yes
$\mathcal{Y}^-$	No	Yes	No	Yes

One should note that while the sub and super solution theorem can be used to verify the *existence* of solutions to the Lichnerowicz equation, it tells us nothing about *uniqueness*. To show uniqueness, we have relied upon the following result (Isenberg [1995]): Let $f : \Sigma^3 \times \mathbb{R}^+ \to \mathbb{R}$ [e.g., $f(x,\psi) = \frac{1}{8}R\psi - \frac{1}{8}\sigma^2\psi^{-7} + \frac{1}{12}\tau^2\psi^5$] satisfy the condition $\frac{\partial f}{\partial \psi}(x,\psi) \geq 0$ for all ψ in some interval I in $\mathbb{R}^+$. If ψ_1 and ψ_2 are both solutions of the PDE $\nabla^2\psi = f(x,\psi(x))$ and if $\psi_1(x)$ and $\psi_2(x)$ both take values in I for all $x \in \Sigma^3$, then $\psi_1(x) = \psi_2(x)$ for all $x \in \Sigma^3$. We prove this result in Isenberg [1995], and then use it there to show that for any given set of conformal data for which a solution to the Lichnerowicz equation exists[6], the solution is unique.

We also wish to point out that while there are six classes for which the conformal method works and six classes for which it doesn't, in a real sense the conformal method does work for *generic* CMC conformal data. This is because it works for the classes $\{\mathcal{Y}^+, \sigma^2 \not\equiv 0, \tau \neq 0\}$, $\{\mathcal{Y}^0, \sigma^2 \not\equiv 0, \tau \neq 0\}$ and $\{\mathcal{Y}^-, \sigma^2 \not\equiv 0, \tau \neq 0\}$. These are the generic classes; all of the others are, roughly speaking, sets of measure zero.

Non Constant Mean Curvature Data

Until recently, many workers believed the conjecture that every globally hyperbolic spacetime with compact Cauchy surfaces admits at least one Cauchy surface with constant mean curvature. The conjecture is now known to be false (Bartnik [1988], Eardley and Witt [1989]), at least for certain topologies and at least for solutions of Einstein's equations with matter sources. Hence it is useful to consider non CMC initial data (γ, K). The study is also motivated by the increased mathematical interest of the non CMC constraint equations.

The non CMC constraint equations are more difficult to handle than the CMC version of these equations because, with $\nabla\tau \neq 0$, one must work with four fully coupled partial differential equations (7) for four unknown functions (ϕ, W) rather than just one PDE (11) for one unknown function. While much less is known about the non CMC system, there has been considerable recent progress in understanding which sets of conformal data do lead to solutions of (7) and hence to non CMC solutions of the constraint equations (2). What we know is the following[7].

Theorem 1 (Choquet-Bruhat et al. [1992], Isenberg and Moncrief [1995]) *If we choose conformal data with $\lambda \in \mathcal{Y}^-$, with $\tau^2 > 0$, and with $|\nabla\tau|$ less than a constant depending upon λ and σ, then there exists a unique solution to equation* (7).

Theorem 2 (Isenberg and Moncrief [1996]) *If we choose conformal data with $\lambda \in \mathcal{Y}^+$ and with $|\nabla\tau|$ less than a constant depending upon λ and σ (τ^2 may have zeroes) then there exists a unique solution to equation* (7).

These results are of course quite limited. For most sets of non CMC conformal data, we don't know if equation (7) admits a solution. However the method which Moncrief and I have used[8] to prove Theorems 1 and 2 shows considerable promise, and it should be a useful tool for demonstrating that the conformal method works for much wider classes of conformal data.

The key to our method is the study of the semi-decoupled sequence of PDEs

$$\nabla_a(L \underset{n}{W})^a{}_b = \frac{2}{3} \underset{n-1}{\phi}{}^6 \nabla_b\tau \tag{22a}$$

and

$$\nabla^2 \underset{n}{\phi} = \frac{1}{8} R \underset{n}{\phi} - \frac{1}{8}(\sigma^{ab} + L \underset{n}{W}{}^{ab})(\sigma_{ab} + L \underset{n}{W}{}_{ab})\phi^{-7} + \frac{1}{12}\tau^2 \underset{n}{\phi}{}^5 . \tag{22b}$$

These equations are semi-decoupled in the sense that if one knows $\underset{n-1}{\phi}$, then equation (22a) is a (linear, elliptic) PDE for $\underset{n}{W}$ alone; and then once one knows $\underset{n}{W}$,

---

[6]excepting conformal data in the (trivial) class $\{\mathcal{Y}^0, \sigma^2 \equiv 0, \tau = 0\}$.

[7]Again, we leave out details concerning function spaces.

[8]Choquet-Bruhat uses a different (Leray-Schauder type) method to prove Theorem 1. Her proof is discussed in Choquet-Bruhat et al. [1992].

equation (22b) is a (quasilinear, Lichnerowicz-type) PDE for $\underset{n}{\phi}$ alone. The idea then is to (a) show that there is a sequence of solutions $\{(\underset{n}{\phi}, \underset{n}{W})\}$ to the sequence of equations (22); (b) show that the sequence $\{(\underset{n}{\phi}, \underset{n}{W})\}$ converges to some $(\underset{\infty}{\phi}, \underset{\infty}{W})$; and (c) show that $(\underset{\infty}{\phi}, \underset{\infty}{W})$ is a solution to equation (7) for the chosen set of conformal data.

We wish to discuss the steps of this method in a bit more detail, both to show how Theorems 1 and 2 are proven, and to indicate what one needs to do if one hopes to loosen some of the conditions in the hypotheses of these theorems.

First, we consider how one shows that the sequence $\{(\underset{n}{\phi}, \underset{n}{W})\}$ exists. The choice of $\underset{0}{\phi}$ is free (within certain upper and lower bounds; see Isenberg and Moncrief [1995]). Once $\underset{0}{\phi}$ is chosen, we obtain $\underset{1}{W}$ from equation (22b) with $n = 1$. This is a linear elliptic PDE, and standard theory (see, e.g., Besse [1987]) shows that for sufficiently nice λ and τ on a closed manifold Σ^3, a unique solution $\underset{1}{W}$ exists. Besides existence, we learn from standard theory that $\underset{1}{W}$ is controlled by certain regularity estimates. Most useful for our work is the sup inequality

$$|L\underset{1}{W}| < C \max_{\Sigma^3} |\nabla\tau| \max_{\Sigma^3} |\underset{0}{\phi}|^6 \tag{23a}$$

which holds for some constant $C(\lambda)$. The same existence and regularity results hold for other values of n; we have

$$|L\underset{n}{W}| < C \max_{\Sigma^3} |\nabla\tau| \max_{\Sigma^3} |\underset{n-1}{\phi}|^6 . \tag{23b}$$

These inequalities play a crucial role in later steps of this method.

For a given fixed value of $\underset{n}{W}$, equation (22b) is essentially the Lichnerowicz equation. We can therefore show that $\underset{n}{\phi}$ exists by applying the same sort of sub and super solution analysis as we used for CMC conformal data. For Theorem 1, in which case the conformal data has $\lambda \in \mathcal{Y}^-$ and $\tau^2 > 0$, equation (22b) takes the form

$$\nabla^2 \underset{n}{\phi} = -\frac{1}{8}\underset{n}{\phi} - \frac{1}{8}(\sigma + L\underset{n}{W})^2 \underset{n}{\phi}^{-7} + \frac{1}{12}\tau^2 \underset{n}{\phi}^5 \tag{24}$$

and we are essentially dealing (for each integer n) with the CMC conformal data class $(\mathcal{Y}^-, \sigma^2 \not\equiv 0, \tau \neq 0)$. We recall from our CMC study that for data in this class, we can always find constant sub and super solutions. The same is true here, for each n. This is to a large extent why Theorem 1 was our first result for non CMC data.

For Theorem 2, we have $\lambda \in \mathcal{Y}^+$, and now equation (22b) takes the form

$$\nabla^2 \underset{n}{\phi} = \frac{1}{8}\underset{n}{\phi} - \frac{1}{8}(\sigma + L\underset{n}{W})^2 \underset{n}{\phi}^{-7} + \frac{1}{12}\tau^2 \underset{n}{\phi}^5 . \tag{25}$$

Here, we are essentially dealing with CMC class $(\mathcal{Y}^+, \sigma^2 \not\equiv 0, \tau \neq 0)$, and we recall that for this class there is no constant sub solution. The same is true for equation (25). But again, with work we can show that there is always a non constant sub solution, and existence follows.

Note that for CMC class $(\mathcal{Y}^-, \sigma^2 \not\equiv 0, \tau = 0)$ there is no solution, while for CMC class $(\mathcal{Y}^+, \sigma^2 \not\equiv 0, \tau = 0)$ there is. This fact is closely related to our present requirement in Theorem 1 that $\tau \neq 0$, and to our ability to dispense with this requirement in Theorem 2.

Once we obtain the sequence $\{(\underset{n}{\phi}, \underset{n}{W})\}$ we need to show that the sequence converges. We do this by first establishing the existence of an (n-independent) upper and lower bound for the sequence of sub and super solutions $\{\underset{n}{\phi_-}, \underset{n}{\phi_+}\}$, and then using a contraction mapping argument to show that $\underset{n}{\phi}$ converges in C^0. The convergence of $\underset{n}{W}$ easily follows. These upper and lower bounds seem to be crucial to our analysis thus far. It is to establish them that we first need to impose the requirement that $|\nabla\tau| < C(\lambda, \sigma)$ for some constant $C(\lambda, \sigma)$. We note that the regularity estimates (23) are very important both in establishing the upper and lower bounds, and in carrying through the contraction mapping argument.

The bulk of the work is done once we know that $\{(\underset{n}{\phi}, \underset{n}{W})\}$ converges to some C^0 limit $\{\underset{\infty}{\phi}, \underset{\infty}{W}\}$. Standard bootstrap arguments show that $\{\underset{\infty}{\phi}, \underset{\infty}{W}\}$ are sufficiently differentiable, and it readily follows that $\{\underset{\infty}{\phi}, \underset{\infty}{W}\}$ solves equation (7). One can show uniqueness using an argument much like that which we use for CMC data.

There are three important restrictions on the conformal data appearing in the hypotheses of Theorems 1 and 2 which we would like to loosen. The first, appearing in Theorem 1, is the requirement that τ have no zeroes. This condition does not appear in Theorem 2, partly for reasons noted above. It would be particularly interesting to remove this condition from Theorem 1, because we would then have a straight path to an alternate proof of a result claimed (but not yet proved) by Eardley and Witt [1989]: That there exists spacetimes with compact Cauchy surfaces which are solutions of the vacuum Einstein equations but do not contain a CMC Cauchy surface. To accomplish the removal of the condition $\tau \neq 0$, we will need to devise a way to produce non constant super solutions, since if τ has zeroes and if $\lambda \in \mathcal{Y}^-$, there cannot be a constant super solution. Moncrief and I have made progress in recent months on a scheme for doing this.

The second restriction we would like to loosen appears in both Theorem 1 and Theorem 2: The requirement that $|\nabla\tau|$ be bounded by some constant $C(\lambda, \sigma)$. This requirement is not needed for establishing the existence of the sequence $\{(\underset{n}{\phi}, \underset{n}{W})\}$ in either theorem. But as noted above, it does play a big role in the proof that the sequence $\{(\underset{n}{\phi}, \underset{n}{W})\}$ converges. We do not yet see a promising way to remove the $|\nabla\tau| < C(\lambda, \sigma)$ condition. But we note that we have begun to examine an alternative, simpler, proof of convergence that could possibly help.

One might wonder whether in fact we should *expect* to be able to find solutions to equation (7) for conformal data with uncontrolled $|\nabla\tau|$. Certainly, in a given globally hyperbolic spacetime, it makes sense that one can locate a "very wrinkly" Cauchy surface $\hat{\Sigma}^3$, with $\max_{\hat{\Sigma}^3} |\nabla\tau|$ arbitrarily large. This suggests that perhaps we shouldn't need very tight controls on $|\nabla\tau|$. But the issue is far from resolved. We note that we do not yet know of any sets of non CMC conformal data for which there is no solution of equation (7).

The third restriction, made in the hypotheses of both theorems, is the condition that Σ^3 be closed. While this condition makes the mathematics simpler and is appropriate for studies of cosmological spacetimes, it is certainly important for gravitational studies of astrophysical processes to look at spacetimes with nonclosed

Cauchy surfaces, especially with asymptotically flat or asymptotically hyperbolic boundary conditions. There has been much research on the implementation of the conformal method for finding solutions of the constraints with such boundary conditions (Andersson and Chrusciel [1995], Brill and Cantor [1981], Cantor [1979], Choquet-Bruhat [1993], Choquet-Bruhat and York [1980]). We won't review that work here. We simply note that the conformal method has proven to be very useful: For CMC and maximal data, one understands fairly well which sets of conformal data map to asymptotically flat or asymptotically hyperbolic solutions, and which do not. For non CMC data, there has recently been small but reasonable progress for both asymtotically flat and asymptotically hyperbolic boundary conditions, and we expect to see further progress in the near future using methods similar to those discussed here.

References

Andersson, L. and Chrusciel, P. [1995] *Solutions of the Constraint Equations in General Relativity Satisfying 'Hyperboloidal Boundary Conditions'*, preprint.

Bartnik, R. [1988] *Remarks on Cosmological Spacetimes and Constant Mean Curvature Surfaces*, Comm. Math. Phys. **117**, 615–624.

Besse, A. [1987] *Einstein Manifolds*, Springer-Verlag, New York.

Brill, D. and Cantor, M. [1981] *The Laplacian on Asymptotically Flat Manifolds and the Specification of Scalar Curvature*, Compositio Mathematica **43**, 317–325.

Bruhat, Y. [1962] *The Cauchy Problem*, Gravitation: An Introduction to Current Research (Witten, L., ed.), Wiley, New York.

Cantor, M. [1979] *A Necessary and Sufficient Condition for York Data to Specify an Asymptotically Flat Spactime*, J. Math. Phys. **20**, 1741–1744.

Choquet-Bruhat, Y. [1993] *Solutions of the Coupled Einstein Constraints on Asymptotically Euclidean Manifolds*, Directions in General Relativity, Vol. 2 (Hu, B. L. and Jacobson, T. A., eds.) Cambridge Univ. Press, Cambridge.

Choquet-Bruhat, Y., Isenberg, J. and Moncrief, V. [1992] *Solutions of Constraints for Einstein Equations*, C. R. Acad. Sci. Paris **315**, 349–355.

Choquet-Bruhat, Y. and Geroch, R. [1969] *Global Aspects of the Cauchy Problem in General Relativity*, Comm. Math. Phys. **14**, 329–335.

Choquet-Bruhat, Y. and York, J. [1980] *The Cauchy Problem*, General Relativity and Gravitation (Held, A., ed.), Plenum, New York.

Eardley, D. and Witt, D. [1989], unpublished.

Isenberg, J. [1987] *Parametrization of the Space of Solutions of Einstein's Equations*, Phys. Rev. Lett. **59**, 2389–2392.

Isenberg, J. [1995] *Constant Mean Curvature Solutions of the Einstein Constraint Equations on Closed Manifolds*, Class. Quant. Grav. **12**, 2249–2273.

Isenberg, J. and Moncrief, V. [1996] *A Set of Non Constant Mean Curvature Solutions of the Einstein Constraint Equations on Closed Manifolds*, Class. Quant. Grav., **13**, 1819-1847.

Isenberg, J. and Moncrief, V. [1996] *A Set of Positive Yamabe Class Non Constant Mean Curvature Solutions of the Einstein Constraint Equations*, unpublished.

Lee, J. and Parker, I. [1987] *The Yamabe Problem*, Bull. A.M.S. **17**, 37–91.

Schoen, R. [1984] *Conformal Deformation of a Riemannian Metric to Constant Curvature*, J. Diff. Geom. **20**, 479–495.

Fields Institute Communications
Volume **15**, 1997

Covariant Double-Null Dynamics and the Internal Geometry of Aspherical Black Holes

Werner Israel
Canadian Institute for Advanced Research
Cosmology Program
Theoretical Physics Institute
University of Alberta
Edmonton, Alberta, Canada T6G 2J1
israel@euclid.phys.ualberta.ca

Abstract. In the 1970's, a series of mathematical theorems established that astrophysical black holes are surprisingly simple, "hairless" objects. It appeared that their external geometries are fully characterized by no more than three parameters: mass, electric charge and angular momentum. In contrast to the quiescent exterior, the interior of the hole is highly dynamical. Yet, here too, recent analytical and numerical studies seem to be converging to the conclusion that, asymptotically (at late advanced times), the internal geometry assumes a universal, well-defined form. Observers falling into the hole at late times encounter a singularity which is lightlike. It occupies the surface of a three-dimensional cylinder, which contracts to eventually form a spacelike crushing singularity.

1 Introduction

No-one more surely invites ridicule as a simpleton, charlatan or crackpot than he who expresses serious interest in the inner workings of a black hole. No matter what happens inside the hole, we can never observe it, nor will it affect anything we can observe. Mathematical theorems of Penrose and Hawking assure us of the presence of a singularity within the hole, signalling a breakdown of all the laws of physics. And if that were not enough, irregularities swept into the hole (along with the collapsing star that formed it) could be expected to pile up internally, producing a condition of complete internal chaos. On the face of it, this hardly seems a fit subject for a serious scientist of sound views.

1991 *Mathematics Subject Classification.* Primary 83C57; Secondary 35Q75.

Much of the work reported here is based on a stimulating collaboration with Alfio Bonanno, Pat Brady, Serge Droz, Sharon Morsink and Eric Poisson. It is a pleasure to express acknowledgment to them, and more specifically to Pat Brady for permission to reproduce Figures 3 and 4. This work was supported by the Canadian Institute for Advanced Research and by NSERC of Canada.

© 1997 American Mathematical Society

I should therefore begin by explaining why I believe the outlook is not quite as bleak as I have painted it. I have so far neglected to mention a key property of the black hole interior: *descent into a black hole is fundamentally a progression in time.* (As the simplest example, in the case of the Schwarzschild metric

$$ds^2 = \frac{dr^2}{1 - 2m/r} + r^2 d\Omega^2 - (1 - 2m/r)dt^2$$

for an uncharged, spherical hole, the radial co-ordinate r becomes timelike inside the Schwarzschild radius: $r < 2m$.) This means we are confronted with an evolutionary, not a structural problem. Therefore we do not need an understanding of physics in the Planck (quantum-gravity) regime to follow the evolution up to the stage where curvatures begin to approach Planck levels near the (classically-predicted) singularity. Simple causality prevents our ignorance of the inner, high-curvature regions from infecting the description that present-day theories provide of the outer and *preceding* layers.

Moreover, initial conditions for the evolution are known with precision, thanks to the "no-hair" theorem. (Contrast the situation in cosmology.) Initially, near the outer boundary (event horizon) of the hole, we have a Kerr-Newman geometry, perturbed by a tail of gravitational waves left in the wake of the collapse and decaying according to an inverse power v^{-p} of advanced time v. The task, then, is to evolve these initial data, using the Einstein field equations, up to the point where a singularity forms or curvatures reach Planck levels.

Exploration of the outer (sub-Planckian) layers of the hole is thus seen to reduce to a standard evolutionary problem of Cauchy's type, no more speculative than following the motion of a fluid, using Euler's equations, up to the onset of turbulence or a shock. Technically, however, it is considerably more complex!

In order to simplify the mathematics, the first studies of this problem (Poisson and Israel [1990] and Ori [1991]), employed idealized spherical models. The results which emerged were sufficiently bizarre to arouse general unease as well as specific reservations. During the past year, however, the dust seems to have settled to the extent that one can claim at least the classical structure of spherical holes to be well understood (Bonanno et al. [1994a], [1995b] and Brady and Smith [1995]), and the focus of attention has shifted to the development of efficient techniques for handling the generic case.

In this brief report, I shall concentrate on these latest developments. For introductory treatments and broader surveys the reader may consult Droz et al. [1996], Bonanno et al. [1994b] and [1995a], or Israel [1994].

2 Structure of Charged Spherical Black Holes

In 1989, Eric Poisson and I examined the effect of a spherical (e.g., scalar-field) wave tail on the internal structure of a charged spherical hole.

We found it convenient to employ a two-dimensional reduction of the spherisymmetric Einstein field equations. Any spherisymmetric 4-metric can be expressed in the form

$$ds^2 = g_{AB}(x^C)\, dx^A\, dx^B + r^2\, d\Omega^2, \qquad (A, B, C, \dots = 0, 1) \tag{1}$$

where $d\Omega^2$ is the metric on the unit 2-sphere, (x^0, x^1) are any pair of co-ordinates which label the privileged 2-spheres, and $r(x^A)$ is a function of these co-ordinates.

Its gradient defines a "mass function" $M(x^A)$:

$$g^{AB}(\partial_A r)(\partial_B r) = 1 - 2M/r.$$

Physically, one may think of $M(x^A)$ as the effective gravitational mass interior to the 2-sphere labelled by x^A. Somewhat more accurately, one can say that $\Psi = M/r^3$ is an invariant measure of the Weyl conformal curvature. (For spherical symmetry the Weyl tensor has only one linearly independent component.)

The Einstein field equations can now be reformulated as two-dimensionally covariant equations for the scalars $r(x^A)$, $M(x^A)$. The key result is a $(1+1)$-dimensional wave equation for M, which takes the form (leaving aside some unessential terms, see Poisson and Israel [1990])

$$\Box M = -16\pi^2 r^3 T^{AB} T_{AB}. \tag{2}$$

Effects of the nonlinearity of the Einstein equations are made manifest in this formula.

The crucial internal feature—which a spherical charged hole shares with the generic (rotating) black hole—is an inner, "Cauchy" horizon, a lightlike hypersurface corresponding to infinite external advanced time v, but reachable in finite proper time by observers who fall into the hole. Wave-tails and other time-dependent disturbances propagating into the hole experience infinite blueshift at the Cauchy horizon, causing the source term on the right side of (2) to diverge inside the hole as $v \to \infty$.

This produces a "mass-inflationary" singularity at the Cauchy horizon (CH, see Figures 1, 2). For a wave tail whose flux decays externally like v^{-p}, M internally diverges like

$$M \approx C|uv|^{-p} e^{\kappa_0(u+v)}$$

while externally it remains nearly constant. Here, u is an internal retarded time, calibrated so that $u = -\infty$ on the event horizon EH, and C and κ_0 are positive constants, κ_0 being the inner surface gravity of the static (Reissner-Nordström) black hole to which the external geometry tends asymptotically.

This conclusion hinges on the presence of at least a nominal outflux (transverse to CH). (For a pure influx, no matter how blueshifted, T_{ab} is lightlike, and its square on the right-hand side of (2) vanishes.) In a realistic setting, an outflux is inevitably produced by backscatter of incoming waves off the curvature inside the hole.

One could imagine that this uncontrollable backscatter might lead to problems, since it implies that the transverse flux across CH falls off only as an inverse power $|u|^{-p}$ of retarded time u toward the past of CH ($u \to -\infty$). Conceivably, this might not be fast enough to allow the Cauchy horizon to begin from an asymptotically static, Reissner-Nordström configuration and might lead to its pre-emption by a spacelike singularity. Two years ago, a numerical integration of scalar wave scattering by a charged spherical hole aroused suspicions that this would actually happen (Gnedin and Gnedin [1993]).

However, analytical investigations do not support these suspicions (Bonanno et al. [1994a] and [1995b]). Instead of entering into the gory details of this, I shall rely on the maxim that seeing is believing, and reproduce in Figures 3 and 4 some typical results of recent numerical integrations by Brady and Smith [1995], which fully confirm the picture of a lightlike, mass-inflationary singularity along the Cauchy horizon.

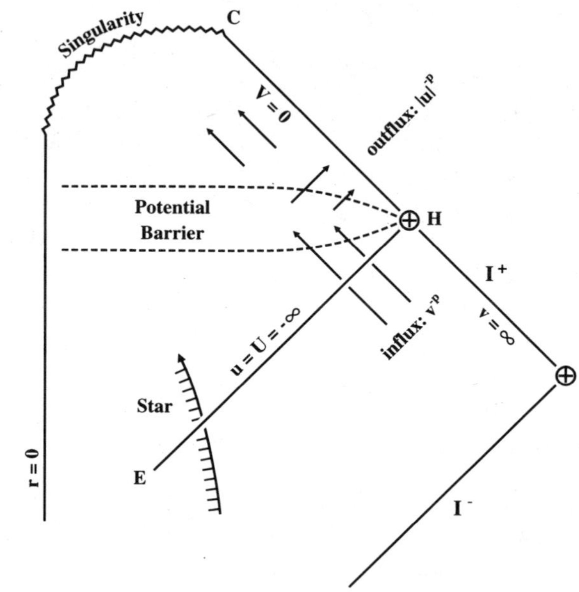

Figure 1 Waves propagating into a charged spherical hole are scattered off a potential barrier situated between the event horizon EH and the Cauchy horizon CH. This is a Penrose conformal map: tracks of radial photons are straight lines which are inclined at $\pm 45°$, and past and future lightlike infinity have been "compactified" to the lines $\mathcal{I}^-$ and $\mathcal{I}^+$ by a conformal transformation. "Points" such as H, marked by circled crosses, are singularities of the conformal mapping. In reality, EH and CH are 3-cylinders of different radii which extend indefinitely without intersecting.

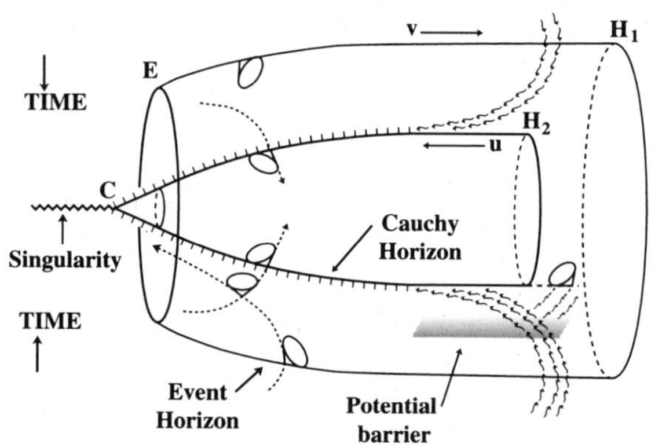

Figure 2 Geometrically faithful rendering of the scenario of Figure 1. One angular co-ordinate has been suppressed. Inside the hole, increasing time coincides with diminishing radius. The points H_1 and H_2 of this figure were telescoped into a single "point" H in Figure 1.

3 Structure of Aspherical Holes

In the aspherical case, one expects the mass-inflationary singularity to be overlaid by a gravitational shock wave imploding along the Cauchy horizon. There is a

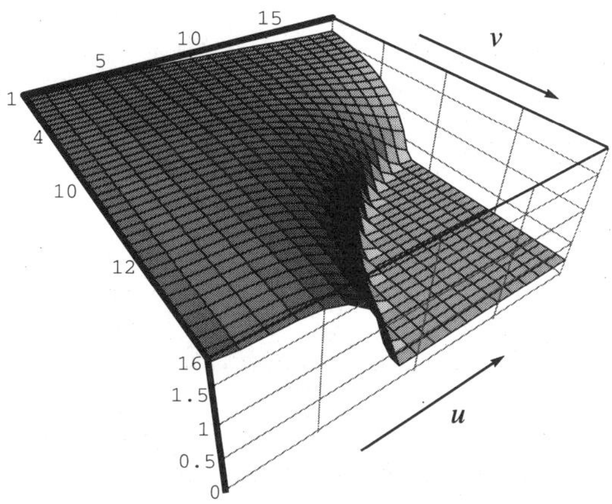

Figure 3 Effect of scalar wave-tail on the internal geometry of a charged spherical black hole. The u- and v-axes mark the Cauchy and event horizons respectively. This three-dimensional plot of the radial co-ordinate $r(u, v)$ shows the radius of the Cauchy horizon decreasing continously to zero with increasing retarded time u.

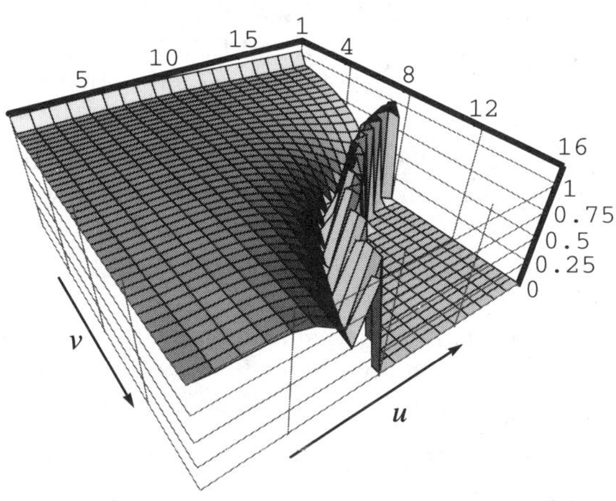

Figure 4 A plot of $\exp(1 - 2M/r)$, showing the effect of mass inflation near the Cauchy horizon. Figures 3 and 4 are from numerical integrations by P. R. Brady and J. D. Smith [1995], reproduced by kind permission of the authors.

growing body of evidence that, apart from this, the spherical picture remains qualitatively valid (Ori [1992], Bonanno et al. [1994b] and [1995a], Brady and Chambers [1995], and Brady et al. [1995]).

Again, doubts have been expressed. Yurtsever [1993] has pointed out that Cauchy horizons are generically spacelike, at least for plane-wave spacetimes. However, the plane-wave analogy appears misleading (Bonanno et al. [1994b] and [1995a] and Flanagan and Ori [1996]). (The concept of "Cauchy horizon" is not necessarily an absolute one, but relative to the hypersurface on which the initial data for the Cauchy problem are set. In the black hole case, the Cauchy horizon for externally set initial data precedes what one would normally think of as the plane-wave

Cauchy horizon in a homogeneous space.)

A comprehensive general treatment of the aspherical case requires development of techniques for studying the gravitational field in the neighbourhood of an asymmetric lightlike singularity. In the remainder of this report, I shall introduce the bare outlines of a versatile mathematical technique built for this purpose.

4 Covariant Double-Null Dynamics

The classic analysis of Arnowitt et al. [1962] formulates gravitational dynamics in terms of the evolution of a spatial 3-geometry. The geometrical framework is the imbedding formalism of Gauss and Codazzi for the foliation of spacetime by spacelike hypersufaces.

Quite often, however, —as in the analysis of Cauchy horizon singularities, but also in other contexts—one encounters circumstances where a lightlike foliation is especially suitable. Because of various degeneracies that arise in the lightlike case, the imbedding relations are quite different and the situation not quite as familiar and under control. To bypass the degeneracies, one is forced to fall back to a foliation of co-dimension 2, by spacelike 2-surfaces.

Several such (2 + 2)-formalisms are extant, notably Geroch et al. [1973], d'Inverno and Smallwood [1980] and Hayward [1993]. Of course, all have basically the same content, but they take very different forms.

The distinguishing feature of the approach I shall present here is that it maintains manifest two-dimensional covariance and operates directly with geometrically significant objects. Two-dimensional covariance permits reduction of the Einstein field equations to an especially concise and geometrically transparent form: the ten Ricci components are embraced in a set of just three compact, two-dimensionally covariant expressions.

(2 + 2)-formalisms are a versatile tool having a range of applications which go far beyond the specific problem of Section 3. They are useful in the analysis and numerical integration of the characteristic initial-value problem of general relativity, the dynamics of strings, of real and apparent horizons, the study of Planck-energy collisions and light-cone quantization. The covariant formalism has been applied by our group (Brady et al. [1995]) to elucidate the singular geometry of the black hole Cauchy horizon.

5 Basic Metric Notions

I shall briefly sketch the elements of the formalism.

We consider a double foliation of spacetime by a net of two intersecting families of lightlike hypersurfaces Σ^0 (with equations u^0 = const.) and Σ^1 (given by u^1 = const.), where $u^A(x^\alpha)$ $(A, B, \ldots, = 0, 1;\ \alpha, \beta, \ldots, = 1, \ldots, 4)$ are a given pair of scalar fields over spacetime, with lightlike gradients:

$$\nabla u^A \cdot \nabla u^B = e^{-\lambda} \eta^{AB},$$

and where the matrix

$$\eta^{AB} = \eta_{AB} = \begin{pmatrix} 0 & -1 \\ -1 & 0 \end{pmatrix}$$

will be used to raise and lower upper-case Latin indices, and $\lambda(x^\alpha)$ is a scalar function. The generators $\ell^{(A)}_\alpha$ of Σ^A are conveniently defined as

$$\ell^{(A)} = e^\lambda \nabla u^A.$$

Two hypersurfaces Σ^0 and Σ^1 intersect in a 2-surface S, with parametric equations

$$x^\alpha = x^\alpha(u^A, \theta^a) \qquad (a, b, \cdots = 2, 3)$$

where (θ^2, θ^3) are intrinsic co-ordinates of S. Both generators $\ell^{(A)}$ are orthogonal to S.

Holonomic basis vectors $e_{(a)}$ and the intrinsic metric of S may now be defined:

$$e^\alpha_{(a)} = \frac{\partial x^\alpha}{\partial \theta^a}, \qquad g_{ab} = e_{(a)} \cdot e_{(b)}.$$

The matrix g_{ab} and its inverse g^{ab} are used to lower and raise lower-case Latin indices, so that $e^{(a)} = g^{ab} e_{(b)}$ are the dual basis vectors tangent to S.

Two-dimensional shift vectors s^a_A are defined by

$$s^a_A = \frac{\partial x^a}{\partial u^A} e^{(a)}_\alpha = -\ell^\alpha_{(A)} \frac{\partial \theta^a}{\partial x^\alpha}.$$

As in the Arnowitt-Deser-Misner formalism, the shift vector s^a_A measures how much one has to deviate from the normal direction $\ell_{(A)}$ to connect points on different 2-surfaces having the same intrinsic co-ordinates θ^a. An infinitesimal four-dimensional displacement dx^α can be decomposed as

$$dx^\alpha = \ell^\alpha_{(A)} du^A + e^\alpha_{(a)} (d\theta^a + s^a_A du^A).$$

Together with the completeness relation

$$g_{\alpha\beta} = e^{-\lambda} \eta_{AB} \ell^{(A)}_\alpha \ell^{(B)}_\beta + g_{ab} e^{(a)}_\alpha e^{(b)}_\beta$$

for the basis $(\ell^{(A)}, e_{(a)})$, this implies that the spacetime metric is decomposable as

$$g_{\alpha\beta} dx^\alpha \, dx^\beta = e^\lambda \eta_{AB} du^A \, du^B + g_{ab}(d\theta^a + s^a_A du^A)(d\theta^b + s^b_B \, du^B).$$

6 Covariant Geometrical Objects Embodying First Derivatives of Metric

Associated with its two normals $\ell_{(A)}$, a 2-surface S has two extrinsic curvatures defined by

$$K_{Aab} = (\nabla_\beta \ell_{(A)\alpha}) e^\alpha_{(a)} \, e^\beta_{(b)}$$

and easily shown to be symmetric in a, b. (Since we are free to rescale the null vectors $\ell_{(A)}$, a certain scale-arbitrariness is inherent in this definition.)

A further basic geometrical property of the double foliation is given by the Lie bracket of $\ell_{(0)}$ and $\ell_{(1)}$. One finds

$$[\ell_{(B)}, \ell_{(A)}] = \epsilon_{AB} \omega^a e_{(a)} \tag{3}$$

where

$$\omega^a = \epsilon^{AB} (\partial_B s^a_A - s^b_B s^a_{A;b}),$$

the semi-colon indicates two-dimensional covariant differentiation associated with metric g_{ab}, and ϵ_{AB} is the two-dimensional permutation symbol.

The geometrical significance of the "twist" ω^a can be read off from (3): the curves tangent to the generators $\ell_{(0)}$, $\ell_{(1)}$ mesh together to form 2-surfaces (orthogonal to the surfaces S) if and only if $\omega^a = 0$. In this case, it would be consistent to allow the co-ordinates θ^a to be dragged along both sets of generators, and thus to gauge both shift vectors to zero.

I shall denote by D_A the two-dimensionally invariant operator associated with differentiation along the normal direction $\ell_{(A)}$. Acting on any two-dimensional geometrical object $X^{a\cdots}_{b\ldots}$, D_A is formally defined by

$$D_A X^{a\ldots}_{b\ldots} = (\partial_A - \mathcal{L}_{s^d_A}) X^{a\ldots}_{b\ldots}.$$

Here, ∂_A is the partial derivative with respect to u^A and $\mathcal{L}_{s^d_A}$ the Lie derivative with respect to the 2-vector s^d_A. As an example:

$$D_A g_{ab} = \partial_A g_{ab} - 2 s_{A(a;b)} = 2K_{Aab}.$$

Geometrically, $D_A X^{a\cdots}_{b\ldots}$ is the projection onto S of the Lie derivative with respect to $\ell_{(A)}$ of the equivalent tangential 4-tensor $X^{\alpha\cdots}_{\beta\ldots}$.

The objects K_{Aab}, ω^a and D_A are all simple projections onto S of four-dimensional geometrical objects. Consequently, they transform very simply under two-dimensional co-ordinate transformations. Under the arbitrary reparametrization

$$\theta^a \to \theta^{a'} = f^a(\theta^b, u^A) \tag{4}$$

(which leaves u^A and hence the surfaces Σ^A and S unchanged), ω_a and K_{Aab} transform cogrediently with

$$e_{(a)} \to e'_{(a)} = e_{(b)} \partial\theta^b / \partial\theta^{a'}.$$

By contrast, the shift vectors s^a_A undergo a more complicated gauge-like transformation, arising from the u-dependence in (4).

7 Ricci Tensor

This geometrical groundwork is already sufficient to allow me to display the simple form that the Ricci components take in this formalism. (Notation for the tetrad components is typified by $R_{aA} = R_{\alpha\beta} e^\alpha_{(a)} \ell^\beta_{(A)}$.)

The results are

$$\begin{aligned}
{}^{(4)}R_{ab} &= \frac{1}{2}\,{}^{(2)}R g_{ab} - e^{-\lambda}(D_A + K_A)K^A_{ab} \\
&\quad + 2e^{-\lambda} K_{A(a}{}^d K^A_{b)d} - \frac{1}{2} e^{-2\lambda}\omega_a\omega_b - \lambda_{;ab} - \frac{1}{2}\lambda_{,a}\lambda_{,b} \\
R_{AB} &= -D_{(A}K_{B)} - K_{Aab}K_B{}^{ab} + K_{(A}D_{B)}\lambda \\
&\quad - \frac{1}{2}\eta_{AB}\left[(D^E + K^E)D_E\lambda - e^{-\lambda}\omega^a\omega_a + (e^\lambda)^{;a}{}_a\right] \\
R_{Aa} &= K^b_{Aa;b} - \partial_a K_A - \frac{1}{2}\partial_a D_A\lambda + \frac{1}{2}K_A\partial_a\lambda \\
&\quad + \frac{1}{2}\epsilon_{AB}e^{-\lambda}\left[(D^B + K^B)\omega_a - \omega_a D^B\lambda\right],
\end{aligned}$$

where ${}^{(2)}R$ is the curvature scalar associated with the 2-metric g_{ab}, and $K_A \equiv K^a_{Aa}$.

It should be clear that this provides a remarkably economical and geometrically transparent generalization of the $(2+2)$ spherical formalism, whose application to spherical black holes was outlined in Section 2. For further details and developments, I must refer to the original papers (Brady et al. [1995] and [1996]).

8 Summary and Conclusion

It appears that the inside of a black hole is, after all, not completely inscrutable. The investigations I have described lead to a rather definite working picture of at least the outer, classical layers of the hole, illustrated in Figures 1 and 2.

The crushing spacelike singularity has a milder, lightlike precursor, situated on the Cauchy horizon CH. In effect, CH serves as a lightlike bridge, linking the quiescent, asymptotically hairless, final phases of the hole's history (at late advanced times v) to the "hairy" crunch near C associated with its formation.

Observers entering the hole long after it is formed fall toward the Cauchy horizon and enter the Planck layer close to CH. What happens then? On this question, contemporary theory is silent. We have only charted a coastline; we do not know whether it heralds a narrow strip of land or a continent. Its further exploration is a task for physics of the millenium.

References

Arnowitt, R., Deser, S. and Misner, C. W. [1962] *Gravitation: an Introduction to Current Research,* Chapter 7 (Witten, L., ed.), Wiley, New York, pp. 227–265.

Bonanno, A., Droz, S., Israel, W. and Morsink, S. M. [1994a] *Structure of the inner singularity of a spherical black hole,* Phys. Rev. D **50**, 7372–7375.

______ [1994b] *The internal evolution of black holes,* Can. J. Phys. **72**, 755–759.

______ [1995a] *erratum,* Can. J. Phys. **73**, 251.

______ [1995b] *Structure of the charged spherical black hole interior,* Proc. Roy. Soc. A **450**, 553–567.

Brady, P. R. and Chambers, C. M. [1995] *Nonlinear instability of Kerr-type Cauchy horizons,* Phys. Rev. D **51**, 4177–4186.

Brady, P. R., Droz, S., Israel, W. and Morsink, S. M. [1996] *Covariant double-null dynamics: (2+2) - splitting of the Einstein equations,* Class. Quantum Grav. **13**, 2211-2230.

______ [1995], in preparation.

Brady, P. R. and Smith, J. D. [1995] *Black hole singularities: a numerical approach,* Phys. Rev. Lett. **75**, 1256–1259.

d'Inverno, R. A. and Smallwood, J. [1980] *Covariant 2+2 formulation of the initial-value problem in general relativity,* Phys. Rev. D **22**, 1233–1247.

Droz, S., Israel, W. and Morsink, S. M. [1996] *Black holes: the inside story,* Physics World **9**, 34–37.

Flanagan, E. and Ori, A. [1996] *How generic are null spacetime singularities?,* Phys. Rev. D **53**, R1754–R1758.

Geroch, R., Held, A. and Penrose, R. [1973] *A space-time calculus based on pairs of null directions,* J. Math. Phys. **14**, 874–881.

Gnedin, M. L. and Gnedin, N. Y. [1993] *Destruction of the Cauchy horizon in the Reissner-Nordström black hole,* Class. Quantum Grav. **10**, 1083–1102.

Hayward, S. A. [1994] *Dual-null dynamics of the Einstein field,* Class. Quantum Grav. **10**, 779–790.

Israel, W. [1994] *The internal geometry of black holes,* Contemp. Math. **170**, 155–165.

Ori, A. [1991] *Inner structure of a charged black hole: an exact mass-inflation solution,* Phys. Rev. Lett. **67**, 789–792.

______ [1992] *Structure of the singularity inside a realistic rotating black hole,* Phys. Rev. Lett. **68**, 2117–2120.

Poisson, E. and Israel, W. [1990] *Internal structure of black holes,* Phys. Rev. D **41**, 1796–1809.

Yurtsever, U. [1993] *Comments on the instability of black-hole inner horizons,* Class. Quantum Grav. **10**, L17–L24.

Fields Institute Communications
Volume **15**, 1997

Charged Black Holes in Generic 2-D Dilaton Gravity

Domingo Louis-Martinez
Department of Physics and
Winnipeg Institute for Theoretical Physics
University of Winnipeg
Winnipeg, Manitoba, Canada R3B 2E9
and
Physics Department
University of Manitoba
Winnipeg, Manitoba, Canada R3T 2N2

Gabor Kunstatter
Department of Physics and
Winnipeg Institute for Theoretical Physics
University of Winnipeg
Winnipeg, Manitoba, Canada R3B 2E9
gabor@theory.uwinnipeg.ca

Abstract. We summarize recent results concerning the classical mechanics, thermodynamics and quantum mechanics of charged black holes in generic 2-D dilaton gravity. In particular, the general solutions are explicitly found and used to derive the Hawking temperature and entropy of the corresponding black holes. The temperature is proportional to the surface gravity as expected, and vanishes in the case of extremal black holes. The generic theory is then quantized in the functional Schrödinger representation. The quantum constraints are solved to yield the physical charge and energy eigenstates in a WKB approximation. An intriguing relationship is found between the imaginary part of the WKB phase and the thermodynamical entropy for all black holes.

1 Introduction

There is growing astronomical and astrophysical evidence for the existence of black holes, both in binary systems and at the center of some galaxies. It is well known that black holes emit thermal radiation (Hawking [1975]) and appear to have thermodynamical entropy (Bekenstein [1973]). However, these facts pose fundamental theoretical problems concerning the statistical mechanical source of black

1991 *Mathematics Subject Classification.* Primary 83C57; Secondary 83C45.

Some of the work presented here was done in collaboration with Jack Gegenberg and Patrick Sutton, to whom we are grateful for many useful discussions. This work was supported in part by the Natural Sciences and Engineering Research Council of Canada.

© 1997 American Mathematical Society

hole entropy and the endpoint of gravitational collapse (see Page [1994], for example). It seems clear that in order to address these questions it is necessary to go beyond the usual semiclassical approximation. Since this is a formidable task in 3+1-dimensional general relativity, much work has focussed on field theoretic models for black holes in 1+1 spacetime dimensions. These models have the advantage of being simple enough to be solvable, but complicated enough to contain the key features relevant to issues surrounding black hole thermodynamics. Most researchers consider specific theories, such as Jackiw-Teitelboim gravity (Jackiw [1984] and Teitelboim [1984]) or the string inspired model (see for example Frolov [1992]). Recently, however, it has been shown (Gegenberg et al. [1995] and Louis-Martinez and Kunstatter [1995]) that it is possible to discuss within a unified framework the thermodynamics and quantum mechanics of black holes in generic dilaton gravity, which contains as special cases all models previously analyzed. Moreover, if one considers black holes in $n + 2$ dimensions with exact n-dimensional spherical symmetry, the effective theory obtained by truncating the non-spherically symmetric modes is also a special case of the generic model. One of the advantages of studying all such theories in a unified framework is that one can look for model independent features, which have a better chance of being physically relevant.

In the present work we will summarize recent results concerning the quantum mechanics and thermodynamics of black holes in generic dilaton gravity coupled to a U(1) gauge field[1]. The charge associated with the U(1) gauge field can be thought of as electric charge (as in spherically symmetric Einstein-Maxwell theory), but other physical interpretations are possible. For example, the Jackiw-Teitelboim model can be obtained by imposing axial symmetry on Einstein gravity with cosmological constant in 2+1 dimensions, in which case the U(1) gauge field is associated with rotations about the axis of symmetry, and the corresponding charge is the angular momentum (Achucarro et al. [1993]). It is important to understand the effects of such charges in general, since they may in principle provide a mechanism for stopping the radiation process: the Hawking temperature of extremal black holes is zero and recent arguments suggest that the entropy is also zero (Hawking et al. [1995] and Teitelboim [1995]).

The paper is organized as follows. In Section 2 we write down the family of actions that we wish to consider and describe the covariant space of solutions. The Hamiltonian analysis is summarized in Section 3, while Section 4 derives the classical thermodynamical properties of charged black holes. In Section 5 the theories are Dirac quantized using the functional Schrödinger representation. It is here that our main result concerning the relationship between the imaginary part of the WKB phase and the thermodynamic entropy is discussed. Finally, Section 6 closes with conclusions and prospects for future work.

2 Generic Maxwell-Dilaton Gravity

The most general coordinate invariant action functional in two spacetime dimensions depending on the metric tensor $g_{\mu\nu}$, scalar field ϕ and vector potential A_μ that contains at most second derivatives of the fields can be written (Banks et al.

[1]The perturbative quantization of this model has recently been considered by Elizalde and Odintsov [1995].

[1991], Mann [1993] and Elizalde and Odintsov [1995])[2],

$$S[g,\phi,A] = \int d^2x\sqrt{-g}\left[\frac{1}{2G}\left(\phi R(g) + \frac{1}{l^2}V(\phi)\right)\right.$$
$$\left. -\frac{1}{4}W(\phi)F^{\mu\nu}F_{\mu\nu}\right], \tag{2.1}$$

where $R(g)$ is the Ricci curvature scalar and $F_{\mu\nu} = \partial_\mu A_\nu - \partial_\nu A_\mu$ is the Abelian field strength tensor[3]. $V(\phi)$ and $W(\phi)$ are arbitrary functions of the dilaton field. It should be noted that in the above, the fields ϕ, $g_{\mu\nu}$ and A_μ are dimensionless, as is the 2-D Newton constant, G. This requires the inclusion of a coupling constant, l, of dimension length in the potential term. The field equations that follow from the action (2.1) are:

$$R + \frac{1}{l^2}\frac{dV}{d\phi} - \frac{G}{2}\frac{dW}{d\phi}F^{\alpha\beta}F_{\alpha\beta} = 0 \tag{2.2}$$

$$\nabla_\mu\nabla_\nu\phi - \frac{1}{2l^2}g_{\mu\nu}V(\phi) + G\left(\delta^\alpha_\mu\delta^\beta_\nu - \frac{3}{4}g_{\mu\nu}g^{\alpha\beta}\right)W(\phi)F_\alpha{}^\gamma F_{\beta\gamma} = 0 \tag{2.3}$$

$$\nabla_\nu\left(W(\phi)F^{\mu\nu}\right) = 0. \tag{2.4}$$

One important aspect of the field equations is that all solutions have (at least) one Killing vector that leaves the metric, dilaton and vector potential invariant. It is:

$$k^\mu = \frac{l}{\sqrt{-g}}\epsilon^{\mu\nu}\partial_\nu\phi \tag{2.5}$$

The most general solution (Louis-Martinez and Kunstatter [1995]) can be written in terms of two coordinate invariant constants of motion, C and q, which can be interpreted as the energy and electric charge, respectively. This fact, plus the existence of the Killing vector (2.5), can be thought of as constituting the generalized Birkhoff Theorem for 2-D dilaton gravity. In "Reissner-Nordstrom gauge", defined by $x = \phi l$ and $g_{01} = 0$ these solutions take the form:

$$\begin{aligned} ds^2 &= -f(\frac{x}{l};C,q)dt^2 + f(\frac{x}{l};C,q)^{-1}dx^2 \\ F &= \frac{q}{W(\frac{x}{l})} \end{aligned} \tag{2.6}$$

where we have defined:

$$f(\phi;C,q) = -(l^2C - j(\phi) + l^2Gq^2K(\phi)) \ , \tag{2.7}$$

with

$$\begin{aligned} j(\phi) &= \int d\phi\, V(\phi) \\ K(\phi) &= \int \frac{d\phi}{W(\phi)}. \end{aligned}$$

[2]One can also add a term of the form $H(\phi)\epsilon^{\mu\nu}F_{\mu\nu}$ but we will not discuss this here. We are grateful to Roman Jackiw for pointing this out.

[3]The non-Abelian case has recently been quantized using slightly different techniques by Strobl [1994].

The constants of motion C and q can be written as scalar functions of the fields:

$$C = -g^{\mu\nu}\phi_{;\mu}\phi_{;\nu} + \frac{j(\phi)}{l^2} - Gq^2K(\phi) \tag{2.8}$$

$$q = \frac{1}{\sqrt{-g}}W(\phi)F\ . \tag{2.9}$$

In terms of these physical observables, the norm of the Killing vector is:

$$\begin{aligned} |k|^2 &= -l^2 g^{\alpha\beta}\partial_\alpha\phi\,\partial_\beta\phi \\ &= l^2C - j(\phi) + l^2Gq^2K(\phi) \\ &= -f(\phi; C, q). \end{aligned} \tag{2.10}$$

We now discuss the existence of black hole solutions. For a given solution to have an event horizon the norm of the Killing vector must vanish along a null surface. By inspection of (2.6) and (2.10), such surfaces are defined by $\phi(x,t) = \phi_0 = constant$ where:

$$f(\phi_0; C, q) = 0. \tag{2.11}$$

As shall be shown below, this provides an effective equation of state from which the thermodynamic properties of the black hole can be easily derived.

We can also determine the conditions under which extremal black holes occur in the generic theory. If $f(\phi; C, q)$ is a monotonic function of ϕ, as shown in Figure 1, then there will exist at most a single event horizon, and the black hole will be Schwarzschild-like.

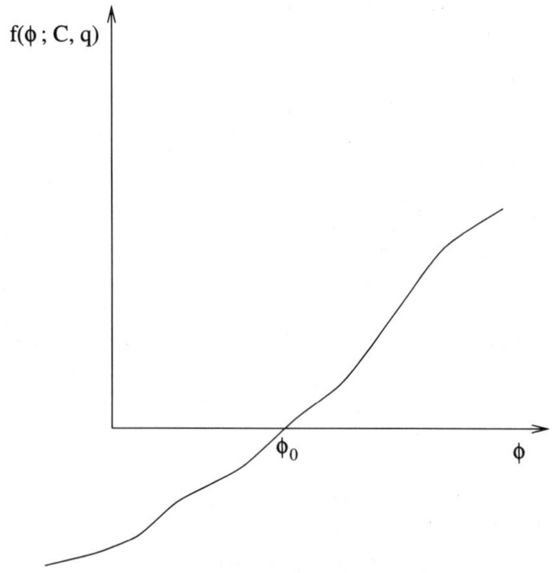

Figure 1 "Schwarzschild" Black Hole: One Event Horizon

If the first derivative of f is monotonic, then it is possible to have two event horizons, one horizon or no horizon, as illustrated in Figure 2.

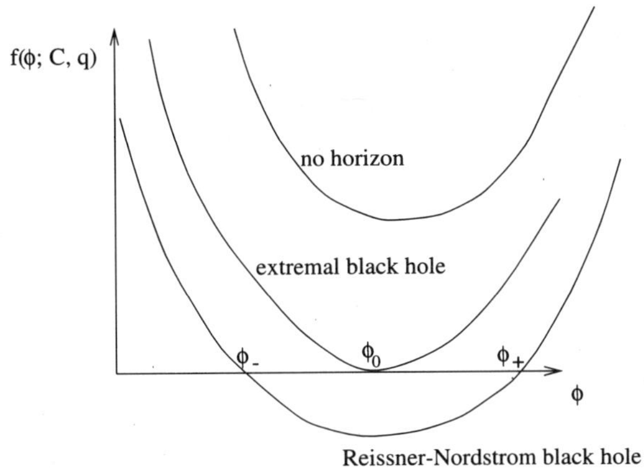

Figure 2 "Reissner-Nordstrom"-type solutions

As seen from Figure 2, the condition for the existence of an extremal black hole is clearly:

$$\left.\frac{df(\phi; C, q)}{d\phi}\right|_{\phi_0} = 0. \tag{2.12}$$

Equation (2.12) will play an important role in the discussion of the thermodynamics of black holes later on.

3 Hamiltonian Formulation

Spacetime is assumed to be locally a direct product $R \times \Sigma$, where the spatial manifold, Σ, can at this stage be either open or closed. The metric can be parametrized as follows:

$$ds^2 = e^{2\rho}\left[-\sigma^2 dt^2 + (dx + M dt)^2\right] \tag{3.1}$$

where x is a local coordinate for the spatial section Σ while ρ, σ and M are functions of spacetime coordinates (x, t). A straightforward calculation leads to the canonical Hamiltonian (up to surface terms that will be discussed below):

$$H_c = \int dx(M\mathcal{F} + \frac{\sigma}{2G}\mathcal{G} + A_0\mathcal{J}) \tag{3.2}$$

where

$$\mathcal{F} = \rho'\Pi_\rho + \phi'\Pi_\phi - \Pi'_\rho \sim 0 \tag{3.3a}$$

$$\mathcal{G} = 2\phi'' - 2\phi'\rho' - 2G^2\Pi_\phi\Pi_\rho - e^{2\rho}\frac{V(\phi)}{l^2} + G\frac{e^{2\rho}}{W(\phi)}\Pi^2_{A_1} \sim 0 \tag{3.3b}$$

$$\mathcal{J} = -\Pi'_{A_1} \sim 0 \tag{3.3c}$$

are secondary constraints. In the above the notation ~ 0 refers to the fact that the constraints must vanish weakly in the sense of Dirac: they can only be set to zero after the evaluation of Poisson brackets. The constraints $\mathcal{F}$ and $\mathcal{G}$ generate spacetime diffeomorphisms. The remaining constraint $\mathcal{J}$ is the analogue of Gauss's law, and generates U(1) gauge transformations. Since there are three first class constraints and six phase space degrees of freedom at each point in space, there are no propagating modes in the theory. However, Equation (3.3c) does not constrain the zero mode of $q := \Pi_{A_1}$, which is therefore a physical observable in the theory.

There is another unconstrained mode hidden in the gravitational phase space variables. The most direct way to find it is to note that the following linear combination of constraints:

$$\tilde{\mathcal{G}} := -e^{-2\rho}(\phi'\mathcal{G} + 2G^2\Pi_\rho\mathcal{F}) + 2GK(\phi)\Pi_{A_1}\mathcal{J} \sim 0 \tag{3.4}$$

is also a total divergence:

$$\tilde{\mathcal{G}} = \frac{\partial C}{\partial x} \tag{3.5}$$

where we have defined:

$$C = e^{-2\rho}(G^2\Pi_\rho^2 - (\phi')^2) + \frac{j(\phi)}{l^2} - GK(\phi)\Pi_{A_1}^2. \tag{3.6}$$

It turns out that C is the conserved quantity associated with translations along the Killing time direction. It therefore can be interpreted as the energy of the black hole. If standard boundary conditions are imposed, C is proportional to the ADM energy. In fact, one can write the complete canonical Hamiltonian, including surface term as follows:

$$H_c = \int dx \left[\frac{\dot{\phi}}{\phi}\mathcal{F} - \frac{1}{2G}\left(\frac{\sigma c^{2\rho}}{\phi'}\right)\tilde{\mathcal{G}} - A_0 q'\right] + H_{surface} \tag{3.7}$$

where

$$H_{surface} = H_{ADM} + \int (A_0 q)'$$

and we have defined:

$$H_{ADM} = \frac{l}{2G}\int dx \left(\frac{\sigma e^{2\rho}}{l\phi'}C\right)'.$$

The resulting ADM energy, which is by definition equal to H_{ADM} evaluated at spatial infinity is:

$$M = \frac{lC}{2G}.$$

It is straightforward to show that C and q as defined above have vanishing Poisson brackets with all the constraints: they are physical observables in the Dirac sense. Moreover, they coincide on shell with the corresponding constants of motion defined in Equations (2.8) and (2.9). Since C and q are independent and have vanishing Poisson bracket with each other, we need to find another pair of phase space observables. The momenta conjugate to C and q can be found by inspection (Louis-Martinez et al. [1994] and Louis-Martinez and Kunstatter [1995]):

$$p_C = -\frac{1}{2}\int dx \frac{e^{2\rho}\Pi_\rho}{((G\Pi_\rho)^2 - (\phi')^2)}$$

$$p_q = -\int dx \left(A_1 + \frac{GK(\phi)e^{2\rho}\Pi_\rho\Pi_{A_1}}{((G\Pi_\rho)^2 - (\phi')^2)}\right).$$

One interesting feature of these expressions is that both integrands have a pole at

$$((G\Pi_\rho)^2 - (\phi')^2) = 0.$$

On shell, we have:

$$|k|^2 = e^{-2\rho}((G\Pi_\rho)^2 - (\phi')^2)$$

so that, on any spacelike slice described by a non-singular coordinate system ($e^{2\rho} > 0$), the pole occurs whenever the norm of the Killing vector changes sign (i.e., at event horizons).

Finally, we note that while the observables C and q are completely invariant under both spacetime diffeomorphisms and U(1) gauge transformations, their conjugates are only invariant under "small" transformations that vanish at the boundaries of the system. This is consistent with the generalized Birkhoff's Theorem which states that, up to diffeomorphisms and U(1) gauge transformations there is only a two parameter family of solutions.

4 Thermodynamics

We now briefly review the thermodynamics of charged black holes in generic dilaton gravity (Gegenberg et al. [1995] and Louis-Martinez and Kunstatter [1995])[4]. For 2-D black holes, we can proceed as follows. First of all, the existence of the Killing vector in Equation (2.5) immediately yields the surface gravity at any exterior horizon ϕ_0 as defined in Section 2. In particular (Wald [1984]),

$$\kappa^2 = -\frac{1}{2}\nabla^\mu k^\nu \nabla_\mu k_\nu \bigg|_{\phi_0} . \tag{4.1}$$

A straightforward calculation yields the on-shell result:

$$\kappa = \frac{V(\phi_0)}{2l} - \frac{lq^2G}{2W(\phi_0)} = \frac{df(\phi;C,q)}{d\phi}\bigg|_{\phi=\phi_0} . \tag{4.2}$$

Note that the surface gravity vanishes for extremal black holes.

The Hawking temperature for black holes can be derived in a number of ways. The most straightforward method is to analytically continue the solutions to Euclidean spacetime and impose a suitable periodicity condition in the imaginary time direction so as to make the corresponding spacetime nonsingular, without boundary. In keeping with the imaginary time formulation of thermodynamics (see for example Kapusta [1989]), the temperature of the black hole is then equal to the period of the imaginary time coordinate. This calculation can be carried through for generic charged black holes as follows. The Euclidean solutions take the general form:

$$ds^2 = f(\frac{x}{l};M,q)dt_E^2 + \frac{1}{f(\frac{x}{l};M,q)}dx^2 \tag{4.3}$$

where $M = lC/2G$, and $f(\frac{x}{l};M,q)$ is the square of the norm of the Killing vector. By defining a dimensionless time coordinate $\alpha := t_E/a$ and a new radial coordinate, R, such that

$$R^2 = a^2 f(\frac{x}{l};M,q) \tag{4.4}$$

the metric can be written:

$$ds^2 = R^2 d\alpha^2 + H(R;M,q)dR^2 \tag{4.5}$$

[4]The thermodynamics of charged black holes in 4-D dilaton gravity has recently been described within a quasilocal formalism by Creighton and Mann [1995].

where

$$H(R;q,M) := \frac{4}{a^2(f'(\frac{x}{l};M,q))^2} \tag{4.6}$$

and the prime denotes differentiation with respect to x. Suppose that there exists an exterior horizon at $R^2 = a^2 f(x_0/l;M,q) = 0$. The solution will be regular at $R = 0$, if and only if $H(R) \to 1$ as $R \to 0$ and the coordinate α is periodic with period 2π. This determines the constant $a = f'(x_0/l;M,q)/2$, and yields a Hawking temperature of:

$$T = (2\pi a)^{-1} = \frac{f'(x_0/l;M,q)}{4\pi}. \tag{4.7}$$

Note that the standard relationship $\kappa = 2\pi T$ is valid for charged black holes in the generic theory, and that the Hawking temperature tends towards zero in the extreme black hole limit.

The black hole entropy can be calculated via Wald's general formalism (Wald [1993]), or more simply by using the black hole "equation of state" (2.11). If we vary the black hole solution infinitesimally, but stay on the event horizon determined by $f(\phi_0;C,q) = 0$ then we obtain the following condition:

$$\begin{aligned} 0 &= \delta f(\phi_0;M,q) \\ &= -2Gl\delta M + 2l\kappa\delta\phi_0 - 2ql^2GK(\phi_0)\delta q \end{aligned} \tag{4.8}$$

where κ is the surface gravity defined in (4.2). This gives:

$$\delta M = \left(\frac{V(\phi_0)}{l} - \frac{lq^2G}{W(\phi_0)}\right)\frac{\delta\phi_0}{2G} - lqK(\phi_0)\delta q \tag{4.9}$$

which is of the desired form:

$$\delta M = \kappa\delta S - \mathcal{P}\delta q \tag{4.10}$$

with entropy

$$S = \frac{2\pi}{G}\phi_0 \tag{4.11}$$

and generalized force $\mathcal{P} := qlK(\phi_0)$ associated with the charge q. The entropy is proportional to the value of the scalar field evaluated on the event horizon. This result justifies *a posteriori* calling $f(\phi_0;C,q) = 0$ the black hole equation of state, since it relates the thermodynamic variables (energy, charge and entropy) of a charged black hole in equilibrium.

5 Quantum Theory

The thermodynamical description given above is purely classical. One would like to be able to derive the thermodynamics from quantum statistical mechanics. However, the Hamiltonian analysis in Section 3 suggests that the reduced phase space for black holes (once all gauge degrees of freedom are removed) is only four dimensional. If this is true, one is left with a riddle concerning the statistical source of black hole entropy: where do the large number of states come from? The generalized Birkhoff Theorem proven in Section 2 implies that there are only two observables that are invariant under arbitrary spacetime diffeomorphism. As shown above, their conjugate variables are only invariant under a restricted class of diffeomorphisms. It is therefore difficult to rule out the existence of additional pairs of conjugate phase space variables that are invariant under a suitably restricted class of diffeomorphisms.

Most quantum mechanical treatments of gravity work with reduced quantization, in which the gauge modes are discarded prior to quantization. Unless this is done very carefully, one may lose important physical information. It is therefore useful to work within a quantization scheme in which all modes are kept until the end of the calculation. This is precisely the point of Dirac quantization: one first constructs a space of states containing all field theoretic modes, including the gauge modes. The constraints are then imposed as operator constraints on physical states. This is the approach that will be described in the present section. It is based on work by Henneaux [1985] who used it to quantize Jackiw-Teitelboim gravity (JT). In order to quantize the generic dilaton gravity theory, it is necessary to work within a geometrical formulation, in which the fundamental fields are the spatial metric and the dilaton field (and U(1) vector potential). In the case of string inspired gravity (SIG) and JT, gauge theoretical formulations exist, and have been used to quantize these theories both with and without matter (Cangemi and Jackiw [1994a,b] and Cangemi et al. [1995]). For SIG, the method described here has been shown to be equivalent to the gauge theoretic version (Benedict [1994]).

The starting point for Dirac quantization in the functional Schrödinger representation is the wave functional $\Psi[\rho, \phi, A]$ defined on the space of fields $\{\rho, \phi, A\}$. The boundary conditions on the fields will ultimately play an important role in the considerations, but for the time being we will leave them unspecified. Most of the results that follow are in fact independent of the boundary conditions. Similarly there are important questions concerning the choice of Hilbert space measure and operator ordering that we cannot address here. Since our analysis will stop at lowest order in the WKB approximation[5], these questions will not be relevant. To this order, the momenta conjugate to the fields will be represented as functional derivatives:

$$\begin{aligned} \hat{\Pi}_\rho &= -i\hbar \frac{\delta}{\delta\rho(x)} \\ \hat{\Pi}_\phi &= -i\hbar \frac{\delta}{\delta\phi(x)} \\ \hat{\Pi}_{A_1} &= -i\hbar \frac{\delta}{\delta A_1(x)}. \end{aligned} \tag{5.1}$$

As stated above, physical quantum wave functionals $\Psi[\rho, \phi, A_1]$ are defined to be those states annihilated by the quantized constraints:

$$\begin{aligned} \hat{\mathcal{F}}\Psi &= 0 \\ \hat{\tilde{\mathcal{G}}}\Psi &= 0 \\ \hat{\mathcal{J}}\Psi &= 0. \end{aligned} \tag{5.2}$$

Since $\hat{\mathcal{J}} = \hat{q}'$ and $\hat{\tilde{\mathcal{G}}} = \hat{C}'$, the last two constraints will be automatically satisfied by eigenstates of the two physical operators in the theory, namely $\hat{C}$ and $\hat{q}$. Thus

[5] For recent discussions of the WKB approximation in gauge theories and quantum gravity see Barvinsky [1993] and Lifschytz et al. [1994].

the equations that must be satisfied by physical energy and charge eigenstates are:

$$\begin{aligned}\hat{C}\Psi &= C\Psi \\ \hat{q}\Psi &= \hat{\Pi}_{A_1}\Psi = q\Psi \\ \hat{\mathcal{F}}\Psi &= 0.\end{aligned} \tag{5.3}$$

If C and q are a complete set of physical observables, it should be possible in principle to construct all physical states out of such eigenstates.

If we expand the phase of the wave functional in powers of $\hbar$:

$$\psi[\rho, \phi, A_1] = exp\frac{i}{\hbar}[S_0[\rho, \phi, A_1] + \hbar S_1[\rho, \phi, A_1] + ...] \tag{5.4}$$

then the lowest WKB order contribution S_0 obeys the following coupled functional differential equations:

$$\left[e^{-2\rho}\left(G^2\left(\frac{\delta S_0}{\delta\rho}\right)^2 - (\phi')^2\right) + \frac{j(\phi)}{l^2} - GK(\phi)\left(\frac{\delta S_0}{\delta A_1}\right)^2\right] = C \tag{5.5}$$

$$\rho'\frac{\delta S_0}{\delta\rho} + \phi'\frac{\delta S_0}{\delta\phi} - \left(\frac{\delta S_0}{\delta\rho}\right)' = 0 \tag{5.6}$$

$$\frac{\delta S_0}{\delta A_1} = q. \tag{5.7}$$

In the present case, there is no anomaly in the algebra of quantum constraints (Cangemi et al. [1995]) so that these equations can in principle be solved. Remarkably, the equations can be solved exactly in closed form in the generic theory. The solution for the WKB phase is (Louis-Martinez and Kunstatter [1995])[6]

$$S_0[C, q; \rho, \phi, A_1] = \int dx\left[Q + \frac{\phi'}{2G}\ln\left(\frac{\phi' - GQ}{\phi' + GQ}\right) + qA_1\right] \tag{5.8}$$

where

$$Q := \frac{1}{G}\sqrt{(\phi')^2 + e^{2\rho}\left(C - \frac{j(\phi)}{l^2} + Gq^2K(\phi)\right)}. \tag{5.9}$$

The WKB phase given above has several important properties. First of all, it is invariant under spatial diffeomorphisms and U(1) gauge transformations that vanish at the boundaries. Secondly, it is possible for the phase to become imaginary. Such imaginary contributions can arise in one of two ways:

1) $\phi'^2 - G^2Q^2 = -e^{2\rho}|k|^2 < 0$

This corresponds to regions along spacelike slices where the Killing vector is spacelike. As mentioned in Section 3, in these regions the physical observables p_C and p_q can become imaginary as well due to the pole in their integrands. Let us

[6]Only one boundary condition had to be imposed in finding this solution, namely that the value of the scalar field ϕ remain fixed at the boundaries of the system.

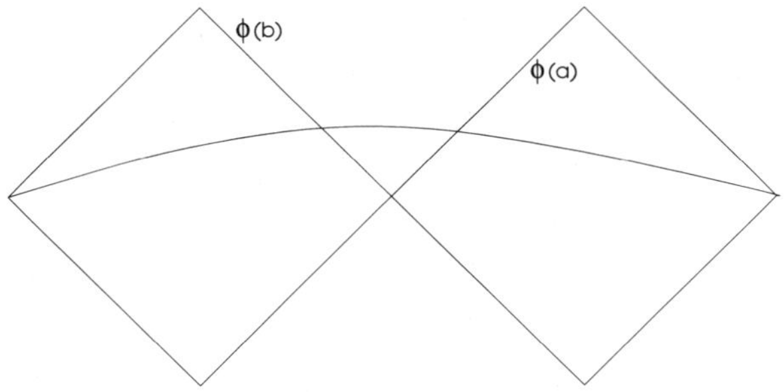

Figure 3 Cauchy Surface for eternal black hole, in Penrose Coordinates

evaluate ImS_0 for a non-extremal black hole, along a spacelike slice in Kruskal coordinates (X, t_k) joining two asympotic regions, as shown in the figure above.

The imaginary contribution comes from the logarithm in the integrand:

$$
\begin{aligned}
ImS_0 &= \frac{1}{2G} Im \int dX \frac{\partial \phi}{\partial X} \ln\left(\frac{X - t_k}{X + t_k}\right) \\
&= \frac{i\pi}{2G}(\phi(a) \pm \phi(b)).
\end{aligned} \tag{5.10}
$$

Note that it depends on the values of the scalar field at the two horizons in the left and right asymptotic regions. Given the expression for entropy in Equation (4.11) this in turn implies that the imaginary part of the WKB phase is proportional to the "entropy" associated with the two horizons. The relative sign of the contributions depends on the nature of the analytic continuation. The most natural choice appears to give a minus sign, in which case the imaginary part of the WKB phase vanishes. The above result suggests an interesting relationship between the entropy of the horizon and the imaginary part of the WKB phase. It is also consistent with a recent calculation by Martinez [1995] who showed that the entropy of eternal non-extremal black holes is zero. It is important to note that for extremal black holes, the Killing vector is never spacelike, so that there is no imaginary contribution. This agrees with recent work (Hawking and Horowitz [1995] and Teitelboim [1995]) claiming that the entropy of extremal black holes vanishes identically.

In the case of non-eternal black holes formed from collapsing matter, a spacelike surface only intersects the horizon once, and one would expect to pick up a contribution to the imaginary part of the phase proportional to the entropy:

$$ImS_0 \approx i\frac{\pi}{2G}\phi_0.$$

2) $\Pi_\rho^2 = Q^2 < 0$.

To see how this may arise, let us temporarily restrict to uncharged black holes ($q = 0$) and consider static "off-shell" configurations defined by:

$$e^{2\rho} = f(\phi; \tilde{C}) \tag{5.11}$$

where "off-shell" means that we allow values of the "mass" $\tilde{C}$ that differ from the energy eigenvalue C in the quantum wave function. For such configurations,

$$Q^2 = \frac{C - \tilde{C}}{f(\phi; \tilde{C})} \tag{5.12}$$

so that imaginary contributions can arise when either $\tilde{C} > C$ or when $f(\phi; \tilde{C}) < 0$. One can now define the probability $P_C(\tilde{C})$ of finding the system in such a configuration, given that it is in an eigenstate of the energy operator with eigenvalue C:

$$P_C(\tilde{C}) := \Psi_C(\tilde{C})\Psi^*_C(\tilde{C}). \tag{5.13}$$

As shown by Gegenberg and Kunstatter [1993], the resulting probability distribution has the intriguing properties illustrated in Figure 4.

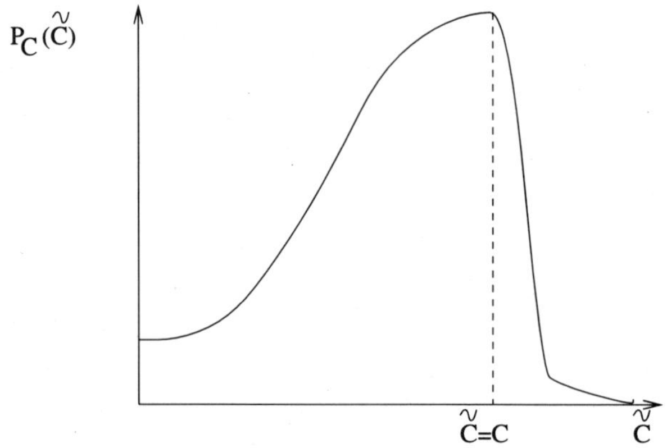

Figure 4 Spherically symmetric gravity

For a suitable choice of slicing, the WKB probability is peaked at the classical value $C = \tilde{C}$ and goes to zero very rapidly for $\tilde{C} > C$. It is non-zero at $\tilde{C} = 0$, and one finds the ratio:

$$\frac{P_C(\tilde{C} = C)}{P_C(\tilde{C} = 0)} = \exp(2\pi C^2/l^2) = \exp(S/2) \tag{5.14}$$

where S is the black hole entropy. Before one can attach any real significance to this result, however, one must determine whether or not $\tilde{C}$ as defined above corresponds to a true physical observable in the theory. This question is currently under investigation.

6 Conclusion

We have shown that all dilaton gravity theories with arbitrary (non-derivative) coupling to an Abelian gauge field can be treated within a single unified framework. An important model independent feature that arises is a connection between the WKB phase and the black hole entropy. This suggests a possible link between black hole thermodynamics and quantum gravitational tunnelling processes. Of course, in order to go beyond the heuristic analogies presented here, it is necessary to encorporate matter, since there are no field theoretic degrees of freedom in the present model, and hence no radiation. Once matter is encorporated, however, one is faced with anomalies that make the constraints non-integrable (Cangemi and Jackiw [1994b]). It would be interesting to clarify the role played by these anomalies in the context of the present formalism. This presumably requires going beyond the WKB approximation so that one is forced to address questions concerning the choice of functional measure and factor ordering. In any case, the preliminary results for the matter-free case described above seem to indicate that dilaton gravity models may be ideally suited for addressing, and hopefully solving, many important questions that arise in the context of black hole thermodynamics in particular, and quantum gravity in general.

References

Achucarro, A. and Ortiz, M. E. [1993] *Relating black holes in two and three dimensions,* Phys. Rev. D **48**, 3600–3605.

Barvinsky, A. [1993] *Unitarity approach to quantum cosmology,* Phys. Rep. **230**, 237–367.

Bekenstein, J. D. [1973] *Black holes and entropy,* Phys. Rev. D **7**, 2333–2346.

Benedict, E. [1994] *A comparison of two quantization procedures for lineal gravity,* Phys. Lett. B **340**, 43–47.

Cangemi, D. and Jackiw, R. [1994a] *Quantum states of string inspired lineal gravity,* Phys. Rev. D **50**, 3193–3922.

______ [1994b] *Quantal analysis of string-inspired lineal gravity with matter fields,* Phys. Lett. B **337**, 271–278.

Cangemi, D., Jackiw, R. and Zwiebach, B. [1995] *Physical states in matter-coupled dilaton gravity,* MIT preprint MIT-CTP-2436.

Creighton, J. D. E. and Mann, R. B. [1995] *Quasilocal thermodynamics of dilaton gravity coupled to gauge fields,* gr-qc/9505007.

Elizalde, E. and Odintsov, S. D. [1995] *Dilatonic gravity near 2 dimensions and asymptotic freedom of the gravitational coupling-constant,* Phys. Lett. B **347**, 211–216.

Frolov, V. P. [1992] *2-dimensional black hole physics,* Phys. Rev. D **46**, 5383–5394.

Gegenberg, J. and Kunstatter, G. [1993] *Quantum theory of black holes,* Phys. Rev. D **47**, R4192–R4195.

Gegenberg, J., Kunstatter, G. and Louis-Martinez, D. [1995] *Observables for two-dimensional black-holes,* Phys. Rev. D **51**, 1781–1786.

Hawking, S. W. [1975] *Particle creation by black holes,* Comm. Math. Phys. **43**, 199–220.

Hawking, S. W., Horowitz, G. T. and Ross, S. F. [1995] *Entropy, area, and black-hole pairs,* Phys. Rev. D **51**, 4302–4314.

Henneaux, M. [1985] *Quantum gravity in two dimensions: exact solution of the Jackiw model,* Phys. Rev. Lett. **54**, 959–965.

Jackiw, R. [1984] *Quantum Gravity* (Christensen, S., ed.), Adam Hilger, Bristol, p. 403.

Kapusta, J. [1989] *Finite Temperature Field Theory,* Cambridge University Press, Cambridge.

Kuchař, K. V. [1994] *Geometrodynamics of Schwarzschild black holes,* Phys. Rev. D **50**, 3961–3981.

Lifschytz, G., Mathur, S. D. and Ortiz, M. [1994] *A note on the semi-classical approximation in quantum gravity,* Phys. Rev. D, to appear.

Louis-Martinez, D., Gegenberg, J. and Kunstatter, G. [1994] *Exact Dirac quantization of all 2-D dilaton gravity theories,* Phys. Lett. B **321**, 193–198.

Louis-Martinez, D. and Kunstatter, G. [1994] *Birkhoff's Theorem in two-dimensional dilaton gravity,* Phys. Rev. D **49**, 5227–5230.

______ [1995] *Two dimensional dilaton gravity coupled to an Abelian gauge field,* Phys. Rev. D **52**, 3494–3505.

Martinez, E. [1995] *Microcanonical functional integral and entropy for eternal black holes,* Phys. Rev. D **51**, 5732–5741.

Page, D. [1994] *Proceedings of the 5th Canadian Conference on General Relativity and Relativistic Astrophysics* (Mann, R. B. and McLenaghan, R. G., eds.), World Scientific, Singapore.

Strobl, T. [1994] *Dirac quantization of gravity-Yang-Mills systems in 1+1 dimensions,* Phys. Rev. D **50**, 7346–7350.

Teitelboim, C. [1984] *Quantum Gravity* (Christensen, S., ed.), Adam Hilger, Bristol, pp. 327.

______ [1995] *Action and entropy of extreme and nonextreme black holes,* Phys. Rev. D **51**, 4315–4318.

Wald, R. M. [1984] *General Relativity,* University of Chicago Press, Chicago.

______ [1993] *Black hole entropy is the Noether charge,* Phys. Rev. D **48**, R3427–R3431.

Fields Institute Communications
Volume **15**, 1997

Gravitational Waves from Coalescing Compact Binaries

Eric Poisson
Department of Physics
Washington University
St. Louis, Missouri USA 63130
Current address: Department of Physics
University of Guelph
Guelph, Ontario, Canada N1G 2W1
poisson@terra.physics.uoguelph.ca

Abstract. This article is intended to provide a pedagogical account of issues related to, and recent work on, gravitational waves from coalescing compact binaries (composed of neutron stars and/or black holes). These waves are the most promising for kilometer-size interferometric detectors such as LIGO and VIRGO. Topics discussed include: interferometric detectors and their noise; coalescing compact binaries and their gravitational waveforms; the technique of matched filtering for signal detection and measurement; waveform calculations in post-Newtonian theory and in the black-hole perturbation approach; and the accuracy of the post-Newtonian expansion.

1 Introduction

The existence of gravitational waves[1] is an unambiguous prediction of the theory of general relativity. Yet, despite efforts originating in the early nineteen sixties, gravitational waves have not been detected directly. Nevertheless, observation of the binary pulsar PSR 1913+16 convinces us that gravitational waves do exist, and that they are correctly described by Einstein's theory (Hulse and Taylor [1975] and Taylor [1994]). That gravitational waves have not yet been detected on Earth is simply due to their incredible weakness: typical waves would produce in a bulk of matter a strain $\Delta L/L$, where L is the extension of the matter, of order 10^{-21} (Thorne [1987] and [1995]). Needless to say, to measure this effect is a great challenge for experimentalists.

1991 *Mathematics Subject Classification.* 83C35.

This work was supported by the Natural Sciences Foundation under Grant No. PHY 92-22902 and the National Aeronautics and Space Administration under Grant No. NAGW 3874. This article was completed while visiting the Theoretical Physics Institute of the University of Alberta; the author is most grateful to Werner Israel for his warm hospitality.

[1]For broader reviews on gravitational waves, see Thorne [1987] and [1995].

© 1997 American Mathematical Society

2 Interferometric Detectors

There is reasonable hope that gravitational waves will be detected within the next ten years, thanks to a new generation of detectors which use interferometry to monitor the small displacements induced by the passage of a gravitational wave. Two groups are currently involved in building large-scale interferometers: the American LIGO team, and the French-Italian VIRGO team.

The LIGO (Laser Interferometer Gravitational-wave Observatory) project (Abramovici et al. [1992]) involves two detectors, to be built in Hanford, Washington, and in Livingston, Louisiana. Construction has begun at both sites. Each interferometer has an armlength of approximately 4 km. LIGO should be completed by the turn of the century.

The VIRGO (so named after the galaxy cluster) project (Bradaschia et al. [1990]) involves a single interferometer, to be built near Pisa, Italy, with an armlength of approximately 3 km. VIRGO should also be completed by the turn of the century.

The basic idea behind interferometric detectors is the following (Thorne [1987] and [1995]):

The interferometer is composed of two long (4 km for LIGO) vacuum pipes forming the letter **L**. A laser beam is split in two at the corner of the **L**, and is sent into each arm of the interferometer. Each beam then bounces off a mass which is suspended at each end of the **L** (a mirror has been coated onto each mass). The light is finally recombined at the beam splitter, and its intensity is measured by a photodiode.

When no gravitational wave is present at the interferometer, the length of each arm is so adjusted that when measured by the photodiode, the light's intensity is precisely zero (the recombined beams are arranged to be precisely out of phase). However, when a gravitational wave passes through the interferometer, the armlengths are no longer constant, and the recombined beams no longer precisely out of phase. More precisely, during the first half of its cycle the gravitational wave increases the length of one arm, and decreases the length of the other. During the second half cycle, the first arm is now shorter, and the second arm longer. The light's intensity therefore oscillates with the gravitational-wave frequency. The intensity is a measure of $\Delta L/L = h$, where L denotes the interferometer armlength, and h the gravitational-wave field.

3 Detector Noise

Interferometers are subject to various sources of noise which limit the detector's sensitivity to gravitational waves. The relative importance of each source depends on the frequency at which the interferometer oscillates (Abramovici et al. [1992]).

At low frequencies ($f < 10$ Hz) the detector's sensitivity is limited by seismic noise, which is due to the Earth's seismic activity. At frequencies larger than 10 Hz the seismic noise can be eliminated with sophisticated isolation stacks; these fail at low frequencies.

At high frequencies ($f > 100$ Hz) the detector's sensitivity is limited by photon shot noise, which is due to statistical errors in the counting of photons by the photodiode. This source of noise can be reduced by increasing the laser power, or making use of "light recycling" (Abramovici et al. [1992]).

At intermediate frequencies (f between 10 Hz and 100 Hz) the noise is dominated by thermal noise, which is due to spurious motions of thermal origin. For

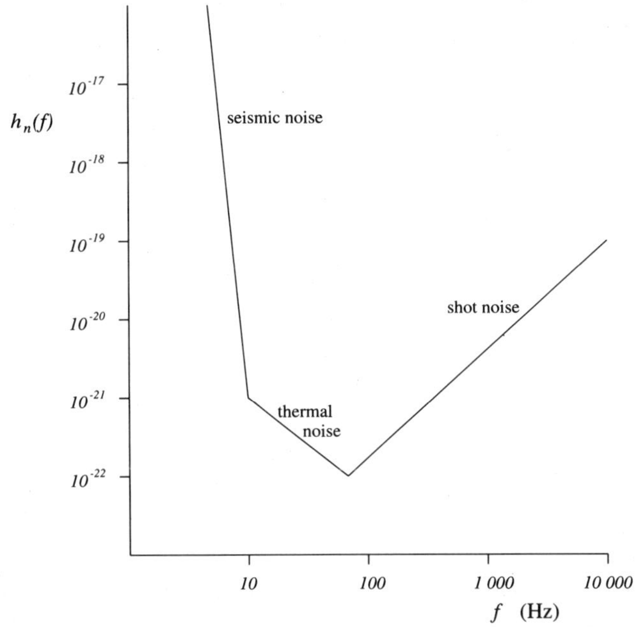

Figure 1 Noise level in an interferometric detector with advanced sensitivity.

example, the suspended masses are thermally excited and vibrate with their normal-mode frequencies; this evidently affects the recombined laser beam.

Interferometers are therefore broad-band detectors, with good sensitivity in the range (Abramovici et al. [1992])

$$10 \text{ Hz} < f < 1\ 000 \text{ Hz}. \tag{3.1}$$

The required sensitivity for full-scale interferometers is approximately $h_n \sim 10^{-22}$ at peak sensitivity — a tall order. [The subscript n stands for "noise level"; we will define $h_n(f)$ precisely below. A plot of $h_n(f)$, appropriate for an interferometric detector with "advanced" sensitivity, is given in Figure 1.] For comparison, we may mention that the Caltech 40 m prototype has already achieved $h_n \simeq 10^{-19}$ at peak sensitivity (f = 450 Hz). It is not implausible that improved technology and a factor of 100 in armlength will permit to reach the desired goal.

4 More About Detector Noise

The detector noise can be measured when no gravitational wave is present at the interferometer, the typical situation. Then the detector output $s(t) = \Delta L(t)/L$ is given by noise alone:

$$s(t) = n(t), \tag{4.1}$$

where $n(t)$ represents the noise. The noise is a random process (see, e.g., Reif [1965]): the function $n(t)$ takes purely random values. Consequently, the noise can only be studied using statistical methods. In the presence of a gravitational wave, Equation (4.1) must be replaced by $s(t) = h(t) + n(t)$, where $h(t)$ is the gravitational-wave field.

The statistical properties of the noise can be determined by careful measurement. For example, the time average

$$\overline{n(t)} = \lim_{T\to\infty} \frac{1}{2T} \int_{-T}^{+T} n(t)\, dt \tag{4.2}$$

can be constructed. This mean value can then be subtracted from $n(t)$ and, without loss of generality, we can put $\overline{n} = 0$. Also from measurements, the noise's *autocorrelation function* $C_n(\tau)$ can be constructed:

$$C_n(\tau) = \overline{n(t)\, n(t+\tau)}; \tag{4.3}$$

$C_n(0)$ gives the mean squared deviation of the noise with respect to the mean value.

In the following we will assume that the noise is *stationary*, in the sense that its statistical properties do not depend on time (Reif [1965]). This means, in particular, that the autocorrelation function does not depend explicitly on the origin of time t, but only on the variable τ, as was expressed in Equation (4.3). We shall also assume that the noise satisfies the *ergodic hypothesis*, so that time averages can be replaced with ensemble averages (Reif [1965]). Here, the noise is imagined to be drawn from a representative ensemble, and the probability that it takes a particular realization $n(t)$ is given by a specified probability distribution. The statistical properties of the noise then refer to this (infinite dimensional) distribution function.

In full generality, the statistical properties of the noise can only be summarized by constructing all the higher moments $\overline{n \cdots n}$. If, however, the noise is assumed to be Gaussian, in the sense that its probability distribution function is an infinite dimensional Gaussian distribution (Helstrom [1968]), then the autocorrelation function contains all the information.

Real detector noise is neither strictly stationary nor strictly Gaussian. However, on a timescale of hours, which is long compared with typical gravitational-wave bursts, the noise appears stationary to a good approximation (Abramovici et al. [1992]). And non-Gaussian components to the noise can be removed, to a large extent, by cross-correlating detector outputs from two widely separated interferometers (Abramovici et al. [1992]). It is therefore a satisfactory approximation to take the noise to be stationary and Gaussian.

Under these assumptions the statistical properties of the detector noise are fully summarized by $C_n(\tau)$. It is convenient to work instead in the frequency domain, and to define the noise's *spectral density* $S_n(f)$ as (Reif [1965])

$$S_n(f) = 2 \int C_n(\tau) e^{2\pi i f \tau}\, d\tau. \tag{4.4}$$

The spectral density is defined for $f > 0$ only; as $C_n(\tau)$ is a real and even function, the negative frequencies only duplicate the information contained in the positive frequencies.

As $n(t)$ is dimensionless, the spectral density has dimensions of time. By multiplying $S_n(f)$ with the frequency and taking the square root (since the spectral density represents the mean squared noise), one obtains the *noise level* $h_n(f)$:

$$h_n(f) = \sqrt{f S_n(f)}. \tag{4.5}$$

This gives the equivalent gravitational-wave amplitude which would make the interferometer oscillate at just the noise level (Thorne [1987] and [1995]); this quantity was introduced in the preceding section (see Figure 1).

5 Coalescing Compact Binaries

Coalescing compact binaries, composed of neutron stars and/or black holes, are the most promising source of gravitational waves for interferometric detectors (Schutz [1986] and [1989] and Thorne [1987] and [1995]).

Consider the binary pulsar PSR 1913+16 (Hulse and Taylor [1975] and Taylor [1994]). This system consists of two neutron stars, each of 1.4 $M_\odot$, in orbital motion around each other. Its present orbital period P is approximately 8 hours, corresponding to orbital separations of about 5×10^5 times the total mass. (Here and throughout we use units such that $G = c = 1$.) Its present eccentricity is approximately equal to 0.6.

The binary's orbital period is observed to decay at a rate $dP/dt = -2 \times 10^{-12}$ corresponding precisely to a loss of energy and angular momentum to gravitational waves (Hulse and Taylor [1975] and Taylor [1994]). In a timescale of approximately 10^8 years the orbital period will have decreased to less than a tenth of a second, corresponding to orbital separations smaller than one hundred times the total mass. In this time, the eccentricity will have been reduced to extremely small values (by the radiation reaction), so that the orbits are practically circular. The gravitational waves produced then have a frequency larger than 10 Hz, and the frequency keeps increasing as the system evolves. *During this late stage of orbital evolution, the gravitational waves sweep through the frequency bandwith of interferometric detectors, and thus become visible.* By the time the orbital separation becomes as small as a few times the total mass, the neutron stars begin to merge. The gravitational waves produced during the final merger cannot be detected by interferometric detectors, at least in the broad-band configuration described above[2]: the gravitational-wave frequency is then larger than 1 000 Hz, for which the detector noise is large.

Of course, it would be foolish to wait 10^8 years in order for PSR 1913+16 to produce gravitational waves with appropriate frequency. Fortunately, interferometric detectors will be sensitive enough to monitor binary coalescences occurring in quite a large volume of the universe, approximately 10^7 Mpc^3 (corresponding to a radius of 200 Mpc (Abramovici et al. [1992])). It has been estimated (Narayan et al. [1991], Phinney [1991] and Tutukov and Yungelson [1993]) that as many as 100 coalescences could occur every year in such a volume (this includes coalescences of black-hole systems as well). This potentially large event rate is one of the factors that make gravitational waves from coalescing compact binaries especially promising.

The other factor comes from the fact that compact binaries are extremely clean astrophysical systems. It can be estimated (Bildsten and Cutler [1992] and Kochanek [1992]) that tidal interactions between the two stars are completely negligible, up to the point where the objects are about to merge. In particular, no mass transfer occurs. The system can therefore be modeled, to extremely good accuracy, as that of two point masses with a limited number of internal properties (such as mass, spin, and quadrupole moment). The challenge in modeling coalescing compact binaries resides in formulating and solving the equations of motion and wave generation for a general relativistic two-body problem. (For an overview, see Damour [1987].)

[2]Other configurations, with narrow-band sensitivity at a chosen frequency, are possible.

6 Waveform According to the Quadrupole Formula

At the crudest level, the gravitational-wave signal corresponding to an inspiraling binary system can be calculated by (i) assuming that the orbital motion is Newtonian (with the effects of radiation reaction incorporated) and (ii) using the standard quadrupole formula for wave generation. (See, for example, Misner et al. [1973]). As motivated above, we may also assume that the orbits are circular.

We define $h(t)$ to be the gravitational-wave signal. This is given by a linear combination, appropriate for interferometric detectors, of the two fundamental polarizations, h_+ and $h_\times$, of the gravitational-wave field. At this level of approximation, the signal is given by (Thorne [1987] and [1995])

$$h(t) = Q(\text{angles})\,(\mathcal{M}/r)\,(\pi\mathcal{M}f)^{2/3}\cos\Phi(t). \tag{6.1}$$

As expected, the signal decays as the inverse power of r, the distance to the source.

In Equation (6.1), Q is a function of all the angles relevant to the problem: position of the source in the sky, orientation of the orbital plane, position and orientation of the detector on Earth. The parameter $\mathcal{M}$ is called the *chirp mass* and represents a particular combination of the masses, given by

$$\mathcal{M} = (m_1 m_2)^{3/5}/(m_1+m_2)^{1/5}. \tag{6.2}$$

The waveform depends on the chirp mass only, and not on any other combination of the masses. The symbol f represents the gravitational-wave frequency, which is equal to *twice* the orbital frequency. Because the system loses energy and angular momentum to gravitational waves, the frequency is not constant, but increases in time according to (Thorne [1987] and [1995])

$$\frac{df}{dt} = \frac{96}{5\pi\mathcal{M}^2}(\pi\mathcal{M}f)^{11/3}. \tag{6.3}$$

As a consequence, Equation (6.1) shows that the amplitude of the signal, which is proportional to $(\pi\mathcal{M}f)^{2/3}$, also increases with time. A signal which increases both in frequency and in amplitude is known as a *chirp*, and this is the origin of the term "chirp mass". Finally, the phase function $\Phi(t)$ is given by

$$\Phi(t) = \int^t 2\pi f(t')\,dt'. \tag{6.4}$$

Because the frequency is not a constant, the phase accumulates nonlinearly with time.

For a system of two neutron stars, the gravitational-wave signal undergoes approximately 16 000 oscillations as it sweeps through the frequency bandwith of an interferometric detector. The timescale for the frequency sweep is approximately 15 minutes. The orbital separation ranges from approximately 180 to 10 times the total mass $M = m_1 + m_2$, and the orbital velocity

$$v \equiv (\pi M f)^{1/3} \tag{6.5}$$

ranges from approximately 0.1 to 0.4. This indicates that relativistic corrections must be inserted in Equations (6.1) and (6.3) in order to obtain a satisfactory degree of accuracy.

7 Matched Filtering

How does one go about finding a gravitational-wave signal in a noisy data stream, when typically the signal is not very strong? And once the signal is found, how does one go about extracting the information it contains? For signals of precisely known form, one goes about this using the technique of *matched filtering* (Wainstein and Zubakhov [1962]). Signals from inspiraling compact binaries, since they can be calculated with high precision, belong to this class. The basic idea behind matched filtering is to use our knowledge about the signal in order to go find it in the data stream, after the noisy frequencies have been filtered out.

Suppose a signal of known form $h(t;\vec{\mu})$ is present in the data stream. Here, the vector $\vec{\mu}$ collectively denotes all the parameters characterizing the signal. In the case of inspiraling binaries, these would be the time of arrival, the initial phase, the distance to the source, the position angles, the chirp mass, and other parameters to be introduced in Section 12. The detector output is given by

$$s(t) = h(t;\vec{\mu}) + n(t), \tag{7.1}$$

where $n(t)$ is the noise, whose statistical properties are fully summarized by the spectral density $S_n(f)$, as discussed in Section 4.

The first step in matched filtering[3] is to pass $s(t)$ through a linear filter which removes the noisy frequencies. The idea here is to use our knowledge about the detector noise contained in $S_n(f)$ in order to discard that part of the detector output for which the detector noise is large. The output is imagined to be decomposed into Fourier modes according to $s(t) = \int \tilde{s}(f)e^{-2\pi ift}\,df$; the filter suppresses the modes $\tilde{s}(f)$ for which $S_n(f)$ is large.

In mathematical terminology, a linear filter is a linear operation on a function $s(t)$. This operation can always be written as (Wainstein and Zubakhov [1962])

$$s(t) \to \int w(t-t')s(t')\,dt', \tag{7.2}$$

where $w(t-t')$ is the filter function. The filter which removes the noisy frequencies is the one such that

$$\tilde{w}(f) = \frac{2}{S_n(|f|)}; \tag{7.3}$$

the factor of 2 is conventional.

The next step in matched filtering consists of computing the overlap integral between the filtered output (7.2) and the known signal $h(t;\vec{\mu}_{\rm trial})$. The true value $\vec{\mu}$ of the source parameters is not known prior to the measurement. This operation must therefore be repeated for a large number of trial values $\vec{\mu}_{\rm trial}$; the corresponding signals $h(t;\vec{\mu}_{\rm trial})$ are known as *templates*. The overlap integral defines the function $S(\vec{\mu}_{\rm trial})$ given by

$$\begin{aligned} S(\vec{\mu}_{\rm trial}) &= \int h(t;\vec{\mu}_{\rm trial})w(t-t')s(t')\,dt'\,dt \\ &= \big\langle h(\vec{\mu}_{\rm trial}) \bigm| s \big\rangle. \end{aligned} \tag{7.4}$$

To obtain the second line we have inserted the Fourier decompositions of h, w, and s, and carried out the integrations over time. The inner product $\langle\cdot|\cdot\rangle$ is defined as

$$\langle a|b\rangle = 2\int_0^\infty \frac{\tilde{a}^*(f)\tilde{b}(f) + \tilde{a}(f)\tilde{b}^*(f)}{S_n(f)}\,df, \tag{7.5}$$

where a and b are arbitrary functions of time, with Fourier transforms $\tilde{a}$ and $\tilde{b}$.

A quantity analogous to $S(\vec{\mu}_{\rm trial})$ can be defined for the noise alone, by replacing s to the right of Equation (7.4) by n. Operationally, this amounts to filtering

[3]This discussion is based on Cutler and Flanagan [1994].

the detector output when a gravitational-wave signal is known *not* to be present. Because the noise is a random function, the integrals $\int h(t; \vec{\mu}_{\rm trial}) w(t-t') n(t')\, dt'\, dt$ are random also. And because the noise is assumed to have zero mean, to take an average over all possible realizations of the noise (by repeating the measurements many times) would yield a zero value. None of these quantities would be especially useful. We are therefore led to consider the root-mean-square average of these integrals,

$$\begin{aligned} N(\vec{\mu}_{\rm trial}) &= {\rm rms} \int h(t; \vec{\mu}_{\rm trial}) w(t-t') n(t')\, dt'\, dt \qquad (7.6) \\ &= \sqrt{\langle h(\vec{\mu}_{\rm trial}) \mid h(\vec{\mu}_{\rm trial}) \rangle} \equiv \rho(\vec{\mu}_{\rm trial}). \end{aligned}$$

To go from the first to the second line requires some machinery which will not be presented in this review. We refer the reader to Wainstein and Zubakhov [1962] for the missing steps.

8 The Signal-to-Noise Ratio

The *signal-to-noise ratio* is defined to be the ratio of S over N:

$$\mathrm{SNR}(\vec{\mu}_{\rm trial}) = \frac{\langle h(\vec{\mu}_{\rm trial}) \mid s \rangle}{\rho(\vec{\mu}_{\rm trial})}. \qquad (8.1)$$

It is clear that $\mathrm{SNR}(\vec{\mu}_{\rm trial})$ is a random function, since $s(t)$ is itself a random function, being the superposition of signal plus noise. Its expectation value, or average over all possible realizations of the noise, is zero in the absence of signal (since $\overline{n} = 0$), and

$$\overline{\mathrm{SNR}(\vec{\mu}_{\rm trial})} = \frac{\langle h(\vec{\mu}_{\rm trial}) \mid h(\vec{\mu}) \rangle}{\rho(\vec{\mu}_{\rm trial})} \qquad (8.2)$$

in the presence of the signal $h(t; \vec{\mu})$. It is important to notice that in Equation (8.2), the numerator is the overlap integral between the true signal $h(t; \vec{\mu})$ and the templates $h(t; \vec{\mu}_{\rm trial})$. In the absence or presence of a signal, the variance in the signal-to-noise ratio is precisely equal to unity (Cutler and Flanagan [1994]), independently of the values of $\vec{\mu}$ and $\vec{\mu}_{\rm trial}$. This fully summarizes the statistical properties of the signal-to-noise ratio, since $\mathrm{SNR}(\vec{\mu}_{\rm trial})$ is a Gaussian random function. [This can be seen from the fact that $s(t)$ itself is Gaussian, being the superposition of signal plus Gaussian noise.]

It is intuitively clear that choosing a template with $\vec{\mu}_{\rm trial} = \vec{\mu}$ will produce the largest possible expectation value of the signal-to-noise ratio. This statement, known as *Wiener's theorem* (Wainstein and Zubakhov [1962]), can easily be shown to be true by applying Schwarz's inequality to the right-hand side of Equation (8.2). We have

$$\max\left\{\overline{\mathrm{SNR}(\vec{\mu}_{\rm trial})}\right\} = \overline{\mathrm{SNR}(\vec{\mu})} = \rho(\vec{\mu}) = \sqrt{\langle h(\vec{\mu}) \mid h(\vec{\mu}) \rangle}. \qquad (8.3)$$

The fact that the signal-to-noise ratio has a variance of unity indicates that a signal can be concluded to be present only if $\rho(\vec{\mu})$ is significantly larger than 1. We will come back to this point in the next section.

The true value of the source parameters can therefore be determined by maximizing $\mathrm{SNR}(\vec{\mu}_{\rm trial})$ over all possible values of the trial parameters. The fact that the signal-to-noise ratio has a variance of unity implies that this determination can only have a limited degree of accuracy. The statistical errors decrease with increasing $\rho(\vec{\mu})$; since this is proportional to the signal's amplitude, a stronger signal gives better accuracy.

Maximizing the signal-to-noise ratio is essentially equivalent to maximizing the overlap integral

$$\big\langle h(\vec{\mu}_{\text{trial}}) \mid h(\vec{\mu})\big\rangle.$$

It is easy to see how the choice of the parameters $\vec{\mu}_{\text{trial}}$ can affect the overlap integral. Consider a toy waveform with three parameters: arrival time, initial phase, and chirp mass. A mismatch in the arrival times clearly reduces the overlap integral: the signal and the template, taken to be functions of time, might have support in entirely different regions of the time axis, leading to a vanishing overlap. Supposing that the arrival times are matched, a mismatch in the initial phases can also reduce the overlap integral: the signal and the template might be out of phase with each other, leading to an oscillating integrand and a vanishingly small overlap. Finally, supposing that both the arrival times and initial phases are matched, a mismatch in the chirp masses would also reduce the overlap integral. This is because the chirp mass governs the rate at which the signal's frequency changes with time; cf. Equation (6.3). Signal and template, starting at the same time with the same phase, might thereafter go out of phase, thereby reducing the overlap.

9 Signal Detection

The first order of business when analysing the output of a gravitational-wave detector is to decide whether or not a signal is present. Here we assume that the signal must be of a specific form, corresponding to a coalescing binary system. In this section we discuss signal detection — how the technique of matched filtering can be employed to find the signal in the noisy data stream. In the next section we will discuss signal measurement — how matched filtering is used to estimate the value of the source parameters once the signal has been found.

As mentioned in the previous section, a signal can be concluded to be present if the maximum value of the signal-to-noise ratio, $\text{SNR} \equiv \max\{\text{SNR}(\vec{\mu}_{\text{trial}})\}$, is significantly larger than unity. In fact, there exists a threshold value SNR^* such that a signal is concluded to present, with a certain confidence level, if $\text{SNR} > \text{SNR}^*$. To figure out how large this threshold must be is a standard application of the statistical theory of signal detection (Helstrom [1968]), which was developed largely for the purpose of detecting radar signals. The theory can easily be taken over to the case gravitational-wave signals (Finn [1992]). To go into the detail of this theory would be outside the scope of this review. We shall simply state that SNR^* is fixed by selecting a small, acceptable value for the probability that a signal would falsely be concluded to be present — the false alarm probability. (This is the Neyman-Pearson criterion (Helstrom [1968]), which is more precisely formulated in terms of the likelihood ratio, the ratio of the probability that a signal is present to the probability that it is absent.) Once SNR^* is fixed, the level of confidence that a signal is indeed present increases with $\text{SNR} > \text{SNR}^*$. A typical ballpark value for the threshold is $\text{SNR}^* = 6$.

In the preceding paragraph it was assumed that the signal-to-noise ratio SNR is computed using template waveforms which are functions of the parameters $\vec{\mu}$ introduced in Section 7. These parameters have direct physical meaning; they include the chirp mass and other meaningful parameters to be introduced in Section

12. However, since these parameters are *not* estimated during the detection stage of the data analysis (they are estimated only *after* a signal has been found), there is no particular need to parametrize the templates with $\vec{\mu}$. In fact, it may be desirable, in order to minimize the computational effort, to parametrize the signal in a completely different way. The new parameters, $\vec{\alpha}$, would then have no particular physical significance. What is required is that the new templates, $h(t;\vec{\alpha})$, reproduce the behaviour of the expected gravitational-wave signal. In other words, the templates $h(t;\vec{\mu})$ and $h(t;\vec{\alpha})$ should span the same "signal space", but $h(t;\vec{\alpha})$ should do so most efficiently. Unlike $h(t;\vec{\mu})$, $h(t;\vec{\alpha})$ need not be derived from the field equations of general relativity. With these new *detection templates*, the signal-to-noise ratio is defined as

$$\text{SNR} = \max\bigl\{\text{SNR}(\vec{\alpha})\bigr\}, \tag{9.1}$$

where $\text{SNR}(\vec{\alpha})$ is defined as in Equation (8.1). As before, a signal is concluded to be present if $\text{SNR} > \text{SNR}^*$.

10 Signal Measurement

Once the detection templates $h(t;\vec{\alpha})$ have been used to conclude that a signal is present in the data stream, they are replaced with the *measurement templates* $h(t;\vec{\mu}_{\text{trial}})$ in order to estimate the value of the physical parameters $\vec{\mu}$.

The procedure to estimate the source parameters was explained in Section 8. As was mentioned, the idea is to maximize over all possible values of the trial parameters the overlap integral

$$\bigl\langle h(\vec{\mu}) \bigm| h(\vec{\mu}_{\text{trial}})\bigr\rangle.$$

We already have discussed the effect of a mismatch in the value of the parameters. What remains to be discussed is the effect of a mismatch in the functional form of the template with respect to that of the true waveform.

The true waveform is governed by the exact laws of general relativity. The template, on the other hand, is necessarily constructed using an approximation to the exact laws. (That approximations must be made in analytic calculations is obvious; in a numerical treatment the approximation resides in the finite differencing of the field equations.) This, clearly, must have an effect on our strategy for extracting the information contained in the gravitational-wave signal. This can be seen simply from the fact that Wiener's theorem, as stated in Section 8, strictly requires signal and template to have the same functional form; they are allowed to differ only in the value of their parameters.

Suppose that gravitational waves coming from a given source are received without noise, so that the true signal $h(t;\vec{\mu})$ is measured accurately. Suppose also that the true value $\vec{\mu}$ of the source parameters is known (God has spoken). Then a computation of the overlap integral with $\vec{\mu}_{\text{trial}} = \vec{\mu}$ but with an approximate template $h(t;\vec{\mu}_{\text{trial}})$ will not yield the maximum possible value for $\text{SNR}(\vec{\mu}_{\text{trial}})$. This is because the approximation differs from the true signal, both in amplitude and in phase. Since both signal and template undergo a large number of oscillations (recall that for a system of two neutron stars, this number is approximately 16 000), the overlap integral is most sensitive to phase differences: a slight phase lag causes the integrand to oscillate, thereby severely reducing the signal-to-noise ratio with respect to its maximum possible value.

A gravitational-wave astronomer doesn't know before the measurement the true value of the source parameters, and must work with an approximation to the true general-relativistic waveform. We have seen that the phase lag occurring between the true signal and the approximate template when the parameters are matched reduces the signal-to-noise ratio from its maximum possible value. It follows that maximizing $\mathrm{SNR}(\vec{\mu}_{\rm trial})$ with approximate templates introduce *systematic errors* into the estimation of the source parameters: evaluating the signal-to-noise ratio with $\vec{\mu}_{\rm trial} = \vec{\mu} + \delta\vec{\mu}$ will return, for some $\delta\vec{\mu}$, a number larger than $\mathrm{SNR}(\vec{\mu})$. The systematic errors are precisely the value of $\delta\vec{\mu}$ for which the signal-to-noise ratio is largest. If the templates are a poor approximation to the true signal, then the systematic errors will be larger than the statistical errors arising because $\mathrm{SNR}(\vec{\mu}_{\rm trial})$ is a random function (see Section 8).

We therefore appreciate the need for constructing measurement templates which are as accurate as possible, especially in phase (Cutler et al. [1993] and Finn and Chernoff [1993]). The requirement is that the systematic errors in the estimated parameters must be smaller than the statistical errors. An estimate for the required degree of accuracy comes from the observation that the overlap integral will be significantly reduced if the template loses phase by as much as one wave cycle with respect to the true signal. Since the total number of wave cycles is approximately 16 000, we have

$$\text{accuracy} \sim \frac{1}{16\,000} \sim 10^{-4}. \tag{10.1}$$

Since the orbital velocity v is of order 10^{-1} when the gravitational-wave frequency is in the relevant bandwidth, relativistic corrections *at least* of order v^4 are required to improve the quadrupole-formula expression given in Section 6. As we shall see, this is an underestimate.

11 Waveform Calculations: Post-Newtonian Theory

We have seen that accurate measurement templates are required to make the most of the gravitational-wave signals originating from coalescing compact binaries. We also have seen that the quadrupole-formula waveform, Equation (6.1), is not sufficiently accurate; relativistic corrections at least of order v^4 are required. How does one go about calculating these?

A possible line of approach is to use a slow-motion approximation to the equations of general relativity. This is based on the requirement that if v is a typical velocity inside the matter source, then

$$v \ll 1. \tag{11.1}$$

We shall call this approximation "post-Newtonian theory". (For an overview, see Will [1994]). We point out that for binary systems, post-Newtonian theory makes no assumption regarding the relative size of the two masses. This is to be contrasted with the perturbation approach, discussed in Section 13, in which the mass ratio is assumed to be small, but no restriction is put on v.

In post-Newtonian theory[4], one starts by defining fields $h^{\alpha\beta}$ as

$$h^{\alpha\beta} = \sqrt{-g}\, g^{\alpha\beta} - \eta^{\alpha\beta}, \tag{11.2}$$

where $g^{\alpha\beta}$ is the inverse of the true metric $g_{\alpha\beta}$ with determinant g, and $\eta^{\alpha\beta}$ is the metric of Minkowski spacetime. When the harmonic gauge conditions

$$\partial_\beta h^{\alpha\beta} = 0 \tag{11.3}$$

[4]This discussion is based on the work of Blanchet, Damour and Iyer, expounded in Blanchet [1987], Blanchet and Damour [1986], [1988] and [1989] and Damour and Iyer [1991].

are imposed, the *exact* Einstein field equations reduce to

$$\Box h^{\alpha\beta} = 16\pi(-g)T^{\alpha\beta} + \Lambda^{\alpha\beta}. \tag{11.4}$$

Here, $\Box = \eta^{\alpha\beta}\partial_\alpha\partial_\beta$ is the flat-spacetime wave operator, $T^{\alpha\beta}$ is the stress-energy tensor of the source, and $\Lambda^{\alpha\beta}$ is nonlinear in $h^{\alpha\beta}$ and represents an effective stress-energy tensor for the gravitational field.

The detailed way in which one solves these equations is quite complicated, and will not be described here. The essential ideas are these (Blanchet [1987], Blanchet and Damour [1986], [1988] and [1989] and Damour and Iyer [1991]):

One first integrates the equations in the near zone ($r < \lambda$, where r is the flat-space radius and λ the gravitational wavelength) assuming slow motion, or $\partial h^{\alpha\beta}/\partial t \ll \partial h^{\alpha\beta}/\partial x^i$. One does this by iterations: the nonlinear terms in Equation (11.4) are first neglected, and the resulting linear equations integrated. These solutions are then used as input for the next iteration. This process is continued until the desired degree of accuracy is obtained. This is the standard post-Newtonian approach (Damour [1987]).

One next integrates the equations everywhere in the vacuum region outside the source. This is done once again by iterations, assuming $h^{\alpha\beta} \ll 1$, but assuming nothing about the relative size of $\partial h^{\alpha\beta}/\partial t$ with respect to the spatial derivatives. This is because the vacuum region contains the wave zone, in which the field propagates with the speed of light; a slow-motion assumption would therefore not do for the field itself. This is the post-Minkowskian approach (Damour [1987]).

Using the post-Minkowskian approach one constructs, by successive approximations, the most general solution to the Einstein equations outside the source. This is characterized by two infinite sets of arbitrary multipole moments (Thorne [1980]), the mass moments $M_{\ell m}(t-r)$ and the current moments $J_{\ell m}(t-r)$, where ℓ and m are the standard spherical-harmonic indices. (In practice, the fields $h^{\alpha\beta}$ are equivalently expressed in terms of symmetric-trace-free moments, not spherical-harmonic moments.) The general solution is then matched to the near-zone solution in the region of common validity, and the multipole moments are thus determined.

Finally, one expresses the radiation field — the time-varying, $O(1/r)$ part of the gravitational field — in terms of the derivatives of the mass and current multipole moments (Thorne [1980]). This gives the gravitational waveform. The gravitational-wave luminosity dE/dt can also be obtained from the radiation field.

12 Waveform to Second Post-Newtonian Order

To date, the post-Newtonian calculation of the waveform has been carried out accurately through second post-Newtonian order — $O(v^4)$ — beyond the leading-order, quadrupole-formula expressions given in Section 6. The complete waveform will not be displayed here. Instead, we will focus solely on the waveform's *phasing*.

The phasing of the waves can be determined from df/dt, the rate of change of the gravitational-wave frequency. This can be expressed as

$$\frac{df}{dt} = \frac{dE/dt}{dE/df}, \tag{12.1}$$

where dE/dt is (minus) the gravitational-wave luminosity, and dE/df relates orbital energy to orbital frequency (f is twice the orbital frequency; the orbits are assumed to be circular). Both these quantities can be expanded in powers of

$$v \equiv (\pi M f)^{1/3}, \tag{12.2}$$

with leading-order terms (Misner et al. [1973])

$$\left(\frac{dE}{dt}\right)_N = -\frac{32}{5}\eta^2 v^{10}, \qquad \left(\frac{dE}{df}\right)_N = -\frac{\pi}{3}\mu M v^{-1}, \tag{12.3}$$

where the subscript N stands for "Newtonian". We have introduced the reduced mass μ, the total mass M, and the mass ratio η as

$$\mu = \frac{m_1 m_2}{m_1 + m_2}, \qquad M = m_1 + m_2, \qquad \eta = \frac{\mu}{M}. \tag{12.4}$$

Notice that η is restricted to the interval $0 < \eta \leq 1/4$, with $\eta = 1/4$ for $m_1 = m_2$. It is easy to check that Equations (12.1) and (12.3) reproduce Equation (6.3) above; the chirp mass can be expressed as $\mathcal{M} = \eta^{3/5} M$.

For simplicity we will, in the following, focus on the quantity dE/dt. To second post-Newtonian order, the luminosity takes the form

$$\begin{aligned}\frac{dE}{dt} &= \left(\frac{dE}{dt}\right)_N \Bigg[1 - \left(\frac{1247}{336} + \frac{35}{12}\eta\right)v^2 + \left(4\pi - \mathrm{SO}\right)v^3 \\ &\quad - \left(\frac{44711}{9072} - \frac{9271}{504}\eta - \frac{65}{18}\eta^2 + \mathrm{SS}\right)v^4 + \cdots\Bigg].\end{aligned} \tag{12.5}$$

Here, the terms SO and SS are due to spin-orbit and spin-spin interactions, respectively (Kidder et al. [1993]); these occur if the masses m_1 and m_2 are rotating. Let $\vec{S}_1$ and $\vec{S}_2$ be the spin angular momentum of each mass, and define the dimensionless quantities $\vec{\chi}_a = \vec{S}_a/{m_a}^2$, for $a = \{1, 2\}$. Let also $\hat{L}$ be the direction of orbital angular momentum. Then (Kidder et al. [1993])

$$\mathrm{SO} = \frac{1}{4}\sum_{a=1}^{2}\Big[11(m_a/M)^2 + 5\eta\Big]\hat{L}\cdot\vec{\chi}_a \tag{12.6}$$

is the spin-orbit term, and

$$\mathrm{SS} = \frac{\eta}{48}\Big(103\vec{\chi}_1\cdot\vec{\chi}_2 - 289\hat{L}\cdot\vec{\chi}_1\,\hat{L}\cdot\vec{\chi}_2\Big) \tag{12.7}$$

is the spin-spin term.

In Equation (12.5), the leading-order term was first calculated in 1963 by Peters and Mathews [1963] using the standard quadrupole formula. The first post-Newtonian correction, at order v^2, was calculated in 1976 by Wagoner and Will [1976]. The $4\pi v^3$ term is due to wave propagation effects: As the waves propagate out of the near zone they are scattered by the curvature of spacetime, and this modifies both the amplitude and the phase of the waveform; this "tail term" was first calculated in 1993 by this author (Poisson [1993]), and then independently by Wiseman [1993] and Blanchet and Schäfer [1993]. The second post-Newtonian correction, at order v^4, was calculated in 1994 by Blanchet, Damour, Iyer, Will, and Wiseman [1995]. Finally, the spin-orbit and spin-spin corrections were calculated in 1993 by Kidder, Will, and Wiseman [1993].

We see from Equation (12.5) that the post-Newtonian corrections bring a number of additional source parameters into the picture. The waveform no longer depends uniquely upon the chirp mass $\mathcal{M}$; it now depends upon the masses m_1 and m_2 separately, and upon the spin-orbit and spin-spin parameters (which stay approximately constant as the system evolves toward coalescence). These new parameters must be included into $\vec{\mu}$ when the signal is analyzed using matched filtering.

What we have at this point is an expression for the waveform which is accurate to second post-Newtonian order. The question facing us is whether this waveform is sufficiently accurate to be used as measurement templates. In other words, are the systematic errors generated by these templates guaranteed to be smaller than the statistical errors?

Evidently, to answer this question is difficult, since we do not have access to the exact waveform in order to make comparisons. In the next section we will consider a model problem for which the waveform *can* be calculated exactly, thereby enabling us to judge the accuracy of the post-Newtonian expansion. We will find, in Section 15, that the answer to this question is, most likely, no: the second post-Newtonian waveform is not sufficiently accurate.

13 Waveform Calculations: Perturbation Theory

A different line of attack for solving Einstein's equations for a compact binary system is to assume that one of the bodies is very much less massive than the other[5]. We therefore demand

$$\mu/M \ll 1, \tag{13.1}$$

where μ is the reduced mass and M the total mass. In this limit μ is practically equal to the smaller mass m_1, and M is practically equal to the larger mass m_2. In contrast with the post-Newtonian approach, we assume *nothing* about the size of the velocity v.

This approach takes advantage of the fact that when Equation (13.1) is valid, the smaller mass creates only a small perturbation in the gravitational field of the larger mass, which can be taken to be the Schwarzschild field. The total gravitational field can therefore be written as

$$g_{\alpha\beta} = g^{(0)}_{\alpha\beta} + h_{\alpha\beta}, \tag{13.2}$$

where $g^{(0)}_{\alpha\beta}$ represents the background Schwarzschild metric, and $h_{\alpha\beta}$ the perturbation.

Writing Einstein's equations for $g_{\alpha\beta}$ and linearizing with respect to $h_{\alpha\beta}$, one finds that the perturbation must satisfy an inhomogeneous wave equation in the Schwarzschild spacetime. Schematically,

$$\Box^{\alpha\beta\mu\nu} h_{\mu\nu} = 8\pi T^{\alpha\beta}, \tag{13.3}$$

where $\Box^{\alpha\beta\mu\nu}$ is an appropriate curved-spacetime wave operator, and $T^{\alpha\beta}$ the stress-energy tensor associated with the orbiting mass.

We will specifically assume that the central body is a Schwarzschild black hole of mass M. This assumption is made for simplicity, and removes the need to model the star's interior. As a matter of fact, the internal structure of the bodies is irrelevant, except during the last few orbital cycles before coalescence; this was discussed in Section 5. Taking advantage of this, we model the orbiting body as a point particle of mass μ, so that its stress-energy tensor is a Dirac distribution with support on the particle's world line. For simplicity, and also because it is physically well motivated (as explained in Section 5), we take the world line to be a circular geodesic of the Schwarzschild spacetime. With $\{t, r, \theta, \phi\}$ as the usual Schwarzschild

[5] This discussion is based on the work of this author and his collaborators. See Poisson and Sasaki [1995] and references therein.

coordinates, r_0 denotes the orbital radius, and $\Omega = d\phi/dt$ is the angular velocity. We have

$$v \equiv \Omega r_0 = (M/r_0)^{1/2} = (M\Omega)^{1/3}, \tag{13.4}$$

and the gravitational-wave frequency f is given by $2\pi f = 2\Omega$. We stress once more that in the perturbation approach, v is not required to be small. The only limitation on v comes from the fact that for $r_0 \leq 6M$, circular orbits are no longer stable; this implies $v < 6^{-1/2} \simeq 0.4082$.

Black-hole perturbations are conveniently treated with the Teukolsky formalism (Teukolsky [1973]), in which gravitational perturbations are represented by the complex-valued function Ψ_4, a particular component of the perturbed Weyl tensor; the tensor $h_{\alpha\beta}$ can be reconstructed from Ψ_4. The equation satisfied by Ψ_4 admits a separation of the variables. When Ψ_4 is expanded in spherical harmonics and decomposed into Fourier modes $e^{-i\omega t}$, one obtains an ordinary second-order differential equation — the Teukolsky equation for the radial function $R_{\omega\ell m}(r)$. Here, ℓ and m are the usual spherical-harmonic indices. Schematically, and omitting the subscript $\omega\ell m$, this equation takes the form

$$\mathcal{D}R(r) = T(r), \tag{13.5}$$

where $\mathcal{D}$ is a second-order differential operator, and $T(r)$ the source, constructed from the particle's stress-energy tensor (Poisson and Sasaki [1995]).

Equation (13.5) can be integrated in the standard way by constructing a Green's function $G(r, r')$ out of two linearly independent solutions to the homogeneous equation. These solutions, $R_<(r)$ and $R_>(r)$, respectively satisfy appropriate boundary conditions at the inner ($r = 2M$) and outer ($r = \infty$) boundaries. Schematically, $G(r, r') = R_<(r_<)R_>(r_>)$, where $r_<$ ($r_>$) is the lesser (greater) of r and r'. The solution to Equation (13.5) can then be expressed as $R(r) = \int G(r, r')T(r')\, dr'$, and Ψ_4 can be reconstructed by summing over all the modes. Finally, the gravitational waveform h and the luminosity dE/dt can be obtained from the asymptotic behaviour of Ψ_4 when $r \to \infty$.

To integrate Equation (13.5) therefore reduces to solving the homogeneous Teukolsky equation, $\mathcal{D}R(r) = 0$, for the functions $R_<(r)$ and $R_>(r)$. This, it turns out, is equivalent to integrating the Regge-Wheeler equation (Regge and Wheeler [1957])

$$\left\{ \frac{d^2}{dr^{*2}} + \omega^2 - \left(1 - \frac{2M}{r}\right)\left[\frac{\ell(\ell+1)}{r^2} - \frac{6M}{r}\right]\right\} X_{\omega\ell}(r) = 0, \tag{13.6}$$

for the functions $X_<(r)$ and $X_>(r)$; here, $d/dr^* = (1 - 2M/r)d/dr$. This comes about because a solution to the homogeneous Teukolsky equation can easily be related to a solution to the Regge-Wheeler equation. The relation is known as the Chandrasekhar transformation (Chandrasekhar [1975]).

14 Luminosity From the Perturbation Approach

The problem of calculating the gravitational waveform for a compact binary system with small mass ratio can therefore be reduced to the simple one of integrating Equation (13.6). The relevant dimensionless parameter entering this equation is $M\omega$, where, for circular orbits, $\omega = m\Omega$ (Poisson and Sasaki [1995]). We therefore have, using Equation (13.4), $M\omega = mv^3$.

For arbitrary values of $M\omega$ the Regge-Wheeler equation must be integrated numerically (Cutler et al. [1993]). This must be done separately for each selected value of v in the interval $0.1 < v < 0.4$ (approximately corresponding, for systems of a few solar masses, to the frequency interval 10 Hz $< f <$ 1 000 Hz).

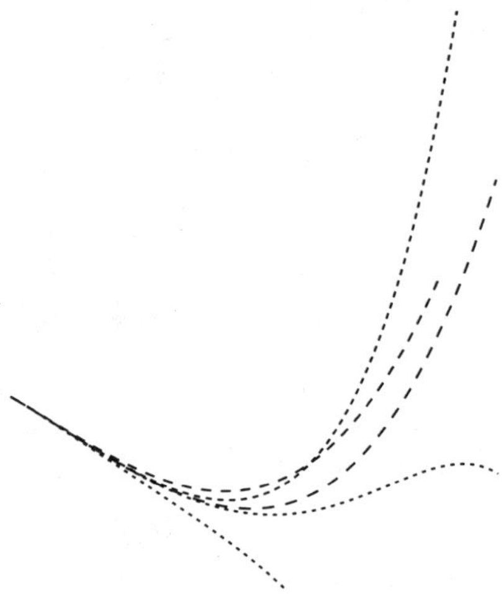

Figure 2 Various representations of $(dE/dt)/(dE/dt)_N$ as a function of orbital velocity v. The solid curve represents the exact results, obtained numerically. The various broken curves represent the various post-Newtonian approximations, as explained in the text.

Once again we will focus on dE/dt, the gravitational-wave luminosity. In Figure 2 we present a plot of $(dE/dt)/(dE/dt)_N$ as a function of v. This is the ratio of the luminosity as calculated exactly (numerically) using the perturbation approach, to the quadrupole-formula expression

$$\left(\frac{dE}{dt}\right)_N = -\frac{32}{5}\left(\frac{\mu}{M}\right)^2 v^{10}. \tag{14.1}$$

The numerical results are depicted as a solid curve. Apart from negligible numerical errors, these results are *exact* to all orders in v; the only assumption used in the calculation concerns the smallness of μ/M. The various broken curves will be described below. The figure shows that $(dE/dt)/(dE/dt)_N$ tends toward unity as $v \to 0$, and stays within 15% of unity everywhere in the interval $0 < v < 6^{-1/2}$.

It is now easy to judge the accuracy of the post-Newtonian expansion for dE/dt in the limit $\mu/M \to 0$. One simply evaluates Equation (12.5) in this limit, with SO = SS = 0, and compare with the numerical results. This post-Newtonian curve is labeled "4" in the figure, and we see that the numerical results are only imperfectly reproduced. The consequences of this will be discussed in the next section.

One can in fact do better than this. The perturbation approach is not only suitable for exact, numerical computations. By combining it with a slow-motion approximation — putting $v \ll 1$ on top of $\mu/M \ll 1$ — it also becomes suitable for approximate, analytical computations. Methods for integrating the Regge-Wheeler equations analytically in the limit $M\omega \ll 1$ were devised by various authors (Poisson [1993], Poisson and Sasaki [1995] and Sasaki [1994]). The basic idea is to proceed by iterations: Equation (13.6) with $M\omega = 0$ is solved in terms of spherical Bessel functions, and this zeroth order solution is used as input for the first iteration.

Using these methods, Tagoshi and Sasaki [1995] were able to calculate dE/dt analytically, accurately through fourth post-Newtonian order, two full orders beyond Equation (12.5). They derive the rather impressive expression

$$\begin{aligned}\frac{dE}{dt} &= \left(\frac{dE}{dt}\right)_N \Bigg[1 - \frac{1247}{336}\,v^2 + 4\pi\,v^3 - \frac{44711}{9072}\,v^4 - \frac{8191}{672}\,\pi\,v^5 \\ &\quad + \left(\frac{6643739519}{69854400} - \frac{1712}{105}\,\gamma + \frac{16}{3}\,\pi^2 - \frac{3424}{105}\,\ln 2 - \frac{1712}{105}\,\ln v\right) v^6 \\ &\quad - \frac{16285}{504}\,\pi\,v^7 + \left(-\frac{323105549467}{3178375200} + \frac{232597}{4410}\,\gamma - \frac{1369}{126}\,\pi^2 + \frac{39931}{294}\,\ln 2 \right. \\ &\quad \left. - \frac{47385}{1568}\,\ln 3 + \frac{232597}{4410}\,\ln v\right) v^8 + \cdots\Bigg]. \qquad (14.2)\end{aligned}$$

Notice the presence of $\ln v$ terms in this expansion, as well as that of the Euler number $\gamma \simeq 0.5772$. Notice also that the first four terms reproduce Equation (12.5) in the limit $\mu/M \to 0$, when SO = SS = 0. This, of course, is as it should be.

The broken curves in Figure 2 represent plots of Equation (14.2) truncated to various orders in v. For example, the curve labeled "6" is a plot of Equation (14.2) with all terms of order v^7 and v^8 discarded. We see from the figure that the post-Newtonian expansion converges poorly. (The suspicion, in fact, is that the series is only asymptotic.) Witness in particular the poor quality of the curve "5" with respect to "4", and compare also "8" to "7".

15 Accuracy of the Post-Newtonian Expansion

We have seen that in the $\mu/M \to 0$ limit, the post-Newtonian expansion for dE/dt converges poorly. A similar statement can be made about dE/df: the post-Newtonian expansion

$$\frac{dE}{df} = \left(\frac{dE}{df}\right)_N \left(1 - \frac{3}{2}\,v^2 - \frac{81}{8}\,v^4 - \frac{675}{16}\,v^6 - \frac{19845}{128}\,v^8 + \cdots\right), \qquad (15.1)$$

where

$$\left(\frac{dE}{df}\right)_N = -\frac{\pi}{3}\mu M v^{-1}, \qquad (15.2)$$

converges slowly to the exact result[6]

$$\frac{dE}{df} = \left(\frac{dE}{df}\right)_N (1 - 6v^2)\left(1 - 3v^2\right)^{-3/2}. \qquad (15.3)$$

Contrary to Equation (14.2), the expansion (15.1) is actually *known* to converge (for all values of v in the interval $0 < v < 6^{-1/2}$).

In this section we address the issue as to how much of an obstacle the poor convergence of the post-Newtonian expansion poses to the construction of accurate measurement templates (Poisson [1995]). We will answer this question for binary systems with small mass ratios, using the results described in the preceding section. Since there is no reason to believe that the convergence of the post-Newtonian

[6]This can be derived simply from the equations for circular geodesic motion in Schwarzschild.

expansion would be much improved for systems of comparable masses, our results should also apply, at least qualitatively, for such systems.

One way to address this question is to ask, given a waveform constructed from the exact numerical results, how much signal-to-noise ratio is lost by matched filtering the exact signal with an approximate post-Newtonian template?

From Section 8 we know that filtering with the exact signal $h(t)$ would give $\mathrm{SNR}|_{\max}$, the largest possible value of the signal-to-noise ratio. On the other hand, filtering with a post-Newtonian template h_n, where n denotes the order in v to which the expansion is taken (for example, $n = 4$ represents a waveform accurate to second post-Newtonian order), gives the smaller value $\mathrm{SNR}|_{\mathrm{actual}}$. From the results of Section 7 and 8 we obtain

$$\mathcal{R}_n \equiv \frac{\mathrm{SNR}|_{\mathrm{actual}}}{\mathrm{SNR}|_{\max}} = \frac{|\langle h|h_n\rangle|}{\sqrt{\langle h|h\rangle\langle h_n|h_n\rangle}}. \tag{15.4}$$

The detail of how to compute $\mathcal{R}_n$ is presented in Poisson [1995].

The calculation described here can only be carried out for binary systems with small mass ratios, because only for these do we have access to the exact waveform $h(t)$. Nevertheless, in the following we shall let μ/M become large, without altering our expressions for the exact and post-Newtonian waveforms. This is done without justification, but reflects the adopted point of view that the quality of the post-Newtonian approximation should not be appreciably affected by the finite-mass corrections.

In doing this extrapolation, some thought must be given as to the interpretation of the ratio μ/M. In the limit $\mu/M \to 0$ this can be taken to be both the ratio of the individual masses or the ratio of reduced mass to total mass. In the case of comparable masses, some choice must be made. We note that as μ/M is allowed to grow large, expressions (14.1) and (15.2) for the Newtonian quantities $(dE/dt)_N$ and $(dE/df)_N$ must be replaced by expressions (12.3), *which are formally identical.* This shows that when extrapolating to the case of comparable masses, μ is to be interpreted as the *reduced mass*, and M as the *total mass*.

n	PN	EXACT
4	0.5796	0.4958
5	0.4646	0.5286
6	0.7553	0.9454
7	0.7651	0.9864
8	0.7568	0.9695

Table 1 Reduction in signal-to-noise ratio incurred when matched filtering with approximate, post-Newtonian templates. The first column lists the order n of the approximation, the second column lists $\mathcal{R}_n$ as calculated using the post-Newtonian approximation for dE/df, and the third column lists $\mathcal{R}_n$ as calculated using the exact expression for dE/df.

We quote the results for $\mathcal{R}_n$ corresponding to a system of two neutron stars, each of 1.4 $M_\odot$; these are displayed in the second column of Table 1. We see that even at quite a high order in the post-Newtonian expansion, only three quarters of the signal-to-noise ratio is reproduced by the post-Newtonian template. This shows that the apparently small discrepancies between the exact and post-Newtonian results for dE/dt and dE/df provide a serious obstacle to the construction of accurate measurement templates.

It is interesting to ask how much of the signal-to-noise ratio would be recovered if dE/df were kept exact instead of being expressed as a post-Newtonian expansion. The third column of the table displays the values of $\mathcal{R}_n$ calculated in this way. We see that most of the signal-to-noise is recovered: $\mathcal{R}_n$ can now be as large as 0.9864 instead of 0.7651. Why does the exact expression for dE/df give such better results? It can be established (Poisson [1995]) that this has to do with the following fact: While the exact expression for dE/df correctly goes to zero at $v = 6^{-1/2}$ (at the innermost circular orbit), its post-Newtonian analogue fails to do so. [For example, the right-hand side of Equation (15.1), truncated to order v^8, goes to zero at $v \simeq 0.4236 > 6^{-1/2}$, corresponding to a radius $r_0 \simeq 5.572M$.] When corrected for this, the post-Newtonian template gives much better results.

16 Conclusion

We therefore see that the poor convergence of the post-Newtonian expansion is a serious obstacle to the construction of accurate measurement templates. Devising ways to extract the information contained in gravitational waves produced during the late inspiral of a compact binary system poses a great challenge to theoretical physicists. It is not clear that "simply" pushing to higher order in post-Newtonian theory will be enough to produce sufficiently accurate measurement templates (such that the systematic errors will be smaller than the statistical errors). There may be a need for theorists to develop alternative ways of dealing with this problem.

It should be stressed that the convergence problem does not arise when constructing *detection templates.* Indeed, as was discussed in Section 9, there is no particular need for these templates to be derived from the equations of general relativity. And since the detection templates need not involve any parameters of direct physical significance, the notion of systematic errors does not apply to them. The only requirement for constructing detection templates is that they should span the appropriate "signal space", and that they should do so the most efficiently.

The poor convergence of the post-Newtonian expansion is therefore not an obstacle for *detecting* gravitational waves from coalescing compact binaries. It is only an obstacle for extracting the information that the waves contain. To overcome this obstacle will undoubtedly be a theorist's challenge for years to come. This state of affairs is highly interesting from a historical point of view: Never before in the history of gravitational physics has experiment demanded such a high degree of sophistication on theoretical calculations.

References

Abramovici, A., Althouse, W. E., Drever, R. W. P., Gürsel, Y., Kawamura, S., Raab, F. J., Shoemaker, D., Sievers, L., Spero, R. E., Thorne, K. S., Vogt, R. E., Weiss, R., Whitcomb, S. E. and Zucker, M. E. [1992] *LIGO: The laser-interferometer gravitational-wave observatory,* Science **256**, 325–333.

Bildsten, L. and Cutler, C. [1992] *Tidal interactions of inspiraling compact binaries,* Astrophys. J. **400**, 175–180.

Blanchet, L. [1987] *Radiative gravitational fields in general relativity. II. Asymptotic behaviour at future null infinity,* Proc. R. Soc. London A **409**, 383–399.

Blanchet, L. and Damour, T. [1986] *Radiative gravitational fields in general relativity. I. General structure of the field outside the source,* Philos. Trans. R. Soc. London A **320**, 379–430.

——— [1988] *Tail-transported temporal correlations in the dynamics of a gravitating system*, Phys. Rev. D **37**, 1410–1435.

——— [1989] *Post-Newtonian generation of gravitational waves*, Ann. Inst. H. Poincaré (Phys. Théorique) **50**, 377–408.

Blanchet, L., Damour, T., Iyer, B. R., Will, C. M. and Wiseman, A. G. [1995] *Gravitational-radiation damping of compact binary systems to second post-Newtonian order*, Phys. Rev. Lett. **74**, 3515–3518.

Blanchet, L. and Shäfer, G. [1993] *Gravitational wave tails and binary star systems*, Class. Quantum Grav. **10**, 2699–2721.

Bradaschia, C., Del Fabbro, R., Di Virgilio, A., Giazotto, A., Kautzky, H., Montelatici, V., Passuello, D., Brillet, A., Cregut, O., Hello, P., Man, C. N., Manh, P. T., Marraud, A., Shoemaker, D., Linet, J. Y., Barone, F., Di Fiore, L., Milano, L., Russo, G., Aguirregabiria, J. M., Bel, H., Duruisseau, J. P., Le Denmat, G., Tourrenc, Ph., Capozzi, M., Longo, M., Lops, M., Pinto, I., Rotoli, G., Damour, T., Bonazzola, S., Marck, J. A., Gourghoulon, Y., Holloway, L. E., Fuligni, F., Iafolla, V. and Natale, G. [1990] *The VIRGO project: A wide band antenna for gravitational wave detection*, Nucl. Instrum. & Methods **A289**, 518–525.

Chandrasekhar, S. [1975] *On the equations governing the perturbations of the Schwarzschild black hole*, Proc. R. Soc. London A **343**, 289–298.

Cutler, C., Apostolatos, T. A., Bildsten, L., Finn, L. S., Flanagan, E. E., Kennefick, D., Markovic, D. M., Ori, A., Poisson, E., Sussman, G. J. and Thorne, K. S. [1993] *The last three minutes: Issues in gravitational-wave measurements of coalescing compact binaries*, Phys. Rev. Lett. **70**, 2984–2987.

Cutler, C., Finn, L. S., Poisson, E. and Sussman, G. J. [1993] *Gravitational radiation from a particle in circular orbit around a black hole. II. Numerical results for the nonrotating case*, Phys. Rev. D **47**, 1511–1518.

Cutler, C. and Flanagan, E. E. [1994] *Gravitational waves from merging compact binaries: How accurately can one extract the binary's parameters from the inspiral waveform*, Phys. Rev. D **49**, 2658–2697.

Damour, T. [1987] *The problem of motion in Newtonian and Einsteinian gravity*, 300 Years of Gravitation (Hawking, S. W. and Israel, W., eds.), Cambridge University Press, Cambridge, pp. 128–198.

Damour, T. and Iyer, B. R. [1991] *Post-Newtonian generation of gravitational waves. II. The spin moments*, Ann. Inst. H. Poincaré (Phys. Théorique) **54**, 115–164.

Finn, L. S. [1992] *Detection, measurement, and gravitational radiation*, Phys. Rev. D **46**, 5236–5249.

Finn, L. S. and Chernoff, D. F. [1993] *Observing binary inspiral in gravitational radiation: One interferometer*, Phys. Rev. D **47**, 2198–2219.

Helstrom, C. W. [1968] *Statistical Theory of Signal Detection*, Pergamon, Oxford.

Hulse, R. A. and Taylor, J. H. [1975] *Discovery of a pulsar in a binary system*, Astrophys. J. **195**, L51–L53.

Kidder, L. E., Will, C. M. and Wiseman, A. G. [1993] *Spin effects in the inspiral of coalescing compact binaries*, Phys. Rev. D **47**, R4183–R4187.

Kochanek, C. S. [1992] *Coalescing binary neutron stars*, Astrophys. J. **398**, 234–247.

Misner, C. W., Thorne, K. S. and Wheeler, J. A. [1973] *Gravitation*, Freeman, San Francisco, Chap. 36.

Narayan, R., Piran, T. and Shemi, A. [1991] *Neutron star and black hole binaries in the galaxy*, Astrophys. J. **379**, L17–L20.

Phinney, E. S. [1991] *The rate of neutron star binary mergers in the universe: Minimal predictions for gravity wave detectors*, Astrophys. J. **380**, L17–L21.

Peters, P. C. and Mathews, J. [1963] *Gravitational radiation from point masses in Keplerian orbit*, Phys. Rev. **131**, 435–440.

Poisson, E. [1993] *Gravitational radiation from a particle in circular orbit around a black hole. I. Analytical results for the nonrotating case*, Phys. Rev. D **47**, 1497–1510.

Poisson, E. [1995] *Gravitational radiation from a particle in circular orbit around a black hole. VI. Accuracy of the post-Newtonian expansion*, Phys. Rev. D **52**, 5719–5723.

Poisson, E. and Sasaki, M. [1995] *Gravitational radiation from a particle in circular orbit around a black hole. V. Black-hole absorption and tail corrections*, Phys. Rev. D **51**, 5753–5767.

Regge, T. and Wheeler, J. A. [1957] *Stability of a Schwarzschild singularty*, Phys. Rev. **108**, 1063–1069.

Reif, F. [1965] *Fundamentals of Statistical and Thermal Physics*, McGraw-Hill, New York.

Sasaki, M. [1994] *Post-Newtonian expansion of the ingoing-wave Regge-Wheeler function*, Prog. Theor. Phys. **92**, 17–36.

Schutz, B. F. [1986] *Determining the Hubble constant from gravitational wave observations*, Nature (London) **323**, 310–311.

——— [1989] *Gravitational wave sources and their detectability*, Class. Quantum Grav. **6**, 1761–1780.

Tagoshi, H. and Sasaki, M. [1994] *Post-Newtonian expansion of gravitational waves from a particle in circular orbit around a Schwarzschild black hole*, Prog. Theor. Phys. **92**, 745–771.

Taylor, J. H. [1994] *Binary pulsars and relativistic gravity*, Rev. Mod. Phys. **66**, 711–719.

Teukolsky, S. A. [1973] *Perturbations of a rotating black hole. I. Fundamental equations for gravitational, electromagnetic, and neutrino-field perturbations*, Astrophys. J. **185**, 635–647.

Thorne, K. S. [1980] *Multipole expansions of gravitational radiation*, Rev. Mod. Phys. **52**, 299–339.

——— [1987] *Gravitational radiation*, 300 Years of Gravitation (Hawking, S. W. and Israel, W., eds.), Cambridge University Press, Cambridge, pp. 330–458.

——— [1995] *Gravitational waves*, Proceedings of the Snowmass 95 Summer Study on Particle and Nuclear Astrophysics and Cosmology (Kolb, E. W. and Peccei, R., eds.), World Scientific, Singapore, in press.

Tutukov, A. V. and Yungelson, L. R. [1993] *The merger rate of neutron star and black hole binaries*, Mon. Not. Roy. Astron. Soc. **260**, 675–678.

Wagoner, R. V. and Will, C. M. [1976] *Post-Newtonian gravitational radiation from orbiting point masses*, Astrophys. J. **210**, 764–775.

——— [1977] *erratum* Astrophys. J. **215**, 984.

Wainstein, L. A. and Zubakov, V. D. [1962] *Extraction of Signals from Noise*, Prentice-Hall, Englewood Cliffs.

Will, C. M. [1994] *Gravitational waves from inspiralling compact binaries: A post-Newtonian approach*, Relativistic Cosmology, Proceedings of the Eighth Nishinomiya-Yukawa Memorial Symposium (Sasaki, M., ed.), Universal Academy Press, Kyoto, pp. 83–98.

Wiseman, A. G. [1993] *Coalescing binary systems of compact objects to (post)$^{5/2}$-Newtonian order. IV. The gravitational wave tail*, Phys. Rev. D **48**, 4757–4770.

Fields Institute Communications
Volume **15**, 1997

Colliding Black Holes with Linearized Gravity

Jorge Pullin
Center for Gravitational Physics and Geometry
Department of Physics, 104 Davey Lab
The Pennsylvania State University
University Park, PA USA 16802
pullin@phys.psu.edu

Abstract. We give a brief summary of results and ongoing research in the application of linearized theory to the study of black hole collisions in the limit in which the holes start close to each other. This approximation can be a valuable tool for comparison and code-checking of full numerical relativity computations. The approximation works quite well for the head-on case and this is motivation to pursue its use in other more interesting contexts. We summarize current efforts towards establishing the domain of validity of the approximation and its use in generation and evolution of initial data for more interesting physical cases.

1 Introduction

The intention of this talk is to summarize the application of linearized gravity, in the specific form of the theory of black hole perturbations, to the study of the collision of black holes. Most of the results are already present in the literature, and the rest of the work is still in progress so I present here only a brief survey.

The motivation for studying black hole collisions is quite clear. In the next few decades gravitational wave detectors will come online that will require "templates" of possible waveforms from different sources. The collision of black holes is one of the main candidates for observable sources of gravitational radiation. Although the initial and advanced LIGO detectors will not quite have the frequency range to detect the waves produced in the final moments of the most common collisions, it is expected that future detectors will, and knowing the waveform for the final moments can also lead to insights into the waveforms emitted earlier on.

The presence of this strong motivation from the experimental side has led to the formation of an alliance of numerical relativity groups (the "binary black hole grand challenge collaboration") with the goal of numerically simulating the collision

---

1991 *Mathematics Subject Classification.* Primary 83C25; Secondary 83C57.

The work described here is in collaboration with Richard Price, John Baker, Reinaldo Gleiser and Oscar Nicasio. I acknowledge support of NSF through grants PHY94-06269 PHY93-96246, funds of the Pennsylvania State University, its Office for Minority Faculty Development, and the Eberly Family research fund.

© 1997 American Mathematical Society

of two black holes using supercomputers. The degree of difficulty of this project is reflected in the fact that several established numerical relativity groups have decided to team efforts in order to tackle it.

Here we will like to offer a much more modest approach, which is based on a simple idea: when a collision of two black holes starts with the holes so close to each other that they are surrounded by a common horizon, the problem looks from the point of view of an external observer as a single distorted black hole. It can therefore be treated with perturbation theory. Although one expects this approach to only yield results in a small range of initial separation, it provides —at least for that range— a benchmark against which one can calibrate numerical codes of the fully numerical approach. In Price and Pullin [1994] an explicit calculation was carried out using this idea. We took the initial data for the head-on collision of two black holes given by the Misner [1960] solution and re-wrote it in such a way that in the case that the two black holes are close to each other it explicitly looks like "Schwarzschild plus something small". We took the "something small" and evolved it using the equations of linearized gravity (the Zerilli equation) and computed the radiated energy. The results are shown in Figure 1, where we plot the energy radiated in the collision as a function of the initial separation and compare with the results of the NCSA group (Anninos et al. [1993]) using a numerical integration of the full Einstein equations. We see that the close approximation works very well until the holes are no longer surrounded by a common apparent horizon ($\mu_0 = 1.3$) and works within the correct order of magnitude up to when the holes are no longer surrounded by an event horizon ($\mu_0 = 2.0$). Also shown is a "far approximation" based on a particle-membrane paradigm (Anninos et al. [1995]). Comparisons of waveforms have also been performed (Anninos et al. [1995]) and they also show very good agreement between the linearized theory and the full numerical simulations.

All this shows that the use of linearized gravity in the close limit can be a valuable aid to full numerical evolutions of the two black hole problem. It is therefore quite tempting to apply the linearized treatment to more interesting situations, specifically the in-spiraling collision of two black holes with angular momentum. There are two main obstacles to doing this computation and we will detail them in the next two sections.

2 Second Order Perturbations: Giving the Formalism Error Bars

Assuming initial data for a black hole collision is given, we can rather easily evolve and compute energies in linearized theory. Why therefore not do it for the in-spiraling collision? The main reason is that for that case there are no numerical results with which to compare and the linearized formalism does not have a measure of error in it: it therefore has little predictive power. There is no consistent way to say when the close approximation breaks down. In fact, this example teaches us a valuable lesson about perturbation theory: when is linearized perturbation valid? The obvious answer "when perturbations are small" is clearly naive. To begin with, "small" should be characterized in a coordinate invariant way. Moreover, as this example shows, perturbations can be "large" and perturbation theory can still be valid: it just needs to happen that the perturbations be large in regions of spacetime that do not contribute in a significant way to the physics of interest. In the two black hole example, such a region is the interior of the horizon and regions close to it, in which perturbations mostly fall into the black hole.

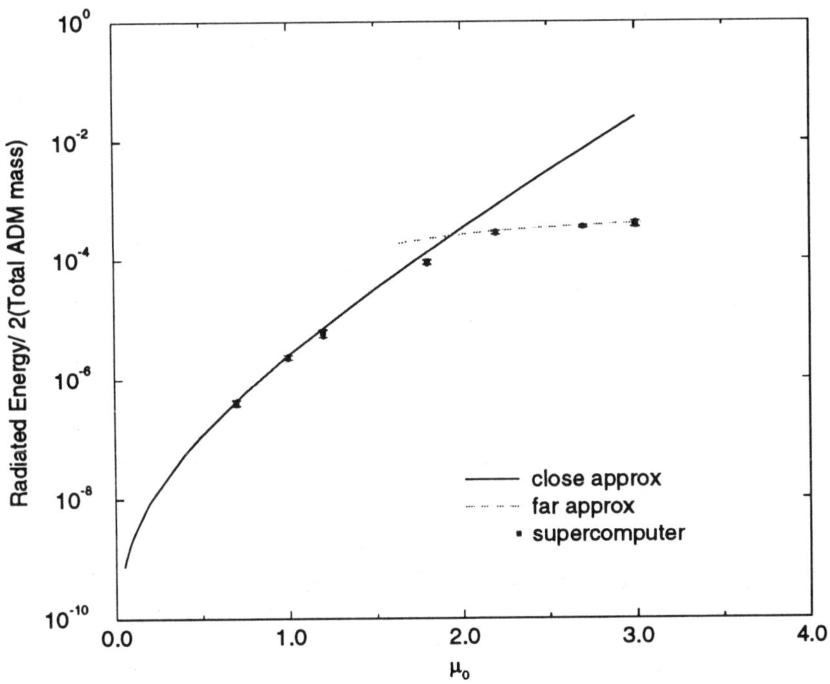

Figure 1 Comparison of results for the radiated energy vs initial separation in a collision for the close approximation, the fully numerical results and the far approximation. Vertical scale is logarithmic.

How is one to characterize when to trust the approximation? The answer is simple: work out the second order perturbations, compute the physical quantities of interest and use how much the first and second order calculation differ as a measure of the accuracy of the first order results. The advantage of this answer is that it is phrased in terms of what one is exactly interested in: the physical quantities. In the case of the collision of two black holes these are the radiated waveforms and energies.

The formalism for second order perturbations of black holes has not been worked out in the past. It can be studied in detail as we do in Nicasio et al. [1995]. Here I just sketch some of the outstanding points. It turns out that all the information can be coded into a single variable, exactly as in the first order perturbation case and that that variable satisfies a "Zerilli equation",

$$-\frac{\partial\psi^{(2)}}{\partial t} + \frac{\partial\psi^{(2)}}{\partial r_*} + V(r)\psi^{(2)} = S \tag{2.1}$$

where $r_* = r + \log(r/2M - 1)$ and the Zerilli function $\psi^{(2)}$ is a coordinate invariant combination of the perturbed metric coefficients. This equation is exactly the same as the one satisfied by the first order perturbations (including the "potential" V(r), which can be seen in Price and Pullin [1994]). However, there is an important difference: the right-hand side is not zero but a "source" term S, which is listed explicitly in Nicasio et al. [1995] and which is a complicated function quadratic in the first order perturbations and their derivatives. The way in which we derived this equation is to compute a particular combination of the Einstein equations, writing the perturbed metric in a particular coordinate system, the so called "Regge-Wheeler" gauge. This, in turn is a way of deriving the original Zerilli equation. The expression we get for $\psi^{(2)}$ is therefore a representation in that gauge of a gauge invariant quantity. The explicitly gauge invariant form of $\psi^{(2)}$ can also be computed.

We therefore are in a position to evolve to second order the problem of black hole collisions and therefore to endow the first order predictions with "error bars". This will be crucial for the inspiralling case, where numerical results are not expected for some time.

3 Initial Data in the Close Approximation

In the head-on collision case we were lucky to have an exact solution to the initial value problem that we could evolve. For the more realistic cases there are no exact solutions available at present and it is unlikely that they will be easily found in the future. There is an immediate alternative at hand. There exist already well tested numerical codes (Cook [1991], [1994]) for solving the initial value problem in general relativity in the context of black hole collisions. One could simply take these initial data evaluated for the case in which the black holes are close and "read off" from them the departures from Schwarzschild to be evolved using the linearized theory. This is certainly possible and has already been illustrated for Brill-Lindquist-type (Brill and Lindquist [1963]) initial data by Abrahams and Price [1995].

Apart from the possibility of using numerical initial data for realistic collisions it is interesting to notice that one can, up to a certain extent, solve the initial value problem analytically if one is only interested in initial data for the close approximation. The idea is simple: in the close approximation the initial data for a black hole collision departs a small amount from the initial data for a Schwarzschild spacetime for a single black hole with mass equal to the sum of the masses of the colliding holes. Therefore one can develop an approximation technique for the initial data starting from the initial data of Schwarzschild and adding small corrections proportional to the separation of the holes. We illustrate here only the zeroth order results, details will be given in a forthcoming paper in collaboration with John Baker.

The initial value problem of general relativity can be conveniently cast in the conformal formalism (see for instance Bowen and York [1980] and references therein). One is interested in solving the momentum and Hamiltonian constraints

$$\nabla^a(K_{ab} - g_{ab}K) = 0 \tag{3.1}$$

$${}^3R - K_{ab}K^{ab} + K^2 = 0 \tag{3.2}$$

where g_{ab} is the spatial metric, K_{ab} is the extrinsic curvature and ${}^3R$ is the scalar curvature of the three metric. One proposes a three metric that is conformally flat $g_{ab} = \psi^4\delta_{ab}$, with ψ^4 the conformal factor and a decomposition of the extrinsic curvature $\widehat{K}_{ab} = \psi^{-2}K_{ab}$.

The constraints become,

$$\widehat{\nabla}^a\widehat{K}_{ab} = 0 \tag{3.3}$$

$$\widehat{\nabla}^2\psi = -\frac{1}{8}\psi^{-7}\widehat{K}_{ab}\widehat{K}^{ab}. \tag{3.4}$$

where $\widehat{\nabla}$ is a derivative with respect to the flat spacetime. Since the momentum constraint is linear, one can propose as a solution for it for the case of two black holes the sum of the solutions for the case of individual holes[1] with momentum P_a,

$$\hat{K}_{ab} = \frac{3}{r^2}\left[P_{(a}n_{b)} - (\delta_{ab} - n_a n_b)P^c n_c\right] \tag{3.5}$$

where n_b is a unit normal in the direction of $\vec{r}$ and all vector fields are defined in the flat background spacetime.

One now can put this solution in the Hamiltonian constraint and one is left with an elliptic, highly non-linear equation for ψ. This is the equation that is usually solved numerically. There exist situations, however, where one can make some progress analytically. Consider the case in which the momenta of the holes is small (Bowen and York [1980]). In that case one can neglect the right-hand side of the Hamiltonian constraint and one only needs to solve a vacuum Laplace equation for ψ. The solution can therefore be very simply found, the difficulty depending on the boundary conditions one chooses for the problem (typically a "symmetrized" boundary condition is imposed, which complicates calculation quite a bit in certain cases, see Cook [1991], [1994] for details).

Another situation in which one can obtain an approximate solution is in the "close approximation". In that case one has two black holes of momenta equal and opposite $P_a^{(1)} = -P_a^{(2)}$, and since the black holes are close, the unit normals appearing in the form for the extrinsic curvature for each hole are approximately equal. That implies that the extrinsic curvature for the problem is approximately zero (as it should, since in the close limit the problem looks like a Schwarzschild black hole at rest.) Therefore one can again neglect the right-hand side of the Hamiltonian constraint and one is again left with a Laplace equation. Let us compare this approximation with the full numerical results. In order to do this we will compare the ADM energy of initial data for a collision of two holes of momentum P. The ADM energy in the conformal formalism is given by

$$E = -\frac{1}{2\pi}\oint_\infty \nabla_i\psi\, d^2S^i \tag{3.6}$$

and we notice that it does not depend explicitly on the extrinsic curvature (it does implicitly via the constraints). Therefore at the approximation we are working, in which the constraints do not couple the conformal factor and the extrinsic curvature, the energy is independent of the extrinsic curvature and therefore independent of the momenta of the holes. We compare this prediction with the full numerical results of Cook in Figure 2.

An interesting aspect is that one can advance this approximation one step further. One can input the extrinsic curvature and the conformal factor found as a fixed "source" in the equation determining the conformal factor and one can obtain a correction through the integration of a Poisson equation. Comparison of this approximation with the numerical data is currently in progress. Details are complicated by the particular boundary conditions that are usually chosen in the numerical computations.

It is evident that the "close approximation" can work in many other cases, apart from the head-on, equal momenta holes we considered here. The only changes will

[1] The particular solution chosen depends on the boundary conditions imposed. This may add other terms to the simple ones we list here for brevity, but they all behave in a similar fashion with respect to the approximations we will consider.

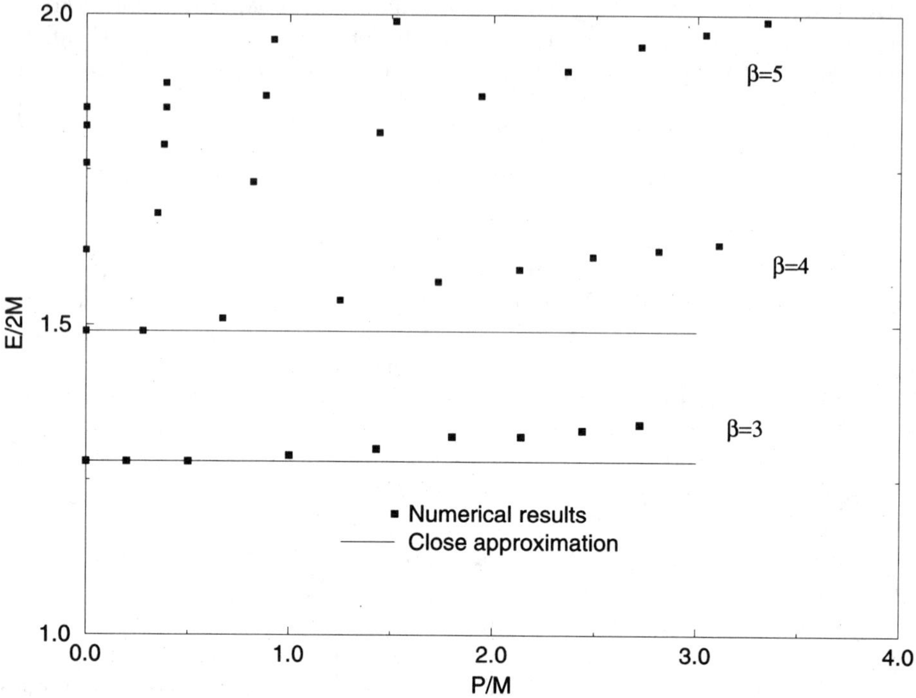

Figure 2 The ADM energy of initial data for collisions of black holes of momenta P. The dots are the full numerical results of Cook, for different values of the initial separation β. We see that for small separations, the energy is approximately independent of the holes momenta, which coincides with the close approximation prediction, depicted by the solid line.

be that the solution one obtains in the "close limit" rather than being a slice of Schwarzschild will be a slice of Kerr or boosted Schwarzschild if the net result of the collision has angular momentum or linear momentum.

4 Summary

We have seen that the use of the "close approximation" can be a valuable aid to full numerical computations of the collision of two black holes. With the introduction of a second order scheme we are now in a position of offering reliable estimates of energies and waveforms that we expect people working on the full numerical simulations will find of use to calibrate codes and design strategies for better integrating the Einstein equations in this problem of great current physical interest.

References

Abrahams, A. and Price, R. [1995] *Black hole collisions from Brill-Lindquist initial data: predictions of perturbation theory*, Phys. Rev. D, to appear.

Anninos, P., Hobill, D., Seidel, E., Smarr, L. and Suen, W.-M. [1993] *Collision of two black holes*, Phys. Rev. Lett. **71**, 2851–2854.

Anninos, P., Price, R., Pullin, J., Seidel, E. and Suen, W.-M. [1995] *Head-on collision of two black holes: comparison of different approaches*, Phys. Rev. D **D52**, 4462–4480.

Bowen, J., York, J. [1980] *Three-dimensional initial data for black holes and black-hole collisions*, Phys. Rev. D **21**, 2047–2056.

Brill, D. and Lindquist, R. [1963] *Interaction energy in geometrostatics*, Phys. Rev. **131**, 471–476.

Cook, G. [1991] *Three dimensional initial data for the collision of two black holes, I*, Phys. Rev. D **44**, 2983–2992.

——— [1994] *Three dimensional initial data for the collision of two black holes, II*, Phys. Rev. D **50**, 5025–5032.

Misner, C. [1960] *Wormhole initial conditions*, Phys. Rev. **118**, 1110–1113.

Nicasio, O., Gleiser, R., Price, R. and Pullin, J. [1995] *Second order perturbations of a Schwarzschild black hole*, preprint gr-qc@xxx.lanl.gov:9510049.

Price, R. and Pullin, J. [1994] *Colliding black holes: the close limit*, Phys. Rev. Lett. **72**, 3297–3300.

Fields Institute Communications
Volume **15**, 1997

Spinors, Jets, and the Einstein Equations

Charles G. Torre
Department of Physics
Utah State University
Logan, UT USA 84322-4415
torre@cc.usu.edu

Abstract. Many important features of a field theory, e.g., conserved currents, symplectic structures, energy-momentum tensors, etc., arise as tensors locally constructed from the fields and their derivatives. Such tensors are naturally defined as geometric objects on the jet space of solutions to the field equations. Modern results from the calculus on jet bundles can be combined with a powerful spinor parametrization of the jet space of Einstein metrics to unravel basic features of the Einstein equations. These techniques have been applied to computation of generalized symmetries and "characteristic cohomology" of the Einstein equations, and lead to results such as a proof of non-existence of "local observables" for vacuum spacetimes and a uniqueness theorem for the gravitational symplectic structure.

1 Introduction

This is a survey of results of work performed largely in collaboration with Ian Anderson (Dept. of Mathematics, Utah State University). The presentation will be somewhat informal; rigorous statements and proofs of our results will be presented elsewhere (Torre and Anderson [1993], Anderson and Torre [1994], [1996a], [1996b] and [1996c]). In the very broadest terms, our efforts are intended to help answer the question: "In what ways are the Einstein equations special?" There is of course a plethora of special features of the Einstein equations which have been uncovered since the advent of general relativity. Here are some examples. The vacuum equations in 4 dimensions are, up to specification of the cosmological constant, the only second-order partial differential equations one can write down for a metric which are "generally covariant" and can be derived from a variational principle. Despite the complexity of the field equations, a large number of exact solutions are known. Indeed, certain reductions of the equations (self-dual equations, stationary-axisymmetric vacuum and electrovac equations) admit transitive symmetry groups and are in some sense "integrable". Special features that are of particular relevance to the physical viability of the Einstein equations include theorems guaranteeing

1991 *Mathematics Subject Classification.* 83C05.

© 1997 American Mathematical Society

that the Cauchy problem is well posed, existence and uniqueness (up to diffeomorphisms) of solutions, existence and positivity of conserved energy-momentum in the asymptotically flat context, etc. Finally, a feature of the Einstein equations which is especially relevant for attempts at quantization is that the equations constitute a constrained Hamiltonian system.

The modern geometric theory of differential equations provides systematic means through which one can characterize certain important structural features of any set of field equations. These structural features arise as geometric objects on the *jet space* of solutions to the field equations (Saunders [1989] and Olver [1993]) and are neatly analyzed in terms of the *variational bicomplex* (Anderson [1992] and Olver [1993]). Such jet space techniques have been used in the analysis of a number of differential equations of applied mathematics and mathematical physics. It is now possible to apply these techniques to study special features of the vacuum Einstein equations; the results of such investigations are the subject of this paper. In particular, I will report on a systematic classification of *generalized symmetries* and *generalized conservation laws* for the vacuum equations. Roughly speaking, generalized symmetries are infinitesimal transformations constructed locally from the metric and its derivatives mapping solutions to solutions. Generalized conservation laws are closed but not exact differential forms constructed locally from solutions to the field equations and/or solutions to the linearized equations. The generalized conservation laws correspond to conserved volume, surface, or line integrals associated to solutions to the Einstein equations. In the mathematical literature, the generalized conservation laws go by the name "characteristic cohomology" of the field equations (Bryant and Griffiths [1993]; for a BRST approach to characteristic cohomology see Barnich et al. [1995]).

To begin our survey of generalized symmetries and conservation laws for the Einstein equations, we should first improve our understanding of the jet space of solutions to the vacuum equations.

2 Jet Space

The jet space of solutions to the Einstein equations can be viewed as the beginning of an answer to the question "What data are freely specifiable *at a point* of a vacuum spacetime?" This question is similar in spirit to the kind of question one asks when investigating the Cauchy problem, where one seeks the data which can be freely specified on a Cauchy surface. Of course the two questions are quite different mathematically but, to carry the analogy a little further, one can think of the jet space of solutions as something of an analog of Cauchy data for analytic solutions. Indeed, to construct an analytic solution to the Einstein equations, one can pick coordinates x^i about some point $x^i = 0$ and write down a power series expansion:

$$g_{ij}(x) = g_{ij}(0) + g_{ij,k}(0)x^k + \frac{1}{2!}g_{ij,kl}(0)x^k x^l + \cdots . \tag{2.1}$$

Evidently, formal power series expansions of a metric about a point x^i are determined by the data

$$(x^i, g_{ij}, g_{ij,k}, g_{ij,kl}, \ldots). \tag{2.2}$$

Such data define a point in the *jet space of metrics* $\mathcal{J}$. Of course, to define a metric via power series about some point, one should enforce some convergence criterion on the metric and its derivatives at the given point; the jet space is, however, defined without such convergence criteria. A more precise way to introduce the jet space of metrics is to view it as a bundle whose base space is spacetime and whose typical fiber is the space of values of the metric and its derivatives at a point. A metric defines a cross section of this bundle. Moreover, Borel's Theorem (see, e.g., Kahn [1980]) implies that given a point $(x^i, g_{ij}, g_{ij,k}, g_{ij,kl}, \ldots) \in \mathcal{J}$, there is always a smooth metric which has that data.

Let us remark that while we have defined $\mathcal{J}$ using coordinates and coordinate derivatives of the metric, the jet space is in fact a coordinate-independent object. It is possible to give a coordinate-free definition of $\mathcal{J}$ as a fiber bundle over spacetime along the lines mentioned above. In particular, it is possible to replace coordinates and coordinate derivatives with globally defined derivative operators (see, e.g., Wald [1990]), but for simplicity we shall stick with an informal local treatment.

Having defined the jet space of metrics we would now like to see what points in jet space are allowed by the Einstein equations. To do this, we need a slightly better parametrization of $\mathcal{J}$. To this end, given coordinates x^i, define variables $\Gamma^{(k)}$ and $Q^{(k)}$, where

$$\Gamma^{(k)} \longrightarrow \Gamma^i_{j_0 j_1 j_2 \cdots j_k} := \Gamma^i_{(j_0 j_1, j_2 \cdots j_k)}, \qquad k = 1, 2, \ldots \tag{2.3}$$

and

$$Q^{(k)} \longrightarrow Q_{ab, c_1 c_2 \cdots c_k} := g_{am}\, g_{bn} \nabla_{(c_3} \cdots \nabla_{c_k} R^m{}_{c_1}{}^n{}_{c_2)} \qquad k = 2, 3, \ldots\,, \tag{2.4}$$

where ∇ is the torsion-free derivative operator compatible with g_{ab} and R_{abcd} is the Riemann tensor. It can be shown that $\Gamma^{(k)}$ and $Q^{(k)}$ are algebraically independent at any given point of spacetime. The variables $\Gamma^{(k)}$ carry the coordinate-dependent information in the kth partial derivatives of the metric. In particular, all of the variables $\Gamma^{(k)}$ vanish at the origin of a geodesic coordinate chart. The variables $Q^{(k)}$ contain all spacetime-geometric information in the kth partial derivatives of the metric. In particular, the curvature tensor and all of its covariant derivatives can be uniquely expressed in terms of the variables $Q^{(k)}$. The tensors $Q^{(k)}$ were apparently introduced by Penrose [1960] and are closely related to Thomas' "normal metric tensors" (Thomas [1934]).

Our first result is that the variables

$$(x^i, g_{ij}, \Gamma^{(1)}, \Gamma^{(2)}, \ldots, Q^{(2)}, Q^{(3)}, \ldots) \tag{2.5}$$

uniquely parameterize points in the jet bundle. In other words, given the data (2.5) one can reconstruct the metric and all of its derivatives at the point labeled x^i. The data (2.5) are freely specifiable at a point of a (pseudo-) Riemannian manifold.

The Einstein tensor can be viewed as a collection of functions on $\mathcal{J}$, and the Einstein equations can be viewed as defining a subspace (actually a submanifold) $\mathcal{E} \hookrightarrow \mathcal{J}$. Because the Einstein equations are geometrically defined, they introduce relations only among the variables $Q^{(k)}$. In fact, the vacuum equations uniquely fix the traces of these tensors in terms of their trace-free parts (Anderson and Torre [1994]). Thus a point in the jet space of Einstein metrics is defined by the variables

$$(x^i, g_{ij}, \Gamma^{(1)}, \Gamma^{(2)}, \ldots, \widetilde{Q}^{(2)}, \widetilde{Q}^{(3)}, \ldots), \tag{2.6}$$

where $\widetilde{Q}^{(k)}$ denotes the completely trace-free part of tensors (2.4) with respect to the metric g_{ij}. The data (2.6) are freely specifiable at a point of a Ricci-flat spacetime. These coordinates on $\mathcal{E}$ can be interpreted in terms of a power series expansion of an Einstein metric as follows. If we are trying to build an Einstein metric by Taylor series we (i) specify the spacetime point x^i around which the series is being developed, (ii) specify the metric components g_{ij} at x^i, (iii) specify the variables $\Gamma^{(k)}$; this fixes the coordinate system in which the metric is being built, (iv) specify the variables $\widetilde{Q}^{(k)}$, which supplies the geometric content of the Einstein metric. Of course, this procedure leaves open the question of convergence of the series.

The parametrization (2.6) of $\mathcal{E}$ turns out to be somewhat unwieldy in applications, primarily because of the need to remove so many traces. A much more useful parametrization, which only exists in four dimensions, uses a spinor representation of the variables $\widetilde{Q}^{(k)}$. Let Ψ_{ABCD} and $\overline{\Psi}_{A'B'C'D'}$ denote the Weyl spinors (Penrose [1960]). Fix a soldering form $\sigma_a^{AA'}$ such that, for a given g_{ij},

$$g_{ij} = \sigma_i^{AA'}\sigma_{jAA'}. \tag{2.7}$$

It can be shown that the variables $\widetilde{Q}^{(k)}$ are uniquely parametrized by the soldering form, the spinor variables

$$\Psi^{(k)} \longleftrightarrow \Psi^{J_1'\cdots J_{k-2}'}_{J_1\cdots J_{k+2}} = \nabla^{(J_1'}_{(J_1}\cdots\nabla^{J_{k-2}')}_{J_{k-2}}\Psi_{J_{k-1}J_kJ_{k+1}J_{k+2})}, \tag{2.8}$$

and their complex conjugates $\overline{\Psi}^{(k)}$. Thus we obtain a spinor parametrization of $\mathcal{E}$ in terms of

$$(x^i, g_{ij}, \Gamma^{(1)}, \Gamma^{(2)}, \ldots, \Psi^{(2)}, \overline{\Psi}^{(2)}, \Psi^{(3)}, \overline{\Psi}^{(3)}, \ldots). \tag{2.9}$$

The spinor aficionado will recognize that our spinor parametrization of $\mathcal{E}$ is closely related to Penrose's notion of an "exact set of fields" (Penrose [1960]).

The parametrization (2.9) of the jet space of solutions to the Einstein equations is an important technical tool needed to classify symmetries and conservation laws. More generally, these variables allow us to address problems of the following type. Suppose we are interested in finding a tensor field $T = T(x, g, \partial g, \ldots)$, locally constructed from the metric and its derivatives to some order, which satisfies some local differential relations,

$$DT = 0, \tag{2.10}$$

when the Einstein equations hold. As an example, suppose we wanted to find a conserved current for the Einstein equations. This would be a vector field $j^a = j^a(x, g, \partial g, \ldots)$ such that

$$\nabla_a j^a = 0 \qquad \text{when } G_{ab} = 0. \tag{2.11}$$

The tensors T and DT can be viewed as a collection of functions on $\mathcal{J}$ and the differential relation (2.10) says that the function DT vanishes on $\mathcal{E}$. If we express the relation (2.10) in the coordinates (2.9), then (2.10) must hold *identically.* The power of spinor analysis can now be brought to bear on classifying all such solutions T to this identity up to terms which vanish on $\mathcal{E}$. As we shall see, classifications of generalized symmetries and generalized conservation laws are precisely problems of this type.

To be honest, it is a bit of an over-simplification to say that these problems can be solved in a straightforward manner given the coordinates (2.9). The results to be given below are in fact intricately related through a mathematical structure on $\mathcal{J}$ (or $\mathcal{E}$) called "the variational bicomplex" (Anderson [1992]). Because my goal here is to emphasize results rather than techniques, I will not be able to say more about the variational bicomplex in this paper. Suffice it to say that the bicomplex is an indispensable tool in the analysis of any set of field equations, and this mathematical structure is playing a vital role "behind the scenes" in all that follows.

3 Generalized Symmetries

A generalized symmetry is an infinitesimal transformation constructed locally from the fields and their derivatives to some order which maps solutions of the field equations to other solutions. Generalized symmetries are of interest because they provide methods of generating new solutions from known solutions, their existence is necessary for the existence of local conservation laws, and because of their role in complete integrability of a variety of partial differential equations (Olver [1993]). Before giving results from our classification of generalized symmetries of the Einstein equations, let us first have a look at an elementary example of a dynamical system that admits a generalized symmetry.

Consider the Kepler problem, which can be described by the non-linear system of ordinary differential equations:

$$\mathbf{r}'' = -k\frac{\mathbf{r}}{r^3}, \tag{3.1}$$

where $\mathbf{r}$ is the relative position of two masses in space, k is a constant, and a prime denotes a time derivative. A point in the jet space $\mathcal{J}$ for this problem is defined by the variables $(t, \mathbf{r}, \mathbf{r}', \mathbf{r}'', \dots)$. The equations (3.1) and their time derivatives define the jet space $\mathcal{E} \hookrightarrow \mathcal{J}$ of solutions. Coordinates on $\mathcal{E}$ are $(t, \mathbf{r}, \mathbf{r}')$; these variables parametrize the extended velocity phase space for the Kepler problem. There are a couple of obvious symmetries of the equations (3.1), namely, rotational symmetry and time translation symmetry. These symmetries are usually called "point symmetries" or "Lie symmetries". The point symmetries are distinguished by the fact that they can be defined as groups of transformations of the underlying space of independent and dependent variables $(t, \mathbf{r})$ only, without involving derivatives of $\mathbf{r}$. By contrast, the following infinitesimal transformation necessarily involves derivatives of $\mathbf{r}$:

$$\delta\mathbf{r} - 2(\lambda \cdot \mathbf{r})\mathbf{r}' - (\lambda \cdot \mathbf{r}')\mathbf{r} - (\mathbf{r} \cdot \mathbf{r}')\lambda \tag{3.2}$$

Here λ is a fixed, time-independent vector. It is straightforward to check that if $\mathbf{r}(t)$ satisfies (3.1) then, to first order in λ, $\mathbf{r}(t) + \delta\mathbf{r}(t)$ also satisfies (3.1). The transformation (3.2) represents a first-order generalized symmetry of the equations (3.1)[1]. I think you will agree that it is somewhat remarkable that the relatively simple system of equations (3.1) admits such a complicated "hidden symmetry". The three-parameter family of symmetries given in (3.2) are responsible for the existence of a conserved vector, known as the Laplace-Runge-Lenz vector $\mathbf{A}$ (see, for example, Goldstein[1990]):

$$\mathbf{A} = \mathbf{r}' \times (\mathbf{r} \times \mathbf{r}') - k\frac{\mathbf{r}}{r}. \tag{3.3}$$

Conservation of the Laplace-Runge-Lenz vector reflects a special feature of the Kepler problem: all its bound orbits are closed. The only other central force that

[1]Note that all generalized symmetries of (3.1) can be expressed as first-order symmetries if we use the equations of motion. Note that this property of generalized symmetries does not generalize to partial differential equations.

has this special property is that of an isotropic oscillator, and in this case there is again a generalized symmetry and (tensor) conservation law which is responsible.

It is natural to wonder if the Einstein equations will admit any hidden generalized symmetries as do many simpler systems of non-linear differential equations. To find such symmetries we must look for an infinitesimal transformation

$$\delta g_{ab} = h_{ab}(x, g, \partial g, \dots) \tag{3.4}$$

mapping solutions to solutions. This means that h_{ab} must satisfy the linearized equations,

$$-\nabla^c\nabla_c h_{ab} - \nabla_a\nabla_b(g^{cd}h_{cd}) + 2\nabla^c\nabla_{(a}h_{b)c} = 0, \tag{3.5}$$

when the metric g_{ab}, out of which h_{ab} is built, satisfies the Einstein equations $G_{ab} = 0$. Because we are only interested in transformations between solutions, any two symmetries that are equal when the field equations hold can be considered equivalent. So, for example, the Einstein tensor defines a generalized symmetry, $h_{ab} = G_{ab}$, but the transformation it generates is trivial, i.e., we identify this symmetry with $h_{ab} = 0$. As described earlier, the classification of on-shell generalized symmetries is exactly the kind of problem that can be fruitfully attacked using the spinor parametrization of $\mathcal{E}$. In detail, the symmetry transformation h_{ab} is viewed as a function on $\mathcal{J}$ (we are actually only interested in the restriction of h_{ab} to $\mathcal{E}$). We view (3.5) as requiring that certain functions on $\mathcal{J}$ built from h_{ab} vanish when restricted to $\mathcal{E}$. That is, as a function on $\mathcal{E}$, h_{ab} should satisfy the identity (3.5). As an identity on $\mathcal{E}$, (3.5) must hold for all values of the coordinates (2.9); analysis of this requirement leads to the following result.

Let $h_{ab} = h_{ab}(x, g, \partial g, \dots)$ be a generalized symmetry of the vacuum Einstein equations in four spacetime dimensions. Then there is a constant c and a covector $V_a = V_a(x, g, \partial g, \dots)$ such that, modulo terms that vanish when $G_{ab} = 0$, the symmetry must take the following form:

$$h_{ab} = c\, g_{ab} + \nabla_a V_b + \nabla_b V_a. \tag{3.6}$$

This form of h_{ab} represents a combination of two types of symmetry transformations. The term $c\, g_{ab}$ corresponds to a scale symmetry admitted by the vacuum equations. If we allowed for a cosmological constant, this symmetry would be absent. The terms $\nabla_a V_b + \nabla_b V_a$ correspond to the infinitesimal change in the metric arising from the pull-back by a 1-parameter family of (local) spacetime diffeomorphisms generated by V^a. Thus the most general symmetry is a combination of a constant scaling and a "gauge transformation". Of course, both of these symmetries are well known and we conclude that the vacuum Einstein equations admit no "hidden local symmetries".

4 Generalized Conservation Laws

The symmetries we have found do not have any non-trivial conservation laws associated with them. Noether's Theorem then suggests that, aside from possible topological conservation laws, there are no conserved currents for the Einstein equations built locally from the metric and its derivatives. This is in fact true, but it does not follow directly from our symmetry classification because it is *a priori* possible to have symmetries that are on-shell trivial and nevertheless generate non-trivial conservation laws. For completely non-degenerate systems of PDE's it is known that every non-trivial conserved current follows from a non-trivial generalized symmetry (Olver [1993]), but the Einstein equations do not qualify as a completely non-degenerate system owing to their general covariance. The problem of rigorously classifying conserved currents for the vacuum Einstein equations can be solved using the variational bicomplex and our spinor techniques. In fact, it is possible to generalize the analysis and classify all closed forms locally built from vacuum metrics as well as a large class of closed forms locally built from vacuum metrics and solutions of the linearized equations. We begin with closed forms built locally from vacuum metrics (see Wald [1990] for a general discussion of identically closed forms locally built from fields).

Let $\omega = \omega(x, g, \partial g, \dots)$ be a p-form locally built from the metric and its derivatives to some order. We assume $p < 4$. If $d\omega = 0$ when the vacuum Einstein equations hold, we say that ω represents a *generalized conservation law* for the vacuum equations. The importance of a generalized conservation law stems from the fact that its integral over a closed[2] p-dimensional submanifold Σ is independent of the choice of Σ (up to homology) when the metric satisfies the field equations. Thus the integral

$$Q[g] = \int_\Sigma \omega(x, g, \partial g, \dots) \tag{4.1}$$

is a conserved charge characterizing vacuum spacetimes. As we are only interested in the values of the conserved charges for solutions of the field equations, we will identify any generalized conservation laws which are equal when the field equations hold. In other words, viewing the generalized conservation laws as functions on $\mathcal{J}$, we are only interested in their restriction to $\mathcal{E}$.

Of course, if (on $\mathcal{E}$) there exists a $(p-1)$-form $\eta = \eta(x, g, \partial g, \dots)$ such that

$$\omega = d\eta, \tag{4.2}$$

then ω is identically closed and $Q[g] = 0$ for any metric g. We will call such exact forms *trivial conservation laws.*

If ω is a closed 3-form, then its Hodge dual is a conserved current; the conserved charge is then a volume integral of a local density. In field theory this is the way in which conserved charges typically arise. However, interesting conserved quantities do sometimes arise not as volume integrals but instead as surface—or even line—integrals. For example, if we restrict our attention to vacuum (regions of) spacetimes admitting a Killing vector field k^a, then the Komar 2-form,

$$\kappa_{ab} = \epsilon_{ab}{}^{cd}\nabla_c k_d, \tag{4.3}$$

defines a generalized conservation law, as does the twist 1-form

$$\tau_a = k^b \kappa_{ab}. \tag{4.4}$$

Is there a field-theoretic explanation for the existence of these generalized conservation laws? Are there any generalized conservation laws admitted by the vacuum Einstein equations which can be defined without assuming the existence of a Killing vector? We shall answer the latter question now and return to the first question toward the end of this article.

We wish to classify closed forms locally built from Ricci-flat metrics. Once again we are looking for some functions on $\mathcal{J}$ satisfying a certain differential relation on

[2]With suitable boundary conditions, the submanifold can be open or have boundaries.

$\mathcal{E}$. Using spinor techniques and basic properties of the variational bicomplex we obtain the following results. If $\omega = \omega(x, g, \partial g, \dots)$ is a p-form locally constructed from the metric, and ω is closed when the Einstein equations hold then, modulo terms which vanish when the field equations are satisfied, there exists a $(p-1)$-form $\eta = \eta(x, g, \partial g, \dots)$ and a constant c such that

$$\begin{aligned} \omega &= \text{const.}, && p = 0; \\ \omega &= d\eta, && p = 1, 2; \\ \omega &= c\,\sigma + d\eta, && p = 3. \end{aligned} \tag{4.5}$$

In (4.5) σ is an *identically* closed 3-form locally built from the metric and its first derivatives. This 3-form is a representative of the non-trivial cohomology class at degree 3 which exists on the bundle of Lorentzian metrics over spacetime (Torre [1995]). The corresponding conserved charge is the "kink-number" of the spacetime, which was discussed by Finkelstein and Misner quite some time ago (Finkelstein and Misner [1959]). Roughly speaking, the kink number counts the number of times light cones tumble as one traverses a hypersurface. Because σ is identically closed, this conservation law exists for any gravitational theory in which one employs a Lorentzian metric. Aside from this single topological conservation law, the vacuum Einstein equations admit no non-trivial generalized conservation laws.

There is an immediate corollary of this result which is relevant for the Hamiltonian formulation of general relativity on closed universes (Torre[1993]). Recall that the Einstein equations can be viewed as defining a constrained Hamiltonian system. The constraints limit the possible range of canonical data and generate canonical transformations which are Hamiltonian expressions of the action of spacetime diffeomorphisms on the Cauchy data. It is of interest, especially in attempts to canonically quantize the theory, to find functions on the gravitational phase space which are invariant under these canonical transformations, i.e., which have vanishing Poisson brackets with the constraint functions. These "gauge invariant" functions on phase space are commonly called the *observables* of the theory. Because, for closed universes, the Hamiltonian of general relativity is a linear combination of the constraint functions, the observables are constants of the motion. But we have, in effect, just classified a large set of constants of motion for the Einstein equations, and so we have a classification of a particular type of observables.

Let us define a *local observable* $\mathcal{O}$ as an observable on the gravitational phase space which is constructed as an integral over a compact p-dimensional manifold Σ of a differential p-form ω, where ω is locally built from the canonical coordinates and momenta and their derivatives to any finite order. Because the canonical variables are related by local formulas to the spacetime metric it follows that, on solutions (in which Σ is embedded in spacetime), $\mathcal{O}$ is a conserved charge of the type (4.1). But there are no such conserved charges except the kink number, which must always vanish in spacetimes that admit a foliation by spacelike hypersurfaces (Finkelstein and Misner [1959]). We conclude that there are no local observables for vacuum spacetimes with compact Cauchy surfaces. This is not to say that there are no observables, but only that the simplest class of observables that one naturally tries to construct for general relativity are simply not there.

Let us now turn to generalized conservation laws built from solutions γ_{ab} to the linearized Einstein equations. An important example is provided by the pre-symplectic current for the vacuum equations. Here we present it as a 3-form:

$$\Omega_{abc} = \left(\gamma^m_{[a}\nabla^n\hat{\gamma}^p_b - \hat{\gamma}^m_{[a}\nabla^n\gamma^p_b\right)\epsilon_{c]mnp}. \tag{4.6}$$

The pre-symplectic current is locally built from the metric, a pair of metric perturbations γ and $\hat{\gamma}$, and their first derivatives. Ω is closed when the metric and perturbations satisfy the Einstein equations and linearized equations respectively. Thus its integral over a compact hypersurface,

$$\Xi(\gamma,\hat{\gamma}) = \int_\Sigma \Omega, \tag{4.7}$$

is independent of the choice of Σ up to homology[2]. Up to a normalization, $\Xi(\gamma,\hat{\gamma})$ is the value of the pre-symplectic form on the space of solutions to the vacuum Einstein equations when acting on a pair of tangent vectors $(\gamma,\hat{\gamma})$ (see, for example, Ashtekar [1990], and references therein).

Integrable systems of PDE's often admit inequivalent symplectic structures (Olver [1993]). In addition, it is known that the existence of generalized conservation laws depending on three or more solutions to the linearized equations is closely related to applicability of Darboux's method of integration (Anderson [1992]). Thus this sort of conservation law—depending on one or more solutions to the linearized equations— can expose important features of a set of field equations, and it is natural to ask if the Einstein equations admit any other generalized conservation laws of this type.

To present our classification of generalized conservation laws depending on solutions to the linearized equations we must introduce some notation. Let $\omega^{(p,q)}$ denote a spacetime p-form locally constructed from the metric and its derivatives as well as from q solutions of the linearized equations and their derivatives. The linearized solutions must appear in a skew q-multilinear fashion. For example, the pre-symplectic current Ω_{abc} would be denoted $\omega^{(3,2)}$ in this notation, and the forms classified in (4.5) would be denoted $\omega^{(p,0)}$. We now look for such forms that are closed when the Einstein equations and their linearization are satisfied by the metric and perturbations. To find such forms, we must generalize our spinor parametrization of $\mathcal{E}$ to include the jet space of solutions to the linearized equations, but this is relatively straightforward. Using techniques from the variational bicomplex we obtain the following results.

Let $\omega^{(p,q)}$, $q > 0$ and $p < 4$, be a generalized conservation law for the vacuum Einstein equations in four spacetime dimensions. Then there exist constants b and c and forms $\eta^{(p-1,q)}$ locally constructed from the metric and perturbations such that, modulo terms which vanish when the Einstein and linearized Einstein equations hold, we have:

$$\begin{aligned}
\omega^{(0,q)} &= 0; \\
\omega^{(p,q)} &= d\eta^{(p-1,q)}, \qquad p = 1,2; \\
\omega^{(3,q)} &= d\eta^{(2,q)}, \qquad q > 2; \\
\omega^{(3,2)} &= b\,\Omega + d\eta^{(2,2)}; \\
\omega^{(3,1)} &= c\,\Theta + d\eta^{(2,1)}.
\end{aligned} \tag{4.8}$$

In this last equation we have denoted by Θ a 3-form Θ_{abc}, which is defined as

$$\Theta_{abc} = \epsilon_{abc}{}^{d}\nabla^{e}\left(\gamma_{de} - g_{de}\, g^{mn}\gamma_{mn}\right). \tag{4.9}$$

Θ is closed when g_{ab} and γ_{ab} satisfy the Einstein equations and their linearization respectively. The integral of Θ over a hypersurface is again independent of the choice of hypersurface and defines the canonical 1-form, or pre-symplectic potential, on the space of solutions to the Einstein equations. This means that, viewing the conserved charge defined by Θ as a 1-form on the infinite-dimensional space of solutions acting on a tangent vector γ, the exterior derivative of the charge is the symplectic 2-form on the space of solutions. Aside from the symplectic current and its associated potential, there are no other non-trivial generalized conservation laws built from solutions of the linearized equations as described above. Note that this result establishes a uniqueness theorem for the gravitational pre-symplectic structure in the sense that any such structure which can be constructed as the spatial integral of a closed, locally constructed 3-form is a multiple of (4.7).

5 Discussion

We have classified the generalized symmetries and generalized conservation laws of the vacuum Einstein equations in four dimensions. Our results indicate that, from the vantage point of geometric structures on the jet space of solutions, one can see only a handful of "special features" of the vacuum equations. Still, let us summarize our results and what they tell us about the vacuum equations.

The generalized symmetries include a constant scale symmetry and a diffeomorphism symmetry. The scale symmetry simply indicates that there are no length scales set by the vacuum equations; this symmetry is absent if one modifies the equations using dimensionful constants, e.g., if one includes a cosmological term in the equations. The diffeomorphism symmetry reflects the general covariance of the Einstein equations. We expect this symmetry to be present in any generally covariant system of field equations. Aside from these well-known transformations, the Einstein equations are devoid of symmetry.

The only closed–not–exact form locally constructed from Ricci-flat metrics corresponds to a topological conservation law—the conservation of "kink number". This conservation law reflects the non-trivial topology of the bundle of Lorentzian metrics over spacetime and will arise in any system of field equations for a Lorentzian metric. If the metric is Riemannian, this conservation law is absent. The absence of any other conserved 3-forms can be traced back to the absence of suitable generalized symmetries. But this does not explain the dearth of lower-degree conservation laws.

Remarkably, it is possible to give a rather simple theory of lower-degree conservation laws in a general field theory (Anderson and Torre [1996a], [1996b] and [1996c]), which can be thought of as somewhat analogous to Noether's theory of conserved currents (Olver [1993]). With some mild technical assumptions it is possible to show that in order for a set of Lagrangian field equations to admit lower-degree conservation laws two conditions must be met. First, the theory must be a "gauge theory", that is, it must admit some form of *gauge transformation*, where we define a gauge transformation as a generalized symmetry built from arbitrary functions of spacetime. Second, the solutions to the field equations must be such that they always allow for *gauge symmetries*, that is, there always exists a gauge transformation that leaves each solution invariant[3]. Thus, the gauge transformation of the Einstein equations is the diffeomorphism symmetry ((3.6) with $c = 0$), and a gauge symmetry of a solution g_{ab} to the field equations would be a diffeomorphism which does not change that solution. The infinitesimal gauge symmetry is then generated by a *generalized Killing vector field*, that is, a vector field locally constructed from the metric and its derivatives to some order which satisfies the Killing equations when the metric is Ricci-flat. The generic solution to the vacuum equations admits no Killing vector fields. More precisely, it is possible to show that there are no generalized Killing vector fields, and so we can say that the absence of lower-degree conservation laws for the Einstein equations reflects the absence of isometries of generic solutions. Conclusions such as these can also be obtained using BRST methods (see Barnich et al. [1995]). If we consider reductions of the Einstein equations obtained by demanding the solutions always admit a Killing vector, then the general theory leads to lower-degree conservation laws such as shown in (4.3) and (4.4).

Finally, we have classified generalized conservation laws built locally from solutions to the linearized equations and found only the symplectic current and its "potential". These conservation laws reflect the variational properties of the Einstein equations. As is well-known (Ashtekar [1990]), a conserved symplectic current arises for any field equations derivable from a Lagrangian. Thus the conserved 3-form (4.6) reflects the fact that the Einstein equations can be derived from the Einstein-Hilbert Lagrangian. The essential uniqueness of the pre-symplectic current leads to a uniqueness result for variational principles for the vacuum Einstein equations, which will be presented elsewhere. Normally, the current defining the symplectic potential for a system of Lagrangian field equations is not conserved. However, it is not hard to show that the current defining the symplectic potential *is* conserved provided the Lagrangian can be chosen to vanish when the field equations hold. Thus the closed 3-form (4.9) reflects the fact that the Einstein-Hilbert Lagrangian vanishes on Ricci-flat metrics.

In this article I have discussed structural features of the vacuum Einstein equations which can be uncovered using spinor–jet space techniques. The techniques that were used can be generalized to analyze related systems of equations, specifically (i) reductions of the Einstein equations such as obtained by restricting attention to solutions with one or more Killing vector fields; and (ii) the Einstein equations with matter couplings, e.g., the Einstein-Maxwell equations. As mentioned above, we expect non-trivial lower-degree conservation laws to arise when one analyzes the Einstein-Killing equations, and a classification of such conservation laws is currently in progress. Matter couplings can also induce lower degree conservation laws; for example, the Einstein-Maxwell equations admit a 2-form conservation law. More intriguing, if perhaps on somewhat less firm physical ground, are matter couplings dictated by Kaluza-Klein reductions of higher-dimensional vacuum relativity. For example, the classical reduction from five to four dimensions corresponds to an Einstein-Maxwell-scalar field theory. This reduction is dictated by the assumption that the five-dimensional vacuum theory admits a Killing vector field, which, on general grounds, indicates the existence of "characteristic cohomology". The details of such investigations will be presented elsewhere.

[3]The gauge symmetry of a solution will in general vary with the choice of solution.

References

Anderson, I. M. [1992] *Introduction to the variational bicomplex,* Cont. Math. **32**, 51–73.

Anderson, I. M. and Torre, C. G. [1994] *Two component spinors and natural coordinates for the prolonged Einstein equation manifolds,* Utah State University Technical Report.

——— [1996a] *Classification of local generalized symmetries of the vacuum Einstein equations,* Comm. Math. Phys. **176**, 479–539.

——— [1996b] *Lower degree conservation laws in field theory,* in preparation.

——— [1996c] *The variational bicomplex for the Einstein equations,* in preparation.

Ashtekar, A., Bombelli, L. and Reula, O. [1990], Analysis, geometry and mechanics: 200 years after Lagrange (Francaviglia, M. and Holm, D., eds.), North-Holland, Amsterdam.

Barnich, B., Brandt, F. and Henneaux, M. [1995] *Local BRST cohomology in the anti-field formalism,* Comm. Math. Phys. **174**, 57–92.

Bryant, R. L. and Griffiths, P. A. [1995] *Characteristic cohomology of differential systems,* J. Am. Math. Soc. **8**, 507-596.

Finkelstein, D. and Misner, C. [1959] *Some new conservation laws,* Ann. Phys. **6**, 230–243.

Goldstein, H. [1980] *Classical mechanics*, Addison-Wesley, Reading.

Kahn, D. W. [1980] *Introduction to global analysis*, Academic Press, New York.

Olver, P. [1993] *Applications of Lie groups to differential equations*, Springer-Verlag, New York.

Penrose, R. [1960] *A spinor approach to to general relativity,* Ann. Phys. **10**, 171–201.

Saunders, D. [1989] *The geometry of jet bundles*, Cambridge University Press, Cambridge.

Thomas, T. Y. [1934] *Differential invariants of generalized spaces*, Cambridge University Press, Cambridge.

Torre, C. G. [1993] *Gravitational observables and local symmetries,* Phys. Rev. D **48**, R2373–R2376.

——— [1995] *Some remarks on gravitational analogues of magnetic charge,* Class. Quantum Grav. **12**, L43–L50.

Torre, C. G. and Anderson, I. M. [1993] *Symmetries of the Einstein equations,* Phys. Rev. Lett. **70**, 3525–3529.

Wald, R. M. [1990] *On identically closed forms locally constructed from a field,* J. Math. Phys. **31**, 2378–2384.

Fields Institute Communications
Volume **15**, 1997

The Laws of Black Hole Mechanics in an Arbitrary Lagrangian Theory of Gravity

Robert M. Wald
Enrico Fermi Institute
and
Department of Physics
University of Chicago
5640 S. Ellis Avenue
Chicago, Illinois USA 60637-1433
rmwa@midway.uchicago.edu

One of the most remarkable developments in the classical theory of black holes in general relativity was the discovery of a complete analogy between certain laws satisfied by black holes and the ordinary laws of thermodynamics (Bardeen et al. [1973]). The discovery by Hawking that black holes radiate as black bodies demonstrated that this analogy is of deep physical significance.

One way to gain more insight into the nature of the laws of black hole mechanics is to examine their validity in a more general context. In the talk, I reviewed recent progress toward deriving the laws of black hole mechanics in the context of an arbitrary theory of gravity derived from a diffeomorphism covariant Lagrangian. The zeroth law (constancy of surface gravity) actually holds for an arbitrary static or stationary-axisymmetric black hole (with $t-\phi$ reflection symmetry), independently of any field equations (Carter [1973], Racz and Wald [1996]). Most significantly, a version of the first law of black hole mechanics (relating differences in mass, angular momentum, and "entropy" of nearby stationary black holes) holds in an arbitrary Lagrangian theory of gravity, with the entropy of a stationary black hole given by the integral of the "Noether charge" form over the horizon (Iyer and Wald [1994], Wald [1993]). (Here the Noether charge form is a potential for the Noether current associated with the local symmetries arising from the diffeomorphism invariance of the theory.) The formula thereby obtained for black hole entropy can be converted to a simple, local, geometrical expression, involving the functional derivative of the Lagrangian with respect to the Riemann curvature (Iyer and Wald [1994]). However, the definition of the entropy of non-stationary black holes and the validity of the second law of black hole mechanics in an arbitrary Lagrangian theory remain open issues, although there are hints that the validity of the second law in a theory may be intimately related to the positivity of energy in that theory.

1991 *Mathematics Subject Classification.* Primary 83C57; Secondary 83D05.

© 1997 American Mathematical Society

Details of the general derivation of the first law of black hole mechanics and the formula for the entropy of stationary black holes can be found in Iyer and Wald [1994].

References

Bardeen, J. M., Carter, B. and Hawking, S. W. [1973] *The Four Laws of Black Hole Mechanics*, Comm. Math. Phys. **31**, 161–170.

Carter, B. [1973] *Black Hole Equilibrium States*, Black Holes (DeWitt, C. and DeWitt, B. S., eds.), Gordon & Breach, New York.

Racz, I. and Wald, R. M. [1996] *Global Extensions of Spacetimes Describing Asymptotic Final States of Black Holes*, Class. Quant. Grav., **13**, 539-552.

Wald, R. M. [1993] *Black Hole Entropy is the Noether Charge*, Phys. Rev. D **48**, R3427–R3431.

Iyer, V. and Wald, R. M. [1994] *Some Properties of the Noether Charge and a Proposal for Dynamical Black Hole Entropy*, Phys. Rev. D **50**, 846–864.

Fields Institute Communications
Volume **15**, 1997

Black Holes, Galaxy Formation and the Large-scale Structure of the Universe

Michael J. West
Department of Astronomy & Physics
Saint Mary's University
Halifax, Nova Scotia, Canada B3H 3C3
west@sisyphus.stmarys.ca

Abstract. A new model for the formation of active galaxies is described. A key feature of this model is the idea that the birth of black holes in the centres of supergiant galaxies is strongly influenced by the large-scale distribution of matter in the universe. This conjecture can successfully explain two observed phenomena: the alignment of the radio, optical and infrared axes of high-redshift radio galaxies, and the alignment of present-day cD galaxies with their environs. Other recent observational tests of this theory are also discussed.

1 Introduction: A Tale of Two Alignment Effects

Black holes are widely believed to be the central engines that power active galaxies such as quasars and radio galaxies. Yet how and when such black holes formed remains one of the great outstanding problems in astrophysics.

In this paper, a truly remarkable link is postulated between the formation of black holes, massive galaxies and the large-scale structure of the universe. Motivation for this work came from two observed phenomena discussed below.

1.1 Observations of high-redshift radio galaxies

Considerable insight has been gained in recent years into the nature of distant galaxies. Because of the finite speed of light, distant galaxies are seen as they were in the past, which allows astronomers to directly probe the earliest stages of galaxy formation. Radio observations have proven to be a particularly effective method of finding high-redshift galaxies. At present the most distant known radio galaxies and quasars are at redshifts $z \geq 4$, which correspond to epochs when the universe was less than 10% its present age.

Optical observations of powerful high-redshift radio galaxies have revealed that they are often very elongated and clumpy in appearance. An example is shown in Figure 1. This appearance is distinctly different from that of most galaxies in

1991 *Mathematics Subject Classification.* Primary 85A40, 83C57; Secondary 85A45.
The author was supported by a grant from the NSERC of Canada.

© 1997 American Mathematical Society

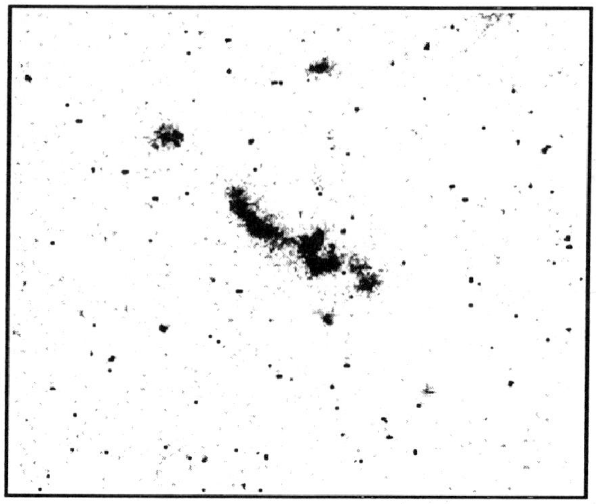

Figure 1 Hubble Space Telescope image of the distant radio galaxy 3C 324 at a redshift $z = 1.21$. Note the very clumpy and elongated appearance of this galaxy, which is characteristic of many powerful radio galaxies at high redshifts.

the present-day universe. The high incidence of double- and multiple-component radio galaxies with separations of only a few thousand light-years (e.g., 3C 65, 114, 265, 324, 356 and 368) strongly suggests that these systems are in the process of merging.

Yet perhaps the most intriguing property of powerful, high-redshift radio galaxies is the observed alignment of their extended optical and infrared structures with their radio jets (e.g., Chambers et al. [1987], McCarthy et al. [1987], Rigler et al. [1992] and Dunlop and Peacock [1993]). This alignment effect is illustrated in Figure 2. This phenomenon is puzzling, since the optical and infrared emission presumably trace the distribution of stellar light (and hence the mass distribution) in these galaxies, whereas the orientation of the radio jet axis is likely determined by physical processes on much smaller scales associated with the black hole central engine. A number of different mechanisms have been proposed for the origin of this alignment effect, however to date none has proven entirely satisfactory.

1.2 Observations of low-redshift cD galaxies

There is a second alignment effect, seemingly unrelated to the first, which involves cD galaxies.

cD galaxies are among the most enigmatic objects in the sky. Found exclusively in the centres of galaxy clusters, cDs are the most massive galaxies in the present-day universe, and the most luminous objects whose light is purely stellar in origin. A typical cD galaxy is shown in Figure 3. These supergiant galaxies are generally quite elongated in appearance, and observational evidence suggests that they are predominantly prolate in shape ($l_1 \gg l_2 \gg l_3$, where l_i are the principal axes of the galaxy mass distribution; Porter et al. [1991] and Ryden et al. [1993]).

One of the most striking features of the large-scale distribution of luminous matter in the universe is its filamentary appearance, with long, quasi-linear ar-

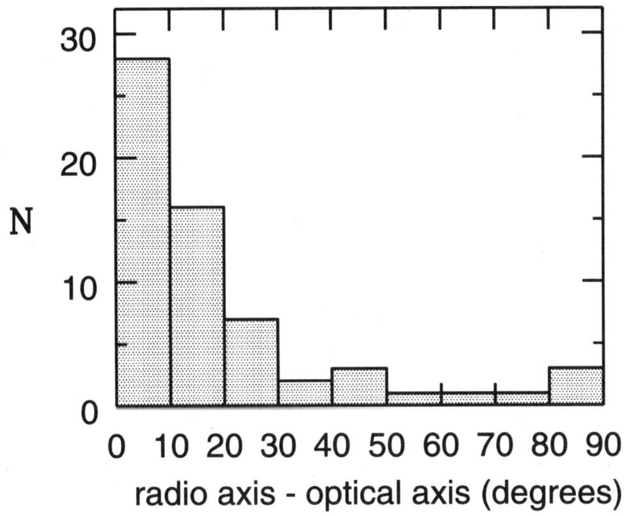

Figure 2 The alignment effect in high-redshift radio galaxies. Shown here is the frequency distribution of the difference between the radio and optical axis orientations of a sample of powerful, high-redshift radio galaxies. If the radio and optical axes were uncorrelated, then a uniform distribution between 0° and 90° would be expected. Instead, a strong tendency is seen for the radio and optical axes to share the same orientation; this alignment tendency is significant at greater than the 99.9% confidence level.

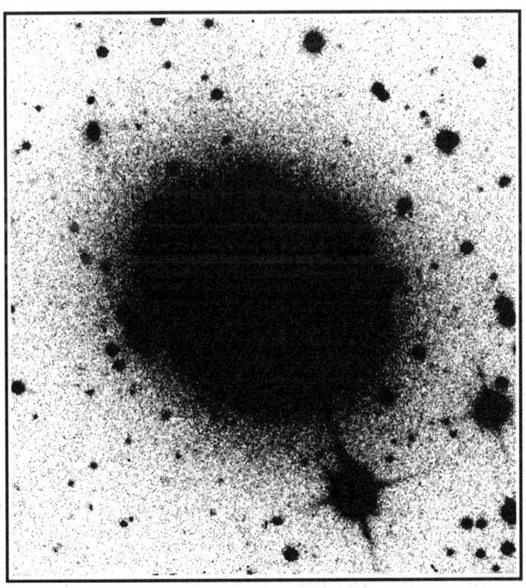

Figure 3 Optical image of a typical cD galaxy, located in the galaxy cluster Abell 3656.

rangements of galaxies extending over hundreds of millions of light-years (see Figure 4). Interestingly, cD galaxies exhibit a strong propensity for their principal

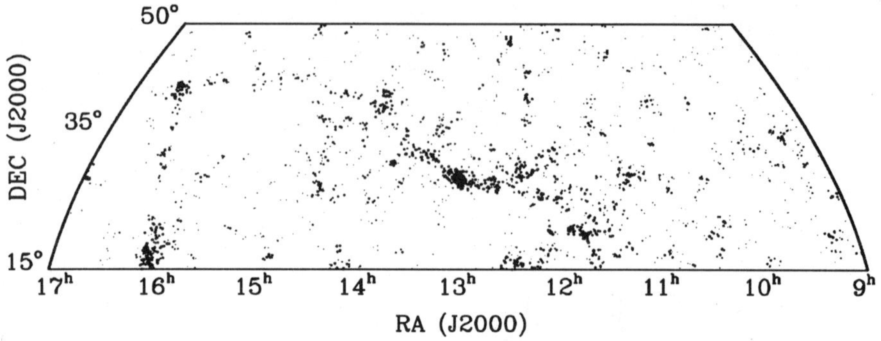

Figure 4 A portion of the observed large-scale galaxy distribution. Each dot represents a galaxy as seen in projection on the plane of the sky. Filamentary features can be seen extending over many millions of light-years in length.

axes to align with the surrounding filamentary matter distribution (e.g., Binggeli [1982], Lambas et al. [1988], van Kampen and Rhee [1990], West [1994] and West and Schombert [1996]). This alignment effect is shown in Figure 5. It is certainly remarkable that the structure of cD galaxies should somehow know about the distribution of matter on scales of tens or hundreds of millions of light-years. No other type of galaxy exhibits this sort of alignment effect, which suggests that cD galaxies must be the result of a unique formation history unlike that of other galaxies.

1.3 Are the two alignment effects related?

To date, the two aligment effects described above have been assumed to be separate and unrelated phenomena. However, as the following sections will endeavour to show, a compelling argument can be made that the two alignment effects are in actuality *related* phenomena that share a common origin.

2 cD Galaxy Formation Via Anisotropic Mergers

In most currently popular models for the evolution of structure in the universe, galaxy formation is envisioned as proceeding in a hierarchical manner. Supergiant galaxies like cDs would likely have formed by a series of mergers of smaller galaxies.

Given the filamentary topology of the large-scale matter distribution, it seems rather unlikely that such mergers would have occurred in a random, isotropic manner. Instead, the mergers which give rise to cD galaxies at the centres of clusters would most likely occur along preferred directions as material infalls anisotropically along filaments. This special formation history produces cD galaxies that are quite *prolate* in shape and that have preferred orientations built into them which reflect the surrounding filamentary pattern of the large-scale matter distribution, even at early epochs (West [1994]). State-of-the-art numerical simulations confirm this claim (see Figure 6). This presumably is the origin of the alignment effect seen in Figure 5.

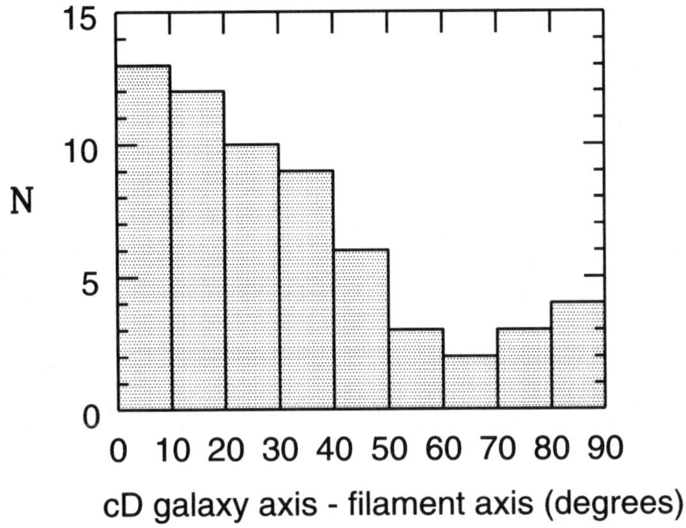

Figure 5 The alignment effect in low-redshift cD galaxies. Shown here is the frequency distribution of the difference between the orientation of the cD galaxy principal axis and the direction defined by the surrounding large-scale filamentary matter distribution. The significance of the alignment effect exceeds the 99.9% confidence level.

3 The Connection Between Powerful, High-Redshift Radio Galaxies and Low-Redshift cD Galaxies

A number of pieces of evidence suggest that powerful radio galaxies and quasars at high redshifts were the precursors of today's cD galaxies. To begin with, observational studies have shown that high-redshift radio galaxies and quasars generally reside in dense environments comparable to those of present-day cD galaxies (e.g., Yee and Green [1987], Yates et al. [1989] and Ellingson et al. [1991]). Furthermore, it is well established that the most powerful radio galaxies in the low-redshift universe are often cD galaxies, and thus it seems likely that powerful radio sources at earlier epochs were also associated with cD galaxies or their progenitors. Additional support for this idea comes from order of magnitude estimates which suggest that the space density of high-redshift radio galaxies and quasars may have been comparable to the present-day space density of cD galaxies (West [1994]). It is worth noting too that the cD galaxy in the simulation shown in Figure 6 has the same sort of clumpy and elongated appearance at early epochs that is characteristic of powerful, high-redshift radio galaxies (compare Figures 1 and 6).

These and other arguments point strongly to an evolutionary link between powerful, high-redshift radio galaxies and low-redshift cD galaxies. Having argued earlier that highly anisotropic mergers will produce prolate cDs with preferred orientations, the effect of such a formation history on the genesis of radio sources at high redshifts is considered in the following section.

4 What Determines the Jet Orientation in Powerful Radio Galaxies?

For several decades now, the the standard model for active galaxies has envi-

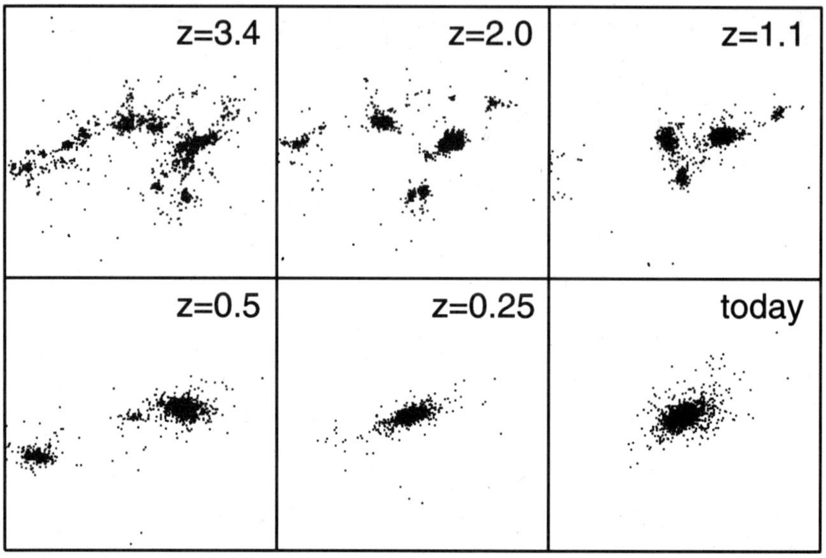

Figure 6 Computer simulation of cD galaxy formation (kindly provided by Ray Carlberg). This has been extracted from a larger million-particle simulation of a universe dominated by cold dark matter (see Carlberg [1994]). Each box is approximately 13 million light-years across. Time sequence is denoted by the redshift z. Only those particles that end up in the final cD galaxy are shown here. Note the highly anisotropic nature of the merger process, and how this is reflected in the shape and orientation of the galaxy at both high and low redshifts.

sioned a massive compact object, most probably a black hole of order $10^7 - 10^9$ solar masses, which acts as a central engine in the production of radio emission (e.g., Lynden-Bell [1969] Zel'dovich and Novikov [1971], Blandford and Rees [1974] and Begelman et al. [1984]). According to this picture, the radio jets are produced by twin beams of relativistic material that are ejected along the black hole spin axis.

At early epochs, frequent mergers of small proto-galactic systems would have provided an abundant source of gas-rich material in burgeoning young cD galaxies (see Figure 6). Cold gas and dust falling into the prolate gravitational potential of a developing cD would quickly settle into a thin rotating disk as dissipative processes damp out random motions. It is straightforward to show from dynamical considerations that a rotating gaseous disk in a prolate gravitational potential will precess until its angular momentum vector is aligned with the principal axis of the galaxy's mass distribution (Kahn and Woltjer [1959] and Gunn [1979]). Angular momentum loss during this disk precession results in an inward flow of gas towards the centre of the galaxy, where its accumulation provides a plentiful source of material for the creation and fuelling of a black hole (Rees [1978] and Gunn [1979]).

In this scenario of black hole formation, one would expect the black hole spin axis to share the same orientation as the angular momentum vector of the fuelling accretion disk, which is ultimately determined by the intrinsic shape and orientation of the host cD galaxy (e.g., Tubbs [1980], Tohline et al. [1981] and Steiman-Cameron and Durisen [1988]). Recent Hubble Space Telescope observations have provided quite convincing evidence of accretion disks associated with black holes in the centres of two galaxies (Jaffe et al. [1993] and Harms et al. [1994]); as expected, in both cases, the radio jet axis is perpendicular to the fuelling accretion disk.

Because cD galaxies are prolate in shape as a result of the anisotropic merger process by which they formed, the black hole spin axis, and hence radio jets, will tend to align with the major axes of these galaxies, even at high redshifts. Hence, the anisotropic merger model proposed here for the origin of the alignment effect in cD galaxies also provides a natural mechanism to explain the alignment of the radio, optical and infrared axes in high-redshift radio galaxies.

5 Low-Redshift Counterparts to High-Redshift Radio Galaxies

It is interesting to note that some of the most powerful radio galaxies at low redshifts exhibit *both* alignment effects, which suggests that they may indeed have formed in the manner proposed here.

A good example is the cD/radio galaxy NGC 708, which is located in the galaxy cluster Abell 262 (see Figure 7) and forms part of the well-known Perseus-Pisces filament. NGC 708 is known to have a gas/dust lane that is oriented nearly perpendicular to its radio source axis. This radio axis shares a similar orientation as the galaxy principal axis, which in turn is aligned with the surrounding large-scale filamentary feature seen in Figure 7.

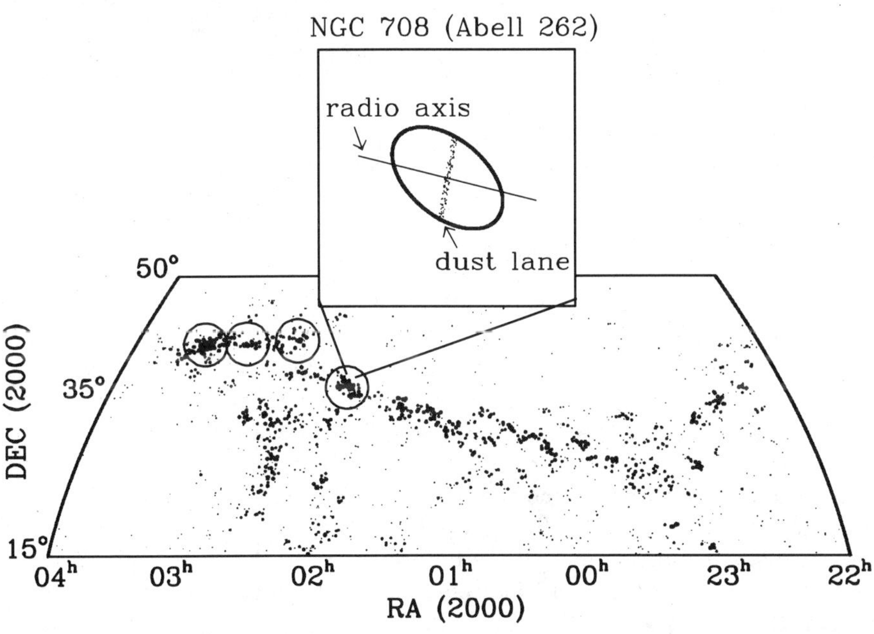

Figure 7 Comparison of the radio, optical and dust-lane orientations in the cD galaxy NGC 708. Also plotted here is the large-scale distribution galaxies in the same region of the sky. Circles denote Abell clusters of galaxies. Note the tendency for the radio and optical axes of NGC 708 to point along the local large-scale filament.

Another example is provided by Hercules A (3C 348), an extremely powerful

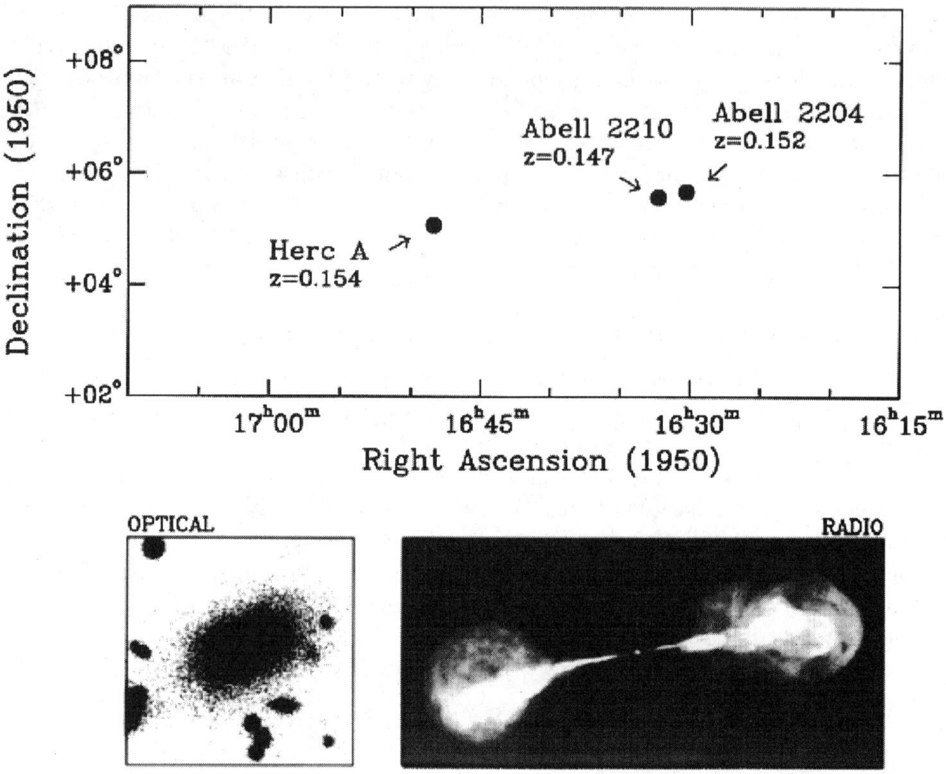

Figure 8. The alignment effect(s) in Hercules A (3C 348). The lower panels show optical and radio images of Hercules A, while the top panels shows Hercules A in relation to its large-scale surroundings. Note the tendency for the radio and optical axes to point towards the neighbouring Abell clusters.

low-redshift ($z = 0.154$) radio source. Hercules A is a very elongated cD galaxy which resides in a cluster of galaxies (Yates et al. [1989]). As Figure 8 shows, Hercules A exhibits the same alignment effect seen in its high-redshift cousins; its radio and optical axes share the same orientation, which indicates that Hercules A is most probably prolate in shape. Although 3C 348 is too far away for the surrounding filamentary galaxy distribution to be seen in existing surveys, Abell clusters can serve as beacons for mapping the surrounding large-scale matter distribution (in Figure 7, for example, the Abell clusters trace the same large-scale structure as the galaxies, albeit more sparsely). As Figure 8 shows, the radio and optical axes of Hercules A appear to "point" along an unseen filament traced by its nearest companion clusters, Abell 2210 and Abell 2204.

Still another example is Cygnus A (3C 405), which is by far the most powerful radio source in the low-redshift universe. Its proximity affords a unique opportunity to study a nearby galaxy whose properties are likely to be quite similar to those of powerful radio galaxies at high redshifts. As shown by West [1994], the orientations of the radio and optical structures of Cygnus A, and the galaxy cluster in which it resides, are remarkably similar. Most importantly, Cygnus A's radio jet clearly "points" towards its nearest neighbouring cluster, Abell 2319, as can be

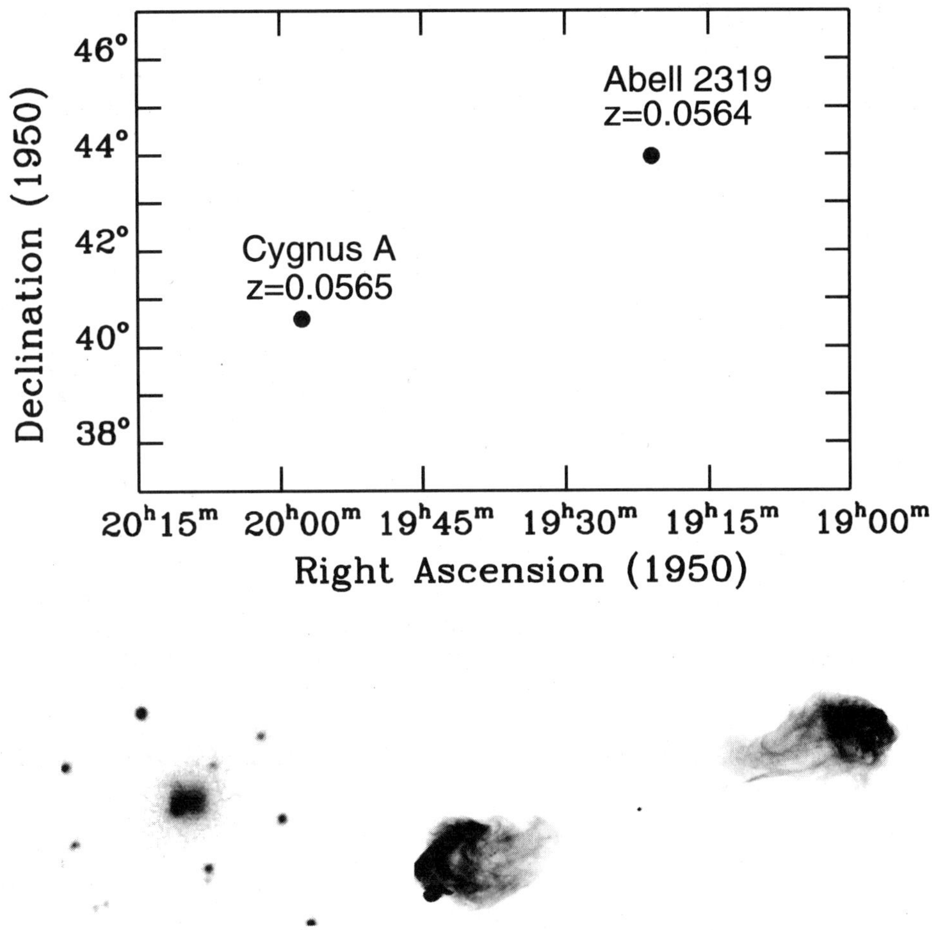

Figure 9 The alignment effect(s) in Cygnus A (3C 405). The top panel shows Cygnus A in relation to its nearest neighbouring cluster of galaxies, Abell 2319. The two bottom panels show the optical (left) and radio (right) structure of Cygnus A. Note the clear alignment of the radio jets with the direction towards the neighbouring cluster.

seen in Figure 9. This strongly suggests that Cygnus A has formed via a process of anisotropic mergers that have influenced not only the orientation of the galaxy, but also the orientation of its radio emission – and by implication, the formation of its central black hole.

6 Predictions of the Anisotropic Merger Model

Any successful theory should strive not only to explain existing observations, but also to make specific predictions that can be tested. In this spirit, West [1994] proposed a number of tests of the anisotropic merger model described here. Recent observational work has confirmed several of these predictions:

(1) *Anisotropic distribution of companion galaxies around high-redshift radio galaxies*. Given the proposed anisotropic distribution of material around proto-

cD galaxies, companion galaxies around high-redshift radio galaxies and quasars should also be anistropically distributed, with a strong tendency for such objects to be found along the direction defined by the radio axis. This means not only clumps associated with the radio galaxy itself, but also other companion galaxies at greater separations. This important prediction of the anisotropic merger model has recently been confirmed by Röttgering et al. [1995].

(2) *Formation of clusters of galaxies via anisotropic mergers of subclusters.* It is widely believed that we are presently in the epoch of cluster formation. According to the anisotropic merger proposed here, clusters of galaxies should form from mergers of smaller subclusters which infall along large-scale filaments. Recent observational work by West et al. [1995] has confirmed this picture of cluster formation.

(3) *Dust lanes in cD galaxies.* If cD galaxies are truly prolate, as the model described here requires, then the dynamical arguments outlined in Sections 2 and 4 lead to the prediction that dust and gas disks which are acquired by such galaxies during mergers should settle into planes that are perpendicular to the major axis of the galaxy's mass distribution. Work is in progress to test this prediction.

7 Conclusions

Although admittedly speculative in nature, the model proposed here is built upon a foundation of widely accepted theoretical ideas regarding black hole and galaxy formation and the physics of radio sources. What is novel about the present work is that it brings together for the first time a number of seemingly disparate points into a cohesive picture. The most attractive feature of this anisotropic merger model is that it unites two alignment effects that had previously been assumed to be completely independent phenomena.

If once accepts the premise that powerful radio galaxies in the early universe were the precursors of today's cD galaxies, then it does not require a great leap of faith to assume that an anisotropic merger process that can produce cD galaxies with preferred shapes and orientations might also communicate these same anisotropies to the very innermost regions of these galaxies, right down to the black hole central engine. Indeed, given the standard accretion disk model for black hole formation and evolution discussed in Section 4, it would be rather surprising if the spin axes of these objects were completely divorced from their surroundings.

In summary, the model proposed here suggests a remarkable connection between the formation of massive black holes, cD galaxies, and the large-scale structure of the universe.

A more detailed account of this work can be found in West [1994].

References

Begelman, M. C., Blandford, R. D. and Rees, M. J. [1984] *Theory of Extragalactic Radio Sources,* Reviews of Modern Physics **56**, 255–351.

Binggeli, B. [1982] *The Shapes and Orientations of Clusters of Galaxies,* Astronomy and Astrophysics **107**, 338–349.

Blandford, R. D. and Rees, M. J. [1974] *A 'Twin-Exhaust' Model for Double Radio Sources,* Monthly Notices of the Royal Astronomical Society **169**, 395–415.

Carlberg, R. G. [1994] *Velocity Bias in Clusters,* Astrophysical Journal **433**, 468–478.

Chambers, K. C., Miley, G. K. and van Breugel, W. [1987] *Alignment of Radio and Optical Orientations in High-redshift Radio Galaxies,* Nature **329**, 604–606.

Dunlop, J. S. and Peacock, J. A. [1993] *Luminosity Dependence of Optical Activity and Alignments in Radio Galaxies,* Monthly Notices of the Royal Astronomical Society **263**, 936–962.

Ellingson, E., Green, R. F. and Yee, H. K. C. [1991] *Clusters of Galaxies Associated with Quasars.* II. *Galaxy Cluster Dynamics,* Astrophysical Journal **378**, 476–486.

Gunn, J. E. [1979] *Feeding the Monster: Gas Discs in Elliptical Galaxies,* Active Galactic Nuclei, Cambridge Univ. Press, Cambridge, pp. 213–226.

Harms, R. J., Ford, H. C., Tsvetanov, Z. I., Hartig, G. F., Dressel, L. L., Kriss, G. A., Bohlin, R., Davidsen, A. F., Margon, B. and Kochhar, A. K. [1994] *HST FOS Spectroscopy of M87: Evidence for a Disk of Ionized Gas Around a Massive Black Hole,* Astrophysical Journal **435**, L35–L38.

Jaffe, W., Ford, H., Ferraresse, L., van den Bosch, F. and O'Connell, R. [1993] *A Large Nuclear Accretion Disk in the Active galaxy NGC 4261,* Nature **364**, 213–215.

Kahn, F. and Woltjer, L. [1959] *Intergalactic Matter and the Galaxy,* Astrophysical Journal **130**, 705–717.

Lambas, D. G., Groth, E. J. and Peebles, P. J. E. [1988] *Alignments of Brightest Cluster Galaxies with Large-scale Structures,* Astronomical Journal **95**, 996–998.

Lynden-Bell, D. [1969] *Galactic Nuclei as Collapsed Old Quasars,* Nature **223**, 690–694.

McCarthy, P. J., van Breugel, W., Spinrad, H. and Djorgovski, S. [1987] *A Correlation Between the Radio and Optical Morphologies of Distant 3CR Radio Galaxies,* Astrophysical Journal **321**, L29–L33.

Porter, A. C., Schneider, D. P. and Hoessel, J. G. [1991] *CCD Observations of Abell Clusters.* V. *Isophotometry of 175 Brightest Elliptical Galaxies in Abell Clusters,* Astronomical Journal **101**, 1561–1594.

Rees, M. J. [1979] *Relativistic Jets and Beams in Radio Galaxies,* Nature **275**, 516–517.

Rigler, M. A., Lilly, S. J., Stockton, A., Hammer, F. and Le Fèvre, O. [1992] *Infrared and Optical Morphologies of Distant Radio Galaxies,* Astrophysical Journal **385**, 61–82.

Röttgering, H. J. A., West, M. J., Miley, G. K. and Chambers, K. C. [1995] *The Optical Counterparts and the Environments of Ultra-steep-spectrum Radio Sources,* Astronomy and Astrophysics, in press.

Ryden, B. S., Lauer, T. R. and Postman, M. [1993] *The Shapes of Brightest Cluster Galaxies,* Astrophysical Journal **410**, 515–519.

Steiman-Cameron, T. Y. and Durisen, R. H. [1988] *Evolution of Inclined Galactic Gas Disks. I. A Cloud-Fluid Approach,* Astrophysical Journal **325**, 26–48.

Tohline, J. E., Simonson, G. F. and Caldwell, N. [1982] *Using Gaseous Disks to Probe the Geometric Structure of Elliptical Galaxies,* Astrophysical Journal **252**, 92–101.

Tubbs, A. D. [1980] *The Dynamical Evolution of NGC 5128,* Astrophysical Journal **241**, 969–980.

van Kampen, E. and Rhee, G. [1990] *Orientation of Bright Galaxies in Abell Clusters,* Astronomy and Astrophysics **237**, 283–295.

West, M. J. [1994] *Anisotropic Mergers at High Redshifts: The Formation of cD Galaxies and Powerful Radio Sources,* Monthly Notices of the Royal Astronomical Society **268**, 79–102.

West, M. J., Jones, C. and Forman, W. [1985] *Substructure: Clues to the Formation of Clusters of Galaxies,* Astrophysical Journal Letters **451**, L1–L4.

West, M. J. and Schombert, J. M. [1996] *Large-scale Alignments of Brightest Cluster Galaxies,* Astrophysical Journal, to appear.

Yates, M. G., Miller, L. and Peacock, J. A. [1989] *The Cluster Environments of Powerful, High-redshift Radio Galaxies,* Monthly Notices of the Royal Astronomical Society **240**, 129–166.

Yee, H. K. C. and Green, R. F. [1987] *Surveys of Fields Around Quasars,* Astrophysical Journal **319**, 28–43.

Zel'dovich, Ya. B. and Novikov, I. D. [1971] *Relativistic Astrophysics. Vol.* I. *"Stars and Relativity",* Univ. of Chicago Press, Chicago.

Fields Institute Communications
Volume **15**, 1997

When Black Holes Collide

Peter Anninos
National Center for Supercomputing Applications
605 E. Springfield Ave.
Champaign, IL USA 61820

David Hobill
National Center for Supercomputing Applications
605 E. Springfield Ave.
Champaign, IL USA 61820
Current address: Department of Physics and Astronomy
University of Calgary
Calgary, Alberta, Canada T2N 1N4
hobill@crag.ucalgary.ca

Edward Seidel
National Center for Supercomputing Applications
605 E. Springfield Ave.
Champaign, IL USA 61820

Larry Smarr
National Center for Supercomputing Applications
605 E. Springfield Ave.
Champaign, IL USA 61820

Wai-Mo Suen
Department of Physics, Washington University
St. Louis, Missouri, USA 63130

Abstract. The head-on collision of two nonrotating axisymmetric equal mass black holes is treated numerically using Misner's time-symmetric initial data where the two holes are represented by Einstein-Rosen bridges that can be placed arbitrarily distant from one another. We have extracted both $\ell = 2$ and $\ell = 4$ gravitational waveforms resulting from the collision. The normal modes of the final black hole dominate the spectrum in all cases studied. The total energy radiated is computed using several independent methods, and is typically less than $0.002M$.

The coalescence of two black holes is considered to be one of the most promising sources of gravitational waves (Schutz [1993]). In a series of papers (Anninos et al.

1991 *Mathematics Subject Classification.* Primary 83C57; Secondary 83C35.
The computations described in this article were carried out on a Cray-YMP (National Center for Supercomputing Applications), a Cray-C90 (Pittsburg Supercomputer Center), a NEC-SX3/44 (Centre de Informatique de Dorval) and a Fujitsu VPX-240 (Calgary High Performance Computing Center). Funding for this project was provided by NCSA and an NSERC (Canada) research grant.

© 1997 American Mathematical Society

[1993], [1994] and [1995]) we investigate numerically a special case of the black hole coalescence problem, namely the head-on collision of two equal mass black holes. Building on the work of DeWitt, Cadez, Smarr and Eppley (Smarr et al. [1976] and Smarr [1977] and [1979]) (henceforth abbreviated as DCSE) and the more recent work involving distorted, single black holes (Abrahams et al. [1992], Anninos et al. [1997] and Bernstein et al. [1994]), a 2D code has been developed that is capable of constructing highly accurate numerical solutions to the vacuum Einstein equations.

The method of solution is based upon the 3+1 (or ADM) formalism (Arnowitt et al. [1962]) where the line element can be written as

$$ds^2 = -(\alpha^2 - \beta^i \beta_i)dt^2 + 2\beta_i \, dx^i dt + \gamma_{ij} \, dx^i dx^j . \tag{1}$$

Here α is the lapse function that foliates the four dimensional spacetime into three dimensional spatial hypersurfaces with 3-metric γ_{ij}, and β^i is the shift vector that specifies three dimensional coordinate transformations from time slice to time slice. Greek indices (ranging from 0 to 3) label four dimensional coordinates, and Latin indices (ranging from 1 to 3) label spatial coordinates. Geometric units ($G = c = 1$) are also used.

The vacuum Einstein equations reduce to four constraint equations

$$R - K_{ij}K^{ij} + K^2 = 0, \tag{2}$$

$$\nabla_j (K^{ij} - \gamma^{ij} K) = 0, \tag{3}$$

and twelve evolution equations

$$\partial_t \gamma_{ij} = -2\alpha K_{ij} + \nabla_i \beta_j + \nabla_j \beta_i, \tag{4}$$

$$\begin{aligned} \partial_t K_{ij} &= -\nabla_i \nabla_j \alpha + \alpha \left(R_{ij} + K_{ij} K - 2K_{ik} K^k_j \right) \\ &\quad + \beta^k \nabla_k K_{ij} + K_{kj} \nabla_i \beta^k + K_{ik} \nabla_j \beta^k . \end{aligned} \tag{5}$$

Here R_{ij} is the Ricci tensor formed from the spatial 3-metric γ_{ij}, K_{ij} is the extrinsic curvature, K is the trace of K_{ij} and ∇_j is the covariant derivative with respect to γ_{ij}.

The initial data we consider consists of two throats connecting two isometric sheets. By construction the throats are spheres, centered on the axis along $\rho = 0$, (where (ρ,ϕ,z) are standard cylindrical coordinates) on which boundary conditions relating the metric across the two sheets may be imposed. Since the natural boundaries (the throats and a sphere surrounding the system far from the throats) do not lie along constant (z,ρ) coordinates, it is useful to introduce the "quasi-spherical" Cadez (Smarr et al. [1976]) coordinates (η,ξ) with η being a logarithmic "radial" coordinate and ξ an "angular" coordinate. Cadez coordinates are related to cylindrical coordinates through a complex conformal transformation

$$\chi(\zeta) = \eta + i\xi = \frac{1}{2}\left[\ln(\zeta + \zeta_0) + \ln(\zeta - \zeta_0)\right] + \sum_{n=1}^{\infty} C_n \left(\frac{1}{(\zeta_0 + \zeta)^n} + \frac{1}{(\zeta_0 - \zeta)^n} \right), \tag{6}$$

where $\zeta = z + i\rho$, and $\zeta_0 = \coth \mu$ is the value of ζ at the throat center. The constant η and ξ coordinate lines of (6) lie along the field and equipotential lines of two equally charged metallic cylinders located at the centers of the two throats $z = \pm \coth \mu$. The coefficients C_n are determined by a least-squares method that set the throats (defined by ${\rho_{th}}^2 + (z_{th} \pm \coth \mu)^2 = 1/\sinh^2 \mu$) on an $\eta = \eta_0 =$ constant coordinate line. Both η_0 and the different C_n are computed on a problem specific basis for different μ using this least-squares procedure. The constant Cadez coordinate lines as viewed in the cylindrical coordinate system are shown in Figure 1a. However the computer views the (η, ξ) coordinates as a rectangular system as shown in Figure 1b.

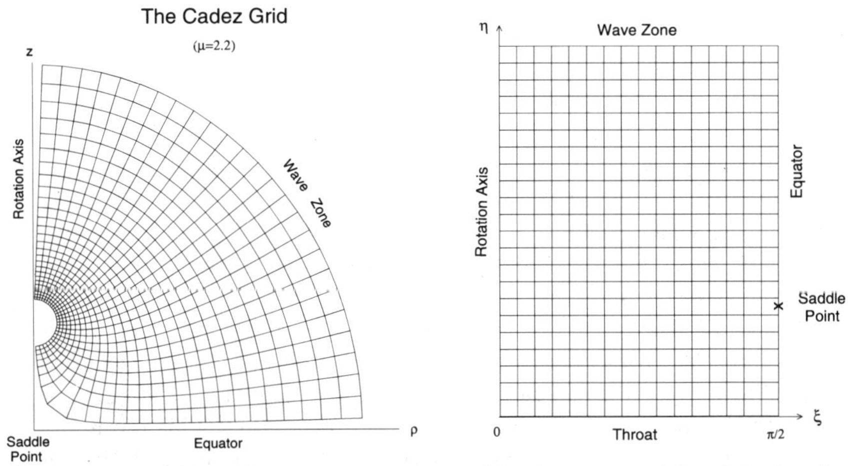

Figure 1 (a) The Cadez grid is constructed for the case $\mu = 2.2$ and displayed in a single quadrant with cylindrical coordinates. The throats are centered on the axis at $z = \pm \coth \mu$. Lines of constant η concentrically surround the throat locally, and become spherical far from the holes. (b) The computational grid is shown in Cadez coordinates.

The most general spacetime four-metric for a non-rotating axisymmetric spacetime may be written as:

$$g_{\mu\nu} = \begin{bmatrix} -\alpha^2 + \beta^i\beta_i & \beta_\eta & \beta_\theta & 0 \\ \beta_\eta & \psi^4 a & \psi^4 c & 0 \\ \beta_\theta & \psi^4 c & \psi^4 b & 0 \\ 0 & 0 & 0 & \psi^4 d \sin^2\theta \end{bmatrix}$$

where (with the exception of $\psi(\eta, \theta)$) the metric coefficients are functions of t, η, and θ.

In 1960 Misner [1960] constructed an analytic time-symmetric solution to the (equal mass) black hole collision initial value problem. The initial 3-metric is conformally flat

$$dl^2 = \psi_M{}^4 \left(d\rho^2 + dz^2 + \rho^2 d\phi^2\right). \tag{7}$$

The conformal factor ψ_M defined by

$$\psi_M = 1 + \sum_{n=1}^{\infty} \frac{1}{\sinh(n\mu)} \left(\frac{1}{\sqrt{\rho^2 + (z+z_n)^2}} + \frac{1}{\sqrt{\rho^2 + (z-z_n)^2}} \right), \tag{8}$$

(where $z_n = \coth(n\mu)$) solves the Hamiltonian constraint (2) with the proper isometry imposed between the upper and lower sheets. This data set which is axisymmetric, reflection symmetric about the equator ($z = 0$) and time symmetric ($K_{ij} = 0$) represents two equal mass black holes with zero rotation. The two black hole centers are aligned along the axis of symmetry (z-axis) so the physical interaction is a head-on collision along the axis. The free parameter μ is related to the physical parameters M (half the total ADM mass and approximately the mass of a single black hole when the holes are "infinitely" separated)

$$M = M_{ADM}/2 = 2\sum_{n=1}^{\infty} \frac{1}{\sinh(n\mu)}, \tag{9}$$

and L (the proper distance along the spacelike geodesic connecting the throats)

$$L = 2\left[1 + 2\mu \sum_{n=1}^{\infty} \frac{n}{\sinh(n\mu)}\right]. \tag{10}$$

The effect of increasing μ is to set the two black holes further away from one another and decrease the total mass of the system.

Solutions are constructed using the ADM evolution Equations (4) and (6) written in a second order time explicit finite differencing scheme. The lapse function is determined from the maximal slicing condition ($\mathrm{Tr}K = 0$) and the condition that the 3-metric is diagonal (i.e., $c = 0$) determines the shift vector. Using a method developed by Abrahams and Evans [1990] and applied in Abrahams et al. [1992] to oscillating black hole spacetimes the gravitational wave-forms have been extracted from the numerical values of the metric functions. This method is based on a Regge-Wheeler perturbation expansion and one can construct the the gauge invariant Zerilli function Ψ (Zerilli [1970]). The function Ψ represents the wave-like part of the metric that is radiative at large distances from the source and is commonly used in semi-analytic calculations of black hole normal mode frequencies (Chandrasekhar [1983]). The asymptotic energy flux carried by gravitational waves can be computed from

$$\frac{dE}{dt} = \frac{1}{32\pi}\left(\frac{\partial\Psi}{\partial t}\right)^2, \tag{11}$$

independently for each ℓ mode contribution for the normalization we use for Ψ.

For all of the simulations we have extracted both the $\ell = 2$ and $\ell = 4$ waveforms at radii of 30, 40, 50, 60 and $70M$. By comparing results at each of these radii we are able to check the propagation of waves and the consistency of our energy calculations.

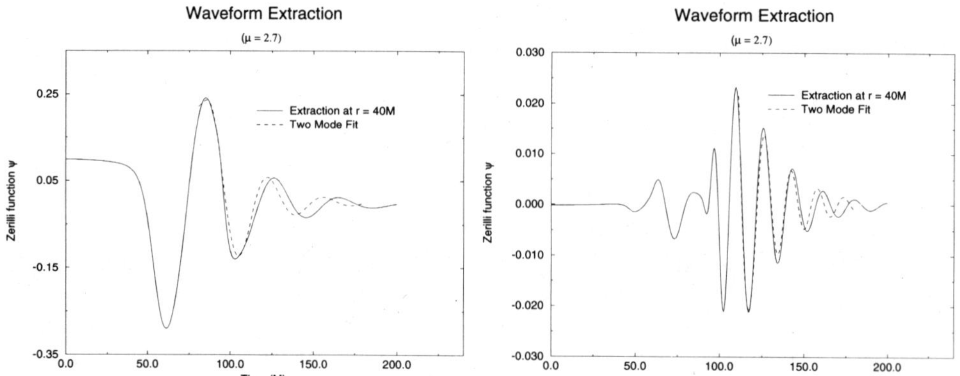

Figure 2 The (a) $\ell = 2$ and (b) $\ell = 4$ waveforms for a black hole collision with $\mu_0 = 2.7$ ($L/M = 12.7$). The solid lines indicate the waveforms measured at $r = 40M$ and the dotted lines are the quasi-normal mode fits to the mass of the final black hole.

Figure 2 shows the waveforms for the $\ell = 2$ and the $\ell = 4$ modes obtained from a single black hole collision. The gravitational waves can be used to measure the mass increase of the single black hole as it absorbs increasing amounts of gravitational wave energy. This explains the difference between the quasi-normal modes and the true wave form.

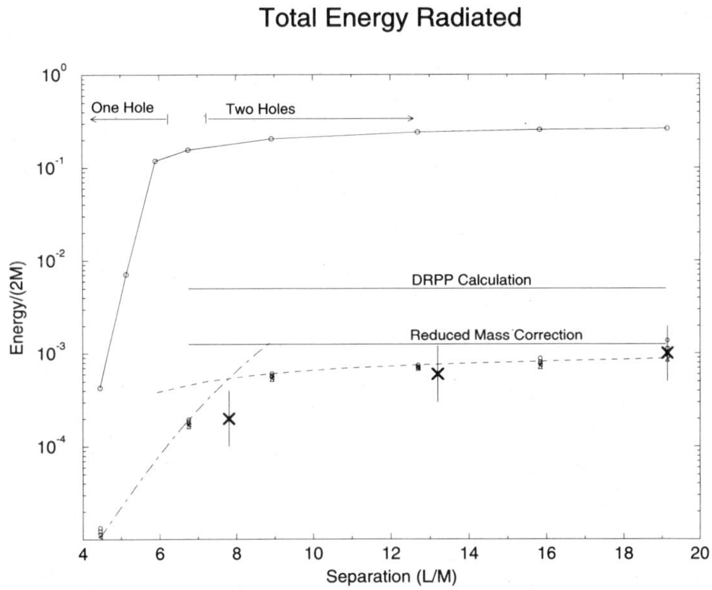

Figure 3 The total gravitational wave energy output is shown for the six parameter studies. The connected circles are the upper limit based on the Hawking area theorem, the clustered symbols show numerical results at various detector locations, and the crosses show early results by Smarr and Eppley with their approximate error bars. The solid line labelled "DRPP Calculation" is the result of a naive application of the point particle result, the solid line labelled "Reduced Mass Correction", takes into account the finite mass of the infalling black hole, and the dashed line is the semi-analytic estimate. Finally, the dot-dashed line is an empirical fit to the data for low values of the separation parameter μ, showing the exponential fall off of the energy in the one black hole regime.

The total radiated energy E can be computed from the Zerilli function using Equation (11). We display these results in Figure 3. The six clusters of unconnected symbols represent six numerical simulations corresponding to the different μ parameter values. Each of the five symbols within a cluster corresponds uniquely to the total integrated $\ell = 2$ energy computed at the five different wave detectors. For reference, the early results of Smarr and Eppley are plotted as large crosses with error bars suggested by Smarr [1979]. Within the large errors quoted, those early results are remarkably consistent with our more accurate results.

Clearly the results in Figure 3 show two distinct regimes, as denoted by the arrows in the upper part of the figure. For $\mu < 1.8$ the initial data contains *one* black hole, and the energy radiated falls off exponentially. We have fit the energy output to an exponential for $\mu \leq 1.8$ and find that the results are approximated well by the formula $E = 3.13 \times 10^{-8} \exp(4.852\mu)$. For $\mu > 1.8$ there are two holes and the energy radiated is somewhat independent of the initial separation. Four lines based on analytic and semi-analytic calculations that treat the system as two black holes are also shown. The top line is the theoretical upper limit obtained using the Hawking area theorem. The curve labelled "DRPP calculation" is based on Davis et al. [1971] (see also Chandrasekhar [1983]), the result labelled "Reduced Mass Correction" takes into account the finite mass of a black hole falling into its partner (see Anninos et al. [1993]), and the dashed line is the complete semi-analytic calculation discussed in Anninos et al. [1995], accounting for other effects such as tidal heating of the event horizon.

References

Abrahams, A., Bernstein, D., Hobill, D., Seidel, E. and Smarr, L. [1992] *Numerically Generated Black Hole Spacetimes: Interaction with Gravitational Waves*, Phys. Rev. D **45**, 3544–3558.

Abrahams, A. and Evans, C. [1990] *Gauge Invariant Treatment of Gravitational Radiation Near the Source: Analysis and Numerical Simulations*, Phys. Rev. D **42**, 2585–2594.

Anninos, P., Bernstein, D., Hobill, D., Seidel, E., Smarr, L. and Towns, J. [1997] *Gravitational Radiation from Oscillating Black Holes*, Computational Astrophysics: Gas Dynamics and Particle Methods (Benz, W., Barnes, J., Muller, E. and Norman, M., eds.), Springer-Verlag, New York, to appear.

Anninos, P., Hobill, D., Seidel, E., Smarr, L. and Suen, W.-M. [1993] *The Two Black Hole Collision*, Phys. Rev. Lett. **71**, 2851–2853.

Anninos, P., Hobill, D., Seidel, E., Smarr, L. and Suen, W.-M. [1994] *The Head-On Collision of Two Equal Mass Black Holes: Numerical Methods*, Technical Report 024, National Center for Supercomputing Applications.

Anninos, P., Hobill, D., Seidel, E., Smarr, L. and Suen, W.-M. [1995] *The Head-On Collision of Two Equal Mass Black Holes*, Phys. Rev. D, **52**, 2044–2058.

Arnowitt, R., Deser, S. and Misner, C. W. [1962] *The Dynamics of General Relativity*, Gravitation: An Introduction to Current Research, (Witten, L., ed.), John Wiley, New York, pp. 227–265.

Bernstein, D., Hobill, D., Seidel, E., Smarr, L. and Towns, J. [1994] *Numerically Generated Axisymmetric Black Hole Spacetimes: Numerical Methods and Code Tests*, Phys. Rev. D **50**, 5000–5024.

Chandrasekhar, S. [1983] *The Mathematical Theory of Black Holes*, Oxford Univ. Press, Oxford.

Davis, M., Ruffini, R., Press, H. and Price, R. H. [1971] *Gravitational Radiation from a Particle Falling Radially into a Schwarzschild Black Hole*, Phys. Rev. Lett. **27**, 1466–1469.

Misner, C. [1960] *Wormhole Initial Conditions*, Phys. Rev. **118**, 1110–1112.

Schutz, B. F. [1993] *Sources of Gravitational Waves and Their Detectability*, General Relativity and Gravitation, 1992 (Gleiser, R., Kozameh, C. and Moreschi, O., eds.), Inst. of Phys. Pub., Bristol, U.K., pp. 191–210.

Smarr, L. [1977] *Spacetimes Generated by Computers: Black holes with Gravitational Radiation*, Ann. N. Y. Acad. Sci. **302**, 569–604.

Smarr, L. [1979] *Gauge Conditions, Radiation Formulae and the Two Black Hole Collision*, Sources of Gravitational Radiation (Smarr, L., ed.), Cambridge University Press, Cambridge, pp. 245–274.

Smarr, L., Čadež, A., DeWitt, B. and Eppley, K. [1976] *Collision of Black Holes: Theoretical framework*, Phys. Rev. D **14**, 2443–2452.

Zerilli, F. J. [1970] *Effective Potential for Even-parity Regge-Wheeler Gravitational Perturbation Equations*, Phys. Rev. Lett. **24**, 737–738.

Fields Institute Communications
Volume **15**, 1997

A New Class of Spherically Symmetric Perfect Fluid Solutions with Kinematic Self-Similarity

Patricia M. Benoit
Department of Mathematics, Statistics and Computing Science
Dalhousie University
Halifax, Nova Scotia, Canada B3H 3J5
pmbenoit@cs.dal.ca

Alan A. Coley
Department of Mathematics, Statistics and Computing Science
Dalhousie University
Halifax, Nova Scotia, Canada B3H 3J5
aac@cs.dal.ca

Abstract. A new class of spherically symmetric, perfect fluid solutions of Einstein's Field Equations admitting a kinematic self-similarity (Carter and Henriksen [1989]) is exhibited. It is explicitly shown that these spacetimes do not admit a homothetic vector.

1 Introduction

Solutions exhibiting self-similarity of the first kind (i.e., those admitting a homothetic vector) have been studied extensively (Cahill and Taub [1971], Wesson [1979], Carr and Yahill [1990], Ori and Piran [1990], Henriksen and Patel [1991] and Carr and Koutras [1993]). Cahill and Taub [1971] presented a comprehensive study of the spherically symmetric perfect fluid case. The aim of this paper is to examine spherically symmetric perfect fluid solutions of Einstein's Field Equations (EFEs) which exhibit self-similarity of a more general type.

A kinematic self-similarity, ξ, considered as a generalization of a homothetic vector, may be defined (Carter and Henriksen [1989] and [1991]) as:

$$\mathcal{L}_\xi h_{ab} = 2h_{ab}, \tag{1.1}$$
$$\mathcal{L}_\xi u_a = \alpha u_a, \tag{1.2}$$

where $\mathcal{L}_\xi$ is the Lie derivative with respect to ξ, $h_{ab} = g_{ab} + u_a u_b$ is the projection tensor, u^a is the fluid velocity vector, and α is a constant. If (and only if) $\alpha = 1$, then ξ is a homothetic vector. In the following we present a new class of solutions, which contains as a subcase the kinematic self-similar Tolman dust solution (Carter

1991 *Mathematics Subject Classification.* Primary 83C20; Secondary 83C55.

This work was supported, in part, by the Natural Sciences and Engineering Research Council of Canada.

© 1997 American Mathematical Society

and Henriksen [1989] and Henriksen [1989]), and show that the resulting spacetimes do not possess a homothetic vector.

We choose coordinates such that the spherically symmetric metric has the form

$$ds^2 = -e^{2\nu}dt^2 + e^{2\lambda}dr^2 + R^2 d\Omega^2, \tag{1.3}$$

where $d\Omega^2 = d\theta^2 + sin^2\theta\, d\phi^2$ and surfaces of constant t are orthogonal to the fluid flow lines; i.e., $u^a = (e^{-\nu}, 0, 0, 0)$. Kinematic self-similarity (Carter and Henriksen [1989]) then implies that the metric functions are of the form

$$\nu(r,t) = \nu(z), \quad \lambda(r,t) = \lambda(z), \quad R(r,t) = r \cdot S(z), \tag{1.4}$$

in terms of the similarity variable ($\alpha \neq 0$)

$$z = r(\alpha t)^{(-1/\alpha)}. \tag{1.5}$$

2 A New Class of Solutions

Using the similarity variables defined above, the following represents a class of solutions of the perfect fluid EFEs;

$$ds^2 = -dt^2 + (S + \dot{S})^2 dr^2 + r^2 S^2 d\Omega^2, \tag{2.1}$$

where it is assumed that $S + \dot{S} \neq 0$, $x \equiv \ln(z)$, $\dot{S} \equiv \dfrac{dS}{dx}$ and $S(x)$ is a solution of the differential equation

$$0 = 2\frac{\ddot{S}}{S} + \frac{\dot{S}^2}{S^2} + 2\alpha\frac{\dot{S}}{S} + k, \tag{2.2}$$

where $\dot{S} \neq 0$ (i.e., the spacetime is not static), k is a dimensional constant, $\alpha \neq 0$ and $\alpha \neq 1$ (i.e., ξ is not a homothetic vector).

The fluid is described by

$$\mu = W(x)t^{-2},$$
$$p = \tfrac{1}{4}k\alpha^{-4}t^{-2}.$$

The dust solution of the Tolman family, described in Henriksen [1989] and Carter and Henriksen [1991], is obtained when $k = 0$. We note that in the solution above there is a dimensional constant appearing (in the pressure), a property that is characteristic of self-similarity of the second kind (Barenblatt and Zel'dovich [1972]). In the above, $W(x)$ is a particular function of x which is given in Benoit and Coley [1996].

3 Non-Existence of a Homothetic Vector

This class of solutions can be shown to have NO homothetic vectors. We recall that we have assumed that $\alpha(\alpha - 1)\dot{S}(S + \dot{S}) \neq 0$. The proof of this statement may be outlined as follows:

We shall assume that the solution admits a homothetic vector, X, so that

$$\mathcal{L}_X g_{ab} = 2g_{ab}, \tag{3.1}$$

whence in the case of a perfect fluid, it follows that

$$\mathcal{L}_X u_a = u_a, \tag{3.2}$$

and we will show that the only possible solutions are such that $\alpha(\alpha - 1)\dot{S}(S + \dot{S}) = 0$ contradicting the conditions of the solution given by (2.1) and (2.2).

Equation (3.2) implies that $X^0{}_{,0} = 1$ and $X^a{}_{,0} = 0$, where $a = 1,2,3$. The (a,0) term of Equation (3.1) then implies that $X^0{}_{,a} = 0$ $(a = 1,2,3)$, and hence $X^0 = t + c_0$. The (1,1) term of Equation (3.1) implies that

$$F(x)(\alpha t X^1 - r X^0) = \alpha r t (X^1{}_{,1} - 1), \tag{3.3}$$

where $F(x) \equiv (\dot{S} + \ddot{S})/(\dot{S} + S)$. Differentiating Equation (3.3) with respect to θ and then t we find that either

$$X^1{}_{,2} = 0, \tag{3.4}$$

or

$$F(x) = a_2, \tag{3.5}$$

where a_2 is a constant.

From Equation (3.5) we find that for $a_2 \neq -1$

$$S(x) = a_3 e^{-x} + \frac{a_1}{1 + a_2} e^{a_2 x}, \tag{3.6}$$

where a_1 and a_3 are constants. In the special case $a_3 = 0$, we have that $\dot{S} = a_2 S$, whence Equation (2.2) implies that $3{a_2}^2 + 2\alpha a_2 + k = 0$. Therefore, for given α and k, if a_2 satisfies this equation then $\dot{S} = a_2 S$ is a particular solution. In this special case we have that

$$ds^2 = -dt^2 + \frac{r^{2a_2}}{(1 + a_2)^2 (\alpha t)^{2a_2/\alpha}} \{[1 + a_2]^2 dr^2 + r^2 d\Omega^2\}, \tag{3.7}$$

which becomes, after the coordinate redefinition $r \to \bar{r} = r^{1+a_2}$,

$$ds^2 = -dt^2 + C^2 t^{-2a_2/\alpha} \{d\bar{r}^2 + \bar{r}^2 d\Omega^2\}. \tag{3.8}$$

This represents a flat, power-law FRW spacetime, which is known to admit a homothetic vector (Carr and Yahill [1990] and Benoit and Coley [1996]). We emphasize that this FRW model is only a very special solution; however, we note that it acts as an asymptote to all solutions of (3.6) as well as to a general class of solutions of (2.2). In the general casc $a_3 \neq 0$, Equation (2.2) implies that $S + \dot{S} = 0$ or $\dot{S} = 0$. Similarly, the special case $a_2 = -1$ also leads to either $S + \dot{S} = 0$ or $\dot{S} = 0$.

Hence, in general, if solutions are to admit a homothetic vector, we must have $F(x) \neq$ constant, and therefore, $X^1{}_{,2} = 0$. Similarly, by differentiating Equation (3.3) with respect to ϕ, we find that $X^1{}_{,3} = 0$. Hence X^1 is a function of r alone. Thus, substituting $X^1 = f(r)$ into Equation (3.3) and twice differentiating with respect to t (assuming $c_0 \neq 0$, since $c_0 = 0$ leads to $\alpha = 1$), we find that

$$\frac{d}{dx}\left(\frac{F(x)}{\dot{F}(x)}\right) = \frac{-\alpha t}{c_0}\left(\frac{f(r)}{r} - \frac{1}{\alpha}\right). \tag{3.9}$$

Therefore,

$$\frac{f(r)}{r} - \frac{1}{\alpha} = c r^{-\alpha}, \tag{3.10}$$

where c is an arbitrary constant. Substituting Equation (3.10) into Equation (3.3) gives

$$F(x) = \frac{(1 - \alpha)(c e^{-\alpha x} + t)}{c e^{-\alpha x} - c_0}, \tag{3.11}$$

so that (for $\alpha \neq 1$)

$$(\frac{ce^{-\alpha x} - c_0}{1-\alpha})F(x) - ce^{-\alpha x} = t, \tag{3.12}$$

which is clearly inconsistent.

Therefore, the solutions in this class which meet the condition $\alpha(\alpha - 1)\dot{S}(S + \dot{S}) \neq 0$, except for the very special flat FRW solution, necessarily do not admit a homothetic vector.

References

Barenblatt, G. E. and Zel'dovich, Ya. B. [1972] *Self Similar Solutions as Intermediate Asymptotes*, Ann. Rev. Fluid Mech. **4**, 285–312.

Benoit, P. M. and Coley, A. A. [1996] *Spherically Symmetric Spacetimes and Kinematic Self-Similarity*, preprint.

Cahill, M. E. and Taub, A. H. [1971] *Spherically Symmetric Similarity Solutions of the Einstein Field Equations for a Perfect Fluid*, Commun. Math. Phys. **21**, 1–40.

Carr, B. J. and Koutras, A. [1993] *Self-Similar Perturbations of a Kantowski-Sachs Model*, Astrophys. J. **405**, 34–50.

Carr, B. J. and Yahill, A. [1990] *Self-Similar Perturbations of a Freidmann Universe*, Astrophys. J. **360**, 330–342.

Carter, B. and Henriksen, R. N. [1989] *A Covariant Characterization of Kinematic Self-Similarity*, Annales de Physique **14**, 47–53.

Carter, B. and Henriksen, R. N. [1991] *A Systematic Approach to Self-Similarity in Newtonian Space-Time*, J. Math. Phys. **32**, 2580–2597.

Henriksen, R. N. [1989] *Spherically Symmetric, Self Gravitating Dust is Always 'Self-Similar'*, Mon. Not. R. Ast. Soc. **240**, 917–929.

Henriksen, R. N. and Patel, K. [1991] *Null Charts and Naked Singularities in Spherically Symmetric, Homothetic Spacetimes*, Gen. Rel. Grav. **23**, 527–581.

Ori, A. and Piran, T. [1990] *Naked Singularities and Other Features of Self-Similar General-Relativistic Gravitational Collapse*, Phys. Rev. D **42**, 1068–1090.

Wesson, P. S. [1979] *Observable Relations in an Inhomogeneous Self-Similar Cosmology*, Astrophys. J. **228**, 647–663.

Fields Institute Communications
Volume **15**, 1997

A Wave-Like Solution in Kaluza-Klein Theory

Andrew Billyard
Department of Physics
Dalhousie University
Halifax, Nova Scotia, Canada B3H 4H8
jaf@cs.dal.ca

Paul S. Wesson
Department of Physics
University of Waterloo
Waterloo, Ontario, Canada N2L 3G1
wesson@astro.uwaterloo.ca

Abstract. Using symbolic software, we derive an exact solution of the 5D gravity equations that describes a wave moving through a de Sitter vacuum.

There has recently appeared significant literature on a new interpretation of Kaluza-Klein theory in which empty geometry in 5D gives rise to curvature and matter in 4D. (See the list of references for a subset of these papers). Many solutions of this theory are known, and it is always possible to interpret any 5D solution in apparent vacuum as a 4D solution of Einstein's equations with an effective or induced energy-momentum tensor (Wesson and Ponce de Leon [1992]). We wish in what follows to give a new type of solution derived using MAPLE symbolic software. It has complex or wave-like metric coefficients for ordinary 3D space. While complex coefficients have occasionally been used in applications of 4D Einstein theory to particle physics, we will concentrate here on giving the solution, and defer a detailed physical interpretation to future work.

We let English indices run 0-4 and Greek indices run 0-3. The coordinates are $x^0 = t$ (time), $x^{123} = xyz$ (space) and $x^4 = \psi$ (the new coordinate). Units are chosen such that the speed of light is $c = 1$ and the gravitational constant is $8\pi G = 1$. The field equations for apparently empty 5D Kaluza-Klein theory are then given in terms of the Ricci tensor by

$$R_{ab} = 0. \tag{1}$$

However, these relations contain the 4D Einstein equations with an effective or induced 4D energy-momentum tensor, given in the general case by Wesson and

1991 *Mathematics Subject Classification.* 83E15.

© 1997 American Mathematical Society

Ponce de Leon [1992]:

$$
\begin{aligned}
{}^{(4)}T_{\alpha\beta} = & \frac{\phi_{\alpha;\beta}}{\phi} - \frac{\epsilon}{2\phi^2}\left\{ \frac{\overset{*}{\phi}\,\overset{*}{g}_{\alpha\beta}}{\phi} - \overset{**}{g}_{\alpha\beta} + g^{\mu\nu}\overset{*}{g}_{\alpha\mu}\overset{*}{g}_{\beta\nu} \right. \\
& \left. - \frac{g^{\mu\nu}\overset{*}{g}_{\mu\nu}\overset{*}{g}_{\alpha\beta}}{2} + \frac{g_{\alpha\beta}}{4}\left[\overset{*}{g}^{\mu\nu}\overset{*}{g}_{\mu\nu} + \left(g^{\mu\nu}\overset{*}{g}_{\mu\nu}\right)^2\right] \right\}.
\end{aligned}
\tag{2}
$$

Here $g_{44} \equiv \epsilon\phi^2$ with $\epsilon = \pm 1$, $\phi_\alpha \equiv \partial\phi/\partial x^\alpha$, an overstar denotes $\partial/\partial x^4$ and a semicolon denotes the usual (4D) covariant derivative. The presence of matter in 4D depends clearly on the extra dimension and derivatives with respect to the extra coordinate.

Consider the x^4-dependent 5D metric

$$
ds^2 = \psi^2 dt^2 - \psi^2 e^{i(wt+k_x x)}dx^2 - \psi^2 e^{i(wt+k_y y)}dy^2 - \psi^2 e^{i(wt+k_z z)}dz^2 + l^2 d\psi^2. \tag{3}
$$

Here w is a frequency, k_x etc. are wave-numbers, and l measures the size of the extra dimension. The symbolic software Maple V was an integral part of our computations; it allowed us to quickly verify that (3) was indeed a vacuum solution (see below) and it enabled us to examine and discard other oscillatory metrics based on required criteria.

For (3), the non-zero components of the Riemann-Christoffel tensor have the form

$$
R_{txtx} = -\left(\frac{\psi^2}{4l^2}\right)\left(w^2 l^2 - 4\right) e^{i(wt+k_x x)} \quad \text{etc.} \tag{4a}
$$

$$
R_{xyxy} = \left(\frac{\psi^2}{4l^2}\right)\left(w^2 l^2 - 4\right) e^{i(2wt+k_x x+k_y y)} \quad \text{etc.}, \tag{4b}
$$

with two similar relations in each case obtained by permutations of the space coordinates. The non-zero components of the Ricci tensor look like

$$
R_{tt} = -\left(\frac{3}{4l^2}\right)\left(w^2 l^2 - 4\right) \tag{5a}
$$

$$
R_{xx} = \left(\frac{3}{4l^2}\right)\left(w^2 l^2 - 4\right) e^{i(wt+k_x x)} \quad \text{etc.} \tag{5b}
$$

It is clear from (5) that the field equations (1) are satisfied for $w = \pm 2/l$, and from (4) that the solution is 5D flat.

However, it is non-trivial in 4D. For example, the 4D Ricci or curvature scalar (Wesson and Ponce de Leon [1992]) is:

$$
{}^{(4)}R = \frac{1}{4l^2}\left[\overset{*}{g}^{\mu\nu}\overset{*}{g}_{\mu\nu} + \left(g^{\mu\nu}\overset{*}{g}_{\mu\nu}\right)^2\right] = \frac{12}{\psi^2 l^2}. \tag{6}
$$

And the non-zero components of the induced energy-momentum tensor (2) are:

$$
T_{00} = -\frac{3w^2}{4} \tag{7a}
$$

$$
T_{11} = \frac{3w^2}{4}\, e^{i(wt+k_x x)} \quad \text{etc.} \tag{7b}
$$

It is possible to match these components to those of a perfect fluid with

$$T_{\alpha\beta} = (p+\rho)\, u_\alpha u_\beta - p\, g_{\alpha\beta} \tag{8}$$

where ρ is density and p is pressure. However, this requires knowledge of the velocities.

These can be obtained by solving the geodesic equations

$$\frac{du^k}{d\lambda} + \Gamma^k_{ij}\, u^i u^j = 0, \tag{9}$$

where $u^k \equiv dx^k/d\lambda$ and λ is a suitable parameter along the trajectory. For metric (3), the non-zero Christoffel symbols of the second kind are:

$$\Gamma^t_{t\psi} = \frac{1}{\psi},\quad \Gamma^t_{xx} = \frac{iw}{2}\, e^{i(wt+k_x x)} \quad \text{etc.} \tag{10a}$$

$$\Gamma^x_{tx} = \frac{iw}{2},\quad \Gamma^x_{xx} = \frac{ik_x}{2},\quad \Gamma^x_{x\psi} = \frac{1}{\psi} \quad \text{etc.} \tag{10b}$$

$$\Gamma^\psi_{tt} = -\frac{\psi}{l^2},\quad \Gamma^\psi_{xx} = \frac{\psi}{l^2}\, e^{i(wt+k_x x)} \quad \text{etc.} \tag{10c}$$

Using these it may be verified by substitution that a solution of (9) is

$$u^0 = \left(\frac{1}{\psi^2}\right)\left(V_t - V_x\, e^{ik_x x/2} - V_y\, e^{ik_y y/2} - V_z\, e^{ik_z z/2}\right) \tag{11a}$$

$$u^1 = \left(\frac{A_x}{\psi^2}\right) e^{-i(wt+k_x x/2)} \quad \text{etc.} \tag{11b}$$

$$u^4 = \left(\pm\frac{w}{2\psi}\right) e^{-iwt/2}\Big[\psi^2 K^2 e^{iwt} - \Big(V_t\, e^{iwt/2} - V_x\, e^{i(wt+k_x x)/2} - V_y\, e^{i(wt+k_y y)/2} - V_z\, e^{i(wt+k_z z)/2}\Big)^2 + A^2\Big]^{1/2}. \tag{11c}$$

Here A_x etc. and V_t are integration constants, $A^2 \equiv A_x^2 + A_y^2 + A_z^2$, $V_x = w/k_x A_x$ etc., and $K = 0$ for the massless case ($ds = 0$) and $K = 1$ for the massive case ($ds \neq 0$). The relations (11) satisfy not only (9) but also, of course, the consistency condition $u_a u^a = K$ set by the metric.

Combining (11), (8) and (7) gives us the properties of matter associated with the metric (3):

$$p = -\rho = \frac{3w^2}{4\psi^2}. \tag{12}$$

Here ψ appears because it is associated with the induction of matter in the same way as it is in the cosmological solutions (Wesson [1992], Wesson et al. [1993], Liu and Wesson [1994] and Wesson [1994]). For the solution (12), the equation of state is clearly that of the classical de Sitter vacuum:

$$p + \rho = 0. \tag{13}$$

Interpreted as vacuum properties related to the cosmological constant, the latter is $\Lambda = -3w^2/4\psi^2$ and is negative for the signature adopted in the metric (3).

This metric can be changed to give insight into the kind of oscillations involved. Thus the oscillatory terms in x, y and z can be absorbed by appropriate coordinate transformations, and w can be made imaginary (which reverses the sign of the last part of the metric since $l^2 = 4/w^2$). The result is a 5D growing-mode de Sitter solution, of a type found previously by Ponce de Leon during a search for 5D cosmological models that reduce to 4D Friedmann-Robertson-Walker ones on hypersurfaces $x^4 = \psi =$ constant. In this regard, it should be noted that the standard 4D de Sitter solution does not necessarily imply a growing (expanding) mode, but also an oscillatory mode. This is most readily seen by considering the Friedmann equations:

$$\rho = \frac{3k}{R^2} + \frac{3\dot{R}^2}{R^2} - \Lambda \tag{14a}$$

$$p = -\frac{k}{R^2} - \frac{\dot{R}^2}{R^2} - \frac{2\ddot{R}}{R} + \Lambda. \tag{14b}$$

Here $R = R(t)$ is the scale factor, k is the curvature constant, and an overdot denotes the derivative with respect to the time t. These equations are satisfied for $\rho = p = 0$, $k = 0$, $R = e^{iwt}$ and $\Lambda = -3w^2$. This oscillatory motion is the analog in 4D of the motion implied by the metric (3) in 5D. Since in both cases there is spherical symmetry in 3D, the oscillations involved are not gravitational waves of the conventional sort. Rather, the metric (3) describes a situation in which the metric coefficients of ordinary 3-space have wave-like behaviour and the associated medium is an unperturbed de Sitter vacuum. We therefore believe it is appropriate to refer to vacuum waves in Kaluza-Klein theory. We will defer a detailed physical interpretation of the solution (3) to future work. However, it is apparent that it is cosmological in nature. It is flat in 5D, and to that extent is similar to other higher-dimensional cosmological models that are 5D flat but provide good cosmological models when viewed on hypersurfaces (Wesson [1992], Liu and Wesson [1994], Wesson [1994] and Coley and McManus [1995]). Indeed, an obvious application of (3) is to the inflationary universe model. We can imagine a situation where a vacuum-dominated universe exists with metric (3), first with w real and in an oscillatory mode and later with w imaginary and in a growing de Sitter mode. This picture is in line with others which interpret the big bang as a quantum-induced phase change, but needs to be investigated further.

References

Coley, A. A. and McManus, D. J. [1995] *A Family of Cosmological Solutions in Higher-Dimensional Einstein Gravity,* J. Math. Phys. **36**, 335–339.

Kalligas, D., Wesson, P. S. and Everitt, C. W. F. [1995] *The Classical Tests in Kaluza-Klein Gravity,* Astrophys. J. **439**, 548–557.

Liu, H. and Wesson, P. S. [1994] *Cosmological Solutions and Their Effective Properties of Matter in Kaluza-Klein Theory,* Int. J. Mod. Phys. D **3**, 627–637.

Liu, H., Wesson, P. S. and Ponce de Leon, J. [1993] *Time-Dependent Kaluza-Klein Soliton Solutions,* J. Math. Phys. **34**, 4070–4079.

Mashhoon, B., Liu, H. and Wesson, P. S. [1994] *Particle Masses and the Cosmological Constant in Kaluza-Klein Theory,* Phys. Lett. B **331**, 305–312.

Ponce de Leon, J. and Wesson, P. S. [1993] *Exact Solutions and the Effective Equation of State in Kaluza-Klein Theory,* J. Math. Phys. **34**, 4080–4092.

Wesson, P. S. [1992] *A Physical Interpretation of Kaluza-Klein Cosmology,* Astrophys. J. **394**, 19–24.

——— [1994] *An Embedding for the Big Bang,* Astrophy. J. **436**, 547–550.

Wesson, P. S., Liu, H., and Lim, P. [1993] *Time-Dependent Kaluza-Klein Solitons and Their Effective Properties of Matter,* Phys. Lett. B **298**, 69–72.

Wesson, P. S. and Ponce de Leon, J. [1992] *Kaluza-Klein Equations, Einstein's Equations and an Effective Energy-Momentum Tensor,* J. Math. Phys. **33**, 3883–3887.

——— [1994] *The Physical Properties of Kaluza-Klein Solitons,* Class. Quant. Grav. **11**, 1341–1351.

Wesson, P. S., Ponce de Leon, J., Lim, P. and Liu, H. [1993] *Physical Properties of Matter Derived from Geometry of Kaluza-Klein Theory,* Int. J. Mod. Phys. D **2**, 163–170.

Fields Institute Communications
Volume **15**, 1997

Mass Inflation in a Rotating Charged Hole

Alfio Bonanno
Canadian Institute for Advanced Research Cosmology Program
Theoretical Physics Institute, University of Alberta
Edmonton, Alberta, Canada T6G 2J1
and
Institute of Astronomy, University of Catania
Viale Andrea Doria 6, 95125 Catania, Italy
abonanno@ct.astro.it

Abstract. The interior of a charged rotating black hole is analyzed in the presence of an ingoing and outgoing flux of gravitational waves. An explicit mass inflation solution is found in the limit of a slowly rotating hole.

The inner structure of a realistic black hole represents an intriguing challenge. Although the evolution of the fields beyond the event horizon is in principle determined by the field equations, the presence at smaller radii of an inner horizon makes any further evolution of the initial data meaningless. The loss of predictability is associated with a lightlike surface, the Cauchy horizon (CH). This lightlike hypersurface is unstable to perturbations since it represents a surface of infinite blueshift for the gravitational waves coming from the collapsing object and infalling in the hole.

The relevant question is to take into account the effect of the backreaction of the fields on the metric. In spherical models the late time behavior of the external problem is characterized by a power law decay of aspherical perturbations $\sim 1/v^{p+1}$ in terms of the standard advanced coordinate, according to Price's result. The internal evolution is therefore determined by setting this boundary condition for the fields outside the event horizon. Israel and Poisson [1990] carried out an analytic model of the interior of a spherical BH where they found that the Hawking quasi-local mass is exponentially divergent at the CH. This singularity turns out to be a scalar singularity where the Ψ_2 component of the Weyl tensor is not regular, although "mild" as noted by Ori [1991], since the mass function is an integrable function of the Kruskalized advanced coordinate. Little progress has been made if the restriction of spherical symmetry is removed. Although the problem has been

1991 *Mathematics Subject Classification.* Primary 83C57; Secondary 83C75.

The author would like to thank Werner Israel for his constant advice and many enlightening discussions, as well Valery Frolov, Sharon Morsink, Charles Torre, David Hobill, Roberto Balbinot and Eric Poisson, for many useful and important comments. The author also thanks the Institute of Astronomy and the University of Catania for financial support.

© 1997 American Mathematical Society

investigated by Ori [1992], Brady and Chambers [1994] and Brady et al. [1995], no explicit solution is available.

We present a specific model of a rotating nonstationary BH that should capture the essential features of the physics of the Cauchy horizon in a rotating BH.

The crossflow region, near the CH, is described by an outgoing lightlike shell - simulating the outgoing flux - embedded in a continuous flow of infalling gravitational waves. This axisymmetric realization of the Ori [1991] model is derived in the slow rotation approximation when $a = \frac{J}{m}$ is small compared to the radius of the inner horizon r_0

$$\epsilon \equiv \frac{a^2}{r_0^2} \ll 1. \tag{1}$$

In particular, as background geometry we consider the non-stationary Vaidya type generalization of the Kerr metric discussed in Carmeli and Kaye [1977]. We extend that model to the charged case in order to retain a finite (nonzero) radius for the inner horizon even when a is small. In the $\{r, \vartheta, \varphi, v\}$ Eddington-Kerr coordinate system the metric reads

$$\begin{aligned} ds^2 &= -\left(1 - \frac{2m\,(v)\,r - e^2}{\Sigma}\right) dv^2 + 2dr\,dv + \Sigma\,d\vartheta^2 - 2a\sin^2\vartheta\,d\varphi\,dr \\ &+\mathcal{R}^2 \sin^2\vartheta\,d\varphi^2 - \frac{2a(2m\,(v)\,r - e^2)\sin^2\vartheta}{\Sigma} dv\,d\varphi \end{aligned} \tag{2}$$

where

$$\mathcal{R}^2 = \frac{(r^2+a^2)^2 - \Delta a^2 \sin^2\vartheta}{\Sigma}, \quad \Sigma = r^2 + a^2\cos^2\vartheta, \quad \Delta = r^2 + a^2 + e^2 - 2m(v)r. \tag{3}$$

As in the spherical Vaidya model, the mass parameter $m(v)$ is a function of the advanced coordinate v. Its functional dependence near the CH, located at $v = +\infty$, is assumed to be of the form

$$m = m_0 + \delta m(v) \qquad \delta m(v) = -\frac{A}{v^{p-1}} \qquad v \to +\infty \tag{4}$$

to reproduce Price's tail of the gravitational waves. Since we shall work in the slow rotation approximation (4) is the only relevant contribution. We introduce the null complex tetrad $\{l_\mu, n_\mu, m_\mu, \bar{m}_\mu\}$, with $-l_\mu n^\mu = m_\mu \bar{m}^\mu = 1$ and l_μ is the repeated principal null direction associated with the infalling radiation. The energy-momentum tensor can be expressed as

$$T_{\mu\nu} = 2\phi_{22} l_\mu l_\nu - 4\phi_{12} l_{(\mu}\bar{m}_{\nu)} - 4\bar{\phi}_{12} l_{(\mu} m_{\nu)} + 4\phi_{11}(l_{(\mu}n_{\nu)} + m_{(\mu}\bar{m}_{\nu)}) \tag{5}$$

where $\phi_{22}, \phi_{12}, \phi_{11}$ are the only non-vanishing tetrad components of the trace-free part of the Ricci tensor $S_{\mu\nu} = R_{\mu\nu} - 1/4 g_{\mu\nu} R$ ($= R_{\mu\nu}$ in our case),

$$\begin{aligned} \phi_{22} &= r(2\dot{m}(v) - a^2\sin^2\vartheta\,\ddot{m}(v))/4\Sigma^2, \\ \phi_{11} &= e^2/2\Sigma^2, \qquad \phi_{12} = -ia\sin\vartheta\,\dot{m}(v)\rho/2\sqrt{2}\Sigma \end{aligned} \tag{6}$$

and $\rho = -(r - ia\cos\vartheta)^{-1}$. The radiation field consists of a pure null part and a residual term. In particular ϕ_{11} represents the contribution of a static source field generated by a charge of strength e. The spacetime is algebraically special of Petrov type II since the only non-zero Weyl invariants are Ψ_2, Ψ_3, Ψ_4. Let us now consider the equation of motion $r = r(v, \vartheta)$ of an axisymmetric outgoing null hypersurface.

Close to the CH, where the functional dependence of the mass function on the advanced coordinate is given by (4), it reads

$$-\frac{\delta m(v)r}{r^2+a^2} - 2\kappa_0(r-r_0) - 2\partial_v r + \frac{a^2\sin^2\vartheta}{r^2+a^2}(\partial_v r)^2 + \frac{(\partial_\vartheta r)^2}{r^2+a^2} = 0 \tag{7}$$

where κ_0 is the surface gravity of the inner horizon located at $r = r_0$. To first order in the effective rotation parameter ϵ, the solution is found to be

$$r - r_0 = f(v)\left(1 - \epsilon\left(\frac{\sin^2\vartheta(p-1)}{2\kappa_0 v} + O(1/v^2)\right)\right) + O(\epsilon^2) \tag{8}$$

where

$$f(v) = \frac{A}{\kappa_0 r_0} v^{-(p-1)}(1 + \frac{p-1}{\kappa_0 v} + \dots). \tag{9}$$

This shows that the angular dependence is suppressed by a factor $1/v$ to first order in ϵ as we approach the CH.

Now we consider the spacetime to be divided into two regions $\mathcal{M}_\pm$ separated by an outgoing shell $\mathcal{S}$ whose equation of motion is assumed to be of the form (9) near the Cauchy horizon (see Bonanno [1995] for details). We assume that the "past" side of the shell is described by the radiating Kerr-Newman metric (2). It is reasonable to think that in the slow rotation regime the production of gravitational waves can be neglected and the structure of the resulting "glued" manifold is still Kerr-like and we assume that the spacetime in the future sector of the shell can be represented with a line element of the type (2). Although a_+ is in general a function of the coordinates in $\mathcal{M}^-$, we shall see that $\partial a_+/\partial v_-$ is exponentially small close to the CH. In this model the presence of the outgoing lightlike shell simply serves to start the contraction of the null generators of the inner horizon. Thus we consider a pressureless shell so that the soldering of the two geometries is affinely conciliable (Israel and Barrabés [1991]). It is convenient to define $m_+ = \bar{m} + M(v_+)$ where, by definition $2\bar{m}\bar{r} \equiv \bar{r}^2 + \bar{a}^2 + e^2$ and $\bar{r}$, $\bar{a}$ are the values of r_+ and a_+ at the CH. In order to determine the spacetime in the past sector of the shell we impose the continuity condition for the area of the intrinsic metric across the shell

$$[\mathcal{A}] = 0 \tag{10}$$

where $[\mathcal{A}] = \mathcal{A}_+ - \mathcal{A}_-$. It is straightforward to see that, to first order in the rotation parameter this implies, if $M(v_+)/r_0 \gg 1$,

$$a_+^2 M_+ \simeq \epsilon r_0^3, \qquad v_- \to +\infty. \tag{11}$$

This equation contains the essential physics. As observed in the beginning, it explicitly shows that the effective Kerr parameter a_+ approaches zero as the mass function grows. The asymptotic geometry close to the CH is not spherically symmetric because the $a_+^2 M_+$ and $a_+ M_+$ terms in the metric are not negligible. The energy-momentum tensor in the future sector of the shell contains further contributions arising from the fact the a_+ is not constant. However, it is possible to show that the leading, divergent behavior is still given by the optical geometric contribution in (5). The missing link to relate the dynamics of the spacetime follows from the condition for matching of normal stresses across the shell. In particular, in the slow rotation approximation it reads

$$[\phi_{22} l_\mu l_\nu s^\mu s^\nu] = 0 \tag{12}$$

where s_μ are generators of $\mathcal{S}$ and the scalar products have to be calculated from the spherically symmetric contribution. This equation contains the relevant couplings between the angular momentum and mass function. By using (11) in (12), after some simplifications one explicitly finds

$$\left[\frac{d^2M_+}{dv_-^2}+\frac{1}{M_+}\left(\frac{dM_+}{dv_-}\right)^2\right]\frac{\epsilon r_0\sin^2\vartheta}{2\partial_v f_-}-\frac{dM_+}{dv_-}=-\kappa_0\left(1-p\frac{2+\kappa_0 r_0\epsilon\sin^2\vartheta}{2\kappa_0 v_-}\right)M_+ \tag{13}$$

and note that $\vartheta_-=\vartheta_+$ in this approximation. Hence we write

$$M_+=m_++\delta m_+ \tag{14}$$

and we have from (13)

$$m_+=\frac{1}{v_-^p}e^{\kappa_0 v_-},\qquad\qquad v_-\to+\infty. \tag{15}$$

This has the Israel-Poisson behavior, and

$$\delta m_+\sim\epsilon\sin^2\vartheta e^{\kappa_0 v_-}. \tag{16}$$

This result shows the effective mass parameter exponentially inflates with a residual angular dependence

$$M_+\sim e^{\kappa_0 v_-}(1+\epsilon\sin^2\vartheta) \tag{17}$$

to first order in ϵ. The geometry in the mass-inflated sector is asymptotically dominated by the large mass term, and the metric, to a good approximation, explicitly reads

$$ds_+^2\simeq\frac{2m_+}{r}(1+\epsilon\sin^2\vartheta)\,dv_+^2+2dr\,dv_++r^2d\vartheta^2+r^2(1-\epsilon\sin^2\vartheta)\sin^2\vartheta\,d\varphi^2$$
$$-4\sqrt{\epsilon m_+ r}\sin^2\vartheta\,dv_+d\varphi. \tag{18}$$

The following "mild" twist of the φ coordinate (since m_+ is an integrable function of the advanced coordinate v_+)

$$d\varphi=d\Phi+\frac{2\sqrt{\epsilon m_+ r_0}}{r_0^2}dv_+ \tag{19}$$

brings the metric into the final form

$$ds_+^2\simeq\frac{2m_+}{r}(1+\epsilon\sin^2\vartheta)\,dv_+^2+2dr\,dv_++r^2d\vartheta^2+r^2(1+\epsilon\sin^2\vartheta)\sin^2\vartheta\,d\Phi^2 \tag{20}$$

to linear order in ϵ. It is straightforward to check the mathematical consistency of the approximations. In fact had we *started* with the line element (20), with m_+ given by (15) we would have found

$$\mathcal{A}_+=4\pi r_0^2(1+\epsilon)=\mathcal{A}_-+O(1/v^p) \tag{21}$$

for the equality of the intrinsic area across the shell. Also it is straightforward to verify that the stress-energy tensor computed from (20) close to the CH satisfies the matching condition for the normal stresses, equation (12). The Komar invariant quantity associated with the rotational Killing vecor field ξ^μ is not conserved since matter is flowing into the system. It is indeed divergent in our model, but with a much slower rate

$$\frac{1}{8\pi}\oint\xi^{\mu;\nu}d\sigma_{\mu\nu}\sim\sqrt{m_+} \tag{22}$$

where the integral is taken over the two-dimensional boundary of any $v_-=$ const. hypersurface, with $v_-\to+\infty$. It is not clear whether or not this "angular momentum inflation" would occur in a more realistic model than ours. In this case the divergence of the $R_{\Phi\Phi}$ component of the Ricci tensor drives the divergence in (22). This component is finite in spherical models, but one might argue that in the presence of the Killing vector field ξ_μ associated with the axial symmetry, the quantity

$$\int R_{\mu\nu}\xi^\nu d\Sigma^\mu \tag{23}$$

is also divergent if $R_{\mu\nu}$ diverges close to the CH. The resulting singularity at the CH is still "mild" since the only tetrad-invariant component of the Weyl tensor, Ψ_2 diverges as in the spherically symmetric case

$$\Psi_2 \sim \frac{1}{v_+ \ln^p |v_+|} \tag{24}$$

The result we have presented explicitly shows that deviations from spherical symmetry at the Cauchy horizon are present in the mass-inflated part of the geometry. This phenomenon should thus characterize the dynamics of a spacetime with a nonspherical inner horizon and should help to understand the more complicated case of a realistic generic rotating black hole.

References

Brady, P. R. and Chambers, C. M. [1994], preprint NCL94-TP13.

Brady, P., Droz, S., Israel, W. and Morsink, S. M. [1995], in preparation.

Bonanno, A. [1995], University of Alberta, preprint Th-16-95.

Carmeli, M. and Kaye, M. [1977] *Gravitational field of a radiating rotating body*, Annals of Physics **103**, 97–120.

Israel, W. and Barrabés, C. [1991] *Thin shells in general relativity and cosmology: the lightlike limit*, Physical Review D **43**, 1129–1142.

Ori, A. [1991] *Inner structure of a charged black hole: an exact mass-inflation solution*, Phys. Rev. Lett. **67**, 789–792.

Ori, A. [1992] *Structure of the singularity inside a realistic rotating black hole*, Phys. Rev. Lett. **68**, 2117–2120.

Poisson, E. and Israel, W. [1990] *Internal structure of black holes*, Phys. Rev. D **41**, 1796–1809.

Fields Institute Communications
Volume **15**, 1997

Quantum Black Holes and Hypertime

Stephen P. Braham
Department of Mathematics and Statistics
University of New Brunswick
Fredericton, N. B., Canada E3B 5A3
Current Address: Centre for Experimental and Constructive Mathematics
Simon Fraser University
Burnaby, B. C., Canada V5A 1S6
warp@lizardo.cecm.sfu.ca

Abstract. It is possible to solve the problem of time for spherically symmetric spacetimes by the use of hypertime variables. This leads to functional Schrödinger equations in the quantum theory. The possible use of this approach in numerical studies of quantum gravity is discussed.

I would like to address some of the questions that arise out of my recent work (Braham [1994] and [1995]) on hypertime approaches to the so-called *problem of time* in quantum gravity, and some possible applications of that work. It has long been a hope of researchers to model the full quantum dynamics of black holes and wormholes. Even in semiclassical theory, these systems display strange properties, from Hawking-Unruh radiation (Hawking [1975] and Unruh [1976]), with its potential destruction of our ability to make predictions in quantum theory, to mass inflation (Hiscock [1981] and Poisson and Israel [1990]), and the corresponding wormhole stability problem. To understand these objects, we need to combine general relativity, a theory which is *invariant* to the choice of time, and quantum theory, which *needs a time coordinate.* For such a theory to be useful, we need:

1. A dynamical quantum description, i.e., a wavefunction.
2. An ontological relationship, i.e., an interpretation of that description.
3. A way of calculating the first two!

The latter two points are very important: It would be very useful if we could formulate numerical models of the full quantum dynamics of black holes, if we did it in a way that truly corresponded to the *construction of a quantum theory.* A quantum theory, however, needs more than just a wavefunction: it needs *observables.* Observables are operators that describe the statements we can make from

---

1991 *Mathematics Subject Classification.* Primary 83C45, 83C57; Secondary 83-08, 83C05.
I would like to thank Bill Unruh and Jack Gegenberg for their insightful comments on canonical quantum gravity and black hole dynamics. This work is supported by the National Sciences and Engineering Council of Canada.

© 1997 American Mathematical Society

the quantum theory, without which any result is useless. In particular any valid numerical approach to quantum gravity must seemingly construct these observables. Unfortunately, this has not been done, to date, in any of the numerical studies of the field. This is a major problem, due to the fact that it seems improbable that any full theory of quantum gravity will lend itself to easy analysis, and thus computers will be central to understanding it. Indeed, we can consider the ability to construct a given quantum theory on a computer as being the razor edge that divides pure speculative theories from ones in which hard calculation can be made. If a theory cannot be represented in a computer calculation, it surely cannot produce numbers than can be compared to experiment.

The contrast between the invariance of general relativity under choices of time, and the need for a choice of time in quantum mechanics leads to the famous above-mentioned problem of time. We can see how this problem arises by looking at the Arnowitt, Deser and Misner (ADM) (Arnowitt et al. [1964]) formulation of geometrodynamics. In this formalism, the spacetime 4-metric is decomposed in terms of the familiar 3-metric g_{ij} and the corresponding lapse N, and shift N^i variables, all introduced on a specific foliation (prescribed by N and N^i), labeled by a time parameter t, and a spatial hypersurface position $\mathbf{x}$. The action for the vacuum part of the theory can then be written in the manifestly canonical form

$$S = \int dt \int d^3x \left(\pi^{ij}(\mathbf{x})\dot{g}_{ij}(\mathbf{x}) - N(\mathbf{x})\mathcal{H}(\mathbf{x}) - N_i(\mathbf{x})\mathcal{H}^i(\mathbf{x})\right), \tag{1}$$

with the overdot representing partial differentiation with respect to t, and π^{ij} being the momenta conjugate to g_{ij}. The $\mathcal{H}$ and $\mathcal{H}^i$ represent vanishing constraint functions. This allows us to view the theory as describing the evolving 3-geometry (and corresponding momenta) on a hypersurface, but with a Hamiltonian that is constrained to *vanish* (Arnowitt et al. [1964] and Braham [1995]),

$$\mathbf{H} = \int d^3x \left(N(\mathbf{x})\mathcal{H}(\mathbf{x}) + N_i(\mathbf{x})\mathcal{H}^i(\mathbf{x})\right) = 0. \tag{2}$$

The corresponding quantum equations describing the system are then obtained using the usual canonical quantization rules for replacing phase-space variables with operators,

$$g_{ij} \to \hat{g}_{ij} = g_{ij}\times, \ \pi^{ij} \to \hat{\pi}^{ij} = -i\frac{\delta}{\delta g_{ij}}, \tag{3}$$

operating on some wavefunction $\psi[g_{ij}]$. The vanishing of the Hamiltonian leads us to

$$\hat{\mathcal{H}}(\mathbf{x})\psi = 0, \ \hat{\mathcal{H}}^i(\mathbf{x})\psi = 0. \tag{4}$$

These quantum constraints mean that the wavefunction is *constant*, the observables are constants of the motion and we seem to have lost quantum dynamics! How can we hope to make any predictions in quantum gravity if we have no dynamics? This is a well-known problem in quantum gravity, and has been discussed by many authors, but a discussion of some of the various problems can be found in a paper by Kuchař presented during the *Fourth Canadian Conference on General Relativity and Relativistic Astrophysics* (Kuchař [1992]).

There is a second problem that arises in the usual ADM formalism, and that is the sheer *size* of the conventional numerical calculations needed to solve Equation (4). From a numerical standpoint, we are interested in calculating $\psi[g_{ij}]$, which is then regarded as being a function of the possible values that g_{ij} can takes on at all points on a given three-dimensional hypersurface. Let us imagine that we wish to solve this on a hypersurface divided into N cells, and that we wish to divide the ranges that each metric component g_{ij} takes into n cells. There are six components, so that means that the numerically approximated wavefunction is defined on n^{6N} variables! For even low values of n and N, we quickly arrive at an impossibly large numbers of cells, and move beyond the ability to computationally analyze the quantum theory. We therefore end up in a situation where a potentially valid quantum theory is rendered useless because it does not lend itself to actual realistic scientific study. To be fair, it should be pointed out that there are better ways than simple brute force to solve equations like Equation (4): There are Monte Carlo techniques which allow one to produce 'probable' field configurations, and sketch out the structure of ψ. One such method is the dynamical triangulation algorithm discussed by Kristin Schleich and Donald Witt in this book. Unfortunately, it is hard to see how one can construct the observables purely from such an approach, and thus something else is needed to provide a valid quantum theory with an equally valid interpretation. Luckily, there may be potentially valid formulations of quantum gravity that avoid this problem.

It is possible to construct an alternative formulation of geometrodynamics that both produces classical and quantum dynamics, and tells us the correct quantum observables, at least for model theories. This is the so-called *hypertime* approach, pioneered by Isham and Kuchař (Kuchař [1972] and Isham and Kuchař [1985a] and [1985b]) and with recent small contributions by myself (Braham [1993], [1994] and [1995]). In this approach, we view the Hamiltonian given by Equation (2) as representing evolution with respect to other variables, constructed out of the gravitational and matter phase-space variables themselves, instead of with respect to the simple label t. These latter are called *embedding variables* and they can be considered as representing the location of a given spacelike hypersurface in an embedding into a surrounding spacetime.

The method starts by finding a canonical transformation in which we can divide the standard local gravitational phase-space variables into three classes (Kuchař [1972] and Braham [1995])

$$g_{ij}, \pi^{ij} \rightarrow T^{\mu}, \Pi_{\mu}, Y_{\mu} \equiv \{g^A, \pi_A\}, \tag{5}$$

where μ ranges from one to four and $A \in \{1, 2\}$. g^A and π_A represent the 'true' dynamical variables of the theory discussed above (which can easily be extended to include matter degrees of freedom without changing our results (Braham [1995]). T^{μ} represent the corresponding internal embedding coordinates. The latter describe the location of the hypersurface in spacetime, commonly dubbed the hypertime. Π_{μ} are then the corresponding momenta (i.e., the energy-momentum densities). The approach is then to simply write down a solution, for Π_{Λ}, to the constraints

$$\Pi_{\Lambda} = -P_{\Lambda}[T^{\Sigma}, Y_{\mu}], \tag{6}$$

where Λ represents an index over position, tensor indices, and potential boundary terms (Braham [1995]). Functional differentiation of this gives us a functional form for the classical dynamics of gravity, *relative to hypertime*,

$$\frac{\delta Y_{\mu}}{\delta T^{\Lambda}} = [Y_{\mu}, P_{\Lambda}]_P, \tag{7}$$

where $[,]_P$ denotes a normal Poisson bracket. Solving these equations corresponds to a path integration in the space of all hypertime-labelled embeddings, and there is a corresponding integrability condition,

$$\frac{\delta P_\Lambda}{\delta T_\Sigma} - \frac{\delta P_\Sigma}{\delta T_\Lambda} + [P_\Lambda, P_\Sigma]_P = 0, \tag{8}$$

which is an extension of a condition discussed by Kuchař [1972]. This condition turns out to be automatically satisfied, even when boundary terms are used to correct for asymptotic behaviour (Braham [1995]). This tells us that, classically, the evolution is independent of how the spacetime is foliated by a choice of embedding. These equations can then be quantized to give the hypertime functional Schrödinger equation

$$i\frac{\delta}{\delta T^\Lambda}\psi[T^\Lambda, Y_\mu] = \hat{P}_\Lambda \psi[T^\Lambda, Y_\mu], \tag{9}$$

and corresponding quantum consistency condition,

$$[\hat{\Pi}_\Lambda + \hat{P}_\Lambda, \hat{\Pi}_\Sigma + \hat{P}_\Sigma] = 0. \tag{10}$$

These equations, both classical and quantum, represent the basic information to solve the problem of time. The first equation tells us that we can represent any set of physical variables at hypertime T_1^Λ by their 'initial' values at hypertime T_0^Λ. This is done by solving the *classical* Equation (7) to find a set of classical "observables",

$$Y_{(\mathrm{in})\mu}(T_1^\Lambda) \equiv Y_\mu(T_0^\Lambda), \tag{11}$$

considered as functionals $Y_{(\mathrm{in})_\mu}[Y_\mu(T^\Lambda)]$, and converting them into quantum observables (or *perennials*) by replacing $Y_\mu(T^\Lambda)$ by $\hat{Y}_\mu(T^\Lambda)$, to define the effective initial condition operators, for a hypertime T_0^Λ, derived from the operators at hypertime T_1^Λ. Torre has detailed this construction in his paper on the observables for cylindrically symmetric spacetime (Torre [1991]). These are then considered to be the operators representing the physical observables. This represents the dynamics in terms of constants of the motion, thus explaining why the original ADM formalism seems frozen in time. Equation (9) tells us how to get the wavefunction at hypertime T_1^Λ, which can then be interpreted using the observables we've constructed. The first-order form of everything makes computer implementation reasonably straightforward. There is one big problem, though, in that Equation (10), although true classically, is not automatically satisfied for an arbitrary ordering of the operators in $\hat{P}_\Lambda$. This point is discussed by Kuchař [1972]. There is also the second problem that operator ordering in the $Y_{(\mathrm{in})\mu}(T^\Lambda)$ operators will usually be non-trivial. This is discussed by Torre [1991], but also, in detail, in the conceptually similar paper on the Gaussian reference fluid interpretation of quantum geometrodynamics by Kuchař and Torre [1991], where they also discuss the problems of constructing an appropriate unitary evolution.

How can this be realized in practice? I have recently developed a hypertime formulation of spherically symmetric black holes and wormholes (Braham [1995]), coupled to matter, that expresses the hypertime evolution in terms of the effective mass $m(r)$ measured by an observer at a varying radius r from a hole with charge q,

$$\Pi^+ + \Pi^- = \frac{1}{4}\frac{\partial}{\partial r}\ln\left(\rho - 2m + \frac{q^2}{\rho}\right). \tag{12}$$

This reduces the calculation down to something like n^{2N} cells. This is a great improvement, but still beyond most computational techniques. However, in the case of collapsing shells of matter, important for black hole formation, and also important in inflationary models of the universe, it turns out that Equations (7) and (9) are simple enough to lend themselves to computer solution: The classical evolution corresponds to only two *ordinary differential equations*, and the quantum one is a simple partial differential equation in two variables, first order in the hypertime variables. This follows from the limited number of free dynamical variables, which are just those describing the shell position. This reduces the calculation down to n cells. We can therefore see that these equations can be solved on the computer to tell us the wavefunction of these simple models, as well as the observables for the complete quantum formalism. This will be the first fully quantum modelling of black hole collapse with actual calculation of the observables. We can hope that this will allow us to understand some crucial matters in quantum gravity, and also produce some new results. We can, if we wish, actually perform computer-based 'experiments' on quantum gravity.

It should be realized that this, by no means, solves the problem of time. It is only a starting point. The first big problem can be seen by looking at Equation (12), and then looking at the quantized solution for $m(r)$, which turns out to be given by a rather complicated (although physically understandable) expression involving exponentiated operators (Braham [1995]). This is not just a problem of the formalism, but can be seen in the similarly pretty, but operator-unfriendly, reduced ADM approaches of Unruh [1976] and Hajicek, Thomi and Isaac (Hajicek [1984] and Hajicek et al. [1984]). Obviously, the operator-ordering problem needed to satisfy Equation (10) is highly non-trivial. One benefit of the simple shell model that I am proposing for numerical study, is that the operator-ordering problem is simplified, just because we have a finite number of dynamical variables, the shell position and momentum, instead of an infinite number, as found in the corresponding field-theoretic problem in the spherical case.

A fascinating question comes into the interpretation of Equation (11). That equation tells us that the ontological structure of the theory is determined by our ability to *classically* evolve the equations of motion back in time. What if this is not, effectively, possible? There is evidence from the critical collapse behaviour found by Choptuik [1993] that gravity may be a non-integrable system. The latter would show up in the exact details of the reduction process, as discussed by Torre elsewhere in the proceedings. The effect of this would be that small perturbations of the configuration at time T_1^Λ, evolved back in time, could easily correspond to a large range of initial conditions at time T_0^Λ, thus leading to a large range of possible $\hat{Y}_{(\mathrm{in})\mu}$ operators. The latter implies that the evolution back in time would *not* be unique, once we take experimental limitations into account. Thus, in the hypertime approach, we could then end up with *more than one* possible interpretation of a given wavefunction at a given T^Λ. This would convert the loss of invertibility in the classical problem into a loss of *predictability* in the quantum problem. The information loss would be above and beyond the well-known potential loss caused by the Hawking-Unruh effect. This is an obviously speculative remark, but is something that numerical work can address, and shows us the kind of behaviour that may be studied in this approach.

In this article, I have described why certain approaches to quantum gravity may be effectively useless, due to an inability to study them using conventional computing technology, and how certain other approaches may avoid this problem. Of course, this does not make one theory physically incorrect: we may be unable to study truly correct theories, at least until more advanced computing technology

T_{ab} at a point may be regarded as holding in any coordinate domain of any point of the manifold.

Now, let X be an MC; it is immediate to see that all Killing Vectors (KV), proper Homothetic Vector fields (HVF) and proper Special Conformal Killing Vector fields (SCKV) are, naturally, MCs. However, the converse is not always true. Assuming that $X \in \mathcal{J}(M)$, the results below follow straightforwardly:

1. Assume that (M, g) is a spacetime such that $R \neq 0$, then X is an RC iff X is an SCKV (i.e., $\mathcal{L}_X g_{ab} = 2\lambda g_{ab}$ with $\lambda(x^c)$ such that $\lambda_{;ab} = 0$; see for instance Carot [1990], Coley and Tupper [1989] and Hall [1990]; the conformal factor then being $\lambda \equiv -\frac{1}{2}\mathcal{L}_X(\log R)$. If $R = 0$, then matter and Ricci collineations obviously coincide.
2. Assuming that f is a (smooth) scalar function defined on M, then $Y \equiv fX$ is also an MC iff either f is constant on M or X satisfies $T_{ab}X^b = 0$, in which case T_{ab} is necessarily degenerate (i.e., its rank is less than 4).
3. If T_{ab} is non-degenerate, then a matter collineation X will satisfy $\mathcal{L}_X T^a_{\ b} = 0$ iff X is a KVF.

When studying the Lie algebra $\mathcal{J}(M)$ of matter collineations, two cases arise naturally, according to whether T_{ab} is non-degenerate or degenerate. If T_{ab} is non-degenerate, then rank $T_{ab} = 4$ and $\mathcal{J}(M)$ is finite dimensional, its maximal dimension being 10, 9 being excluded by Fubini's theorem. If, on the other hand, T_{ab} is degenerate, then rank $T_{ab} < 4$ and we cannot guarantee the finite dimensionality of $\mathcal{J}(M)$. In fact, from remark (2) above it is easy to show that in many cases this algebra is infinite dimensional.

From now on, we will mainly deal with the degenerate case; therefore, it will be assumed that rank $T_{ab} < 4$, in which case, the only physically significant spacetimes are either dust fluids (perfect fluids with $p = 0$) or null fluids (radiation and null Einstein-Maxwell fields, Kramer et al. [1980]); the rank of T_{ab} being in both cases 1. As for the Ricci tensor, again only two cases of physical interest arise, both of them also being rank 1; namely: stiff fluids (perfect fluid with $p = \mu$, p and μ denoting pressure and energy density respectively) and, as before, null fluids.

2 Degenerate Case

As it was pointed out above, we need only consider rank 1 tensors (either energy-momentum or Ricci) and in all cases, the tensor under consideration can be simply written as the tensor product of a certain 1-form, say ϕ_a, with itself. Thus, for the energy-momentum tensor one has:

$$T_{ab} = \phi_a \phi_b \tag{2.1}$$

where $\phi_a \equiv \mu^{1/2} u_a$ in the case of dust (u_a denoting the four-velocity of the fluid, which is necessarily geodesic), and $\phi_a \equiv e^{\gamma/2} k_a$ in the case of a null fluid (where e^γ is the energy density and k_a is a geodesic null field). As for the Ricci tensor, one could also write $R_{ab} = \phi_a \phi_b$, but now, in the case of a stiff fluid $\phi_a \equiv (2\mu)^{1/2} u_a$.

It is now straightforward to show that $\mathcal{L}_X T_{ab} = 0$ if and only if $\mathcal{L}_X \phi_a = 0$, that is: the problem of finding the MC in a spacetime whose energy-momentum tensor is rank 1, can be formally reduced to that of finding those vector fields X that leave a given 1-form ϕ invariant. (The same comments obviously apply to the case of RCs).

In order to solve this problem let us consider the following (differentiable) distributions:

$$\begin{aligned}\mathcal{K} \quad &: \quad p \to \mathcal{K}_p \subseteq T_pM : Z_p \in \mathcal{K}_p \text{ iff } \quad < \mathcal{Z}, \phi > |_p = 0 \quad (\mathcal{K} : \text{Kernel of } \phi) \\ \mathcal{J} \quad &: \quad p \to \mathcal{J}_p \subseteq T_pM : X_p \in \mathcal{J}_p \text{ iff } \quad (\mathcal{L}_X \phi)|_p = 0 \\ \mathcal{F} \quad &\equiv \quad \mathcal{K} \cap \mathcal{J}\end{aligned}$$

where T_pM is the tangent space to M at a point p.

Obviously, $\mathcal{J}$ is integrable; and one can show quite easily that so is $\mathcal{F}$. Assuming now that *dim* $\mathcal{J}_p$ and *dim* $\mathcal{F}_p$ are constant over M one has the following different possibilities:

	I	II	III	IV	V	VI	VII
dim $\mathcal{K}_p$	3	3	3	3	3	3	3
dim $\mathcal{J}_p$	0	1	1	2	2	3	3
dim $\mathcal{F}_p$	0	0	1	1	2	2	3
dim $\mathcal{J}$	0	1	∞	∞	∞	∞	∞

Notice that, whenever *dim* $\mathcal{F}_p \neq 0$, the dimension of the Lie algebra of matter collineations $\mathcal{J}(M)$ is necessarily ∞; for given a vector field X in $\mathcal{F}$ (and therefore $X \in \mathcal{J}(M)$) and any function f on M, the vector field $Y \equiv fX$ is also in $\mathcal{F}$ and hence is another matter collineation (this corresponds to remark 2 in the previous section). In the case of dust, some interesting results may be derived for those matter collineations in $\mathcal{F}$. We collect them in the following

Proposition Let $Y \in \mathcal{F}$; therefore it satisfies $\mathcal{L}_Y \mu = \mathcal{L}_Y u_a = 0$ and $\omega_{ac} Y^c = 0$ where ω_{ac} is the vorticity of the fluid.

The proof is straightforward and the converse result also holds, for further details see Carot et al. [1994]. Now, since ω_{ac} is a skew tensor and since $\omega_{ac} u^c = 0$; its rank can only be 0 or 2; therefore if, at a given point p (and hence at every point, according to our hypothesis), dim $\mathcal{F}_p$ happens to be greater than or equal to 2, the fluid is then irrotational. If, on the other hand $\omega_{ac} \neq 0$, then $\mathcal{F}_p$ is at most of dimension 1. In view of all this, we can give the following algebraic recipe for finding $\mathcal{F}$:

1. Solve $\omega_{ac} Z^c = 0$ for Z.
2. Find $Z^\perp$, the projection of Z orthogonal to u (i.e., $(Z^\perp)^a = h^a_{.c} Z^c$ with $h^a_{.c} = \delta^a_{.c} + u^a u_c$).
3. Consider the algebraic system of equations given by $\mu_{,c} Y^c = 0$ and $(Z^\perp)^a = Y^a$, and solve it for Y^c.

Then, the dimension of $\mathcal{F}_p$ will be equal to the number of linearly independent solutions to the last equation; and any linear combination of them with smooth functions as coefficients, will be a vector field in $\mathcal{F}$.

As for a null fluid, some specific conditions can also be derived for those matter collineations in $\mathcal{F}$. In order to work them out, let us choose a real null tetrad adapted to the null eigenvector of the Ricci tensor; i.e., $\{k_a, l_a, x_a, y_a\}$ satisfying $-k_a l^a = x^a x_a = y^a y_a = 1$ and all other inner products zero. Then, with the above notation, we can state:

Proposition Let $Y \in \mathcal{F}$, therefore it satisfies $\omega x_{[a} y_{c]} Y^c = 0$ (ω being the vorticity of k^a.

As in the previous case the proof is straightforward. For further information see Carot et al. [1994].

3 Examples

Szekeres' Solutions. Using coordinates (t, r, y, z) and according to Bonnor et al. [1977], the general form of these metrics can be written as follows:

$$ds^2 = -dt^2 + e^{\lambda} dr^2 + e^{\sigma}(dx^2 + dy^2) \tag{3.1}$$

where λ and σ depend on all four coordinates. This family of solutions satisfies Einstein's Field Equations for a dust perfect fluid with 4-velocity $u^a = (1, 0, 0, 0)$, and therefore $\omega_{ac} = 0$. It is a well known fact that they do not admit KVF's (Bonnor et al. [1977]) and it can also be proven that they do not admit proper HVF, CKV, affine collineations or curvature collineations, however, they do admit infinitely many matter collineations. Following the procedure outlined in the previous section for finding those MCs $Y \in \mathcal{F}$ and since $\omega_{ac} = 0$; one sees that it suffices for Y to satisfy $Y^a u_a = 0$ and $\mu_{,a} Y^a = 0$. These conditions define at each point $p \in M$ a 2-plane (as it follows from the functional dependence of the energy density μ, see Bonnor et al. [1977] for further details). Hence dim $\mathcal{F}_p = 2$ so that there exist two linearly independent vector fields spanning $\mathcal{F}$ and therefore any combination of them with arbitrary (smooth) functions will also be a proper matter collineation in $\mathcal{F}$.

Null fluids. Using coordinates (z, v, u, w) we now consider a null Einstein-Maxwell field with the following metric (Carot et al. [1992]):

$$ds^2 = dz^2 + dv^2 + 2A\, dv\, dw + 2D\, du\, dw + B\, dw^2 \tag{3.2}$$

Here $A = A(v, w)$; $B = S(v, w)u + T(v, w)$ and $D = D(w)$. The energy-momentum corresponding to this metric is that of a null fluid with $k^a = \partial_u$, and then $\omega = 0$; thus, the energy density of the electromagnetic field becomes: $e^{\gamma} = -1/(2D^3)\{2D_{,w}A_{,v} - 2DA_{,vw} - SA_{,v} + DT_{,vv}\}$ and $Y^a = k^a$ becomes then a non-trivial matter (or Ricci) collineation in $\mathcal{F}$.

4 RCs in Spherically Symmetric Space-Times

Next, we give without proof a few results concerning RCs for non-degenerate Ricci tensors in the special case of spherically symmetric space-times. Proofs can be found in Carot et al. [1995]. The line element for this class of space-times can be written as (Kramer et al. [1980]): $ds^2 = -e^{\nu(t,r)} dt^2 + e^{\lambda(t,r)} dr^2 + r^2 d\Omega^2$, and since rank $R_{ab} = 4$, $\mathcal{R}(M)$ is finite dimensional, and one then has:

1. In the local coordinate chart in which the metric is written, a proper RC (i.e., non-homothetic, non curvature collineation) is of the form: $X = X^t(t, r)\partial_t + X^r(t, r)\partial_r$. This result can be easily generalized to a wider class of space-times, namely "Class B warped space-times" (Carot and da Costa [1993]) which include the spherically symmetric ones as a special case.
2. The maximal dimension of $\mathcal{R}(M)$ in one such space-time is 6, and it contains a 3-dimensional Lie subalgebra of isometries.
3. If in addition to spherical symmetry, one has staticity (i.e., ∂_t is also a KV), then the component of the Ricci tensor R_{rr} must be necessarily constant. As an example of this, consider the above line-element with Carot et al. [1995]: $\nu(r) = \dfrac{ar^4}{8r_0^2} + h \log r + k$ and $\lambda(r) = \nu(r) + 2 \log \dfrac{r}{r_0}$, where a, h, k and r_0 are constants. Then $X = -\dfrac{1}{a}\sqrt{2(h+1)} \log r\, \partial_t + \dfrac{t + t_0}{\sqrt{2(h+1)}} \partial_r$ is a proper RC, and so is $Y = \partial_r$.

References

Bonnor, W. B., Sulaiman, A. H. and Tomimura, N. [1977] *Szekeres' space-times have no Killing vectors*, Gen. Rel. Grav. **8**, 549–559.

Carot, J. [1990] *Exact solutions for space-times admitting non-null special conformal Killing vectors*, Gen. Rel. Grav. **22**, 1135–1145.

Carot, J. and da Costa, J. [1993] *On the geometry of warped spacetimes*, Class. Quantum Grav. **10**, 461–481.

Carot, J., da Costa, J. and Vaz, E. G. L. R. [1994] *Matter collineations: the inverse 'symmetry inheritance' problem*, J. Math. Phys. **35**, 4832–4838.

Carot, J., Mas, L., Rago, H. and da Costa, J. [1992] *Affine collineations in Einstein-Maxwell space-times*, Gen. Rel. Grav. **24**, 959–972.

Carot, J., Nunez, L. A. and Percoco, U. [1995] *Remarks on Ricci Collineations of Static Spherically Symmetric Spacetimes*, preprint U.L.A.

Coley, A. A. and Tupper, B. O. J. [1989] *Special conformal Killing vector space-times and symmetry inheritance*, J. Math. Phys. **30**, 2616–2625.

Hall, G. S. [1990] *Special conformal symmetries in General Relativity*, Gen. Rel. Grav. **22**, 203–215.

Hall, G. S. and Rendall, A. D. [1984] *Local and Global algebraic structures in General Relativity*, Int. J. Theor. Phys. **28**, 365–375.

Kramer, D., Stephani, H., MacCallum, M. A. H. and Herlt, E. [1980] *Exact Solutions of Einstein's Field Equations*, VEB Deutscher Verlag der Wissenschaften, Berlin.

Fields Institute Communications
Volume **15**, 1997

Topology and Causal Structure

Andrew Chamblin
Department of Applied Mathematics and Theoretical Physics
University of Cambridge
Cambridge CB3 9EW, England
h.a.chamblin@damtp.cam.ac.uk

Abstract. We investigate the causal structure of spacetimes (M, g) for which the metric g is singular on a set of points.

1 Metric Topology

Statements about the kinematical aspects of causal structure have long been of interest to those working in classical and quantum gravity (Bass and Witten [1957], Geroch [1967], Horowitz [1991] and Chamblin and Penrose [1992]). These studies have mostly focussed on the causal structure of spacetimes (M, g) for which the metric g is globally non-singular. However, the work of a number of authors (Hawking [1978], Ashtekar [1986], Witten [1988] and Sorkin [1990]) suggests that degenerate metrics may be a generic feature of any quantum theory of gravity which incorporates topology-changing processes. With this in mind, it is natural to investigate the causal properties of spacetimes with degenerate Lorentzian structures. In order to fully appreciate this issue, we need to first recall, in a detailed way, what a 'singularity' in the metric is, and how the 'index' of a singularity is associated with the topology of the metric.

Let us first assume that (M, g) is n-dimensional and *time-orientable* (Hawking and Ellis [1973]). It follows (Steenrod [1951]) that we can express the Lorentz metric, g^L_{ab}, in terms of some auxiliary Riemannian metric g^R_{ab} and some timelike vector field V, in the usual way:

$$g^L_{ab} = g^R_{ab} - 2V_a V_b \tag{1}$$

where in equation (1), g^R_{ab} is some (arbitrary) Riemannian metric on M, and V has been normalised with respect to g^R_{ab}, that is, $g^R_{ab}V^a V^b = 1$. Thus, there is a correspondence between vector fields and Lorentz metrics on M.

1991 *Mathematics Subject Classification.* Primary 83C45; Secondary 83C99.

The author would like to thank Dr. L. J. Alty and Dr. G. W. Gibbons for discussions which helped to formulate this work. Also eternal gratitude and thanks go to Jo Chamblin (Piglit) for loving support and help with the preparation of this paper.

This work was supported by NSF Graduate Fellowship No. RCD-9255644.

© 1997 American Mathematical Society

Now, a natural question is: Given two Lorentz metrics, g and g' (on M), when can g' be obtained from g by a continuous deformation? This question should be relevant in any Lorentzian sum-over-histories version of quantum gravity, since, if two metrics g and g' are related by some element ϕ of the identity-connected component of the diffeomorphism group, it follows that the two metrics must also be homotopic. In light of the above correspondence between timelike vector fields and Lorentz metrics, we see that this question reduces to the question: Given two vector fields, V and V' (on M), when can V' be obtained from V by continuous deformation? This question is well understood (Hector and Hirsch [1981]); it turns out that the answer depends upon the degrees of maps (defined by the vector fields V and V') from certain $(n-1)$-dimensional submanifolds of M, into S^{n-1}. In the context of General Relativity, the degree of such a map is called the 'kink number' (see Finkelstein and Misner [1959] and Gibbons and Hawking [1992]), and one often writes $\mathrm{kink}(\Sigma; V)$ to denote the degree of the map from Σ to S^{n-1} (for Σ some $(n-1)$-manifold). We adopt the usual definition that $\mathrm{kink}(\Sigma; V) = \mathrm{kink}(\Sigma; g)$, where V is any timelike vector relative to g. We now need to define precisely what we mean by a 'singularity'.

In this paper, we shall only consider the situation where the metric g_L fails to be invertible on a collection of points, $p_i \in M$; we shall refer to such a point where the metric is not invertible as a *singularity*. We should point out that this sort of singularity is highly non-generic, since in general the metric will fail to be invertible on higher-dimensional hypersurfaces in the spacetime. We adopt this definition of a singularity purely as a simplifying assumption. Using equation (1), we see that a metric singularity is thus present at any point $p \in M$ at which the timelike vector field, V, becomes singular. The topological nature of vector field singularities is well understood (Hector and Hirsch [1981]). To each singularity, p, of a vector field V on M, one assigns an *index*, defined as follows:

Let $S^{n-1}_\varepsilon(p)$ be a tiny $(n-1)$-dimensional sphere (of radius ε) at p. Since the vector field V is assumed to vanish only at p (and not on a neighbourhood of p), V is non-vanishing on $S^{n-1}_\varepsilon(p)$. V thus defines a map from $S^{n-1}_\varepsilon(p)$ to S^{n-1}, and we define the *index of* V *at* p (denoted $\mathrm{ind}(V,p)$) by the formula

$$\mathrm{ind}(V,p) = 1 - \mathrm{kink}(S^{n-1}_\varepsilon(p); V). \tag{2}$$

In light of equation (1), we therefore define the index of a Lorentz metric, g_L, at a metric singularity p by the relation

$$\mathrm{ind}(g_L,p) = 1 - \mathrm{kink}(S^{n-1}_\varepsilon(p); g_L). \tag{3}$$

2 Causality and Singularities

To begin with, we recall the precise statement of Geroch's result:

Theorem 1 (Geroch [1967]) *Let (M, g_L) be a compact, time-orientable spacetime with globally non-singular Lorentz metric g_L. Suppose that the boundary of M is equal to the disjoint union of two compact spacelike three-manifolds, $\partial M \cong \Sigma_1 \cup \Sigma_2$, and that there are no closed timelike curves in (M, g_L). Then Σ_1 and Σ_2 are diffeomorphic, and M is topologically of the form $\Sigma_1 \times [0,1]$.*

Thus, if Σ_1 and Σ_2 are not even *homeomorphic*, there must exist CTCs in (M, g_L) (we shall assume time-orientability throughout this paper). If we allow the metric g_L to be singular, then the above result will not in general hold. As Geroch noted in his original proof, the timelike flow (induced by g_L) will induce a diffeomorphism from Σ_1 to Σ_2 *unless* the timelike curves close back on themselves *or* they hit singularities.

A natural question, then, is whether or not it is always possible to arrange for all the CTCs to hit singularities. Given a compact causality-violating spacetime (M, g_L) (with or without boundary), such that g_L is singular at some finite collection of points $\{p_1, p_2, ...p_n\}$, can we always 'deform' all of the CTCs so that they 'hit' the singularities? The answer is yes. By slightly modifying the arguments of Chamblin and Penrose [1992], we can prove

Theorem 2 *Let M be a smooth compact n-manifold, without boundary, and let g_L be some Lorentz metric on M such that g_L is singular on a finite collection of points $\{p_1, p_2, ...p_m\}$ in M and (M, g_L) is causality violating. Then there always exists a new metric, g'_L, which can be obtained by a continous deformation of the original metric g_L such that (M, g'_L) is causal.*

Proof Given a point, $p_i \in M$, at which g_L is singular, let $B^n_\varepsilon(p_i)$ denote the *closed* n-ball of radius ε about p_i and $S^{n-1}_\varepsilon(p_i) \cong \partial B^n_\varepsilon(p_i)$, as above. Let N be obtained from M by removing all of the $B^n_\varepsilon(p_i)$:

$$N \cong M - (B^n_\varepsilon(p_1) \cup B^n_\varepsilon(p_2) ... \cup B^n_\varepsilon(p_m)).$$

By construction, the spacetime (N, g_L) is non-singular and *open* (since the boundary, ∂N, does not lie in N; this is the main technical difference from the construction in Chamblin and Penrose [1992]). We wish to 'cut' all of the CTCs in (N, g_L) by deforming them onto ∂N, as outlined in Chamblin and Penrose [1992]. We need to briefly pass to $(\bar{N}, g_L)$, where $\bar{N}$ denotes the *closure* of N. $(\bar{N}, g_L)$ is then a *compact* spacetime with boundary ∂N. As in Chamblin and Penrose [1992], we can construct a continuous deformation of the metric g_L on $\bar{N}$ to a new metric g'_L, such that all of the CTCs are pushed onto the boundary, ∂N. We now pass back to (N, g'_L). We have pulled all of the CTCs arbitrarily close to the singular points.

Furthermore, since the deformation is continuous, the kink number on each $S^{n-1}_\varepsilon(p_i)$ is unchanged, that is, we have not altered the index of any singularity. We then continue this construction in the limit $\varepsilon \longrightarrow 0$, so that all CTCs are continuously forced onto the singular points. We have therefore constructed a continuous deformation, $g_L \longrightarrow g'_L$, such that (M, g'_L) is causal. □

Thus, given a singular spacetime with CTCs, we can always 'destroy' the CTCs using the singularities. Furthermore, by construction, the deformation does not move any of the singularities together.

Of course, one could take the view that *spacelike* topology change is too restrictive and that the topology of the metric should also be allowed to fluctuate. In the context of a 'Lorentzian' quantum gravity theory obtained by a path integral prescription, this would mean that instead of calculating the probability amplitude to pass from some initial *spacelike* three-manifold to some final *spacelike* three-manifold, one would instead seek an 'amplitude' describing the probability of transition 'from' a three-manifold Σ_1, such that the Lorentz metric g_L has kink number n_1 on Σ_1, 'to' a three-manifold Σ_2, such that the Lorentz metric g_L has kink number n_2 on Σ_2. In this way, the 'boundary data' (or asymptotic states) of such a theory would be purely topological. We stress this viewpoint (even though it goes against the spirit of canonical quantum gravity) since we feel it nonsensical to study strictly *spacelike* topology change via spacetimes which have metric singularities. To see why, let (M, g_L) be any spacetime with spacelike boundary ∂M and let g_L be singular at p. We then obtain a spacetime (M, g'_L), with a *kink* in g'_L on ∂M, simply be deforming the singular point p *across* ∂M. The point is, the continuous deformation could perhaps represent the 'flow of time'. The creation/annihilation of *singularities* corresponds to creation/annihilation of *kinks* (note that in the above example, one would have $\text{kink}(\partial M; g'_L) = \text{ind}(g_L, p)$). The two concepts are dual in this sense (see also Gibbons [1992] for ideas along these lines).

References

Ashtekar, A. [1986] *New Variables for Classical and Quantum Gravity*, Phys. Rev. Lett. **57**, 2244–2247.

Bass, R. W. and Witten, L. [1957] *Remark on Cosmological Models*, Rev. Mod. Phys. **29**, 452–453.

Chamblin, A. and Penrose, R. [1992] *Kinking and Causality*, Twistor Newsletter **34**, 13–18.

Finkelstein, D. and Misner, C. [1959] *Some New Conservation Laws*, Annals of Physics **6**, 230–243.

Geroch, R. P. [1967] *Topology in General Relativity*, J. Math. Phys. **8**, 782–786.

Gibbons, G. W. [1992] *Topology Change: Kinks and Spinors*, Proc. of Intl. Workshop of Theor. Phys., 6th Session, Erice, Sicily, 1992 (eds.), pp. 13–18.

Gibbons, G. W. and Hawking, S. W. [1992] *Kinks and Topology Change*, Phys. Rev. Lett. **69**, 1719–1721.

Hawking, S. W. [1978] *Spacetime Foam*, Nucl. Phys. B **144**, 349–362.

Hawking, S. W. and Ellis, G. [1973] *The Large-Scale Structure of Spacetime*, Cambridge University Press, Cambridge.

Hector, G. and Hirsch, U. [1981] *Introduction to the Geometry of Foliations*, Aspects of Mathematics, Friedr. Vieweg & Sohn Verlagsgesellschaft mbH.

Horowitz, G. T. [1991] *Topology Change in Classical and Quantum Gravity*, Class. Quant. Grav. **8**, 587–602.

Sorkin, R. [1990] *Consequences of Spacetime Topology*, Proc. of 3rd Canadian Conf. on Gen. Rel. and Rel. Astrophys., May, 1989 (Coley, A., Cooperstock, F. and Tupper, B., eds.). pp. 39–59.

Steenrod, N. [1951] *Topology of Fibre Bundles*, Princeton Univ. Press, Princeton.

Witten, E. [1988] 2 + 1 *Dimensional Gravity as an Exactly Soluble System*, Nucl. Phys. B **311**, 46–78.

Fields Institute Communications
Volume **15**, 1997

Charged Dilaton Black Holes with Unusual Asymptotics

Kevin C. K. Chan
Department of Physics
University of Waterloo
Waterloo, Ontario, Canada N2L 3G1
kckchan@astro.uwaterloo.ca

Abstract. A new class of black hole solutions in Einstein-Maxwell-dilaton gravity in $3+1$ dimensions is briefly reviewed. These solutions have regular horizons and a singularity only at the origin. Their asymptotic behavior is neither asymptotically flat nor (anti-) de Sitter. Similar solutions exist for certain Liouville-type potentials for the dilaton.

1 Introduction

It is always interesting and important to investigate how the properties of black holes are modified when a dilaton is present. In this paper, we briefly review (see (Chan et al. [1995] for details) a new class of electrically charged static spherically symmetric (SSS) black hole solutions to $(3+1)$-dimensional dilaton theories of gravity. We consider the following action in which gravity is coupled to a dilaton ϕ and Maxwell field $F_{\mu\nu}$

$$S = \int d^4x\sqrt{-g}(R - 2(\nabla\phi)^2 - V(\phi) - e^{-2a\phi}F^2). \tag{1.1}$$

Three special cases are considered: (i) $V(\phi) = 0$, (ii) $V(\phi) = 2\Lambda e^{2b\phi}$ and (iii) $V(\phi) = 2\Lambda_1 e^{2b_1\phi} + 2\Lambda_2 e^{2b_2\phi}$. In Case (i), $a = 1$ corresponds to low energy string theory. Case (ii) corresponds to a Liouville-type gravity. It was shown in Poletti and Wiltshire [1994] that with the exception of a pure cosmological constant potential (i.e., $b = 0$), no asymptotically flat ($\Lambda = 0$) or asymptotically anti-de Sitter SSS solutions to the field equations associated with Case (ii) exist. The form of potential in Case (iii) was previously investigated by a number of authors (Ozer and Taha [1992] and Easther [1993]) in the context of FRW scalar field cosmologies.

The solutions we derived are not vacuum solutions since the action Equation (1.1) contains a static "dilaton fluid" whose energy-momentum tensor is nowhere vanishing. Here we adopt special dilaton fluid models in which the metric approaches neither the Minkowski nor anti-de Sitter metrics at spatial infinity.

1991 *Mathematics Subject Classification.* 83C57.

© 1997 American Mathematical Society

of possible causal structures depending on the values of a and Λ. We leave this investigation as an exercise for readers.

Finally, we consider the sum of two Liouville-type terms in Case (iii). If $b_1 = b_2$, then this case reduces to Case (ii), so we will not repeat these solutions. Requiring $b_1 \neq b_2$ again gives three classes of solutions. The first class is identical to Equation (2.7), except for the modifications

$$U(r) = U_1(r) - \frac{\Lambda_2(1+a^2)^2 e^{2a\phi_0}}{(3-a^2)} r^{\frac{2}{1+a^2}}, \tag{2.15}$$

with $b_2 = a$. The causal structures associated with this case can be obtained as in the previous cases. The general form of the temperature is

$$T = \frac{r_O^{-\frac{3+a^2}{1+a^2}}}{2\pi\gamma^2}\left(r_O{}^2 - M(a^2-3)r_O - 2\frac{Q^2 e^{2a\phi_0}}{\gamma^2}\right), \tag{2.16}$$

where r_O denotes the location of the outer horizon when it exists. In general r_O must be solved for numerically, although closed form solutions can be obtained for special values of a.

The second class of solutions for Case (iii) is identical to Equation (2.10), except for the modifications

$$U(r) = U_2(r) + \frac{\Lambda_2(1+a^2)^2 e^{-\frac{2\phi_0}{a}}}{a^2(1-3a^2)} r^{\frac{2a^2}{1+a^2}}, \tag{2.17}$$

with $b_2 = -\dfrac{1}{a}$. For arbitrary a, the extra power of r in Equation (2.17) makes it impossible to find explicitly the positions of the horizons, but we can determine the number and type of horizons by graphical methods. The temperature in this case is given by

$$T = \frac{r_O^{-\frac{2a^2}{1+a^2}}}{2\pi\gamma^2}\left[\left(\frac{2Q^2 e^{2a\phi_0}}{\gamma^2} - 1\right) r_O - \frac{3a^2-1}{a^2} M\right], \tag{2.18}$$

where again r_O denotes the location of the outer horizon (when it exists).

A third class of solutions exists that has no counterpart in the solutions with the Liouville potential in Case (ii). These solutions exist for $0 < N < 1$ with $N \neq \frac{1}{2}$, and satisfy

$$U(r) = \frac{1}{(2N-1)\gamma^2} r^{2(1-N)} - r_0 r^{1-2N} + \frac{2Q^2 e^{2a\phi_0} r^{2(1-2N\pm 2\tilde{a})}}{(1-2N\pm 2\tilde{a})(1-3N\pm\tilde{a})\gamma^4}, \tag{2.19}$$

with

$$\tilde{a} = a\sqrt{N(1-N)},\ b_1 = \mp\frac{N}{\sqrt{N(1-N)}},\ b_2 = a \mp \frac{2N}{\sqrt{N(1-N)}},$$

$$\Lambda_1 = \frac{(N-1)}{(2N-1)\gamma^2} e^{\pm\frac{2N\phi_0}{\sqrt{N(1-N)}}},\ \Lambda_2 = -\frac{Q^2(1-N\pm\tilde{a})}{(1-3N\pm\tilde{a})\gamma^4} e^{\pm\frac{4N\phi_0}{\sqrt{N(1-N)}}} \tag{2.20}$$

$$\text{and } \phi_1 = \pm\sqrt{N(1-N)}.$$

r_0 is a constant related to the mass. We leave the derivation of the temperature and causal structures as an exercise for readers. Finally note that the solutions given in Cases (i)-(iii) can be generalized to dimensions $n > 4$ (see Chan et al. [1995]).

3 Summary

To sum up, we have obtained several one-parameter families of non-asymptotically flat ((anti-) de Sitter) electrically charged SSS black hole solutions for the action Equation (1.1) with a zero potential, Liouville potential and a two-term exponential potential. We briefly review their varies properties including their temperatures. We close by commenting on further possible extensions of our work. It would be interesting to add angular momentum to the present solutions. Generalization of all the solutions derived in this paper to arbitrary $n > 4$ with both electric and magnetic charge present is another problem of interest. One can also attempt to construct any asymptotically flat or (anti-) de Sitter black holes for the two-term exponential potential.

References

Brown, J. D., Creighton, J. and Mann, R. B. [1994] *Temperature, Energy, and Heat Capacity of Asymptotically Anti-de Sitter Black Holes*, Phys. Rev. D **50**, 6394–6403.

Brown, J. D. and York, J. W. [1993] *Quasilocal Energy and Conserved Charges Derived from the Gravitational Action*, Phys. Rev. D **47**, 1407–1419.

Chan, K. C. K., Horne, J. H. and Mann, R. B. [1995] *Charged Dilaton Black Holes with Unusual Asymptotics*, Nucl. Phys. B **447**, 441–461.

Chan, K. C. K. and Mann, R. B. [1994] *Static Charged Black Holes in* $(2+1)$*-dimensional Dilaton Gravity*, Phys. Rev. D **50**, 6385–6393.

Easther, R. [1993] *Exact Superstring Motivated Cosmological Models*, Class. Quant. Grav. **10**, 2203–2215.

Ozer, M. and Taha, M. O. [1992] *Exact Solutions in String-Motivated Scalar-field Cosmology*, Phys. Rev. D **45**, R997–R999.

Poletti, S. J. and Wiltshire, D. L. [1994] *Global Properties of Static Spherically Symmetric Charged Dilaton Spacetimes with a Liouville Potential*, Phys. Rev. D **50**, 7260–7270.

——— [1995] *erratum*, Phys. Rev. D **52**, 3753–3754.

Fields Institute Communications
Volume **15**, 1997

General Relativity Calculations in Maple

Kenneth C-K Chu, Cesar E. Farell and Greg J. Fee
Department of Applied Mathematics
University of Waterloo
Waterloo, Ontario, Canada N2L 3G1

R. G. McLenaghan
Department of Applied Mathematics
University of Waterloo
Waterloo, Ontario, Canada N2L 3G1
rgmclena@daisy.uwaterloo.ca

Abstract. A new component tensor package for Maple is described. Some of the functions in the package are illustrated by calculations with the Kerr-Newman and Schwarzschild solutions.

1 Introduction

A new package for performing calculations in general relativity and differential geometry in Maple is described. The package includes the functions for the computation of the connection and curvature including the NP spin coefficients and curvature components and the determination of Petrov type that are currently available in Maple V Release 3. It also includes new functions for performing the following component calculations: (i) tensor addition, multiplication, contraction, symmetrizations, and raising and lowering indices; (ii) complex conjugation of quantities; (iii) tensor transformations; (iv) covariant, exterior, Lie, Pfaffian, and ordinary derivatives; (v) Lie bracket; (vi) Killing's equations; (vii) geodesic equations; (viii) the scalar invariants of the Riemann tensor. The package will be available in the Library of the next version of Maple. Descriptions of the functions in the package will be available on line. (See also McLenaghan [1994].) The package is *distinct* from the package described in Kavian, McLenaghan and Geddes [1997], which also performs *indicial* tensor calculations. In the following sections we illustrate the use of some functions of the package with calculations for the Kerr-Newman (KN) metric by means of a Maple worksheet.

2 Computations in the Natural Basis

The calculation is begun with the following commands:

```
> with(tensor);
```

1991 *Mathematics Subject Classification.* Primary 83-08; Secondary 83C15.

© 1997 American Mathematical Society

[*Christoffel1*, *Christoffel2*, *Einstein*, *display_allGR*, *displayGR*, *tensorsGR*, *Jacobian*, *Killing_eqns*, *Levi_Civita*, *Lie_diff*, *Ricci*, *Ricciscalar*, *Riemann*, *RiemannF*, *Weyl*, *act*, *antisymmetrize*, *change_basis*, *commutator*, *compare*, *conj*, *connexF*, *contract*, *convertNP*, *cov_diff*, *create*, *d1metric*, *d2metric*, *directional_diff*, *dual*, *entermetric*, *exterior_diff*, *exterior_prod*, *frame*, *geodesic_eqns*, *get_char*, *get_compts*, *get_rank*, *init*, *invars*, *invert*, *lin_com*, *lower*, *npcurve*, *npspin*, *partial_diff*, *permute_indices*, *petrov*, *prod*, *raise*, *symmetrize*, *transform*]

```
> coord:= [t,r,theta,phi];
```

$$coord := [t, r, \theta, \phi]$$

```
> Delta:=r^2 - 2*m*r + a^2 + Q^2: Sigma:=r^2 + a^2*cos(theta)^2:
> g_components:= array (1..4, 1..4, symmetric,
[(1,1)=(Delta-a^2*sin(theta)^2)/Sigma, (1,2)=0, (1,3)=0,
(1,4)=a*sin(theta)^2*(r^2+a^2-Delta)/Sigma,
(2,2)=-Sigma/Delta, (2,3)=0, (2,4)=0, (3,3)=-Sigma, (3,4)=0,
(4,4)=((a^2*sin(theta)^2*Delta)-(r^2+a^2)^2)*sin(theta)^2/(Sigma)]):
> g:= create ([-1,-1], eval(g_components)):
```

The coordinates are stored as a Maple list. The metric tensor is stored using a Maple table, known as a `tensor_type`, which contains two fields: `compts`, an array of components, and `index_char`, a list representing the covariant / contravariant character of the tensor component indices (positive 1's represent contravariant indices, negative 1's represent covariant indices). The `create` utility is used to define a `tensor_type`.

3 Computation and Simplification of Results

Each routine of the tensor package uses its own simplification routine. To customize the way simplification is done throughout the computation, the user can directly redefine these routines as needed. Initially, these routines are all set equal to a generic simplifier for the entire package called `tensor/simp`, which is itself initialized to use the Maple routine `normal`. Here we customize the simplfication to something appropriate for the KN metric:

```
> `tensor/simp` := proc(a) local n,d;
n:=factor(simplify(a,'trig')); d:=denom(n); n:=numer(n);
collect(n,Q,proc(b) frontend(factor,[b],[{`+`,`*`,`^`},{}]); end)/
collect(d,Q,proc(b) frontend(factor,[b],[{`+`,`*`,`^`},{}]); end);
end:
```

The routine `tensorsGR` computes the following quantities given the coordinates and covariant metric tensor components: contravariant metric tensor components, determinant of the metric components, Christoffel symbols of the first and second kind, covariant Riemann tensor components, covariant Ricci tensor components, Ricci scalar, covariant Einstein tensor components, and covariant Weyl tensor components, denoted by `contra_g`, `det_g`, `C1`, `C2`, `Rm`, `Rc`, `R`, `G`, and `C`, respectively.

```
> tensorsGR(coord, g, contra_g, det_g, C1, C2, Rm, Rc, R, G, C):
```

The `displayGR` routine displays the non-zero Einstein tensor components:

```
> displayGR('Einstein',G);
```

The Einstein Tensor non-zero components:

$$G11 = \frac{-Q^4 + \left(a^2\,\cos(\theta)^2 - 2\,a^2 + 2\,m\,r - r^2\right)\,Q^2}{\left(r^2 + a^2\,\cos(\theta)^2\right)^3}$$

$$G14 = (-a\,(\cos(\theta) - 1)(\cos(\theta) + 1)\,Q^4$$
$$-2a(\cos(\theta) - 1)(\cos(\theta) + 1)\,(a^2 + r^2 - m\,r)\,Q^2)/\left(r^2 + a^2\,\cos(\theta)^2\right)^3$$

$$G22 = \frac{Q^2}{(r^2 + a^2\,\cos(\theta)^2)\,Q^2 + (r^2 + a^2\cos(\theta)^2)\,(r^2 - 2mr + a^2)}$$

$$G33 = -\frac{Q^2}{r^2 + a^2\cos(\theta)^2}$$

$$G44 = \big(- (\cos(\theta) - 1)^2\,(\cos(\theta) + 1)^2\,a^2\,Q^4 - (\cos(\theta) - 1)(\cos(\theta) + 1)$$
$$\big(- 2a^2\,\cos(\theta)^2 mr - 3\,r^2\,a^2 + r^2\,a^2\,\cos(\theta)^2$$
$$+a^4\,\cos(\theta)^2 - 2a^4 + 2a^2\,m\,r - r^4\big)Q^2/\left(r^2 + a^2\,\cos(\theta)^2\right)^3$$

$$\mathit{character} \;:\; [-1,\,-1]$$

4 Computations Using a Rigid Moving Frame

In the case of the general rigid moving frame, the routines `connexF` and `RiemannF` compute the connection coefficients and the curvature tensor components respectively (McLenaghan [1994]). In the special case of a complex null frame on spacetime, the routines `npspin` and `npcurve` compute the Newman and Penrose (NP) spin coefficients and curvature components respectively (McLenaghan [1994]).

As an example of the latter routines, we perform a computation in the standard null co-frame for the KN metric (Wald [1984]).

```
> e:= (1/2)^(1/2):   T:= Delta^(1/2):   p:= Sigma^(1/2):
> h_compts:=array(1..4,1..4,
   [[e*T/p, -e*p/T, 0, -e*T/p*a*sin(theta)^2],
   [e*T/p,    e*p/T, 0, -e*T/p*a*sin(theta)^2],
   [I*e*sin(theta)/p*a, 0, -e*p, -I*e*sin(theta)/p*(r^2+a^2)],
   [-I*e*sin(theta)/p*a, 0, -e*p, I*e*sin(theta)/p*(r^2+a^2)]]):
> h := create([1,-1],eval(h_compts)): hinv:= invert(h,'det_h'):
```

The NP spin coefficients and curvature components are computed by:

```
> NPspin := npspin(coord,h,'Gdet','c_info'):
> NPcurve := npcurve(NPspin,c_info):
```

The independent non-zero spin coefficients are:

```
> NPspin[epsilon]; NPspin[tau]; NPspin[alpha]; NPspin[rho];
```

$$\frac{1}{4}\,\frac{\sqrt{2}\,(\,-r\,\cos(\,\theta\,)\,a - I\,m\,r + m\,\cos(\,\theta\,)\,a + I\,a^2 + I\,Q^2\,)}{\sqrt{r^2 + a^2\,\cos(\,\theta\,)^2}\,\sqrt{r^2 - 2\,m\,r + a^2 + Q^2}\,(\,I\,r + \cos(\,\theta\,)\,a\,)}$$

$$\frac{1}{2}\,\frac{\sqrt{2}\,a\,\left(a\,\cos(\,\theta\,)^3 - \cos(\,\theta\,)\,a - I\,r\,\cos(\,\theta\,)^2 + I\,r\right)}{(r^2 + a^2\,\cos(\,\theta\,)^2)^{3/2}\,\sin(\,\theta\,)}$$

$$\frac{1}{4}\,\frac{\sqrt{2}\,\left(\cos(\,\theta\,)\,r^2 + \cos(\,\theta\,)\,a^2 + I\,a\,r\,\cos(\,\theta\,)^2 - I\,a\,r\right)}{(r^2 + a^2\,\cos(\,\theta\,)^2)^{3/2}\,\sin(\,\theta\,)}$$

$$\frac{1}{2}\,\frac{I\,\sqrt{2}\,\sqrt{r^2 - 2\,m\,r + a^2 + Q^2}}{\sqrt{r^2 + a^2\,\cos(\,\theta\,)^2}\,(\,I\,r + \cos(\,\theta\,)\,a\,)}$$

The independent non-zero curvature components are:

```
> NPcurve[Phi][1,1];   NPcurve[Psi][2];
```

$$\frac{1}{2}\frac{Q^2}{(r^2+a^2\cos(\theta)^2)^2}$$
$$\frac{-Q^2+I(\cos(\theta)a-Ir)m}{(\cos(\theta)a-Ir)(Ir+\cos(\theta)a)^3}$$

5 Petrov Classification and Algebraic Invariants

The Petrov type of the KN metric is determined by:

```
> petrov(NPcurve[Psi]);
```

$$type\_D$$

The package is also able to compute the algebraic invariants of the curvature tensor, some of which are (Carminati and McLenaghan [1991]):

```
> `tensor/invars/simp`:= proc(a) a; end:
> invar_R1 := invars('r1', NPcurve);
```

$$invar\_R1 := \frac{Q^4}{(r^2+a^2\cos(\theta)^2)^4}$$

```
> invar_W1 := invars('w1', NPcurve);
```

$$invar\_W1 := 6\,\frac{(-Q^2+I(\cos(\theta)a-Ir)m)^2}{(\cos(\theta)a-Ir)^2(Ir+\cos(\theta)a)^6}$$

```
> invar_W2 := invars('w2',NPcurve): invar_M1 := invars('m1',NPcurve);
```

$$invar\_M1 := -2\,\frac{(-Q^2+I(\cos(\theta)a-Ir)m)Q^4}{(\cos(\theta)a-Ir)(Ir+\cos(\theta)a)^3(r^2+a^2\cos(\theta)^2)^4}$$

```
> invar_M2 := invars('m2',NPcurve):
```

Since the principal null directions of the Weyl tensor coincide with those of the Maxwell field tensor, the invariants must satisfy the alignment condition $w_1m_2 - 2w_2m_1 = 0$ (Carminati and McLenaghan [1991]):

```
invar_W1 * invar_M2 - 2*invar_W2 * invar_M1;
```

$$0$$

6 Tensor Operations

The package also provides facilities for the following general tensor operations: linear combination, product, contraction, symmetrization, raising and lowering of indices, covariant, partial, and Lie derivatives, exterior product and derivative, dual, coordinate transformation, change of basis, and various other tools.

We illustrate some of these functions by computing the Killing tensor field of the KN metric (Wald [1984]). Define two principal contravariant null vectors by:

```
> l:=create([1],array(1..4,[(r^2+a^2)/Delta, 1, 0, a/Delta])):
> n:=create([1],array(1..4,[(r^2+a^2)/2/Sigma, -Delta/2/Sigma;
0, a/2/Sigma])):
```

Obtain the covariant components of these vectors:

```
> l_lower:=lower(g,l,1): n_lower:=lower(g,n,1):
```

The Killing 'tensor' is obtained by symmetrizing the product of the covariant null vectors obtained above and taking the appropriate linear combination with the metric tensor:

```
> l_n:=prod(l_lower,n_lower): l_n_sym:=symmetrize(l_n, [1,2]):
> K:= lin_com(2*Sigma,l_n_sym, -r^2,g):
```

Finally, we verify that `K` is a Killing tensor by computing the left-hand-side of $K_{(ab;c)} = 0$. Instead of displaying all of the components of the result, we use the `display` option of the `act` facility (which is generally used to apply a procedure to each component of a `tensor_type`):

```
> LHS_Keqns:= Killing_eqns(K,coord,C2): act(display,LHS_Keqns);
```

No non-zero independent components.

Since all components are zero, we have verified that `K` is indeed a Killing tensor.

References

Carminati, J. and McLenaghan, R. G. [1991] *Algebraic invariants of the Riemann tensor in a four-dimensional Lorentzian space*, J. Math. Phys. **11**, 3135–3140.

Kavian, M., McLenaghan, R. G. and Geddes, K. O. [1997] *MAPLETENSOR: A new system for performing indicial and component tensor calculations by computer*, The 6th Canadian Conference on General Relativity and Relativistic Astrophysics, University of New Brunswick, 1995, The Fields Institute Communications Series (Braham, S. P., Gegenberg, J. D. and McKellar, R. J., eds.), American Mathematical Society, Providence, RI.

McLenaghan, R. G. [1994] *Maple applications to general relativity*, Algebraic computing in general relativity, Lecture notes from the 1st Brazilian School on Computer Algebra, Vol. 2 (Roboucas, M. J. and Roque, W., eds.), Oxford University Press, Oxford.

Wald, R. M. [1984] *General Relativity*, University of Chicago Press, Chicago, p. 313.

Fields Institute Communications
Volume **15**, 1997

Vacuum Solutions of Gauss-Bonnet Theory and Induced Cosmological Models

Alan A. Coley
Department of Mathematics, Statistics and Computing Science
Dalhousie University
Halifax, NS, Canada B3H 3J5
aac@cs.dal.ca

Abstract. The exact, cosmological (flat) solution of the vacuum field equations for the five-dimensional theory of gravity based on the Einstein-Hilbert action plus the Gauss-Bonnet quadratic curvature invariant is obtained. The solution has the following properties: (i) there is an initial singularity, (ii) the three-expansion decreases monotonically, and (iii) the fifth-dimension compactifies. This solution can then be (formally) interpreted as a solution of Einstein's four-dimensional theory with a non-zero cosmological (perfect) fluid material source with a 'non-linear' equation of state (and hence the four-dimensional matter is geometric in origin). This model asymptotically approaches the familiar (four-dimensional) Friedmann-Robertson-Walker radiation model at later times.

There have been many attempts to construct a unified field theory based on a multidimensional space-time and it is generally believed that higher dimensions would play a significant role in the early universe. In addition, theories of gravity with non-linear gravitational contributions to the linear Einstein-Hilbert action are of great interest, motivated primarily by quantum gravity (but the theories are also of interest classically (Stelle [1977])). In particular the Gauss-Bonnet Theory (GBT), in which the D-dimensional action is given by

$$S = \int \sqrt{-g} d^D x \{R + \alpha[R^2 - 4R_{ab}R^{ab} + R_{abcd}R^{abcd}]\}, \tag{1}$$

and so includes quadratic curvature in the Gauss-Bonnet combination (where α is a constant), is motivated by the low energy expansion of supersymmetric string theory (Green et al. [1987]), the fact that such a theory does not have ghosts (Zwiebach [1985]), and that in higher-dimensions the resulting field equations are of second order only.

1991 *Mathematics Subject Classification.* Primary 83E15; Secondary 83F05.

© 1997 American Mathematical Society

In this paper we shall study vacuum solutions of the $5D$ (five-dimensional–$D = 5$) GBT with the cosmological metric

$$ds^2 = -dt^2 + R^2(t)(dx_1{}^2 + dx_2{}^2 + dx_3{}^2) + S^2(t)dx_4{}^2, \tag{2}$$

which is the $5D$ generalization of the $4D$ (flat) Robertson-Walker metric. Varying the action (1) yields the following non-trivial vacuum field equations (Madore [1985]):

$$\frac{3\dot{R}^2}{R^2} = -\frac{3\dot{R}\dot{S}}{RS} - 12\alpha\left[\frac{\dot{R}^3}{R^3}\frac{\dot{S}}{S}\right], \tag{3}$$

$$\frac{2\ddot{R}}{R} + \frac{\dot{R}^2}{R^2} = -\frac{2\dot{R}\dot{S}}{RS} - \frac{\ddot{S}}{S} - 4\alpha\left[2\frac{\ddot{R}}{R}\frac{\dot{R}}{R}\frac{\dot{S}}{S} + \frac{\dot{R}^2}{R^2}\frac{\ddot{S}}{S}\right], \tag{4}$$

$$\frac{\ddot{R}}{R} + \frac{\dot{R}^2}{R^2} = -4\alpha\left[\frac{\ddot{R}}{R}\frac{\dot{R}^2}{R^2}\right]. \tag{5}$$

We shall solve these equations exactly below.

In the spirit of the original Kaluza-Klein theory we shall investigate whether the $4D$ properties of matter may be completely geometric in origin (Einstein [1956] and Salam [1980]) by studying whether the higher-dimensional vacuum field equations reduce (formally) to Einstein's $4D$ theory with a non-zero energy-momentum tensor constituting the material source. We shall interpret the source as a cosmological (perfect) fluid with energy-density, μ, and pressure, p; from the $4D$ Einstein field equations for the Robertson-Walker metric μ and p can then be identified by

$$\mu \equiv \frac{3\dot{R}^2}{R^2}, \qquad p \equiv -\frac{2\ddot{R}}{R} - \frac{\dot{R}^2}{R^2}. \tag{6}$$

It is known that an energy-momentum tensor of a particular form may formally admit a number of physical interpretations (Coley and Tupper [1986]); for example, the source can also be formally interpreted as an electromagnetic field as in the original $5D$ Kaluza-Klein theory (Kaluza [1921], and Klein [1926a] and [1926b]) or (as is more commonly done) as a scalar field (Soleng [1991]).

The (vacuum, Einstein) field equations in Wesson's (Wesson [1984]) $5D$ Space-Time-Matter (STM) theory (or variable-mass theory), in which the fifth coordinate is associated with mass, are formally obtained by setting $\alpha = 0$ in Equations (3-5), which yield the solution (Wesson [1990])

$$R = t^{1/2}, \qquad S = t^{-1/2}, \tag{7}$$

and

$$\mu = \frac{3}{4}t^{-2} = 3p, \tag{8}$$

which is the familiar $4D$ (flat) Friedmann-Robertson-Walker (FRW) radiation cosmological model, but where the properties of the matter (8) are prescribed by the ($5D$) geometry (and hence the matter is interpreted in terms of more fundamental geometric quantities).

Since the theory is purely gravitational the induced matter (see (8)) can only have a simple form; however, the theory does give rise to physically acceptable matter (and matter which is appropriate in the cosmological context). It is, however, of interest to investigate whether models exist with more complex forms of matter (than (8)). For D-dimensional GBT (for a metric which is essentially dependent only on the spacetime coordinates, but general otherwise) it is found that $\mu - 3p = (D-5)F + \alpha G$, where F and G are functions of the metric components and their derivatives (Poisson [1991]). Hence for $5D$ STM theory $\mu = 3p$ (regardless of the 4-geometry). To obtain models with $\mu \neq 3p$ we must therefore investigate models with $D > 5$ (see, e.g., Coley [1994]), or, as will be done here, models with $\alpha \neq 0$ (see, for example, Equation (5)). Thus, the work can be thought of as generalizing the STM model ((7), (8)) by taking Einstein's $4D$ theory with matter to be embedded in a higher-dimensional ($D = 5$) generalized Lagrangian Theory (GBT).

Defining

$$r \equiv \frac{\dot{R}}{R}\ , \quad s \equiv \frac{\dot{S}}{S}, \tag{9}$$

Equation (3) becomes

$$s = \frac{-r}{1 + 4\alpha r^2}, \tag{10}$$

which defines s in terms of r (and the denominator is always non-zero for $\alpha > 0$), and Equation (5) becomes

$$\dot{r} = \frac{-2r^2(1 + 2\alpha r^2)}{1 + 4\alpha r^2}. \tag{11}$$

Integrating this equation then yields

$$2t = \frac{1}{r} + \sqrt{2\alpha}\left[\frac{\pi}{2} - \arctan(\sqrt{2\alpha} r)\right], \tag{12}$$

where the constant of integration has been scaled so that $r \to \infty$ as $t \to 0^+$. Using Equations (10) - (12) it can be shown that Equation (4) is then satisfied identically. Finally, using (9), Equation (10) can be integrated to yield

$$S = S(0)\left[\frac{2\alpha r^2}{1 + 2\alpha r^2}\right]^{1/4}, \tag{13}$$

where $S(0)$ is the constant value of $S(t)$ at $t = 0$. From (9), Equation (12) is a differential equation for $R(t)$.

Equations (12) and (13) represent the exact solution of Equations (3) - (5) for $t \in (0, \infty)$. As $t \to 0^+$, we have that $r = \dfrac{\dot{R}}{R} \to \infty$ (i.e., there is a singularity at $t = 0$) and $S \to S(0) =$ constant. From (12) $r = \dfrac{\dot{R}}{R} > 0$ for all t, and from (11) r is monotonically decreasing, whence from (10) $s = \dfrac{\dot{S}}{S} < 0$ for all t. [We note parenthetically that the time-reversed solution is, of course, also permitted.] As $t \to \infty$, $r \to 0^+$ and $S \to 0^+$; asymptotically we obtain (7) and (8) (which is the exact solution of Equations (3) - (5) with $\alpha = 0$), so that the solution asymptotically approaches the familiar (zero-curvature) radiation FRW model. (However, the radiation FRW model cannot be an exact solution of the GBT with $\alpha \neq 0$).

Finally, from (6) we find that

$$\mu = 3r^2,$$
$$p = \frac{r^2(1-4\alpha r^2)}{1+4\alpha r^2}. \tag{14}$$

We can write the equation of state as

$$p = \frac{\mu(3-4\alpha\mu)}{3(3+4\alpha\mu)}. \tag{15}$$

Therefore, the model gives rise to $4D$ matter with more complex structure (than (8)). In addition, the matter has some (physically) interesting properties. As $t \to 0^+$, $p \to -\frac{1}{3}\mu$ $(\mu \to \infty)$, and as $t \to \infty$, $p \to \frac{1}{3}\mu$ $(\mu \to 0^+)$.

References

Coley, A. A. [1994] *Higher-dimensional vacuum cosmology,* Ap. J. **427**, 585–602.

Coley, A. A. and Tupper, B. O. J. [1986] *Two fluid cosmological models,* J. Math. Phys. **27**, 406–416.

Einstein, A. [1956] *The Meaning of Relativity,* Princeton University Press, Princeton.

Green, M. B., Schwarz, J. H. and Witten, E. [1987] *Superstring Theory,* Cambridge University Press, Cambridge.

Kaluza, Th. [1921] *Zum Unitätsproblem der Physik,* Sitz. Preuss. Akad. Wiss. **33**, 966–972.

Klein, O. [1926a] *Quantentheorie und fünf-dimensionale Relativitätstheorie,* Z. Phys. **37**, 895–906.

——— [1926b] *The atomicity of electricity as a quantum theory law,* Nature (London) **118**, 516.

Madore, J. [1985] *Kaluza-Klein theory with the Lanczos Lagrangian,* Phys. Lett. A **110**, 289–294.

Poisson, E. [1991] *Quadratic gravity as hair tonic for black holes,* Class. Quantum Grav. **8**, 639–650.

Salam, A. [1980] *Gauge unification of fundamental forces,* Rev. Mod. Phys. **52**, 525–538.

Soleng, H. H. [1991] *Trapping the Kaluza-Klein scalar,* Gen. Rel. Grav. **23**, 313–319.

Stelle, K. S. [1977] *Renormalization of higher derivative quantum gravity,* Phys. Rev. D **16**, 955–969.

Wesson, P. S. [1984] *An embedding for general relativity with variable rest mass,* Gen. Rel. Grav. **16**, 193–203.

——— [1990] *Clarification of an extended theory of gravity and a reply to Grøn and Soleng,* Gen. Rel. Grav. **22**, 707–713.

Zwiebach, B. [1985] *Curvature squared terms and string theories,* Phys. Lett. B **156**, 315–317.

Fields Institute Communications
Volume **15**, 1997

Quasilocal Formalism for Dilaton Gravity with Yang-Mills Fields

Jolien D. E. Creighton
Department of Applied Mathematics and Theoretical Physics
Cambridge University, Silver Street
Cambridge, England CB3 9EW
Current address: Department of Physics
University of Waterloo
Waterloo, Ontario, Canada N2L 3G1
jolien@avatar.uwaterloo.ca

Robert B. Mann
Department of Applied Mathematics and Theoretical Physics
Cambridge University, Silver Street
Cambridge, England CB3 9EW
Current address: Department of Physics
University of Waterloo
Waterloo, Ontario, Canada N2L 3G1
mann@avatar.uwaterloo.ca

Abstract. We present a quasilocal formalism, based on the one proposed by Brown and York, for dilaton gravity with Yang-Mills fields. For solutions possessing sufficient symmetry, we define conserved quantities such as mass, angular momentum, and charge. We also present a micro-canonical action and use it to arrive at a quasilocal version of the first law of thermodynamics for static systems containing a black hole.

1 Introduction

Suppose a gravitating system[1] is surrounded by some spatial surface. On this surface, it is possible to define quantities that will yield information about the system contained. These quantities we call quasilocal quantities. In the past it has been the practice to determine analogous quantities in an asymptotic limit. However, this procedure must be adapted for the specific asymptotic behaviour of the spacetime under consideration; it is unclear how to proceed except when the spacetime is asymptotically flat. Furthermore, when we study the thermodynamics of

---

1991 *Mathematics Subject Classification.* Primary 83C40; Secondary 83C57, 81T13.
This work was supported by NSERC.

[1]In this paper, we shall assume a four dimensional spacetime for definiteness.

© 1997 American Mathematical Society

a gravitating system, the quasilocal surface takes on an important physical significance as the thermodynamic wall containing the system. Thus, we find it important to develop a formalism for constructing quasilocal quantities on the boundary of a gravitating system, as such quantities can then be identified directly as the thermodynamic variables of the system.

An excellent quasilocal formalism was developed by Brown and York [1993a] for General Relativity. In this formalism, a surface stress energy-momentum tensor was constructed by a functional analysis of the action for General Relativity, and, from this tensor, a quasilocal energy can be found. In addition, for spacetimes possessing sufficient symmetry, a conserved mass (which differs, in general, from the quasilocal energy) and angular momentum can be defined. In their following paper (Brown and York [1993b]), they used the Feynman path integral technique of statistical mechanics to demonstrate the connection between canonical and thermodynamic conjugacy. They showed how to construct a "microcanonical" action for which the "extensive" variables are fixed on the boundary of the system. From this form of the action, the entropy and the first law of thermodynamics (at the zeroth order of quantum corrections) follow easily.

Our purpose is to extend this formalism to gravity theories involving a dilaton. In addition, we will include a Yang-Mills field in our analysis to demonstrate the rôles of matter in the formalism. This paper is a summary of the formalism; a more complete discussion is given in our recent paper (Creighton and Mann [1995]).

2 The Quasilocal Formalism

Assume that the region of spacetime under consideration, $\mathcal{M}$ (the "gravitating system"), is topologically the direct product of a spacelike hypersurface, Σ, with a real interval. The boundary $\partial\Sigma$ is the quasilocal surface that surrounds the system, and $\mathcal{T}$ is the history of this surface. We take our dynamical fields to be the spacetime metric, $g_{\mu\nu}$, a scalar field we call the dilaton, Φ, and a Yang-Mills potential $\mathfrak{A}_\mu{}^{\mathfrak{a}}$. The gauge covariant derivative operator is $(\mathfrak{D}_\mu)^{\mathfrak{a}}{}_{\mathfrak{b}} = \nabla_\mu \delta^{\mathfrak{a}}{}_{\mathfrak{b}} + \mathfrak{f}^{\mathfrak{a}}{}_{\mathfrak{bc}}\mathfrak{A}_\mu{}^{\mathfrak{c}}$ where $\mathfrak{f}^{\mathfrak{a}}{}_{\mathfrak{bc}}$ are the structure constants of the gauge group[2]. The action we take is

$$\begin{aligned} S = \int_{\mathcal{M}} d^4x\sqrt{-g}\,\bigl(D(\Phi)R(g) + H(\Phi)(\nabla\Phi)^2 + V(\Phi) - \tfrac{1}{4}W(\Phi)\mathfrak{F}^{\mu\nu}{}_{\mathfrak{a}}\mathfrak{F}_{\mu\nu}{}^{\mathfrak{a}}\bigr) \\ - 2\int_{\mathcal{T}} d^3x\sqrt{-\gamma}\,D(\Phi)\mathrm{tr}(\Theta) - 2\int_{\Sigma} d^3x\sqrt{h}\,D(\Phi)\mathrm{tr}(K). \end{aligned} \tag{2.1}$$

Here, $R(g)$ is the Ricci scalar of the metric $g_{\mu\nu}$ and $\mathfrak{F}_{\mu\nu}{}^{\mathfrak{a}}$ is the curvature of the gauge covariant derivative operator. The induced metrics on $\mathcal{T}$ and Σ are γ_{ij} and h_{ij} respectively, while the extrinsic curvatures of these surfaces are Θ_{ij} and K_{ij} respectively. The functions $D(\Phi)$, $H(\Phi)$, $V(\Phi)$, and $W(\Phi)$ are arbitrary functions of the dilaton alone (and none of its derivatives) that specify the theory[3].

Under variations of the field configurations for which the boundary configurations are fixed the induced variation in the action of equation (2.1) is minimized when the equations of motion are satisfied. Under more arbitrary variations, surface terms proportional to the variations of the fields on the boundaries will appear

[2] From these structure constants, we are able to define a Killing metric that will allow us to raise and lower gauge group indices.

[3] The action functional can also possess an arbitrary functional of the boundary fields, but we shall ignore this here.

along with the equation of motion terms. The coefficients of the boundary variations of the fields are identified as the momenta conjugate to those surface fields. The phase space on Σ is given by $\{(p^{ij}, h_{ij}), (\mathfrak{P}^i{}_{\mathfrak{a}}, \mathfrak{A}_i{}^{\mathfrak{a}}), (P, \Phi)\}$ so that, for example, $\pi^{ij} = -(\delta S/\delta h_{ij})_{c\ell}$ (where the "$c\ell$" indicates evaluation on a classical solution). There will also be momenta conjugate to the fields on the $\mathcal{T}$ boundary; its phase space coördinates are $\{(\pi^{ij}, \gamma_{ij}), (\mathfrak{K}^i{}_{\mathfrak{a}}, \mathfrak{U}_i{}^{\mathfrak{a}}), (\Pi, \phi)\}$. (We have written the gauge and dilaton fields on $\mathcal{T}$ as $\mathfrak{U}_i{}^{\mathfrak{a}}$ and ϕ respectively.)

Our primary interest will be in the $\mathcal{T}$ boundary which is the history of our quasilocal surface. Parameterize the time interval with t whose corresponding affine vector is $t^\mu = Nu^\mu + N^\mu$ where u^μ is the unit normal to the leaves, Σ_t, of the foliation, and N and N^μ are the lapse and shift vectors. On the quasilocal surfaces, $\partial\Sigma_t$, there are induced metrics $\sigma_{ij} = \gamma_{ij} + u_i u_j$. We can decompose the fields and conjugate momenta on $\mathcal{T}$ into portions parallel to the unit normals, u^i (which we assume lie along $\mathcal{T}$), and projections onto the surfaces $\partial\Sigma$. The metric, γ_{ij}, is thus decomposed into functions of the lapse, the shift, and the induced metric σ_{ab} in the usual way. The gauge field, $\mathfrak{U}_i{}^{\mathfrak{a}}$ is decomposed into a scalar, $\mathfrak{V}^{\mathfrak{a}}$, and a vector on $\partial\Sigma$, $\mathfrak{W}_a{}^{\mathfrak{a}}$. Similarly we decompose the momenta. Define, on $\partial\Sigma$, $\mathcal{E} = \sqrt{\sigma} u_i u_j \tau^{ij}$, and, $\mathcal{S}^{ab} = \sqrt{\sigma}\sigma_i^a \sigma_j^b \tau^{ij}$ as the quasilocal surface energy and stress density respectively. Here, $\tau^{ij} = 2\pi^{ij}/\sqrt{-\gamma}$ is the surface stress energy-momentum tensor on $\mathcal{T}$. Similarly, define $\mathcal{Q}_{\mathfrak{a}} = \sqrt{\sigma} W(\Phi) n_i \mathfrak{E}^i{}_{\mathfrak{a}}$ and $\mathcal{I}_a{}^{\mathfrak{a}} = \sqrt{\sigma} W(\Phi) \epsilon_a{}^{ij} n_i \mathfrak{B}_j{}^{\mathfrak{a}}$ as the quasilocal Yang-Mills surface charge and current densities respectively. Here, n^i is the unit normal to Σ and $\mathfrak{E}_i{}^{\mathfrak{a}}$ and $\mathfrak{B}_i{}^{\mathfrak{a}}$ are the usual electric and magnetic (spatial) vectors constructed from the Yang-Mills field strength. The surface momentum density is given by $\mathcal{J}^a = -\sqrt{\sigma} u_i \sigma_j^a \tau^{ij} + \mathcal{Q}^{\mathfrak{a}} \mathfrak{W}^a{}_{\mathfrak{a}}$; note that it includes an EMF term from the Yang-Mills field. Finally, a surface dilaton density is given by $\mathcal{F} = \Pi/N$. Explicit expressions for these quantities are given elsewhere (Creighton and Mann[1995]).

With these definitions, the variation of the action on the boundary $\mathcal{T}$ is

$$\delta S|_{\mathcal{T}} = \int_{\mathcal{T}} d^3x \big(-\mathcal{E}\delta N + \mathcal{J}_a \delta N^a - \mathcal{Q}_{\mathfrak{a}} \delta(N\mathfrak{V}^{\mathfrak{a}}) + N(\tfrac{1}{2}\mathcal{S}^{ab}\delta\sigma_{ab} + \mathcal{I}^a{}_{\mathfrak{a}} \delta\mathfrak{W}_a{}^{\mathfrak{a}} + \mathcal{F}\delta\phi)\big). \tag{2.2}$$

Any variable that is a function of the phase space coördinates (on Σ) only is called an *extensive* variable. The variables $\mathcal{E}$, $\mathcal{J}_a$, $\mathcal{Q}_{\mathfrak{a}}$, σ_{ab}, $\mathfrak{W}_a{}^{\mathfrak{a}}$, and ϕ are all extensive. Conversely, a variable that is not a function on the phase space of Σ is known as an *intensive* variable. Functions of the lapse and the shift are examples of intensive variables. We see in equation (2.2) that the first three terms of the integrand involve variations of intensive variables while the last three terms involve variations of extensive variables. We shall return to this important point in the next section.

The quasilocal surface energy, momentum, and charge densities yield valuable information about the enclosed spacetime when it possesses sufficient symmetries. If the boundary, $\mathcal{T}$, admits a timelike, surface forming, Killing vector (such that the Lie derivative of all the fields along this vector vanish), and if the matter stress energy tensor is negligible on the quasilocal boundary, then the quantity $\mathbb{M} = \oint d^2x N\mathcal{E}$ (where $\oint$ represents an integral over the quasilocal surface, $\partial\Sigma$) is a conserved quantity representing the mass contained within the boundary. By "conserved" we mean that the mass is independent of the foliation of $\mathcal{T}$. Similarly, a conserved angular momentum can be associated with the interior of the boundary under similar conditions, but where the Killing vector, φ^a, must now be spacelike and azimuthal. If the quasilocal boundary contains the orbits of this vector, then the angular momentum is $\mathbb{J} = \oint d^2x \mathcal{J}_a \varphi^a$. Finally, a conserved Yang-Mills charge can be associated with each Lie-algebra-valued scalar field, $\mathfrak{k}^{\mathfrak{a}}$, that is covariantly constant: $(\mathfrak{D}_\mu)^{\mathfrak{a}}{}_{\mathfrak{b}}\mathfrak{k}^{\mathfrak{b}} = 0$. The conserved charge is then $\mathbb{Q}[\mathfrak{k}] = \oint d^2x \mathcal{Q}_{\mathfrak{a}}\mathfrak{k}^{\mathfrak{a}}$. This is the analog of Gauss' law. By *quasilocal energy*, we shall mean the quantity $\mathbb{E} = \oint d^2x \mathcal{E}$. Notice that this differs from the quasilocal mass and is not necessarily a conserved quantity. However, it will be interpreted as the thermodynamic internal energy of the system.

3 The Microcanonical Action

The action that we have considered thus far is the appropriate action for the generation of the equations of motion when, under variations, the fields on the initial and final spacelike hypersurfaces are fixed as are various quantities on the quasilocal boundary. However, we have seen that the quantities that must be fixed on the quasilocal boundary are a mixture of extensive and intensive variables. The *microcanonical action* will be that action for which only extensive variables need be fixed on the quasilocal boundary under variations. It differs from the original action by a boundary term:

$$S_{\rm m} = S + \int_{\mathcal{T}} d^3x (\mathcal{E}N - \mathcal{J}_a N^a + \mathcal{Q}_{\mathfrak{a}} N \mathfrak{V}^{\mathfrak{a}}) \tag{3.1}$$

so that

$$\delta S_{\rm m}|_{\mathcal{T}} = \int_{\mathcal{T}} d^3x N(\delta\mathcal{E} - \omega^a \delta\mathcal{J}_a + \mathfrak{V}^{\mathfrak{a}}\delta\mathcal{Q}_{\mathfrak{a}} + \tfrac{1}{2}\mathcal{S}^{ab}\delta\sigma_{ab} + \mathcal{I}^a{}_{\mathfrak{a}}\delta\mathfrak{W}_a{}^{\mathfrak{a}} + \mathcal{F}\delta\phi) \tag{3.2}$$

where $\omega^a = N^a/N$ is interpreted as the angular velocity of observers with zero vorticity on $\mathcal{T}$. The covariant form of the microcanonical action is just the same as equation (2.1), but now the $\mathcal{T}$ boundary term is:

$$S_{\rm m}|_{\mathcal{T}} = \int_{\mathcal{T}} d^3x\sqrt{-\gamma}\left(D(\Phi)t_\mu\Theta^{\mu\nu}\partial_\nu t - 2n^\mu\partial_\mu D(\Phi) - W(\Phi)(n_\mu \mathfrak{F}^{\mu\nu}{}_{\mathfrak{a}}\partial_\nu t)(t^\alpha \mathfrak{A}_\alpha{}^{\mathfrak{a}})\right). \tag{3.3}$$

We now use the microcanonical action to find an expression for the entropy of systems containing a stationary[4] black hole. Since we will use path integral methods, it will be useful to adopt a "Euclidean" notation. In the Euclideanization prescription, we shall construct a new microcanonical action functional, $I_{\rm m} = -{\rm i}S_{\rm m}$, so that the phase of the path integral is $\exp(-I_{\rm m})$. The imaginary factor is absorbed into the intensive variables (specifically, the lapse) so that the extensive variables are invariant under the Wick rotation.

The quantum mechanical microcanonical density matrix is defined as the path integral over all fields in $\mathcal{M}$ with the phase weighting $\exp(-I_{\rm m})$. It is a functional of the variables held fixed on the initial and final spatial hypersurfaces, $\Sigma_{\rm initial}$ and $\Sigma_{\rm final}$, as well as the extensive variables held fixed on $\mathcal{T}$. The most important statistic of this density matrix is the partition function which, in the microcanonical ensemble, is known as the density of states. This is obtained by identifying the initial and the final spatial hypersurfaces (with some period, Δt) and "tracing over" the field configurations on these identified surfaces. The density of states, then, is a functional of the extensive variables on the qusilocal surface.

[4]Stationarity of all the fields is a necessary condition of thermodynamic equilibrium.

Because the system contains a black hole, the foliation becomes degenerate on the event horizon. When the initial and final spatial hypersurfaces have been identified, the complex manifold has the topology of a cone crossed with S^2 (we assume that it is possible to foliate from the event horizon to the quasilocal boundary with parameter r). The conical singularity at the event horizon is removed by fixing the period of identification to the value $\Delta t = 2\pi/\kappa_{\mathrm{H}}$ where $\kappa_{\mathrm{H}}^2 = h^{ij}(\partial_i N)(\partial_j N)$ (evaluated on the event horizon) is the surface gravity of the event horizon.

We now find the entropy of the gravitating system to "zeroth order" in quantum corrections. In this approximation, the density of states is just the path integral phase evaluated on a classical solution, so the entropy is $\mathbb{S} \approx -I_{\mathrm{m}}|_{c\ell}$. We thus need to evaluate the Euclideanized microcanonical action at its (complex) extremal value. The event horizon is not included in the system as it acts as a one-way membrane for observers orbiting on $\mathcal{T}$. This is accomplished most easily by performing a canonical decomposition of the microcanonical action with the surface term given in equation (3.3). No boundary term is appended to the action on the event horizon. We find that:

$$
\begin{aligned}
I_{\mathrm{m}} = \mathrm{i}\int dt \Big(& \int_{\Sigma} d^3x(-p^{ij}\dot{h}_{ij} - \mathfrak{P}^i{}_{\mathfrak{a}}\dot{\mathfrak{A}}_i{}^{\mathfrak{a}} - P\dot{\Phi} + \mathcal{H}N + \mathcal{H}_i N^i + \mathcal{G}_{\mathfrak{a}}\mathfrak{A}_t{}^{\mathfrak{a}}) \\
& + \int_H d^2x\Big(2\sqrt{\sigma}\,(D(\Phi)n^i\partial_i N - Nn^i\partial_i D(\Phi)) - N^a\mathcal{J}_a + N\mathcal{Q}_{\mathfrak{a}}\mathfrak{V}^{\mathfrak{a}}\Big)\Big)
\end{aligned}
\tag{3.4}
$$

where the over-dot indicates a Lie derivative with respect to time and $\mathcal{H}$, $\mathcal{H}_i$, and $\mathcal{G}_{\mathfrak{a}}$ are the Hamiltonian, momentum, and Gauss constraints respectively. Since the solution is stationary, all the time derivatives vanish and since the action is to be evaluated on a classical solution, the constraints also vanish. All that is left then, is the integral over the event horizon. Here we have $N = 0$, $N^a = 0$, and, from the regularity conditions, $\int dt\, n^i\partial_i N = 2\pi\mathrm{i}$. Thus we find that the entropy is $\mathbb{S} = \int d^2x\sqrt{\sigma}\,D(\Phi)$ where the integral is over the event horizon. Notice that, in the case of General Relativity, we have $D(\Phi) = (16\pi)^{-1}$ and the entropy is the usual value of one quarter of the event horizon area (in units of the rationalized Planck's constant and Newton's constant).

The first law of thermodynamics follows immediately from equation (3.2). Define $\beta = \mathrm{i}\int dt\, N$ evaluated on the quasilocal boundary, $\partial\Sigma$. It is the reciprocal temperature of the system. Notice that it is not necessarily constant over the quasilocal surface unless the quasilocal surface is so chosen. The first law of thermodynamics for the quasilocal system is:

$$
\delta\mathbb{S} = \int_{\partial\Sigma} d^2x\,\beta(\delta\mathcal{E} - \omega^a\delta\mathcal{J}_a + \mathfrak{V}^{\mathfrak{a}}\delta\mathcal{Q}_{\mathfrak{a}} + \tfrac{1}{2}\mathcal{S}^{ab}\delta\sigma_{ab} + \mathcal{I}^a{}_{\mathfrak{a}}\delta\mathfrak{W}_a{}^{\mathfrak{a}} + \mathcal{F}\delta\phi). \tag{3.5}
$$

When the quasilocal surface is chosen to be an isotherm, the first term of the integral becomes the usual "$\beta\delta\mathbb{E}$" term. However, since the angular velocity, for example, will not necessarily be constant over an isotherm, it is necessary to write the first law in integral form.

References

Brown, J. D. and York, J. W. [1993a] *Quasilocal energy and conserved charges derived from the gravitational action,* Phys. Rev. D. **47**, 1407–1419.

______ [1993b] *Microcanonical functional integral for the gravitational field,* Phys. Rev. D. **47**, 1420–1431.

Creighton, J. D. E. and Mann, R. B. [1995] *Quasilocal Thermodynamics of Dilaton Gravity Coupled to Gauge Fields,* Phys. Rev. D **52**, 4569–4587.

Fields Institute Communications
Volume 15, 1997

Einstein-Yang-Mills Cosmology with Taub-Type Symmetry

Bahman Darian
Theoretical Physics Institute
University of Alberta
Edmonton, Alberta, Canada T6G 2J1
darian@phys.ualberta.ca

H. P. Künzle
Department of Mathematics
University of Alberta
Edmonton, Alberta, Canada T6G 2G1
hp.kunzle@ualberta.ca

Abstract. Using a general scheme to construct Yang-Mills connections in homogeneous spaces, we construct a Yang-Mills connection for the $SU(2)$ gauge group (referred to as $SU(2)$-gauge) in "axially symmetric" Bianchi IX cosmologies, namely space-times which posses the symmetries of Taub space. After finding some numerical solutions, we discuss briefly the effect of anisotropy on the general behavior of the space-time geometry and the Yang-Mills fields.

1 Introduction

Due to the highly non-linear nature of the Yang-Mills (YM) equations and strong gravity, Einstein-Yang-Mills (EYM) systems (in which a source free YM field is the source of gravity) are notoriously difficult to analyze. However in recent years there has been significant progress on two sides. On one side there are exact general solutions to non-thermal YM fields in Friedmann-Robertson-Walker cosmologies with positive, zero or negative curvature for which $SO(4)$, $E(3)^1$, or $SO(3,1)$ are the isometry groups, respectively (Galt'sov [1991] and Moniz and Mourão [1991]). The high space-time symmetry and conformal invariance of YM equations in 4D result in a decoupling of Einstein and YM equations in conformal time such that one can easily find exact solutions to the system of ODEs.

On the other hand there has been much activity on the static asymptotically flat, spherically symmetric solutions of EYM equations known as Bartnik-McKinnon (BM) particles (Bartnik and McKinnon [1988]) and $SU(2)$-black holes (Bizoń [1990], Künzle and Masood-ul-Alam [1990] and Volkov and Galt'sov [1990]).

1991 *Mathematics Subject Classification.* Primary 83C20, 83F05; Secondary 53C30.
Supported in part be NSERC grant A8059.
[1] $E(3)$ is the six-dimensional Euclidean group.

© 1997 American Mathematical Society

Due to the smaller symmetry group $SU(2) \otimes R$, this system is more difficult to analyze than the cosmological solutions. Despite this fact, there have been extensive investigations of numerical solutions (Kleihaus et al. [1995] and Künzle [1994]).

The present work which is basically an investigation of the cosmological solution of the EYM equations with a $SU(2) \otimes U(1)$ isometry group, was motivated by the intensive activity on static $SU(2) \otimes R$ symmetric solutions mentioned in the above paragraph. One rightfully expects that any analysis of such solutions encounters similar problems as that of the BM particles and $SU(2)$ black holes. However, this analysis is aided by the fact that for all homogenous 3-spaces with a 4-dimensional isometry groups, there is always a 3-dimensional subgroup which acts simply transitively on the homogeneous hypersurface (Kramer et al. [1980]). Therefore the problem reduces to that of a special case of Bianchi cosmologies, namely Bianchi IX with YM matter fields.

This article is a brief account of how homogeneous YM fields in Bianchi cosmologies are constructed. Then the EYM equations for "axially symmetric" Bianchi cosmologies are derived followed by brief comments on the space-time geometry and the general behavior of the fields.

2 Invariant YM Fields in Bianchi Cosmologies

In Bianchi cosmologies with three nowhere vanishing and pointwise linearly independent space-like Killing vector fields (kvfs) $\xi_i, (i = 1, 2, 3)$ a left-invariant vector basis $(e_\perp, e_i)$ and a one-form basis $(\theta^\perp, \theta^i)$ are defined such that

$$[\xi_i, e_j] = 0,\ \theta^j(e_i) = \delta_i^j,\ [e_\perp, e_j] = 0,\ \theta^\perp(e_\perp) = 1,\ \theta^\perp(e_i) = 0,\ \theta^i(e_\perp) = 0, \quad (2.1)$$

where e_i are tangent to the homogeneous space-like hypersurface Σ. In this basis any left-invariant metric can be written as (Ryan and Shepley [1975])

$$ds^2 = -\theta^\perp \otimes \theta^\perp + \overset{(3)}{g}_{ij}(t)\, \theta^i \otimes \theta^j, \quad (2.2)$$

which is a consequence of $\mathcal{L}_{\xi_i} \overset{(4)}{g} - \mathcal{L}_{\xi_i} \overset{(3)}{g} = 0$.

An additional isometry can further restrict $\overset{(3)}{g}_{ij}$. For example, in a Taub cosmology with a $SU(2) \otimes U(1)$ isometry group and kvfs (ξ_i, ξ_ϕ) the commutation relations are as follows

$$[\xi_i, \xi_j] = \epsilon_{ijk}\xi_k,\ [\xi_1, \xi_\phi] = -\xi_2,\ [\xi_2, \xi_\phi] = \xi_1,\ [\xi_3, \xi_\phi] = 0. \quad (2.3)$$

At the point in which $\xi_\phi = 0$, ξ_ϕ generates the isotropy subgroup $U(1)$. Using the torsion free condition $\nabla_{\xi_i}\xi_\phi - \nabla_{\xi_\phi}\xi_i = \nabla_{\xi_i}\xi_\phi = [\xi_i, \xi_\phi]$, and $\mathcal{L}_{\xi_\phi} \overset{(3)}{g}_{ij} = 0$ we derive the canonical form of the metric in Misner parametrization

$$\overset{(3)}{g}_{ij} = R^2\, \mathrm{diag}(e^{2\beta_+}, e^{2\beta_+}, e^{-4\beta_+}). \quad (2.4)$$

The invariance of the $SU(2)$-gauge fields is another story. It is possible that gauge Lie algebra valued fields are not constant in a left invariant basis but the change from one point to another point on Σ in a left invariant frame is merely a gauge transformation. Of course such fields give rise to invariant physical quantities. Therefore in the above basis a Lie algebra valued one-form, the YM connection

$$A^A = A_\perp^A(x,t)\theta^\perp + A_i^A(x,t)\theta^i, \quad (2.5)$$

has to satisfy the further conditions

$$\begin{aligned} \mathcal{L}_{\xi_i} A_\perp^A &= 0, & (2.6)\\ \mathcal{L}_{\xi_i} A_j^A &= D_j W_i^A(x,t). & (2.7) \end{aligned}$$

Equation (2.6) simply requires $A_\perp^A = A_\perp^A(t)$ whereas condition (2.7), in which $W_i^A(x)$ are Lie algebra valued scalar functions and D is the gauge covariant derivative, guarantees that there is a gauge such that

$$A_i^A = \Lambda_i^A(t), \tag{2.8}$$

which is equivalent to setting the right hand side of (2.7) to zero. However, it can be shown that for any invariant connection (2.7), one can always choose a gauge such that (2.8) holds[2].

There are three important facts that should be mentioned regarding the YM connection constructed so far:

1. There is no local gauge freedom left in (2.8) and the remaining gauge freedom is global. Only
$$A^A \mapsto \gamma(t) A^A \gamma^{-1}(t), \qquad (\gamma \in SU(2)) \tag{2.9}$$
are allowed.
2. Such global gauge transformations are generated by the Gauss law constraint, and if these gauge transformations are time dependent, then necessarily $A_\perp^A \neq 0$. One can use this global gauge freedom to triangularize Λ. The six remaining components of Λ are the dynamical degrees of freedom of $SU(2)$-YM field.
3. Any additional kvf ξ_ϕ has a non-trivial action on the tangent space via the commutation relations (2.3). The invariance of the metric under this action via the Killing equation, gave rise to (2.4). The invariance of the YM connection requires the action of ξ_ϕ on Λ_i^A to be equivalent to a gauge transformation. Wang's theorem states that such invariant connections are in one-to-one correspondence with the conjugacy classes of homomorphisms
$$\lambda : (\text{isotropy group generated by } \xi_\phi) \to SU(2), \tag{2.10}$$
such that the YM connection Λ as a gauge-Lie algebra valued one-form satisfies (Kobayashi and Nomizu [1963])
$$\Lambda([\xi_\phi, \xi_i]) = [\lambda_*(\xi_\phi), \Lambda(\xi_i)]. \tag{2.11}$$

Conjugacy classes of homomorphisms when the isotropy group is $U(1)$ are well known for $SU(2)$ black holes (Künzle [1991]). If $\tau_3 \in$ su(2) is the generator of $U(1) \subset SU(2)$-gauge, then the only non-trivial connection corresponds to the conjugacy class of the homomorphisms

$$\lambda : U(1) \to SU(2),\ z \mapsto \begin{pmatrix} z^1 & 0 \\ 0 & z^{-1} \end{pmatrix}. \tag{2.12}$$

Inserting this relation into (2.11) and using (2.3) yields the linear system

$$\epsilon_{3j}{}^k \Lambda_k^A \tau_A = [\tau_3, \Lambda_j^A \tau_A], \tag{2.13}$$

[2]The gauge transformation $W_i^A \mapsto \gamma(t) W_i^A \gamma^{-1}(t) + \gamma(t) d\gamma(t)^{-1}$ with $\gamma(t) = \mathcal{P}\exp\{-\int_0^x W_i^A(y,t)dy\}$ in which the path-ordered line integral is taken along the integral curves of ξ_i, sets the right hand side of (2.8) to zero (Forgács and Manton [1980]). A global proof which is more elegant is given in Brodbeck and Straumann [1993].

Bizoń, P. [1990] *Colored black holes*, Phys. Rev. Lett. **64**, 2844–2847.

Brill, D. R. [1964] *Electromagnetic fields in a homogeneous, nonisotropic universe*, Phys. Rev. **133**, 845–848.

Brodbeck, O. and Straumann, N. [1993] *A generalized Birkhoff theorem*, J. Math. Phys. **34**, 2412–2423.

Forgács, P. and Manton, N. S. [1980] *Spacetime symmetries in gauge theories*, Commun. Math. Phys. **72**, 15–35.

Galt'sov, D. V. and Volkov, M. S. [1991] *Yang-Mills cosmology. Cold matter for a hot universe*, Phys. Lett. B **256**, 17–21.

Kleihaus, B., Kunz, J. and Sood, A. [1995] *SU(3) Einstein-Yang-Mills sphalerons and black holes*, Phys. Lett. B **354**, 240–246.

Kobayashi, S. and Nomizu, K. [1963] *Foundations of differential geometry II*, Wiley, New York.

Kramer, D., Stephani, H., MacCallum, M. and Herlt, E. [1980] *Exact solutions of Einstein's field equations*, Cambridge University Press, Cambridge.

Künzle, H. P. [1991] *SU(n)-Einstein-Yang-Mills fields with spherical symmetry*, Classical Quantum Gravity **8**, 2283–2297.

——— [1994] *Analysis of the static spherically symmetric SU(n)-Einstein-Yang-Mills equations*, Commun. Math. Phys. **162**, 371–397.

Künzle, H. P. and Masood-ul-Alam, A. K. M. [1990] *Spherically symmetric static SU(2) Einstein-Yang-Mills fields*, J. Math. Phys. **31**, 928–935.

Moniz, P. V. and Mourão, J. M. [1991] *Homogeneous and isotropic closed cosmologies with a gauge sector*, Classical Quantum Gravity **8**, 1815–1831.

Ryan, M. and Shepley, L. [1975] *Homogeneous relativistic cosmologies*, Princeton University Press, Princeton.

Sudarsky, D. and Wald, R. M. [1992] *Extrema of mass, stationarity and staticity, and solutions to the Einstein-Yang-Mills equations*, Phys. Rev. D **46**, 1453–1474.

Volkov, M. S. and Gal'tsov, D. V. [1990] *Black holes in Einstein-Yang-Mills theory*, Soviet J. Nuclear Phys. **51**, 747–753.

Fields Institute Communications
Volume **15**, 1997

Using the Schrödinger Equation for Astrophysical and Cosmological Simulations

George Davies
Department of Physics
Queen's University
Kingston, Ontario, Canada K7L 3N6
davies@astro.queensu.ca

Abstract. A method of modeling structure formation by solving the coupled Schrödinger and Poisson equations is presented. This method can be used to model, in the non-relativistic limit, both collisionless matter and the cosmological evolution of scalar field perturbations. That this method can model the latter follows directly from taking the non-relativistic limit of the coupled Klein-Gordon and Einstein equations. To model collisionless matter one constructs a Schrödinger phase space distribution function by applying a windowed Fourier transform to the usual spatial wave function. This distribution function can be shown to obey the collisionless Boltzmann equation with the addition of higher order terms that are tuned to be small. This allows the evolution of the the initial phase space distribution to be determined by simply evolving the associated spatial wave function.

1 Introduction

In astrophysical and cosmological settings systems of collisionless matter (CM) abound. The importance of understanding the dynamics of these systems, and possessing a method with which to model them, is evident when one considers the vast range of applicability of the collisionless approximation. The spatial scale of CM systems ranges from that of globular clusters and galaxies through to large scale structure formed by cosmological dark matter particles (i.e., massive neutrino, axion, etc.).

Commonly used methods for modeling CM are N-body, hydrodynamic and phase space methods, each of these techniques has its own particular strengths and limitations. A new method of modeling CM, the 'Schrödinger method' (SM), has been proposed recently (Widrow and Kaiser [1993]) and incorporates some of the desirable features of each of the previously mentioned methods.

1991 *Mathematics Subject Classification.* 85-08.

© 1997 American Mathematical Society

2 Collisionless Matter

The central equation governing the dynamics of collisionless systems is the collisionless Boltzmann equation, $\dot{f}(\mathbf{x},\mathbf{p},t)=0$, or,

$$\frac{\partial f}{\partial t}=\sum_{i=1}^{3}\left(\frac{\partial V}{\partial x^i}\frac{\partial f}{\partial v^i}-v^i\frac{\partial f}{\partial x^i}\right)\ ,\ \nabla^2 V=4\pi G\int d^3\mathbf{p}f(\mathbf{x},\mathbf{p}); \tag{1}$$

where V is the Newtonian potential. In some special cases, usually situations with a high degree of symmetry, these coupled equations can be solved analytically. However, for most systems of physical interest they must be solved numerically.

Numerical phase space methods are, in general, computationally costly due to the large number of dimensions typically involved. If a method of encoding the full phase space information in a position space function could be found this problem could be overcome. The problem would then be reduced to mapping an initial phase space distribution $f(\mathbf{x},\mathbf{p},t_0)$ to an initial space function $\psi(\mathbf{x},t_0)$, time evolving this space function and then mapping the final space function back to phase space to obtain the final distribution function. The two objects that need to be supplied for this procedure are an 'invertible' phase space map $\mathcal{M}$, and the dynamical equation of motion (DEM) for the space function. Symbolically one can represent this procedure as,

$$f(\mathbf{x},\mathbf{p},t_0)\overset{\mathcal{M}}{\Longrightarrow}\psi(\mathbf{x},t_0)\overset{DEM}{\Longrightarrow}\psi(\mathbf{x},t)\overset{\mathcal{M}^{-1}}{\Longrightarrow}f(\mathbf{x},\mathbf{p},t) \tag{2}$$

The obvious flaw in this idealized method is the assumption that given an $\mathcal{M}$ one can find $\mathcal{M}^{-1}$. This is problematic, as applying $\mathcal{M}$ to a phase space distribution projects out information about its momentum structure. The viability of this type of method will therefore be determine largely by the extent to which momentum structure can be reconstructed from $\psi(\mathbf{x},t)$, given that one does not expect an exact $\mathcal{M}^{-1}$ to exist. This is one of the problems that the SM was designed to address.

3 The Schrödinger Method

The SM uses as $\mathcal{M}^{-1}$ the modulus squared of a coherent state transformation.

$$\mathcal{F}(\mathbf{x},\mathbf{p},t)=|\Psi(\mathbf{x},\mathbf{p},t)|^2=\left|N\int d^3\mathbf{x}'e^{-(\mathbf{x}-\mathbf{x}')^2/2\eta^2-i\mathbf{p}\cdot\mathbf{x}'/\hbar}\psi(\mathbf{x}',t)\right|^2, \tag{3}$$

where the DEM is taken here to be the coupled Schrödinger and Poisson equations.

$$i\hbar\frac{\partial\psi}{\partial t}=-\frac{\hbar^2}{2\mu}\nabla^2\psi+\mu V\psi\ ,\quad \nabla^2 V=4\pi G\rho\ ,\quad \rho=\int d^3\mathbf{p}\,\mathcal{F} \tag{4}$$

By direct calculation one can verify that $\mathcal{F}$ satisfies the Vlasov equation with the addition of higher order terms (Widrow and Kaiser [1993]), which are proportional to free parameters that are chosen to be small. The construction of $\mathcal{F}$ required taking the modulus of a complex function, so that inverting this map is clearly not possible analytically. Regardless of the method chosen to invert the map to phase space, it will be an approximate one; an *Ansatz.*

The SM can be shown symbolically as,

$$f(\mathbf{x},\mathbf{p},t_0)\overset{Ansatz}{\Longrightarrow}\psi(\mathbf{x},t_0)\overset{Schrödinger}{\Longrightarrow}\psi(\mathbf{x},t)\overset{|\Psi|^2}{\Longrightarrow}\mathcal{F}(\mathbf{x},\mathbf{p},t), \tag{5}$$

where $\mathcal{F}\approx f$, referring to the equivalence of $\mathcal{F}$ and f on scales greater than η in the position coordinates and greater than $(\hbar/\eta)$ in the momentum coordinates. Restricting attention to structures that form above this scale one can use this method to model CM. Putting this in terms of finding an $\mathcal{M} \circ \mathcal{M}^{-1} = I$ pair, we have found a map that is invertible above a chosen spatial scale.

To illustrate both the essential features and the general implementation of this method, a one dimensional cold collapse was simulated. Plots of both the wave functions and corresponding distribution functions appear in the accompanying diagrams. The simulation was run through two dynamical collapse times to illustrate the SM ability to model systems in which multi-streaming has occured. The development of a velocity distribution in phase space can be seen to be encoded in ψ as high momentum modes superposed on the density curve. This is due to the structure of our $\mathcal{M}^{-1}$; the coherent state is just a windowed Fourier transform.

Work is in progress to extend the SM to include spherical systems that possess angular momentum. Equations (1) now become,

$$\frac{\partial f}{\partial t} + v_r \frac{\partial f}{\partial r} + \left(\frac{j^2}{r^3} - \frac{\partial \Phi}{\partial r} \right) \frac{\partial f}{\partial v_r} = 0 \ , \ r^2 \, \nabla_r^2 \, \Phi = 4\pi G \int_{-\infty}^{\infty} dv_r \int_0^{\infty} dj^2 f. \qquad (6)$$

The idea here is to use the angular eigenfunction Schrödinger equation along with a collection of radial functions $\{\phi_j\}$ to represent the angular distribution. One can construct a Schrödinger distribution that obeys the Vlasov equation (plus higher order terms) as follows,

$$\mathcal{F}(r, v_r, j^2) = |\Psi_j(r, v_r)|^2 \ , \ \Psi_j = N \int_0^{\infty} dr' e^{-(r-r')^2/2\eta^2} e^{-imv_r r'/\hbar} \phi_j(r'), \qquad (7)$$

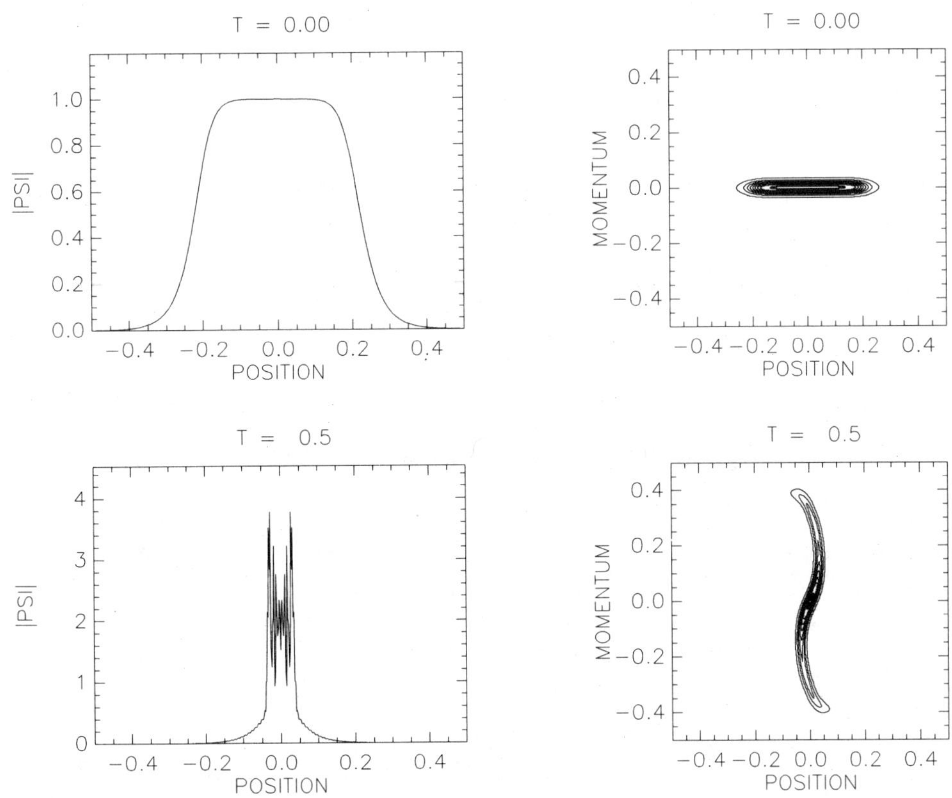

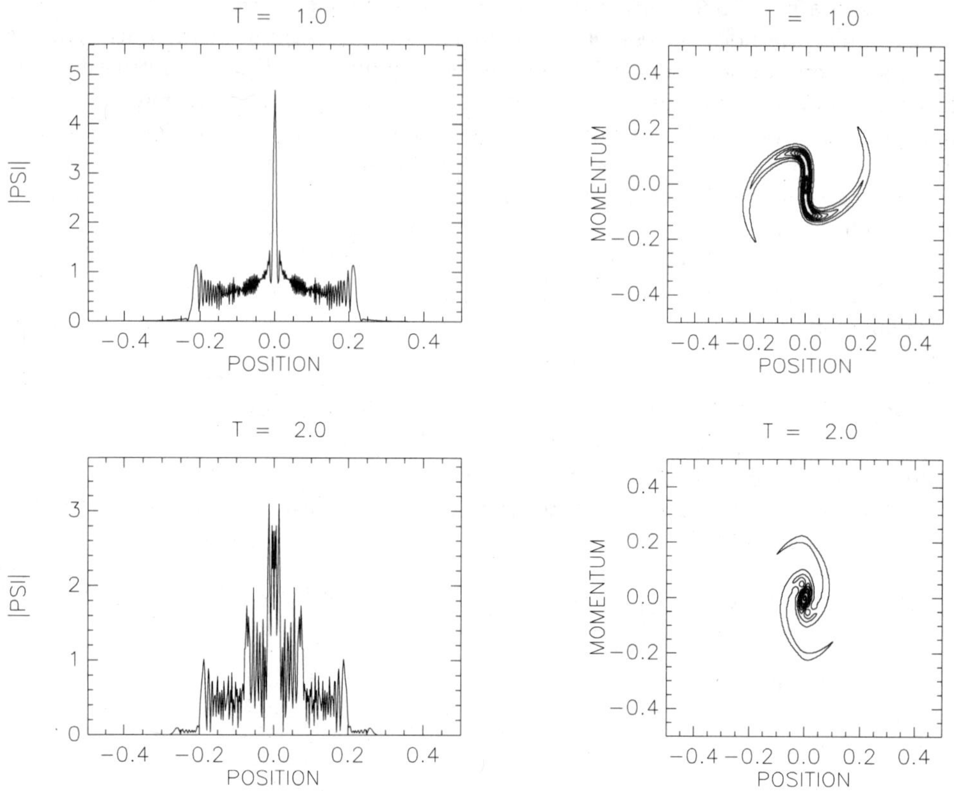

where j^2 is the angular momentum appearing in the Schrödinger equation;

$$i\hbar\frac{\partial\phi_j}{\partial t} = -\frac{\hbar^2}{2m}\frac{\partial^2\phi_j}{\partial r^2} + m\Phi\phi_j + \frac{mj^2}{2r^2}\phi_j\ , \tag{8}$$

and the potential is given by,

$$r^2\,\nabla_r{}^2\,\Phi = 4\pi G\sum_j\int_o^\infty dr' e^{-(r-r')^2/\eta}\,|\phi_j(r')|^2\,. \tag{9}$$

So that solving the coupled equations (8) and (9), taken together with (7), one can model spherical systems that have a 3 dimensional phase space.

References

Widrow, L. M. and Kaiser, N. [1993] *Using the Schrödinger Equation to Simulate Collisionless Matter*, Ap. J. **416**, L71–L74.

Fields Institute Communications
Volume **15**, 1997

Decoherence of the Unruh Detector

Jean-Guy Demers
Department of Physics
McGill University
3600 University St.
Montreal, PQ, Canada H3A 2T8
jgdemers@hep.physics.mcgill.ca

Abstract. As is well known, the Minkowski vacuum appears thermally populated to a quantum mechanical detector on a uniformly accelerating course. We investigate how this thermal radiation may contribute to the classical nature of the detector's trajectory through the criteria of decoherence. In 1+1 dimensions, an uncertainty-type relation is obtained for the detector, involving the fluctuations in temperature, the time of flight and the coupling to the bath.

1 Introduction

In the Copenhagen interpretation of the measurement process, the existence of classical systems is postulated from the start. In quantum cosmology however, the whole universe receives a quantum mechanical description and classicality must be somehow derived. An important mechanism by which some degrees of freedom may behave classically is the environment-induced decoherence. In that case, partial diagonalization of the density matrix in a given basis is obtained by integrating out some unobserved degrees of freedom (Zurek [1981], [1982] and [1991] and Gell-Mann and Hartle [1990]). A simple illustration is given by a system and its environment having each a single degree of freedom described by the actions $S_{\text{sys}}[x]$ and $S_{\text{env}}[y]$, respectively, with interaction described by the action $S_{\text{int}}[x, y]$. Starting with an uncorrelated density matrix at $t = 0$,

$$\rho(x_1, y_1, x_2, y_2; t = 0) = \rho_{\text{sys}}(x_1, x_2; t = 0) \times \rho_{\text{env}}(y_1, y_2; t = 0), \tag{1.1}$$

then at a time t later, the reduced density matrix (where the environment is integrated out) will be given by (Dowker and Halliwell [1992] and Hu and Matacz [1994]):

$$\rho_{\text{red}}(x_1^f, x_2^f; t) = \int dx_1^i \int dx_2^i \, P(x_1^f, x_2^f, t | x_1^i, x_2^i, 0) \, \rho_{\text{sys}}(x_1^i, x_2^i; 0), \tag{1.2}$$

1991 *Mathematics Subject Classification.* Primary 81T20; Secondary 81P15.

I would like to thank the Organizing Committee for putting together this interesting conference. This work was supported in part by funds provided by the U.S. Department of Energy under cooperative agreement # DE-FC02-94ER40818.

© 1997 American Mathematical Society

with the kernel

$$P(x_1^f, x_2^f, t | x_1^i, x_2^i, 0) = \int_{x_1^i}^{x_1^f} \mathcal{D}x_1 \int_{x_2^i}^{x_2^f} \mathcal{D}x_2 \, e^{i(S_{\rm sys}[x_1] - S_{\rm sys}[x_2])} \, e^{i\Gamma[x_1, x_2]}, \tag{1.3}$$

where the boundary conditions on the functional integrals are given for times 0 and t (our units are such that $\hbar = c = 1$). $\Gamma[x_1, x_2]$ is the influence functional (IF). It is a property of the environment and the way it is coupled to the system (Feynman and Vernon [1963]). For an environment in a pure state at $t = 0$, it takes the conceptually simple form

$$e^{i\Gamma[x_1, x_2]} = \langle \psi_2(t) | \psi_1(t) \rangle. \tag{1.4}$$

Here, $|\psi_1(t)\rangle$ is the state that evolved from the initial state at $t = 0$ under the dynamics dictated by $S_{\rm env}[y_1] + S_{\rm int}[x_1, y_1]$ in which x_1 acts as a c-number, time dependent source; likewise $|\psi_2(t)\rangle$ is governed by $S_{\rm env}[y_2] + S_{\rm int}[x_2, y_2]$. For paths $[x_1(t)]$ and $[x_2(t)]$ such that the states in (1.4) are not adiabatically disturbed, Γ will have a (positive) imaginary part Γ_I. It is clear that if Γ_I is large for a pair of paths $(x_1(t), x_2(t))$ that are far apart from one another in spacetime, then the contribution of these pairs will be suppressed in (1.3). As a result, $\rho_{\rm red}(x_1^f, x_2^f; t)$ will be more diagonal. Decoherence can thus be studied through Γ_I.

In recent times, the study of black holes received renewed interest, in part due to the introduction of more tractable two dimensional models. An important issue arising in that context is the limit of the semiclassical approximation, in which the Hawking radiation is derived. As part of the groundwork to that problem, we report here on the related issue of the decoherence of a detector in uniform acceleration. In the next section, we review how the IF predicts the heating of the scalar field forming the environment of a detector on a uniformly accelerating course (Unruh [1976] and Anglin [1993]). In Section 3, the spacetime trajectory is taken to be a *decohered* spacetime path, with a spread around the uniformly accelerating trajectory, and taking now the detector's excitation to be fixed. We then obtain an uncertainty-type relation involving the spread in acceleration involved in constructing the approximately classical path, the time of flight and the coupling to the thermal bath.

2 Unruh Effect

Consider an environment made of a massless scalar field in two spacetime dimensions (Anglin [1993] and Hu and Matacz [1994]). In a finite box of spatial dimension L, the mode expansion reads

$$\phi(x, t) = \sqrt{\frac{2}{L}} \sum_k [y_k^+ \cos kx + y_k^- \sin kx], \tag{2.1}$$

for amplitudes y_k^- and y_k^+ and wave vectors $k = 2\pi n/L$ with $n = 1, 2, \dots$. The kinetic term is

$$L_{\rm env} = \int dx \, \frac{1}{2} \partial_\mu \phi \, \partial^\mu \phi = \frac{1}{2} \sum_{\sigma=+,-} \sum_k [(\dot{y}_k^\sigma)^2 - k^2 (y_k^\sigma)^2], \tag{2.2}$$

where dots denote time derivative. The system is formed by a detector that is linearly and locally sensitive to the matter field with an interacting Lagrangian density

$$\mathcal{L}_{\rm int}(x) = -\varepsilon Q \phi(x, t) \delta(x - X(t)), \tag{2.3}$$

where ε is a coupling constant while $Q(t)$ is the DeWitt monopole moment (DeWitt [1979]) and plays the role of $x(t)$ in Section 1 while $X(t)$ is the trajectory of the detector. For a uniform acceleration a, $X(\tau) = \frac{1}{a}\cosh(a\tau)$ and $t(\tau) = \frac{1}{a}\sinh(a\tau)$ where τ is the proper time and the resulting IF is given by (Hu and Matacz [1994])

$$
\begin{aligned}
\Gamma_R[Q_1, Q_2] &= -\int_0^t ds_1 \int_0^{s_1} ds_2 [Q_1(s_1) - Q_2(s_1)]\mu(s_1, s_2)[Q_1(s_2) + Q_2(s_2)], \\
\Gamma_I[Q_1, Q_2] &= \int_0^t ds_1 \int_0^{s_1} ds_2 [Q_1(s_1) - Q_2(s_1)]\nu(s_1, s_2)[Q_1(s_2) - Q_2(s_2)],
\end{aligned}
\tag{2.4}
$$

where

$$
\begin{aligned}
\mu(\tau_1, \tau_2) &= -\int_0^\infty d\omega \,\frac{\varepsilon^2}{2\pi\omega} \sin\omega(\tau_1 - \tau_2), \\
\nu(\tau_1, \tau_2) &= \int_0^\infty d\omega \,\frac{\varepsilon^2}{2\pi\omega} \coth\frac{\pi\omega}{a} \cos\omega(\tau_1 - \tau_2),
\end{aligned}
\tag{2.5}
$$

are respectively the noise and dissipation kernels arising for the same detector at rest in a bath of thermal oscillators at temperature $T = a/2\pi$ (Feynman and Vernon [1963] and Caldeira and Legget [1983]).

3 Fluctuation in Acceleration

In displaying the Unruh effect in the last section, it was assumed that the position degree of freedom of the detector was not quantum mechanical, as the detector followed a determined trajectory. From a path integral point of view, this treatment would require that the uniformly accelerating trajectory sufficiently decohere from other neighboring paths to form a consistent history. But of course the decoherence is always finite at a given time. We now consider the uncertainty associated with the fluctuations in the detector's paths around the uniformly accelerating trajectory. For simplicity, we assume that the value of the monopole Q is given by its quantum mechanical average, supposed to be constant in time. To study the extent to which the acceleration is well defined, we consider two uniformly accelerating trajectories with slightly different accelerations: $x_i = \sqrt{a_i^{-2} + t^2}$ with $i = 1, 2$. For the scalar field in (2.1) with interaction in (2.3) , the sum over modes can be converted into an integral, and we then find (Demers [1995])

$$
\Gamma_I(a_1, a_2, t) = \frac{\varepsilon^2 Q^2}{8\pi} \int_0^t ds_1 \int_0^{s_1} ds_2 \; P(s_1, s_2, a_1, a_2),
\tag{3.1}
$$

where

$$
P(s_1, s_2, a_1, a_2) \equiv \ln\left(\frac{C^2 - F^2}{A^2 - F^2} \; \frac{D^2 - F^2}{B^2 - F^2}\right)^2 .
\tag{3.2}
$$

and

$$
\begin{aligned}
A &\equiv \sqrt{a_1^{-2} + {s_1}^2} - \sqrt{a_1^{-2} + {s_2}^2}, & B &\equiv \sqrt{a_2^{-2} + {s_1}^2} - \sqrt{a_2^{-2} + {s_2}^2}, \\
C &\equiv \sqrt{a_1^{-2} + {s_1}^2} - \sqrt{a_2^{-2} + {s_2}^2}, & D &\equiv \sqrt{a_2^{-2} + {s_1}^2} - \sqrt{a_1^{-2} + {s_2}^2}, \\
F &\equiv s_1 - s_2 \,.
\end{aligned}
\tag{3.3}
$$

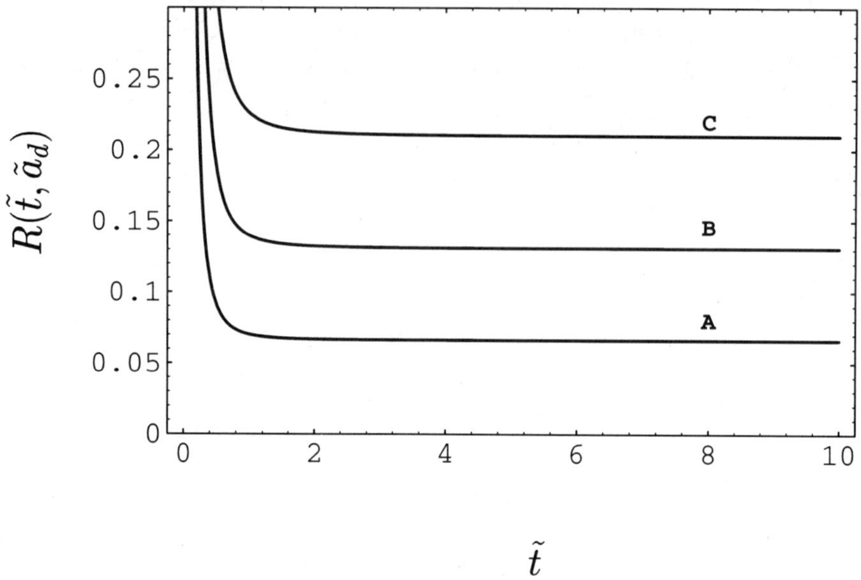

Figure 1 Plot of $R(\tilde{t}, \tilde{a}_d)$ for $\tilde{a}_d = 0.05$ (curve **A**), $\tilde{a}_d = 0.075$ (curve **B**) and $\tilde{a}_d = 0.1$ (curve **C**).

We now wish to extract the features of Γ_I for paths lasting a time large compared with the inverse accelerations ($ta_1 \gg 1, ta_2 \gg 1$). Consider the rate of increase of Γ_I with time. From (3.1),

$$\dot{\Gamma}_I(t, a_1, a_2) = \frac{\varepsilon^2 Q^2}{8\pi} t\ R(t, a_m, a_d), \tag{3.4}$$

with:

$$R(t, a_m, a_d) = \int_0^1 dy\ P(s_1 = t, s_2 = yt, a_1, a_2) = R(\tilde{t}, \tilde{a}_d). \tag{3.5}$$

In (3.5), we made the change of variable $s_2 = yt$ (y is dimensionless), and worked with the average $a_m \equiv \frac{1}{2}(a_1^{-1} + a_2^{-1})$ and difference $a_d \equiv (a_1^{-1} - a_2^{-1})$ in *inverse* accelerations. In the last line of (3.5), a_m was used for the scale factor in P, namely: $\tilde{t} \equiv t/a_m$ and $\tilde{a_d} \equiv a_d/a_m$. Now except for a transient time $\tilde{t} \lesssim 1$, we have $R(\tilde{t}, \tilde{a}_d) \rightarrow f(\tilde{a}_d)$ that is, R is time independent to a good approximation. (Obviously, $f(\tilde{a}_d)$ is an even function.) This can be seen in Figure 1, where numerical evaluations of R are given as a function of $\tilde{t}$ for a few representative values of $\tilde{a}_d$. (The apparently large value for small $\tilde{t}$ is damped to a finite and narrow peak by the factor t in (3.4).) The function $f(\tilde{a}_d)$ may be evaluated numerically, and for small $\tilde{a}_d$ is found to be nearly quadratic.

With (3.4), we thus have for $|\tilde{a}_d| \ll 1$ and $\tilde{t} \gg 1$ the excellent approximation

$$\Gamma_I(t, a_1, a_2) \approx \frac{N}{16\pi} \varepsilon^2 \langle Q \rangle^2 t^2 \left(\frac{a_1 - a_2}{(a_1 + a_2)/2} \right)^2, \tag{3.6}$$

where Q was replaced by its quantum mechanical average (quantum fluctuations of Q are neglected) and N is a numerical constant. We now take the condition $\Gamma_I \approx 1$ as separating the decohering and non-decohering accelerations. Using (3.6) and the relation $T = a/2\pi$ between temperature and acceleration, we conclude that a detector on a decohered accelerating trajectory for a time t will be subject to thermal fluctuations of the order:

$$\left|\frac{\Delta T}{T}\right| \approx \frac{1}{|\varepsilon|\langle Q\rangle t}, \tag{3.7}$$

where we dropped a constant of order one.

4 Discussion

Equation (3.7) can be understood as an uncertainty relation between the 'time of flight' of the detector and the spread in temperature, with the uncertainty given by $\frac{1}{|\varepsilon|\langle Q\rangle}$ and valid for $t/T > 1$. In Equation (3.6), the quadratic dependence of the decoherence on the coupling to the bath is general for a system coupled linearly to the bath (by construction, the amplitude of the detector's excitations act to modify the effective coupling). It is interesting to see that for the class of paths considered here, the dependence in time is also quadratic. This shows that the time is also making the coupling constant stronger or inversely, that a larger coupling constant is identical to a longer time evolution. Qualitatively, this is in line with the result of Calzetta and Mazzitelli [1990]. As the coupling to the bath gets stronger and that one waits longer, clearly more particles are created and decoherence is increased. This can also be seen through (1.4). When particles are copiously produced, a slight difference in the two paths will easily make the two states orthogonal.

Issues of decoherence similar to the ones considered here could well prove to be relevant in the context of black holes. As is well known, one of the distinguishing features of black holes is their ability to Hawking radiate, in close connection with the uniformly accelerating detector. The environment given by these radiated particles could then serve in the study of the validity of the semi-classical approximation. This issue is currently under investigation (Demers and Kiefer [1995]).

References

Anglin, J. R. [1993] *Influence Functionals and the Accelerating Detector*, Phys. Rev. D **47**, 4525–4537.

Caldeira, A. O. and Legget, A. J. [1983] *Path Integral Approach to Quantum Brownian Motion*, Physica A **121**, 587–616.

Calzetta, E. and Mazzitelli, D. [1990] *Decoherence and Particle Creation*, Phys. Rev. D **42**, 4066–4069.

Demers, J.-G. [1995] *Fluctuations of the Unruh Temperature*, Mod. Phys. Lett. A **10**, 1745–1753.

Demers, J.-G. and Kiefer, C. [1995], Phys. Rev. D.

DeWitt, B. S. [1979] *General Relativity: An Einstein Centennial Survey* (Hawking, S. W. and Israel, W., eds.), Cambridge University Press, Cambridge, England, p. 680.

Dowker, H. F. and Halliwell, J. J. [1992] *Quantum Mechanics of History - The Decoherence Functional in Quantum Mechanics*, Phys. Rev. D **46**, 1580–1609.

Feynman, R. and Vernon, F. L. [1963] *The Theory of a General Quantum System Interacting with a Linear Dissipative System*, Ann. Phys. **24**, 118–173.

Gell-Mann, M. and Hartle, J. B. [1990] *Complexity, Entropy and the Physics of Information*, Santa Fe Institute Studies in the Sciences of Complexity, Vol. VIII (Zurek, W. H., ed.), Addison-Wesley, Reading, MA.

Hu, B. L. and Matacz, A. [1994] *Quantum Brownian Motion in a Bath of Parametric Oscillators - A Model for System-Field Interactions*, Phys. Rev. D **49**, 6612–6635.

Unruh, W. G. [1976] *Notes on Black-Hole Evaporation*, Phys. Rev. D **14**, 870–892.

Zurek, W. H. [1981] *Pointer Basis of Quantum Apparatus: Into What Mixture does the Wave Packet Collapse?*, Phys. Rev. D **24**, 1516–1525.

Zurek, W. H. [1982] *Environment-Induced Superselection Rules*, Phys. Rev. D **26**, 1862–1880.

is no longer given by the Bekenstein-Hawking formula (Visser [1993], Wald [1993], Iyer and Wald [1994], Jacobson et al. [1994]). However, the entropy is given by the integral of a geometric density over a cross section of the horizon. For example, in a theory described by the effective gravitational action

$$I = \int d^4x \sqrt{-g} \left[\frac{R}{16\pi G} + \frac{\gamma}{4\pi} R^{abcd} R_{abcd} \right] , \tag{1.1}$$

the entropy of a black hole is

$$S = \oint_H d^2x \sqrt{h} \left[\frac{1}{4G} - \frac{\gamma}{4\pi} R^{abcd} \hat{\epsilon}_{ab} \hat{\epsilon}_{cd} \right] . \tag{1.2}$$

Here H is a space-like cross section of the horizon, h_{ab} is the induced metric on H, and $\hat{\epsilon}_{ab}$ is the binormal to H.

Our present understanding of black hole entropy is limited to a thermodynamic framework. Many attempts have been made to define black hole entropy within a statistical mechanical context, but despite a great deal of effort, such a microphysical interpretation is still lacking. A common feature of many of the attempts is that they yield a black hole entropy proportional to the area, but with a divergent coefficient. A recent suggestion by Susskind and Uglum [1994] is that these divergences can be absorbed in the Bekenstein-Hawking formula as a renormalization of Newton's constant. We investigate this possibility by using a common regulator for a calculation of the renormalization of the gravitational coupling constants, and a statistical calculation of black hole entropy. This allows a precise comparison of the divergences appearing in these two calculations. Our results support the suggestion of Susskind and Uglum [1994]. A more complete description of this work appears in Demers et al. [1995a].

2 Renormalization of the Gravitational Action

In the study of the one-loop effective action (see, e.g., Birrell and Davies [1982]), one may start with the gravitational action

$$I_g = \int d^4x \sqrt{-g} \left[-\frac{\Lambda_B}{8\pi G_B} + \frac{R}{16\pi G_B} + \frac{\alpha_B}{4\pi} R^2 + \frac{\beta_B}{4\pi} R_{ab} R^{ab} + \frac{\gamma_B}{4\pi} R_{abcd} R^{abcd} + \ldots \right] \tag{2.1}$$

where Λ_B is the cosmological constant, G_B is Newton's constant, while α_B, β_B and γ_B are dimensionless coupling constants for the interactions quadratic in the curvature. The subscript B indicates that these are all bare coupling constants. The ellipsis indicates that the action may also include other covariant, higher-derivative interactions, but no terms beyond those shown will be of interest in the present analysis. We also include the action for a minimally coupled neutral scalar field,

$$I_m = -\frac{1}{2} \int d^4x \sqrt{-g} \left[g^{ab} \nabla_a \phi \nabla_b \phi + m^2 \phi^2 \right] . \tag{2.2}$$

We wish to determine the effective action for the metric which results when in the path integral the scalar field is integrated out. This integration is simply gaussian, yielding the square root of the determinant of the propagator. Thus the contribution to the effective gravitational action, which is the logarithm of this result, is given by $W(g) = -\frac{i}{2} \mathrm{Tr} \log[-G_F(g, m^2)]$ (Birrell and Davies [1982]). Of course, as it stands, this expression is divergent and it must first be regulated to be properly defined.

The divergences of this one-loop effective action, as well as its metric dependence, are then easily identified using an adiabatic expansion for the DeWitt-Schwinger proper time representation of the propagator. To regulate our calculations, we choose the Pauli-Villars method. In this scheme, one introduces fictitious regulator fields with very large masses. Some of these fields are also quantized with the "wrong" statistics. For the present four-dimensional scalar field theory, one needs five regulator fields, and the total action for the matter fields becomes

$$I_m = -\frac{1}{2}\sum_{i=0}^{5}\int d^4x\,\sqrt{-g}\left[g^{ab}\nabla_a\phi_i\nabla_b\phi_i + m_i^2\phi_i^2\right]\,. \tag{2.3}$$

Here, $\phi_0 = \phi$ is the physical scalar with mass $m_0 = m$, ϕ_1 and ϕ_2 are two anticommuting fields with mass $m_{1,2} = \sqrt{\mu^2+m^2}$, ϕ_3 and ϕ_4 are two commuting fields with mass $m_{3,4} = \sqrt{3\mu^2+m^2}$, and ϕ_5 is an anticommuting field with mass $m_5 = \sqrt{4\mu^2+m^2}$. Now, each field makes a contribution to the effective action as discussed above, except that as a result of their anticommuting statistics, the contributions of ϕ_2, ϕ_3 and ϕ_5 have the opposite sign, i.e., $W(g) \simeq +\frac{i}{2}\mathrm{Tr}\log[-G_F(g,m_i^2)]$. We will focus on the divergent terms of the one-loop effective action:

$$W_{div} = \frac{1}{32\pi^2}\int d^4x\,\sqrt{-g}\,[-C\,a_0(x) + B\,a_1(x) + Aa_2(x)]\,. \tag{2.4}$$

In this expression, the coefficients a_0, a_1, a_2 are functionals of the local geometry

$$\begin{aligned} a_0 &= 1 \qquad\qquad a_1 = \frac{1}{6}R \\ a_2 &= \frac{1}{180}R^{abcd}R_{abcd} - \frac{1}{180}R^{ab}R_{ab} + \frac{1}{30}\Box R + \frac{1}{72}R^2 \end{aligned} \tag{2.5}$$

and A, B and C are constants which depend on the masses, m and μ, and which diverge for $\mu \to \infty$:

$$\begin{aligned} A =& \ln\frac{4\mu^2+m^2}{m^2} + 2\ln\frac{\mu^2+m^2}{3\mu^2+m^2} \\ B =& \mu^2\left[2\ln\frac{3\mu^2+m^2}{\mu^2+m^2} + 4\ln\frac{3\mu^2+m^2}{4\mu^2+m^2}\right] + m^2\left[\ln\frac{m^2}{4\mu^2+m^2} + 2\ln\frac{3\mu^2+m^2}{\mu^2+m^2}\right] \\ C =& \mu^4\left[8\ln\frac{3\mu^2+m^2}{4\mu^2+m^2} + \ln\frac{3\mu^2+m^2}{\mu^2+m^2}\right] + 2m^2\mu^2\left[\ln\frac{3\mu^2+m^2}{\mu^2+m^2} + 2\ln\frac{3\mu^2+m^2}{4\mu^2+m^2}\right] \\ &+ \frac{m^4}{2}\left[\ln\frac{m^2}{4\mu^2+m^2} + 2\ln\frac{3\mu^2+m^2}{\mu^2+m^2}\right]\,. \end{aligned} \tag{2.6}$$

For large values of μ, the constants A, B and C grow to leading order as $\ln(\mu/m)$, μ^2 and μ^4, respectively, but they also contain subleading and finite contributions. Combining the scalar one-loop action with the original bare action in Equation (2.1), we can identify the renormalized coupling constants in the effective gravitational action

$$
\begin{aligned}
I_{eff} &= I_g \;+\; W \\
&= \int d^4x\sqrt{-g}\Big[-\frac{1}{8\pi}\left(\frac{\Lambda_B}{G_B}+\frac{C}{4\pi}\right)+\frac{R}{16\pi}\left(\frac{1}{G_B}+\frac{B}{12\pi}\right)+\frac{R^2}{4\pi}\left(\alpha_B+\frac{A}{576\pi}\right) \\
&\qquad +\frac{1}{4\pi}R_{ab}R^{ab}\left(\beta_B-\frac{A}{1440\pi}\right)+\frac{1}{4\pi}R_{abcd}R^{abcd}\left(\gamma_B+\frac{A}{1440\pi}\right)+\dots\Big]
\end{aligned}
\tag{2.7}
$$

where in the action we discard the total derivative term $\Box R$ occurring in a_2. In particular from Equation (2.7), we obtain the renormalized Newton's constant:

$$
\frac{1}{G_R}=\frac{1}{G_B}+\frac{B}{12\pi}\ . \tag{2.8}
$$

In Equation (2.7), divergent renormalizations also occur for the cosmological constant Λ_B and the quadratic-curvature coupling constants α_B, β_B and γ_B. The higher order bare coupling constants (beyond those explicitly shown) would receive finite renormalizations but they will play no role in the present analysis.

3 Statistical Entropy

't Hooft's statistical mechanical calculation of black hole entropy ('t Hooft [1985]) involves counting the states for a thermal ensemble of a scalar field propagating just outside a black hole. Following 't Hooft's calculation, we introduce a fixed background metric, which is a Schwarzchild black hole

$$
ds^2=-\left(1-\frac{r_0}{r}\right)dt^2+\left(1-\frac{r_0}{r}\right)^{-1}dr^2+r^2(d\theta^2+\sin^2\theta\,d\varphi^2)\ . \tag{3.1}
$$

In this background, we consider a minimally coupled scalar field satisfying the Klein-Gordon equation

$$
(\Box-m^2)\phi(x)=0\ . \tag{3.2}
$$

Because of the infinite blue-shift at the horizon, 't Hooft introduced a "brick wall" cut-off: $\phi(x)=0$ for $r\le r_0+h$, with $h\ll r_0$. To eliminate infrared divergences, a second cut-off is introduced at a large radius $L\gg r_0$: $\phi(x)=0$ for $r\ge L$.

To calculate the entropy, we consider the free energy of a thermal ensemble of scalar particles at inverse temperature β

$$
\beta F=\int_0^\infty dE\,\frac{dg}{dE}\,\ln(1-e^{-\beta E})\ , \tag{3.3}
$$

where $dg(E)/dE$ is the density of modes with energy E. In the WKB appproximation, the number of states $g(E)$ is given by

$$
g(E)=\frac{2}{3\pi}\int_{r_0+h}^{L}dr\,r^2\left(1-\frac{r_0}{r}\right)^{-2}\left[E^2-\left(1-\frac{r_0}{r}\right)m^2\right]^{3/2}\ . \tag{3.4}
$$

From Equation (3.4), it is apparent that F will diverge if one takes the limit $h\to 0$. However, the divergences are removed if we add the same Pauli-Villars fields that were introduced in Section 2. With the regulator fields and $h \to 0$, the free energy becomes

$$\bar{F} = -\frac{2}{3\pi}\int_0^\infty \frac{dE}{e^{\beta E}-1}\int_0^L dr\, r^2\left(1-\frac{r_0}{r}\right)^{-2}\sum_{i=0}^{5}\Delta_i\left[E^2-\left(1-\frac{r_0}{r}\right)m_i^2\right]^{3/2}, \quad (3.5)$$

where $\Delta_0 = \Delta_3 = \Delta_4 = 1$ for the commuting fields and $\Delta_1 = \Delta_2 = \Delta_5 = -1$ for the anticommuting fields. Now the divergent contributions at the horizon are:

$$\bar{F} = -r_0^3\left[\frac{\pi}{6\beta^2}B + \frac{8\pi^3}{45\beta^4}A\right], \quad (3.6)$$

where A and B are the same constants given by Equation (2.6).

The corresponding entropy is then:

$$S_q = \beta^2\frac{\partial F}{\partial\beta}\bigg|_{\beta=4\pi r_0} = \frac{\mathcal{A}}{4}\frac{B}{12\pi} + \frac{A}{90}. \quad (3.7)$$

Using Equation (1.2), the bare entropy for the Schwarzchild black hole related to the gravitational action (2.1) is

$$S_B = \frac{\mathcal{A}}{4G_B} + 16\pi\gamma_B. \quad (3.8)$$

Adding this result to S_q yields:

$$\begin{aligned} S = S_B + S_q &= \frac{\mathcal{A}}{4}\left(\frac{1}{G_B}+\frac{B}{12\pi}\right)+16\pi\left(\gamma_B+\frac{A}{1440}\right) \\ &= \frac{\mathcal{A}}{4G_R} + 16\pi\gamma_R. \end{aligned} \quad (3.9)$$

Thus this final entropy expression contains precisely the renormalized coupling constants appearing in Equation (2.7).

4 Conclusion

Through the use of a Pauli-Villars regulator, we have found evidence for the suggestion by Susskind and Uglum [1994] — i.e., the divergences appearing in 't Hooft's statistical mechanical calculation of black hole entropy can be absorbed by a renormalization of the gravitational coupling constants. (Similar results were found by Fursaev and Solodukhin [1994] and Solodukhin [1995a] and [1995b].) We have extended these calculations to a Reissner-Nordström background, and the same conclusions follow (Demers et al. [1995a]). In the case of a non-minimally coupled scalar field, naively it appears that the divergences appearing in the statistical mechanical entropy will be different from those in the renormalization of Newton's constant. However, one obtains the correct renormalization after taking into account the appropriate degrees of freedom at the horizon (Demers et al. [1995b]).

References

Bekenstein, J. D. [1972] *Black Holes and the Second Law,* Lett. Nuov. Cim. **4**, 737–740.

——— [1973] *Black Holes and Entropy,* Phys. Rev. D **7**, 2333–2346.

Birrell, N. D., and Davies, P. C. W. [1982] *Quantum Fields in Curved Space*, Cambridge University Press, Cambridge.

Demers, J.-G., Lafrance, R., and Myers, R. C. [1995a] *Black hole entropy without brick walls,* Phys. Rev. D **52**, 2245–2253, gr-qc/9503003.

——— [1995b], in preparation.

Fursaev, D. V., and Solodukhin, S. N. [1994] *On one-loop renormalization of black hole entropy,* Preprint E2-94-462, hep-th/9412020.

Hawking, S. W. [1975] *Particle Creation by Black Holes,* Commun. Math. Phys. **43**, 199–220.

Iyer, V., and Wald, R. M. [1994] *Some Properties of Noether Charge and a Proposal for Dynamical Black Hole Entropy,* Phys. Rev. D **50**, 846–864, gr-qc/9403028.

Jacobson, T., Kang, G., and Myers, R. C. [1994] *On Black Hole Entropy,* Phys. Rev. D **49**, 6587–6598, gr-qc/ 9312023.

Solodukhin, S. [1995a] *The Conical Singularity and Quantum Corrections to Entropy of Black Hole,* Phys. Rev. D **51**, 609–617, hep-th/9407001.

——— [1995b] *On 'Nongeometric' Contributions to the Entropy of Black Holes due to Quantum Corrections,* Phys. Rev. D **51**, 618–621, hep-th/9408068.

Susskind, L., and Uglum, J. [1994] *Black Hole Entropy in Canonical Quantum Gravity and Superstring Theory,* Phys. Rev. D **50**, 2700–2711, hep-th/9401070.

't Hooft, G. [1985] *On the Quantum Structure of a Black Hole,* Nucl. Phys. B **256**, 727–745.

Visser, M. [1993] *Dirty Black Holes: Entropy as a Surface Term,* Phys. Rev. D **48**, 5697–5705, gr-qc/9307194.

Wald, R. M. [1993] *Black Hole Entropy as Noether Charge,* Phys. Rev. D **48**, 3427–3431, gr-qc/9307038.

Fields Institute Communications
Volume **15**, 1997

Matching Conditions for the 1+1 de Sitter Kinked Spacetime

Kenneth A. Dunn
Department of Mathematics, Statistics and Computing Science
Dalhousie University
Halifax, Nova Scotia, Canada B3H 3J5
dunn@cs.dal.ca

Tina A. Harriott
Department of Mathematics
Mount Saint Vincent University
Halifax, Nova Scotia, Canada B3M 2J6
tina.harriott@msvu.ca

J. G. Williams
Department of Mathematics and Computer Science
(and the Winnipeg Institute for Theoretical Physics)
Brandon University
Brandon, Manitoba, Canada R7A 6A9
williams@brandonu.ca

Abstract. Unlike the usual de Sitter spacetime, the *kinked* de Sitter spacetime requires more than one coordinate patch for its description. The matching at the boundary between two patches is checked in 1+1 dimensions by computing the extrinsic curvature. This is found to be discontinuous at the boundary so that physical consistency will require the presence of a layer of matter.

A spacetime with Finkelstein-Misner kinks (Finkelstein and Misner [1959]) requires two or more coordinate patches for its description and Finkelstein and McCollum [1975] have outlined a general procedure for describing such patches in terms of Kruskal UV-coordinates. Dunn et al. [1994] recently studied de Sitter spacetime with a kink present and constructed the different coordinate patches explicitly. The question to be answered by this present paper is whether or not these patches of de Sitter spacetime can be smoothly joined.

Because of the symmetry of the de Sitter solution, it is sufficient to pose the above question in 1+1 dimensions. A (1+1)-dimensional time-orientable one-kink spacetime is given by Dunn et al. [1992]

$$ds^2 \;=\; -\cos 2\alpha\, dt^2 - 2\,\sin 2\alpha\, dt\, dx + \cos 2\alpha\, dx^2,$$

1991 *Mathematics Subject Classification.* Primary 83C15; Secondary 83C05.
It is a pleasure to thank Greg Burnett, David Finkelstein and Peter Musgrave for helpful conversations. This work was supported by grants from the Natural Sciences and Engineering Research Council of Canada and by grant #2661 from the Brandon University Research Committee.

© 1997 American Mathematical Society

where the angle $\alpha(x)$ of orientation of the light cones is illustrated in Figure 1. When a specific functional form is chosen for $\alpha(x)$, four coordinate patches will be needed: $0 \leq x \leq A$ (which will be called the first patch), $A \leq x$ (which will be called the second patch) and similar patches for negative x. For present purposes, it will be enough to consider the $x = A$ boundary between the first and second patches.

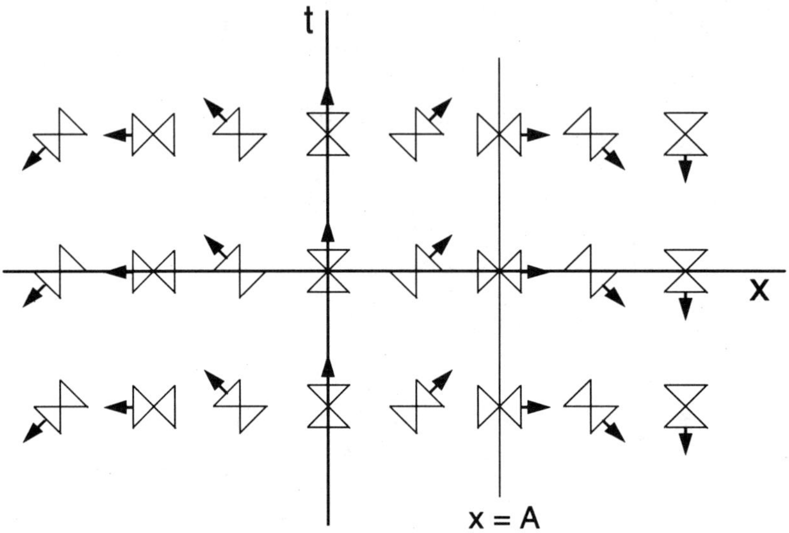

Figure 1 Kink in 1+1 dimensions

In 1+1 dimensions an appropriate field equation for empty space is simply $R = \Lambda$ (Mann et al. [1990]) and, since $R = -\partial_x^2(\cos 2\alpha)$ (Dunn et al. [1992]), it follows that for de Sitter spacetime the first patch corresponds to $0 \leq x \leq A$ with $\sin\alpha = x/A$ (Dunn et al. [1992]), and the second patch corresponds to $A \leq x \leq 2A$ with $\sin\alpha = 2 - x/A$, whence $R = \Lambda = 4A^{-2}$. The Lichnerowicz junction conditions (Lichnerowicz [1955]) require the metric tensor and its first derivatives to be continuous at the $x = A$ boundary. That the latter condition is violated is clear from Figure 2 and the fact that $\cos 2\alpha$ occurs explicitly in the metric. However, the Lichnerowicz conditions are coordinate dependent and the above failure of the second condition may be due to having chosen "bad" coordinates. The corresponding coordinate *independent* condition is that computation of the extrinsic curvature (i.e., Gauss' second fundamental form) of the $x = A$ boundary give the same answer in both patches (Israel[1966] and Lake [1987]). Matching conditions in 1+1 dimensions have been discussed by Mann and Ross [1992] who point out that the extrinsic curvature of a curve C is a *scalar*,

$$K := t^\mu t^\nu \nabla_\mu n_\nu,$$

where t^μ is the unit tangent to C and n^μ is the unit normal. For the curve $x = A$, the above metric and its Christoffel symbols (Dunn et al. [1994]) lead to

$$\begin{aligned}
(t^0, t^1) &= (-\cos 2\alpha)^{-1/2}\,(1, 0), \\
(n^0, n^1) &= (-\cos 2\alpha)^{-1/2}\,(\sin 2\alpha, -\cos 2\alpha), \\
K &= \frac{1}{2}\,(-\cos 2\alpha)^{-1/2}\,\partial_x(-\cos 2\alpha) \;=\; \partial_x(-\cos 2\alpha)^{1/2},
\end{aligned}$$

which implies $K = 2/A$ for the first patch and $K = -2/A$ for the second patch.

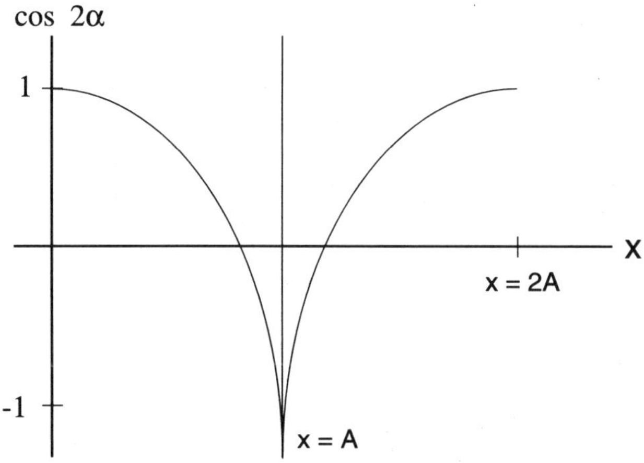

Figure 2 Behaviour of $\cos 2\alpha$ for the de Sitter kink

It is instructive to perform the above calculation in Kruskal UV-coordinates, which were used by the authors in their study of the extension of geodesics in the kinked de Sitter spacetime (Dunn et al. [1994]). In these coordinates, the metric becomes

$$ds^2 \;=\; -4a^2\kappa\,(\kappa - UV)^{-2}\,dU\,dV,$$

where $\kappa = +1$ for the first patch and $\kappa = -1$ for the second patch. The curve forming the common boundary between these patches has unit tangent vector, (appropriately directed) unit normal vector and matrix $\|\nabla n\|$ given by

$$(t^U, t^V) \;=\; (2a)^{-1}(\kappa - UV)\,(-(\sqrt{2}-1)V^{-1},\ \kappa V\,(\sqrt{2}-1)^{-1}),$$

$$(n^U, n^V) \;=\; (2a)^{-1}(\kappa - UV)\,(\kappa(\sqrt{2}-1)V^{-1},\ V\,(\sqrt{2}-1)^{-1}),$$

$$\|\nabla n\| \;=\; \frac{a}{(\sqrt{2}-1)^2}\begin{pmatrix} \kappa\,(\sqrt{2}-1)^{-1}\,V^2 & -(\sqrt{2}-1) \\ -(\sqrt{2}-1)^{-1} & \kappa\,(\sqrt{2}-1)V^{-2} \end{pmatrix}.$$

It follows that the extrinsic curvature of the boundary curve is $K = 2\kappa/A$, in agreement with the previous computation in tx-coordinates. For this lack of continuity in K to be physically acceptable, a layer of matter must be included at the interface between the two patches (Israel [1966] and Lake [1987]).

References

Dunn, K. A., Harriott, T. A. and Williams, J. G. [1992] *Toy Model for Gravitational Kinks*, J. Math. Phys. **33**, 1437–1444.

——— [1994] *Extension of the deSitter Kink Metric*, J. Math. Phys. **35**, 4145–4156.

Finkelstein, D. and McCollum, G. [1975] *Kinks and Extensions*, J. Math. Phys. **16**, 2250–2259.

Finkelstein, D. and Misner, C. W. [1959] *Some New Conservation laws*, Ann. Phys. (NY) **6**, 230–243.

Israel, W. [1966] *Singular Hypersurfaces and Thin Shells in General Relativity*, Nuovo Cimento B **44**, 1–14.

Lake, K. [1987] *Some Notes on the Propagation of Discontinuities in Solutions to the Einstein Equations*, Fifth Brazilian School of Cosmology and Gravitation (Novello, M., ed.), World Scientific, Singapore, pp. 6–15.

Lichnerowicz, A. [1955] *Théories Relativistes de la Gravitation et de l'Electro-magnétisme*, Masson, Paris.

Mann, R. B. and Ross, S. F. [1992] *Matching Conditions and Gravitational Collapse in Two Dimensional Gravity*, Class. Quantum Grav. **9**, 2335–2350.

Mann, R. B., Shiekh, A. and Tarasov, L. [1990] *Classical and Quantum Properties of Two-dimensional Black Holes*, Nucl. Phys. B **341**, 134–154.

Fields Institute Communications
Volume **15**, 1997

COBE Constraints on Kaluza-Klein Cosmologies

Valerio Faraoni
Department of Physics & Astronomy
University of Victoria
P. O. Box 3055
Victoria, B. C., Canada V8W 3P6
faraoni@uvphys.phys.uvic.ca

Fred I. Cooperstock
Department of Physics & Astronomy
University of Victoria
P. O. Box 3055
Victoria, B. C., Canada V8W 3P6
cooperstock@uvphys.phys.uvic.ca

James M. Overduin
Department of Physics & Astronomy
University of Victoria
P.O. Box 3055
Victoria, B. C., Canada V8W 3P6
overduin@uvastro.phys.uvic.ca

Abstract. A recently proposed class of Kaluza-Klein cosmologies is compared with observations of the cosmic microwave background. Some models are ruled out, while others turn out to be viable, with the number of extra dimensions being constrained by the *COBE* and Tenerife experiments.

1 Introduction

Kaluza-Klein (KK) theory (Appelquist et al. [1987], Bailin and Love [1987], Collins et al. [1989] and Duff [1994]) has found a new lease on life in supergravity (Duff et al. [1986]) and superstring theories (Green et al. [1987]), among others. In general, such theories involve scalar fields which can give rise to inflation (Hu et al. [1994] and Liddle [1994]) in some cases. Here we consider several recent KK models (Cho [1990] and [1992] and Cho and Yoon [1993]). Where they are compatible with inflation, we examine their implications for the spectral index of density perturbations, and constrain them using *COBE* and Tenerife data.

1991 *Mathematics Subject Classification.* Primary 83E15, 83F05; Secondary 57P05, 81T13.
This paper is a preliminary version of one that appeared in Int. J. Mod. Phys. D **4**, World Scientific Publishing, Singapore (1995), 387–396. This research was supported in part by a grant from the National Sciences and Engineering Research Council (NSERC) of Canada.

© 1997 American Mathematical Society

In KK theory one begins with a Lagrangian density of the usual form, but in $(4+d)$ dimensions:

$$\mathcal{L}^{(4+d)} = -\frac{m_p^2}{16\pi}(\hat{R}+\hat{\Lambda})\sqrt{-\det(\hat{g}_{AB})}, \tag{1.1}$$

where a hat means "$(4+d)$-dimensional" $(A, B = 0, 1, ..., (3+d))$. Once $\hat{g}_{AB}$ is specified, an effective theory in four dimensions follows. For example, Kaluza and Klein's original choice of $d = 1$ and:

$$(\hat{g}_{AB}) = \begin{pmatrix} g_{\mu\nu} + \phi A_\mu A_\nu & \phi A_\mu \\ \phi A_\nu & \phi \end{pmatrix}, \tag{1.2}$$

where $g_{\mu\nu}$ is the four-dimensional metric $(\mu, \nu = 0, 1, 2, 3)$ and A_μ and ϕ are an arbitrary vector and massless scalar respectively, leads to the Lagrangian density $\mathcal{L} = \mathcal{L}_{GR}(g_{\mu\nu}) + \mathcal{L}_{EM}(A_\mu) + \mathcal{L}_{KG}(\phi)$. This is the famous result that five-dimensional general relativity contains Maxwell's electromagnetism and massless Klein-Gordon scalar field theory as well as Einstein's theory.

We are interested here in the case where $d \geq 1$ and:

$$(\hat{g}_{AB}) = \begin{pmatrix} \hat{g}_{\mu\nu} & 0 \\ 0 & \hat{\phi}_{ab} \end{pmatrix}, \tag{1.3}$$

where $\hat{g}_{\mu\nu}$ is a 4-dimensional Friedmann-Lemaitre-Robertson-Walker (FLRW) metric and $\hat{\phi}_{ab}$ is a d-dimensional Riemannian metric $(a, b = 1, 2, ..., d)$, with all extra dimensions spacelike. We use units in which $\hbar = c = 1$ so that $G = m_{Pl}^{-1/2}$. The lack of off-diagonal terms in (1.3) reflects the cosmological assumption of isotropy and the belief that, during inflationary times, the universe is dominated by the scalar field(s).

2 Model of Cho

In Cho's KK cosmology (Cho [1990] and [1992]) the scalar field responsible for inflation is introduced via $\phi = \left|\det(\hat{\phi}_{ab})\right|$, and there is a *conformal transformation* (Bombelli et al. [1987], Sokolowski and Golda [1987] and Sokolowski [1989]) of the 4-dimensional metric: $g_{\mu\nu} = \sqrt{\phi}\,\hat{g}_{\mu\nu}$. The "dilaton field" σ is then defined by:

$$\sigma = \frac{1}{2}\sqrt{\frac{d+2}{d}}\ln\phi\,. \tag{2.1}$$

With these definitions, the metric (1.3) in Equation (1.1) leads to the following Lagrangian density (Cho [1990] and [1992]) in four dimensions:

$$\mathcal{L} = -\frac{m_{Pl}^2}{16\pi}\left[R + \frac{1}{2}\nabla_\mu\sigma\nabla^\mu\sigma + V(\sigma) + \lambda\left(|\det(\rho_{ab})| - 1\right)\right]\sqrt{-g}\,, \tag{2.2}$$

where $\rho_{ab} \equiv \phi^{-1/d}\hat{\phi}_{ab}$, $g \equiv \det(g_{\mu\nu})$, ∇_μ and R are the covariant derivative and the Ricci curvature of the FLRW metric $g_{\mu\nu}$, and the dilaton potential is:

$$V(\sigma) = R_K \exp\left(-\sqrt{\frac{d+2}{d}}\,\sigma\right) + \hat{\Lambda}\exp\left(-\sqrt{\frac{d}{d+2}}\,\sigma\right). \tag{2.3}$$

Here R_K is the Ricci curvature of the metric ρ_{ab} and λ is a Lagrange multiplier introduced to satisfy the constraint equation $\det(\rho_{ab}) = 1$. (For simplicity, we have omitted the gauge fields of Cho [1990] and [1992] and Cho and Yoon [1993] in this Lagrangian since the universe is dominated by the dilaton (Cho [1990]).)

The kinetic energy term in Equation (2.2) does not have the canonical form. For this we have to replace σ by a *renormalized* scalar field $\bar{\sigma}$:

$$\bar{\sigma} = \frac{\sigma}{\sqrt{16\pi G}} \,. \tag{2.4}$$

This point is crucial because the arguments of the exponentials in the scalar field potential, Equation (2.3), are changed by this renormalization. Without it, one finds (incorrectly) that Cho's model can be compatible with observation in many cases, whereas in fact it cannot.

We also introduce the renormalized quantities:

$$\bar{R}_K = \frac{R_K}{16\pi G} \,, \quad \bar{\Lambda} = \frac{\hat{\Lambda}}{16\pi G} \,, \quad \bar{\lambda} = \frac{\lambda}{16\pi G} \,. \tag{2.5}$$

The Lagrangian density (with the *correct* coupling of the scalar field) then becomes:

$$\mathcal{L} = -\left[\frac{m_{Pl}^2}{16\pi} R + \frac{1}{2} \nabla_\mu \bar{\sigma} \nabla^\mu \bar{\sigma} + W(\bar{\sigma}) + \bar{\lambda}\left(|\det(\rho_{ab})| - 1\right)\right] \sqrt{-g} \,, \tag{2.6}$$

and the correct scalar field potential is:

$$W(\bar{\sigma}) = \bar{R}_K \exp\left(-\sqrt{\frac{16\pi(d+2)}{d}} \frac{\bar{\sigma}}{m_{Pl}}\right) + \bar{\Lambda} \exp\left(-\sqrt{\frac{16\pi d}{d+2}} \frac{\bar{\sigma}}{m_{Pl}}\right) . \tag{2.7}$$

It is to be noted that the cosmological constant $\hat{\Lambda}$ has been absorbed in the potential of the new scalar field as a result of the conformal transformation and hence there is no cosmological constant in the 4-dimensional spacetime.

It is suggested by Cho [1990] that this model might describe an inflationary universe. Knowing the form of the inflaton potential, Equation (2.7), we can ask: (i) Does it actually lead to inflation? (ii) If so, is the inflationary spectrum of density perturbations consistent with *COBE* and Tenerife observations? We distinguish five different cases:

Case 1) Negative or Vanishing Potential: When $\bar{R}_K \leq 0$ and $\bar{\Lambda} \leq 0$ (as in many theories with compact or Ricci-flat internal spaces (Duff et al. [1986], Sahdev [1984], Sunahara et al. [1990] and Randjbar-Daemi and Wetterich [1986])), then $W(\bar{\sigma}) \leq 0$ and there can be no inflation, since the Friedmann equation:

$$\frac{\ddot{a}}{a} = -\frac{4\pi G}{3}(\rho + 3P) \,, \tag{2.8}$$

combined with the expressions for the energy density ($\rho = \dot{\bar{\sigma}}^2/2 + W$) and pressure ($P = \dot{\bar{\sigma}}^2/2 - W$) of the scalar field implies that:

$$\ddot{a} = \frac{8\pi G a}{3}\left(W - \dot{\bar{\sigma}}^2\right) . \tag{2.9}$$

Thus inflation ($\ddot{a} > 0$) requires $W > 0$, and Cho's model is not inflationary in these cases.

Case 2) Power-Law Inflation with $p < 1$: When $\bar{\Lambda} = 0$ and $\bar{R}_K > 0$ then the potential has the form:

$$W(\bar{\sigma}) = |\bar{R}_K| \exp\left(-\sqrt{\frac{16\pi(d+2)}{d}} \frac{\bar{\sigma}}{m_{Pl}}\right) , \tag{2.10}$$

which has the form associated with power-law inflation ($a \propto t^p$) or PLI (Abbott and Wise [1984], Lucchin and Matarrese [1985] and Liddle and Lyth [1993]):

$$V_{PLI}(\phi) = V_0 \exp\left(-\sqrt{\frac{16\pi}{p}}\,\frac{\phi}{m_{Pl}}\right). \tag{2.11}$$

By comparing Equations (2.10) and (2.11) we see that:

$$p \;=\; \frac{d}{d+2} \;<\; 1 \qquad (\text{for all } d \geq 1). \tag{2.12}$$

This is incompatible with inflation since, on differentiating $a(t)$ twice with respect to time, we find that $\ddot{a} \propto p(p-1)t^{p-2}$. Therefore Cho's model is also not inflationary in this case.

Case 3) Power-Law Inflation with $n \leq 0.9$: When $\bar{R}_K = 0$ and $\bar{\Lambda} > 0$ then the potential has the form:

$$W(\bar{\sigma}) = |\bar{\Lambda}| \exp\left(-\sqrt{\frac{16\pi d}{(d+2)}}\,\frac{\bar{\sigma}}{m_{Pl}}\right), \tag{2.13}$$

which again has the form (2.11) associated with PLI. The power-law scenario can be solved exactly, and the density perturbations are described by a power-law spectrum with index $n-1$, where $n = 1 - 2/p$ (see Liddle and Lyth [1993] for a review). Comparison of Equations (2.11) and (2.13) yields:

$$n \;=\; 1 - \frac{2d}{d+2} \;\leq\; \frac{1}{3} \qquad (\text{for all } d \geq 1). \tag{2.14}$$

The index of density perturbations is tightly constrained by experiment: *COBE* 1σ results (Smoot et al. [1992]) are $n = 1.1 \pm 0.5$ and a combined statistical analysis of 1σ *COBE* and Tenerife data (Hancock et al. 1994]) produces the limit $n \geq 0.9$. Cho's model is thus incompatible with observation in this case.

Case 4) Easther-type Model (Cho): When $\bar{R}_K < 0$ and $\bar{\Lambda} > 0$ then Cho's potential is of the type studied by Easther [1994]:

$$V(\phi) = \mathcal{A} \exp\left(-\xi\sqrt{8\pi}\,\frac{\phi}{m_{Pl}}\right) - \mathcal{B} \exp\left(-m\xi\sqrt{8\pi}\,\frac{\phi}{m_{Pl}}\right), \tag{2.15}$$

where comparison with Equation (2.7) gives:

$$\mathcal{A} = \bar{\Lambda}\;,\quad \mathcal{B} = -\bar{R}_K\;,\quad \xi = \sqrt{\frac{2d}{d+2}}\;,\quad m = \frac{d+2}{d}\,. \tag{2.16}$$

According to Easther, no values of m and ξ can produce a spectrum of perturbations consistent with microwave background anisotropies if $\xi \geq 0.5$. In the present case, $d \geq 1$ implies $\xi \geq (2/3)^{1/2} \simeq 0.82$. Therefore Cho's model is ruled out by observation in this case.

Case 5) "Pseudo-Easther"-type Model (Cho): When $\bar{R}_K > 0$ and $\bar{\Lambda} \neq 0$, then the potential has the same form as Easther's potential (2.15), except that the parameter m is less than one (which was not considered in Easther's study (Easther [1994])). Preliminary work along the same lines indicates, however, that this case too violates experimental constraints on microwave background anisotropies.

One can in any case consider the interesting limiting case $d \gg 1$, since $d \approx 40$ has been advocated by many authors (Alvarez and Belèn Gavela [1983], Abbott et al. [1984] and [1985], Barr and Brown [1984], Kolb et al. [1984] and Tomita and

Ishihara [1985]) in order to generate a large amount of entropy in the 4-dimensional universe. In the limit $d \to \infty$ we find:

$$W(\bar{\sigma}) \approx \left(\bar{R}_K + \bar{\Lambda}\right) \exp\left(-\sqrt{16\pi}\,\frac{\bar{\sigma}}{m_{Pl}}\right) . \tag{2.17}$$

This is PLI (2.11) with $p = 1$, which describes a "coasting universe" and is not inflationary.

Taking all five cases together, we conclude that Cho's theory is probably not a viable inflationary model. It is important to note that, without the renormalization of Equation (2.4), the theory would have appeared compatible with observation in Cases 3) and 4).

3 Model of Cho and Yoon

We consider now the cosmologies introduced by Cho and Yoon [1993]. In these models the starting point is Einstein-Cartan theory in $(4 + d)$ dimensions with a non-vanishing torsion tensor. This introduces a new scalar field φ, which must be renormalized in exactly the same manner as σ in Equation (2.4): $\bar{\varphi} = \varphi/\sqrt{16\pi G}$. (The potential $V(\sigma, \varphi)$ in Equation (5.2) of Cho and Yoon [1993] has to be corrected accordingly.) The particular cases given by Equations (5.6) and (5.7) of Cho and Yoon [1993] are particularly interesting, since it is suggested that they might give rise to inflation. When $\bar{\varphi} = 0$, Cho and Yoon's potential is exactly like that of Cho above, and the same conclusions follow (the model is probably not viable). When, however, one keeps $\bar{\varphi} \neq 0$ (but sets $\bar{\sigma} = 0$ for simplicity), then one obtains the following new potential:

$$\begin{aligned} W(\bar{\varphi}) &= \frac{m_{Pl}^2}{16\pi}\left\{\frac{\bar{A}}{4}\left[\exp\left(-\sqrt{\frac{32\pi}{d(d-1)}}\,(d+1)\,\frac{\varphi}{m_{Pl}}\right)\right.\right. \\ &\quad \left.\left. - (d+1)\exp\left(-\sqrt{\frac{32\pi}{d(d-1)}}\,\frac{\varphi}{m_{Pl}}\right)\right] - \Lambda\right\} , \end{aligned} \tag{3.1}$$

where $\bar{A}$ is defined by Cho and Yoon [1993]. Setting $\Lambda = 0$ for simplicity, we distinguish two cases:

Case 1) Negative or Vanishing Potential: If $\bar{A} \geq 0$ then $W(\bar{\varphi}) \leq 0$ and there is no inflation, as in Case 1) of Cho's model.

Case 2) Easther-type Model (Cho and Yoon): When $\bar{A} < 0$, on the other hand, we have a potential of the Easther form (2.15) with:

$$\mathcal{A} = \frac{m_{Pl}^2}{64\pi}\,|\bar{A}|\,(d+1) \ , \ \mathcal{B} = \frac{m_{Pl}^2}{64\pi}\,|\bar{A}| \ , \ \xi = \frac{2}{\sqrt{d(d-1)}} \ , \ m = d+1 \, . \tag{3.2}$$

Applying Easther's constraints (Easther [1994]), we find that this potential is incompatible with observation if $d < 7$, and compatible with observation if $d > 28$. For intermediate numbers of compact dimensions, $7 \leq d \leq 28$, the theory is viable for a *limited range* of values of the scalar field, $\bar{\varphi} > \bar{\varphi}_{min}$.

It is remarkable that observations of the microwave background can constrain the dimensionality of spacetime, and this is the first such case of which we are aware. Once again, we stress that without the renormalization of φ as discussed above, inaccurate conclusions would be reached about the viability of the theory. Finally, we note that observational constraints similar to the ones obtained here can probably be placed on other (possibly) inflationary KK theories such as those of Berezin et al. [1989], Yoon and Brill [1990], Chatterjee and Banerjee [1993], Fabris and Martin [1993], Sarkar and Banerji [1993], and Van den Burgh and Tavakol [1993].

References

Abbott, L. F. and Wise, M. B. [1984] *Constraints on Generalized Inflationary Cosmologies*, Nucl. Phys. B **244**, 541–548.

Abbott, R. B., Barr, S. M. and Ellis, S. D. [1984] *Kaluza-Klein Cosmologies and Inflation*, Phys. Rev. D **30**, 720–727.

______ [1985] *Kaluza-Klein Cosmologies and Inflation. II*, Phys. Rev. D **31**, 673–680.

Alvarez, E. and Belén Gavela, M. [1983] *Entropy from Extra Dimensions*, Phys. Rev. Lett. **51**, 931–934.

Appelquist, T., Chodos, A. and Freund, P. G. O. [1987] *Modern Kaluza-Klein Theories*, Addison-Wesley, Reading.

Bailin, D. and Love, A. [1987] *Kaluza-Klein Theories*, Rep. Prog. Phys. **50**, 1087–1170.

Barr, S. M. and Brown, L. S. [1984] *Entropy from Extra Dimensions*, Phys. Rev. D **29**, 2779–2782.

Berezin, V. A. et al. [1989] *Kaluza-Klein Cosmology with N Dilaton Fields*, Gen. Rel. Grav. **21**, 1177–1192.

Bombelli, L. et al. [1987] *On Energy in 5-dimensional Gravity and the Mass of the Kaluza-Klein Monopole*, Nucl. Phys. B **289**, 735–756.

Chatterjee, S. and Banerjee, A. [1993] *Kaluza-Klein Type of Inhomogeneous Cosmological Model*, Class. Quant. Grav. **10**, L1–L5.

Cho, Y. M. [1990] *Unified Cosmology*, Phys. Rev. D **41**, 2462–2471.

______ [1992] *Reinterpretation of Jordan-Brans-Dicke Theory and Kaluza-Klein Cosmology*, Phys. Rev. Lett. **68**, 3133–3136.

Cho, Y. M. and Yoon, J. H. [1993] *Geometric Symmetry Breaking and Cosmological Potential in Kaluza-Klein Theory*, Phys. Rev. D **47**, 3465–3473.

Collins, P. D. B., Martin, A. D. and Squires, E. J. [1989] *Particle Physics and Cosmology*, Wiley, New York.

Deruelle, N. and Madore, J. [1987] *A Smooth Oscillating Cosmological Solution*, Phys. Lett. B **186**, 25–28.

Duff, M. J. [1994] *Kaluza-Klein Theory in Perspective*, SISSA preprint hep-th/9410046 (1994).

Duff, M. J., Nilsson, B. E. W. and Pope, C. N. [1986] *Kaluza-Klein Supergravity*, Phys. Rep. **130**, 1–142.

Easther, R. [1994] *Inflationary Perturbations and Exponential Potentials*, preprint NZ-CAN-RE-94/1.

Fabris, J. C. and Martin, J. [1993] *Growth of Scalar Perturbations in Kaluza-Klein Theories*, Phys. Lett. B **316**, 476–482.

Green, M. B., Schwarz, J. H. and Witten, E. [1987] *Superstring Theory*, Cambridge Univ. Press, Cambridge.

Hancock, S. et al. [1994] *Direct Observation of Structure in the Cosmic Microwave Background*, Nature **367**, 333–337.

Hu, Y., Turner, M. S. and Weinberg, E. J. [1994] *Dynamical Solutions to the horizon and flatness problems*, Phys. Rev. D **49**, 3830–3836.

Kolb, E. W., Lindley, D. and Seckel, D. [1984] *More Dimensions — Less Entropy*, Phys. Rev. D **30**, 1205–1213.

Liddle, A. R. [1994] *Inflation as the unique causal mechanism for generating density perturbations on scales well above the Hubble radius*, preprint SUSSEX-AST 94/10-2.

Liddle, A. R. and Lyth, D. H. [1993] *The Cold Dark Matter Density Perturbation*, Phys. Rep. **231**, 1–105.

Lucchin, F. and Matarrese, S. [1985] *Power-law Inflation*, Phys. Rev. D **32**, 1316–1322.

Randjbar-Daemi, S. and Wetterich, C. [1986] *Kaluza-Klein Solutions with Non-Compact Internal Spaces*, Phys. Lett. B **166**, 65–68.

Sahdev, D. [1984] *Towards a Realistic Kaluza-Klein Cosmology?*, Phys. Lett. B **137**, 155–159.

Sarkar, R. P. and Banerji, S. [1993] *A Scalar Field Influenced Robertson-Walker Model*, Gen. Rel. Grav. **25**, 1107–1111.

Smoot, G. F. et al. [1992] *Structure in the COBE Differential Microwave Radiometer First-Year Maps*, Ap. J. Lett. **396**, L1–L5.

Sokolowski, L. M. [1989] *Uniqueness of the metric line element in dimensionally reduced theories*, Class. Quant. Grav. **6**, 59–76.

Sokolowski, L. M. and Golda, Z. [1987] *Instability of Kaluza-Klein Cosmology*, Phys. Lett. B **195**, 349–356.

Sunahara, K., Kasai, M. and Futamase, T. [1990] *Kaluza-Klein Inflation with Non-minimally coupled scalar field*, Prog. Theor. Phys. Lett. **83**, 353–357.

Tomita, K. and Ishihara, H. [1985] *Entropy Production Due to Gravitational-wave Viscosity in a Kaluza-Klein Inflationary Universe*, Phys. Rev. D **32**, 1935–1941.

Van den Bergh, N. and Tavakol, R. K. [1993] *Sensitivity of Dynamical Dimensional Reduction in Kaluza-Klein Cosmology*, Class. Quant. Grav. **10**, 183–192.

Yoon, J. H. and Brill, D. R. [1990] *Inflation from extra dimensions*, Class. Quant. Grav. **7**, 1253–1260.

Fields Institute Communications
Volume **15**, 1997

Cosmologies with Two-Dimensional Inhomogeneity

A. Feinstein
Dpto. Física Teórica
Universidad del País Vasco
Bilbao, Spain

J. Ibáñez
Dpto. Física Teórica
Universidad del País Vasco
Bilbao, Spain
wtpibmej@lg.ehu.es

Ruth Lazkoz
Dpto. Física Teórica
Universidad del País Vasco
Bilbao, Spain

Abstract. We present a new generating algorithm to construct exact non static solutions of the Einstein field equations with two-dimensional inhomogeneity. Infinite dimensional families of G_1 inhomogeneous solutions with a self interacting scalar field, or alternatively with perfect fluid, can be constructed using this algorithm.

One of the most important issues related to the inflationary scenario is how sensitive are the inflationary models to the change in the initial conditions (Goldwirth and Piran [1992] and Olive [1992]). The dependence of the inflationary scenario on the initial conditions is usually studied either numerically or perturbatively, under some simplifying assumptions, or analytically, within the framework of exact solutions using cosmological models with a high degree of symmetry. Until now, it was only possible to study exact scalar field cosmological models with one dimensional inhomogeneity (Aguirregabiria et al. [1993]).

We present a new generating algorithm to obtain solutions with a single isometry. The algorithm allows one to obtain spacetimes with homogeneity broken in two directions from already known ones with one dimensional inhomogeneity. One starts with a vacuum orthogonal G_2 cosmology which is well studied and some general solutions for certain topologies are known. (For a recent review see Bonnor et al. [1994].) Then, by using a known algorithm the solution is transformed into a

---

1991 *Mathematics Subject Classification.* Primary 83C15; Secondary 83F05.
This work is supported by the Spanish Ministry of Education grant (CICYT) PB93-0507. R. L.'s work is supported by Basque Government fellowship BFI94-094.

© 1997 American Mathematical Society

massless scalar field G_2 cosmology. The second step involves a conformal transformation which breaks the G_2 symmetry leaving one with the solution with the G_1 symmetry but with a self interacting scalar field with a Liouville type of potential.

To this end we consider a vacuum G_2 cosmology with the following line element

$$ds^2 = e^{f(t,z)}(-dt^2 + dz^2) + G(t,z)\left(e^{p(t,z)}dx^2 + e^{-p(t,z)}dy^2\right). \tag{1}$$

If p_v, G_v and f_v represent a vacuum solution of the Einstein field equations then, by using a simplified version of the Charach-Malin algorithm (Charach and Malin [1979]) the following solution

$$\bar{p} = Bp_v + C\log G_v, \quad \bar{f} = f_v + Ep_v + F\log G_v, \quad \bar{\phi} = Ap_v + D\log G_v \tag{2}$$

will be a solution of Einstein-masless scalar field equations, where the scalar field $\bar{\phi}$ is minimally coupled to gravity and its energy momentum tensor is given by

$$T_{\alpha\beta} = \bar{\phi}_{,\alpha}\bar{\phi}_{,\beta} - \frac{1}{2}\bar{g}_{\alpha\beta}\,\bar{\phi}_{,\gamma}\bar{\phi}^{,\gamma}. \tag{3}$$

The constants A, B, C, D, E and F are subject to the following constraints

$$BC - E + 2AD = 0, \quad C^2 + 2D^2 - 2F = 0, \quad B^2 + 2A^2 = 1. \tag{4}$$

Theorem 1 *Given a minimally coupled massless scalar field solution $\bar{g}_{ab}$ described by the equations (1)-(3), a new solution with minimally coupled scalar field with exponential potential can be built by the the conformal transformation $g_{ab} = \Omega(x)^2\,\bar{g}_{ab}$. The new scalar field is given by $\phi = \bar{\phi} + k\log\Omega(x)$ and the potential will take the form $V(\phi) = V_0\,e^{-k\phi}$. For $k^2 = 2$ the conformal factor Ω is given by $\Omega = \exp(\alpha x)$ and $V_0 = 2\alpha$. For any other value of k the conformal factor is given by*

$$\Omega = (\alpha x)^{\frac{2}{k^2-2}}, \quad V_0 = \frac{12 - 2k^2}{(k^2-2)^2}\,\alpha^2 \tag{5}$$

The constants A, B, C, F and E can be expressed in terms of the constants D and k which are related with the dynamical and potential parts of the scalar field respectively:

$$\begin{aligned} A &= \pm\frac{1}{\sqrt{k^2+2}}, \quad B = \pm\frac{k}{\sqrt{k^2+2}}, \quad C = Dk - 1 \\ F &= \frac{1}{2}(Dk-1)^2 + D^2, \quad E = \pm\frac{1}{\sqrt{k^2+2}}\,(k(Dk-1) + 2D) \end{aligned} \tag{6}$$

We will not prove the theorem here, but the results can be verified in straightforward manner by direct substitution into the Einstein field equations.

This theorem gives one a simple way to reduce the symmetry of the problem. Generically, one reduces the symmetry at the cost of introducing a potential term. However in the special case $k^2 = 6$ in the expression (5) one still obtains solutions with a massless scalar field (or stiff fluid.) One does not have to start with the G_2 symmetry but with any larger group which contains the G_2. Also, since we have not specified the character of the transitivity surface area which gradient $G_\mu G^\mu$ can be timelike, spacelike or null one can construct models starting from the metrics describing cylindrical gravitational waves, plane waves, cosmic strings, etc.. The new obtained metrics can be looked at as perfect fluid spacetimes, if one wishes, by identifying the scalar field with the velocity potential of the fluid. Thus this is the first algorithm to obtain perfect fluid solutions with G_1 symmetry.

To demonstrate the working power of the algorithm we present several examples.

Bianchi I as a seed. We start with the G_3 Bianchi I vacuum (Kasner) solution by specifying the line element (1) with the metric functions

$$G_v = t, \;\; p_v = \beta \log t, \;\; f_v = \frac{\beta^2 - 1}{2} \log t \tag{7}$$

The massless scalar field solution will be described by the metric functions

$$\bar{p} = (B\beta + C) \log t, \;\; \bar{f} = \left(\frac{\beta^2 - 1}{2} + E\beta + F \right) \log t, \tag{8}$$

while the scalar field will be given by $\bar{\phi} = (A\beta + D) \log t$, where the constants are subject to conditions (4). And finally, the solution with a self interacting scalar field with an exponential potential is obtained. The new solution represents an inhomogeneous G_2 cosmology. This solution is a particular case of more general solutions obtained in Feinstein et al. [1995] integrating directly the Einstein equations.

Cosmological gravitational waves as a seed. Here one starts with the vacuum cosmological model describing propagation of inhomogeneities in form of gravitational radiation on a spatially flat homogeneous background. To simplify the discussion we concentrate on a single mode perturbation. The vacuum solution is given by

$$\begin{aligned} G_v &= t, \;\; p_v = \beta \log t + A_0 \cos(\omega z)\, J_0(\omega t) \\ f_v &= \frac{\beta^2 - 1}{2} \log t + \beta\, A_0 \cos(\omega z)\, J_0(\omega t) + \frac{1}{4} (\omega t)^2 A_0^2 \left[J_0^2(\omega t) + J_1^2(\omega t) \right] \\ &\quad - \frac{1}{2} \omega t\, A_0^2 \cos^2(\omega z)\, J_0(\omega t)\, J_1(\omega t) \end{aligned} \tag{9}$$

From this solution one constructs the minimally massless scalar field solution and then using the theorem the exponential potential solution. Since the construction is straightforward we will not present the explicit form of the solution. The resulting new solution represents a spacetime with two dimensional inhomogeneity. The interpretation of the solution as representing propagation of gravitational waves is not at all clear now. It would be interesting to consider if the obtained solution represents the interaction of the gravitational waves with the scalar field.

The structure of the solution is much more complicated than that obtained from the Bianchi I seed. Since the new solutions are always conformally related with the seeds the asymptotic behaviour at $t = 0$ or at late times can be easily deduced from the asymptotic behaviour of the seed metric. It was shown by Charach and Malin [1979] that the massless scalar field solutions mentioned above evolve at large times, or equivalently in the high frequency limit, into an anisotropic, but spatially homogeneous model filled with spatially homogeneous scalar field and collisioneless flows of massless scalar particles and gravitons, so that the final state of evolution would be described by the so-called Doroshkevich-Zeldovich-Novikov spatially inhomogeneous anisotropic universe (Doroshkevich et al. [1968]). Near the initial singularity, on the other hand, the structure is more complicated. It is known that the conformal transformations usually change the nature of the singularity.

Also, the presence of the potential term for the scalar field changes the energy balance which can result in violating the strong energy condition. It is known that the breaking of the strong energy condition is intimately related with inflation. If the cosmological model does not deviate strongly from homogeneity, say a slightly perturbed model, then the hypersurfaces of homogeneity perfectly define a preferred observer which can decide as to whether the model inflates. If the scalar field is not very inhomogeneous the hypersurfaces of $\phi = constant$ serve to select again a preferred observer related to the motion of the perfect fluid which is defined from the scalar field. Thus, the notion of inflation in regular spacetimes is well clear. If the spacetime is highly irregular, on the other hand, neither the scalar field nor other geometric considerations help to indicate inflation. In this case one has to use somewhat a weaker way to specify inflationary behaviour. For example, one may look at the fulfillment of the strong energy condition.

In the inhomogeneous models obtained from the Bianchi I seed (the models with one dimensional inhomogeneity) the behaviour to respect inflation is quite simple. In general, at early times and for $k^2 < 2$ the strong energy condition is fulfilled and the cosmological model behaves in the standard way. Later the energy condition breaks down and the model starts to inflate. This also depends on the spatial coordinate x. In the models obtained from the G_2 seed the behaviour presents much more complicated structure: there is a "wave-like" pattern along the z direction, with energy condition broken periodically. As the time passes the model starts to inflate everywhere, and the "wave-like" pattern gets washed away.

References

Aguirregabiria, J. M., Feinstein, A. and Ibáñez, J. [1993] *Exponential-potential scalar field universes II. Inhomogeneous models*, Phys. Rev. D **48**, 4669–4675.

Bonnor, W. B., Griffiths, J. B. and MacCallum, M. A. H. [1994] *Physical interpretation of vacuum solutions of the Einstein's equations. Part II. Time dependent solutions*, Gen. Rel. Grav. **26**, 687–729.

Charach, Ch. and Malin, S. [1979] *Cosmological model with gravitational and scalar waves*, Phys. Rev. D **19**, 1058–1069.

Doroshkevich, A. G., Zeldovich, Ya. B. and Novikov, I. D. [1968] *Weakly interacting particles in the anisotropic cosmological model*, Sov. Phys. JETP **26**, 408–413.

Feinstein, A., Ibáñez, J. and Labraga, P. [1995] *Scalar field inhomogeneous cosmologies*, J. Math. Phys. **36**, 4962–4974.

Goldwirth, D. S. and Piran, T. [1992] *Initial conditions for inflation*, Phys. Rep. **214**, 223–292.

Olive, K. [1992] *Inflation*, Phys. Rep. **190**, 307–403.

Fields Institute Communications
Volume **15**, 1997

Stiff Matter G_2 Cosmologies

Stephen W. Goode
Department of Mathematics
California State University
Fullerton, CA USA 92634
stephen_goode@qmail2.fullerton.edu

Scott Bullock
Department of Mathematics
California State Univerity
Fullerton, CA USA 92634

Abstract. All stiff matter G_2 solutions of the Einstein Field Equations (EFE) are obtained when the line-element relative to comoving coordinates assumes the form

$$ds^2 = e^{2k}(-dt^2 + dx^2) + r(e^{2f}dy^2 + e^{-2f}dz^2),$$

where r and f are separable functions of the t and x coordinates. We then generalize the Wainwright-Ince-Marshman (WIM) algorithm for generating stiff matter G_2 solutions of the EFE from vacuum seed solutions to incorporate heat flow. Finally the extended WIM algorithm is applied to the Mars [1995] solution to obtain new exact G_2 solutions with heat flow.

1 Non-separable Stiff Matter G_2 Solutions of the EFE

Consider the line-element

$$ds^2 = e^{2k}(-dt^2 + dx^2) + r[e^{2f}dy^2 + e^{-2f}dz^2],$$

where r and f are separable functions in the sense that

$$r(t,x) = r_1(t)r_2(x), \quad f(t,x) = f_1(t) + f_2(x),$$

The EFE with stiff matter as source and $u = e^{-k}\frac{\partial}{\partial t}$ yield the following system of linear ODE for r and f:

$$\ddot{r}_1 - 4\epsilon\beta^2 r_1 = 0, \tag{1.1}$$

$$r_2'' - 4\epsilon\beta^2 r_2 = 0, \tag{1.2}$$

$$\ddot{f}_1 + \frac{\dot{r}_1}{r_1}\dot{f}_1 = c_1, \tag{1.3}$$

1991 *Mathematics Subject Classification.* Primary 83C15; Secondary 83C20.

© 1997 American Mathematical Society

$$f_2'' + \frac{r_2'}{r_2} f_2' = c_1, \tag{1.4}$$

where c_1 is an arbitrary constant, β is an arbitrary positive constant, and $\epsilon = \pm 1, 0$. The metric function k is determined from the linear PDE

$$\frac{\partial k}{\partial x}\frac{\dot{r}_1}{r_1} + \frac{\partial k}{\partial t}\frac{r_2'}{r_2} = \frac{1}{2}\frac{\dot{r}_1}{r_1}\frac{r_2'}{r_2} + 2\dot{f}_1 f_2', \tag{1.5}$$

$$\frac{\partial^2 k}{\partial t^2} - \frac{\partial^2 k}{\partial x^2} + \frac{\dot{r}_1}{r_1}\frac{\partial k}{\partial t} + \frac{r_2'}{r_2}\frac{\partial k}{\partial x} = 2{f_2'}^2 - \frac{{r_2'}^2}{2{r_2}^2} + 4\epsilon\beta^2. \tag{1.6}$$

The pressure and energy density are given by

$$8\pi p = 8\pi\mu = e^{-2k}({f_2'}^2 - \dot{f}_1^2 + \frac{\dot{r}_1^2}{4{r_1}^2} - \frac{{r_2'}^2}{4{r_2}^2} - \frac{\partial^2 k}{\partial t^2} + \frac{\partial^2 k}{\partial x^2}). \tag{1.7}$$

The DE (1.1) and (1.2) are easily integrated, and lead to the following distinct classes of solutions:

$$r_1(t) = \begin{cases} \left.\begin{matrix} \sinh(2\beta t), \\ \cosh(2\beta t), \\ e^{2\beta t}, \end{matrix}\right\} \text{if } \epsilon = +1, \\ \sin(2\beta t), \quad \text{if } \epsilon = -1, \\ A_1 t + B_1, \quad \text{if } \epsilon = 0. \end{cases} \qquad r_2(x) = \begin{cases} \left.\begin{matrix} \sinh(2\beta x), \\ \cosh(2\beta x), \\ e^{2\beta x}, \end{matrix}\right\} \text{if } \epsilon = +1, \\ \sin(2\beta x), \quad \text{if } \epsilon = -1, \\ A_2 x + B_2, \quad \text{if } \epsilon = 0. \end{cases}$$

As we now show, in the case when $\epsilon = \pm 1$, a uniform derivation of the solutions can be given. The $\epsilon = 0$ case requires a separate derivation and for brevity the details are omitted. When $\epsilon = \pm 1$ we can solve the linear DE (1.3) and (1.4) to obtain

$$\dot{f}_1 = \frac{1}{4\epsilon\beta^2}(c_1\frac{\dot{r}_1}{r_1} + \frac{d_1}{r_1}), \quad f_2' = \frac{1}{4\epsilon\beta^2}(c_1\frac{r_2'}{r_2} + \frac{d_2}{r_2}),$$

where d_1 and d_2 are constants. It is useful to introduce the constants δ, κ defined as follows:

$$\delta = \begin{cases} +1, \text{ if } r_1(t) = \begin{cases} \sinh(2\beta t), \\ \sin(2\beta t), \end{cases} \\ -1, \text{ if } r_1(t) = \cosh(2\beta t), \\ 0, \quad \text{if } r_1(t) = e^{2\beta t}, \end{cases} \qquad \kappa = \begin{cases} +1, \text{ if } r_2(x) = \begin{cases} \sinh(2\beta x), \\ \sin(2\beta x), \end{cases} \\ -1, \text{ if } r_2(x) = \cosh(2\beta x), \\ 0, \quad \text{if } r_2(x) = e^{2\beta x}, \end{cases}$$

We now solve the PDE (1.5) and (1.6). The key point is that (1.5) is a first order linear PDE for k and so can be solved by the method of characteristics. The general solution to this PDE is

$$k(t,x) = G(\frac{r_2}{r_1}) + \frac{1}{8\beta^4}({c_1}^2 + 4\beta^4)\ln r_1 + \frac{c_1 d_1}{8\beta^4}I_1 + \frac{d_2}{8\beta^2}(c_1 I_2 + d_1 I_3),$$

where G is an arbitrary function of r_2/r_1 and

$$I_1 = \int \frac{1}{r_1(t)}dt; \quad I_2 = \frac{1}{2\beta}\int \frac{\dot{r}_1}{r_1\sqrt{\kappa + \epsilon c^2 {r_1}^2}}dt; \quad I_3 = \frac{1}{2\beta}\int \frac{1}{r_1\sqrt{\kappa + \epsilon c^2 {r_1}^2}}dt.$$

Here c is a constant which must be replaced by r_2/r_1 upon integration. These integrals can be evaluated in closed form for each value of κ and ϵ. Substituting the above expression into (1.6) yields the following ODE for G:

$$8\beta^2(\delta q^2-\kappa)\frac{d^2G}{dq^2}+\frac{8\beta^2}{q}(\delta q^2+\kappa)\frac{dG}{dq}=\frac{1}{4\beta^4q^2}({d_2}^2-16\beta^6\kappa+4\beta^2{c_1}^2\kappa)-\frac{1}{4\beta^4 q}(1-\delta^2)(1-\kappa^2)d_1d_2, \tag{1.8}$$

where $q=r_2/r_1$. When $\delta^2+\kappa^2\neq 0$ this linear DE is easily integrated to yield

$$G(t,x)=\alpha[\delta^2\ln|\frac{{r_2}^2}{{r_1}^2}-\kappa\delta|+(1-\delta^2)(\frac{{r_2}^2}{{r_1}^2})]+\frac{\delta {d_2}^2}{128\beta^6}(1-\kappa^2)(\frac{{r_1}^2}{{r_2}^2})$$

$$+\frac{\kappa}{64\beta^6}(4\beta^2\kappa {c_1}^2+{d_2}^2-16\kappa\beta^6)\ln(\frac{r_2}{r_1}),$$

where α is an arbitrary constant. When $\delta=\kappa=0$ (1.8) reduces to

$$\frac{1}{4\beta^4q^2}{d_2}^2-\frac{1}{4\beta^4q}d_1d_2=0,$$

which implies that $d_2=0$ in this case. However, in this case, G is an arbitrary (sufficiently smooth) function of r_1/r_2. Substitution into (1.7) yields, after some calculation,

$$8\pi p=8\pi\mu=\frac{1}{16\beta^4{r_1}^2}e^{-2k}\left\{4\beta^2\delta(1-\kappa^2){c_1}^2+128\alpha\beta^8[\kappa(1-\delta^2)-\delta]\right.$$

$$\left.+16\beta^6\delta(\kappa^2+3)-{d_1}^2-\kappa\delta {d_2}^2\right\}.$$

Remark 1.1 We have checked the above solutions using Mathematica.

2 Extension of the WIM Algorithm to Include Heat Flow

We now give an extension of the WIM algorithm (Wainwright et al. [1979]) that can be used to generate G_2 solutions of the EFE when the source of the gravitational field is a stiff perfect fluid with heat flow.

Suppose that

$$ds^2=e^{2k}(-dt^2+dx^2)+r[e^{2f}(dy+wdz)^2+e^{-2f}dz^2],\qquad u=e^{-k}\frac{\partial}{\partial t},$$

is a solution of the EFE with stiff perfect fluid as source, and let $\Omega(t,x)$ be any solution to the linear PDE

$$\Omega_{,tt}-\Omega_{,xx}+\frac{1}{r}(\Omega_{,t}r_{,t}+\Omega_{,x}r_{,x})=0. \tag{2.1}$$

Then

$$ds^2=e^{2k+\Omega}(-dt^2+dx^2)+r[e^{2f}(dy+wdz)^2+e^{-2f}dz^2],\qquad u=e^{-k-\Omega/2}\frac{\partial}{\partial t},$$

is a stiff fluid solution of the EFE with heat flow where

$$8\pi p=8\pi\mu=\frac{1}{4}e^{-2k-\Omega}(\Omega_{,xx}-\Omega_{,tt}+\Omega_{,t}\frac{r_{,t}}{r}+\Omega_{,x}\frac{r_{,x}}{r})+8\pi e^{-\Omega}\mu_{\text{seed}}, \tag{2.2}$$

and the only non-zero coordinate component of the heat flow vector is

$$8\pi q_1=\frac{1}{2}e^{-k-\Omega/2}(\Omega_{,t}\frac{r_{,x}}{r}+\Omega_{,x}\frac{r_{,t}}{r}).$$

In (2.2) μ_{seed} denotes the energy density in the seed solution.

As an example we apply the above algorithm to the following singularity free stiff fluid solution in the class $B(i)$ of the Wainwright classification recently obtained by Mars [1995]:

$$ds^2 = e^{\alpha a^2 x^2} \cosh 2at(-dt^2 + dx^2) + x \left[\frac{1}{x \cosh 2at}(dy + ax^2 dz)^2 + x \cosh 2at\, dz^2 \right],$$

$$u = \frac{1}{\cosh^{1/2} 2at} e^{-(1/2)\alpha a^2 x^2} \frac{\partial}{\partial t}, \qquad 8\pi p = 8\pi\mu = \frac{a^2(\alpha - 1)e^{-\alpha a^2 x^2}}{\cosh 2at}.$$

We note that the metric functions have been assumed separable. We now apply the extended WIM algorithm to obtain a corresponding stiff fluid solution with heat flow. For the above perfect fluid solution we have $r(t, x) = x$, so that the PDE (2.1) reduces to

$$\Omega_{,tt} - \Omega_{,xx} + \frac{1}{x}\Omega_{,x} = 0. \tag{2.3}$$

For simplicity, and in keeping with the separability assumptions placed on the original metric functions, we assume that

$$\Omega(t, x) = \Omega_1(t) + \Omega_2(x).$$

It is then straightforward to derive the following solutions to (2.3):

$$\Omega_1(t) = c_1 t^2 + c_2 t + c_3, \qquad \Omega_2(x) = x^2(c_1 \ln x + c_4),$$

where c_1, c_2, c_3, c_4 are arbitrary constants. The density, pressure and heat flow vector are

$$8\pi p = 8\pi\mu = \frac{e^{-(\Omega + \alpha a^2 x^2)}}{2 \cosh 2at} \left[(2c_1 \ln x + c_1 + 2c_4) + a^2(\alpha - 1) \right],$$

$$8\pi q_1 = \frac{(c_1 t + c_2)}{x} e^{-k - \Omega/2}.$$

To the best of our knowledge this gives the first example of a solution with heat flow in the Wainwright $B(i)$ class. A full analysis of this solution and its associated thermodynamics will be given in a subsequent paper.

Obviously the extended WIM algorithm can be applied to the non-separable solutions derived earlier.

References

Agnew, A. F. and Goode, S. W. [1994] *The $p = \mu$ separable G_2 cosmologies with heat flow*, Class. Quantum Grav. **11**, 1725–1742.

Mars, M. [1995] *New nondiagonal singularity-free cosmological perfect-fluid solution*, Phys. Rev. D **51**, R3989–R3991.

Wainwright, J. [1981] *Exact spatially inhomogeneous cosmologies*, J. Phys. A **14**, 1131–1146.

Wainwright, J., Ince, W. C. W., and Marshman, B. J. [1979] *Spatially homogeneous and inhomogeneous cosmologies with equation of state $p = \mu$*, Gen. Rel. Grav. **10**, 259–271.

Fields Institute Communications
Volume **15**, 1997

Projective Symmetry in Space-Times

Graham S. Hall
Department of Mathematical Sciences
University of Aberdeen
Aberdeen AB9 2TY, UK
mth093@aberdeen.ac.uk

Abstract. There has been much recent interest in the symmetry properties of space-times but little has been done regarding the important property of projective collineations–the most general geodesic–preserving transformations. Here a technique is demonstrated to tackle this problem in a fairly general way and which links such symmetries with a type of curvature degeneracy and ultimately with the space-time holonomy group. The approach is straightforward and algebraic and lends itself in a simple way to the construction of specific examples of proper projective collineations. An alternative approach using the projective curvature tensor will also be discussed.

1 Introduction

In differential geometry and, in particular, in general relativity, geodesics play a central role. In view of the recent interest in the study of symmetries in Einstein's theory one should perhaps study the symmetries of geodesics (the so-called *projective collineations*), that is, local transformations of the space-time M with metric g which map geodesics into geodesics and which are represented in the usual way as the local diffeomorphisms associated with a vector field on M. Such a vector field is called *projective* and, probably due to the difficulties involved, there does not seem to be much available literature concerning them (except Barnes [1993] and the case when the associated collineations preserve the geodesics' affine parameters in which case they are *affine collineations* and are well understood (Hall and da Costa [1988] and Hall et al. [1994])) If X is a projective vector field on M then in any coordinate system one writes in a standard notation,

$$X_{a;b} = \frac{1}{2}h_{ab} + F_{ab} \quad (h_{ab} = h_{ba} = \mathcal{L}_X g_{ab},\ F_{ab} = -F_{ba}) \tag{1.1}$$

and X is projective if and only if

$$h_{ab;c} = 2g_{ab}\phi_c + g_{ac}\phi_b + g_{bc}\phi_a \tag{1.2}$$

1991 *Mathematics Subject Classification.* Primary 83C20; Secondary 53A20.

The author thanks David Lonie for many valuable discussions regarding projective symmetry and holonomy.

© 1997 American Mathematical Society

for some 1-form ϕ_a. X is then *affine* if $\phi_a \equiv 0$ ($\iff h_{ab;c} = 0$), *homothetic* if $h_{ab} = cg_{ab}$ (c constant), *Killing* if $h_{ab} = 0$ and *proper affine* if it is affine but neither Killing nor homothetic. If X is not affine it is called *proper projective* and if X is proper projective and $\phi_{a;b} \equiv 0$ on M it is called *special projective*.

2 The Curvature Tensor

Significant progress can be made into the theory and the finding of examples of projective collineations in space-times by classifying the latter in the first instance by the algebraic structure of their curvature tensor R_{abcd}. Such a classification has been employed usefully before (Hall and McIntosh [1983]) and consists of an exhaustive description of the space-time curvature at some $p \in M$ using four types, A, B, C and D.

Class A: The curvature tensor is of class A at $p \in M$ if there exists a unique direction at p spanned by a vector k at p such that

$$R_{abcd}k^d = 0. \tag{2.1}$$

In the usual 6×6 notation for the curvature, the latter's rank is either 2 or 3.

Class B: The curvature tensor is of class B at $p \in M$ if it has rank equal to 2 at p and it is spanned by an orthogonal timelike-spacelike pair of simple bivectors at p. In this case there are no (non-trivial) solutions to (2.1) for k.

Class C: The curvature tensor is of class C at $p \in M$ if there exist exactly two independent solutions for k of (2.1). In this case the curvature rank at p is unity.

Class D: The curvature tensor is of class D at $p \in M$ if it is non-zero and is not of class A, B or C at p. The curvature rank at p in this case is either 2,3,4,5 or 6 and there are no (non-trivial) solutions for k of (2.1).

A space-time will be said to be of *class A* (respectively *B*, *C* or *D*) if the curvature tensor is of class A (respectively B, C or D) at *all* points of M.

3 Preliminary Results

Theorem 3.1 *Let M be a space-time which admits a projective collineation X and suppose that at $p \in M$ the curvature tensor satisfies $R_{abcd}G^{cd} = 0$ for some real bivector G at p (i.e., it is not of maximal rank 6). Then if G is simple its blade consists of eigenvectors of the symmetric tensor $\phi_{a;b}$ at p each with the same eigenvalue and if G is non-simple each of the two principal complementary blades of G consists of eigenvectors of $\phi_{a;b}$ at p with the same eigenvalue.*

The proof of this result is given in Hall and Lonie [1995] and its application to the classification given in the previous section is clear from equations like (2.1). In fact the conditions of Theorem 3.1 lead to the relation

$$\phi_{a;c}G^c{}_b + \phi_{b;c}G^c{}_a = 0 \tag{3.1}$$

from which expressions for $\phi_{a;b}$, given the class of the curvature and the nature of G, follow from results in Hall [1983] and Hall and McIntosh [1983]. It can then be shown that the following theorem holds (Hall and Lonie [1995]).

Theorem 3.2 *There are no proper projective collineations in any space-time of class A or B. If M is of class C then if M admits either two independent global covariantly constant vector fields or no such vector fields then M admits no proper* *projective collineations. If M admits exactly one such vector field then any proper projective collineation admitted by M is special.*

Before proceeding to the final sections of this paper and the introduction of holonomy theory one recalls that if X is a projective collineation and W is the *Weyl projective curvature tensor* given by

$$W^a{}_{bcd} = R^a{}_{bcd} - \frac{1}{3}\left(\delta^a_c R_{bd} - \delta^a_d R_{bc}\right) \tag{3.2}$$

where $R_{ab} \equiv R^c{}_{acb}$ is the Ricci tensor, then

$$\mathcal{L}_X W = 0. \tag{3.3}$$

Now M is an *Einstein space* if and only if W is equal to the Weyl tensor and one may then prove the following result (Hall and Lonie [1995]).

Theorem 3.3 *If M is an Einstein space in which the Weyl tensor does not vanish over a non-empty open subset of M then any projective collineation admitted by M is affine (and it can only be proper affine if M is a vacuum pp-wave).*

4 Projective Symmetry and Holonomy

In this section it will be assumed that the *rank* of the curvature tensor (regarded as a 6×6 matrix) is *constant and non-zero* over M. In addition M will be assumed *simply connected* (or one could merely restrict attention to a simply connected open region of M). Then the holonomy group (Kobayashi and Nomizu [1963]) of the Levi-Civita connection of g is a connected Lie subgroup of the identity component $\mathcal{L}^0$ of the Lorentz group and thus is uniquely determined by the corresponding subalgebra of the Lie algebra A of $\mathcal{L}^0$. These subalgebras can be conveniently classified by a labelling scheme $R_1, \ldots, R_{15}$ (Schell [1961]) with R_1 the trivial subalgebra (which is excluded by the non-zero (constant) curvature rank condition) and $R_{15} = \mathsf{A}$ (the 'general' case). Each of these subalgebras leads to a holonomy group of some space-time except the type R_5 (Hall [1993] and references given therein). A detailed description of these holonomy types in terms of conventional local differential geometry can be found in Hall [1991] and [1993].

There is a close relationship between this holonomy type of M and its curvature class and the results of Section 2 lead to the following theorem (Hall and Lonie [1995]).

Theorem 4.1 *With the assumptions on M as given at the beginning of this section, M can only admit a* proper *projective collineation if its holonomy group is of type $R_9, R_{10}, R_{11}, R_{13}, R_{14}$, or R_{15}. In the R_{10}, R_{11}, and R_{13} cases any proper projective collineation is special.*

In the above theorem the holonomy condition alone is not sufficient for the existence of a proper projective collineation. For example, the R_{10}, R_{11} and R_{13} cases require the extra condition that the curvature rank be unity over M for such a collineation to exist. However, more can be said in these latter three holonomy cases and, in fact, the situation can be completely resolved. Firstly one can show that for the R_{10}, R_{11} and R_{13} cases any (necessarily special) projective collineation admitted is "unique" in the sense that if X and Y are such collineations then some linear combination of them is *affine.* Secondly, and again for these holonomy types, those metrics for which a proper projective is admitted can be completely determined.

For example, in the R_{13} case such a metric can, in appropriate coordinates, be written as

$$ds^2 = -(dx^0)^2 + (dx^1)^2 + (x^1)^2 q_{\alpha\beta}\, dx^\alpha dx^\beta \tag{4.1}$$

where Greek letters take the values 2 and 3, $q_{\alpha\beta} = q_{\alpha\beta}(x^\gamma)$ and $x^1 > 0$. The vector field given in components by $((x^0)^2, x^0x^1, 0, 0)$ is then a proper (special) projective collineation.

5 The Projective Tensor

So far nothing has been said about the projective tensor W defined in (3.2) and its "invariance" property under projective transformations represented by (3.3). The property (3.3) can be utilised to reproduce some of the results discussed earlier in this article. Suppose that g and g' are space-time metrics for M with corresponding curvature tensors R and R' and projective tensors W and W' and are such that *R and R' are both of class A or of class B or of class C at each $p \in M$.* Then if, in some coordinate system, the components of W and W' are equal, $W^a{}_{bcd} = W'^a{}_{bcd}$, it can be shown that the curvature components of g and g' are also equal, $R^a{}_{bcd} = R'^a{}_{bcd}$. Significant information is then available regarding the relation between g and g' (Hall [1983] and Hall and McIntosh [1983]). This can be translated into the language of projective collineations by letting X be a projective vector field on the space-time M (with metric g) and ψ_t the corresponding local diffeomorphisms associated with X. For some t let ψ_t map the open chart domain U of M with coordinates x^a onto the chart domain $U' = \psi_t(U)$ with coordinates $x'^a = x^a \circ \psi_t^{-1}$. Then, since X is projective, Equation (3.3) shows that the components of the projective tensor computed from the metric g in U' in the coordinates x'^a equal those of the projective tensor computed from the metric $\psi_t^{-1*}g$ in the same region and coordinates. Thus, from what was said above, the same applies to the corresponding sets of curvature components and so X is a *curvature collineation.* It is then easily checked from (3.2) that the 1-form ϕ_a associated with X (Equation (1.2)) satisfies $\phi_{a;b} = 0$ on M and so X is either special projective or affine. The calculation is completed using the techniques developed in Hall and Lonie [1995] and Lonie [1995] (together with those in Hall [1983] and Hall and McIntosh [1983]). Full details will be published elsewhere.

References

Barnes, A. [1993] *Projective Collineations in Einstein Spaces*, Class. Quant. Grav. **10**, 1139–1145.

Hall, G. S. [1983] *Curvature Collineations and the Determination of the Metric from the Curvature in General Relativity*, Gen. Rel. Grav. **15**, 581–589.

——— [1991] *Covariantly Constant Tensors and Holonomy Structure in General Relativity*, J. Math. Phys. **32**, 181–189.

——— [1993] *Proc. 5th Int. Conf. Diff. Geom. and its Appl.* ed. O. Kowolski and D. Krupka, Silesian University Press, Opava, 201–210.

Hall, G. S. and da Costa, J. [1988] *Affine Collineations in Space-Time*, J. Math. Phys. **29**, 2465–2472.

Hall, G. S. and Lonie, D. P. [1995] *Projective Collineations in Space-Time*, Class. Quant. Grav. **12**, 1007–1020.

Hall, G. S., Low, D. J. and Pulham, J. R. [1994] *Affine Collineations in General Relativity and their Fixed Point Structure*, J. Math. Phys. **35**, 5930–5944.

Hall, G. S. and McIntosh, C. B. G. [1983] *The Algebraic Determination of the Metric from the Curvature in General Relativity*, Int. J. Theor. Phys. **22**, 469–476.

Kobayashi, S. and Nomizu, K. [1963] *Foundations of Differential Geometry, Vol. 1*, Interscience, New York.

Lonie, D. P. [1995] Ph.D. Thesis, University of Aberdeen.

Schell, J. F. [1961] *Classification of Four-Dimensional Riemannian Spaces*, J. Math. Phys. **2**, 202–206.

Fields Institute Communications
Volume **15**, 1997

Light Repulsion in Higher-Dimensional Relativity

Dimitri Kalligas
GP-B, Hansen Labs.
Department of Physics
Stanford University
Stanford, CA USA 94305

Paul S. Wesson
Department of Physics
University of Waterloo
Waterloo, Ontario, Canada N2L 3G1
and
GP-B, Hansen Labs.
Deptartment of Physics
Stanford University
Stanford, CA USA 94305
wesson@astro.uwaterloo.ca

Abstract. We show that in 5D gravity the extra potential in the one-body problem can lead to light repulsion.

The world appears to have one time and three space dimensions. But the idea that it may have extra dimensions of which we are not directly aware has led to some remarkable results. These include the unification of gravity and electromagnetism in 5D by Kaluza and Klein in the 1920s (Kaluza [1921] and Klein [1926]) and the wider unification offered by 10D superstrings and 11D supergravity in more recent years (Collins et al. [1989]). Intrigued by the possibility of extra dimensions, we take general relativity with one extra dimension and ask if the latter can be observationally detected using light rays and astronomical sources. Specifically, the light-deflection problem shows an unexpected result: while certain values of parameters lead to light attraction as verified for photons passing near the Sun (Will [1992]), other values that involve the extra potential associated with the fifth dimension lead to light *repulsion*. Thus the existence of an extra dimension can be tested for using data on the lensing of light from quasars and microlensing by objects associated with the Milky Way.

In 5D general relativity, there is a class of solutions which is standard for that theory like the Schwarzschild solution is for the 4D theory (Gross and Perry [1983],

1991 *Mathematics Subject Classification.* 83E15.

© 1997 American Mathematical Society

Davidson and Owen [1985], Lim and Wesson [1992] and Wesson [1992]). (There is a class of solutions rather than a unique one because Birkhoff's theorem does not hold.) The solutions are spherically symmetric in ordinary 3D space, static, and have a line element it is convenient to write as

$$\begin{aligned} ds^2 = \left(1-\frac{2M}{r}\right)^{a} dt^2 - \left(1-\frac{2M}{r}\right)^{-(a+b)} dr^2 \\ - \left(1-\frac{2M}{r}\right)^{(1-a-b)} r^2\left(d\theta^2+\sin^2\theta\, d\phi^2\right) - \left(1-\frac{2M}{r}\right)^{b} d\psi^2 . \end{aligned} \tag{1}$$

Here the coordinates are the time (t), spherical polars (r, θ, ϕ) and a new coordinate (ψ). We have for algebraic convenience absorbed the speed of light c into t, and the gravitational constant G into the mass M of the object at the centre of the geometry. The two dimensionless constants a and b are related by a consistency relation:

$$a^2+ab+b^2=1. \tag{2}$$

This follows from the field equations of the theory, which in terms of the Ricci tensor are $R_{ij}=0$ $(i,j=0-4)$ and are satisfied by the metric (1). Some values of the constants a and b appear to be realized in Nature (e.g., the Schwarzschild case), but some do not. There are discussions of this in the literature, but they go far afield and do not seem appropriate for a short paper like this one. The quantity which is commonly called the Lagrangian for the metric (1) is

$$\begin{aligned} L = \left(1-\frac{2M}{r}\right)^{a} \dot{t}^2 - \left(1-\frac{2M}{r}\right)^{-(a+b)} \dot{r}^2 \\ - \left(1-\frac{2M}{r}\right)^{(1-a-b)} r^2\left(\dot{\theta}^2+\sin^2\theta\, \dot{\phi}^2\right) - \left(1-\frac{2M}{r}\right)^{b} \dot{\psi}^2 , \end{aligned} \tag{3}$$

where a dot denotes the derivative with respect to s or another affine parameter for particles with finite or zero rest mass, respectively. We have used (3) to investigate the classical tests of relativity in the solar system (Kalligas et al. [1995]). For these and similar problems, we can without loss of generality confine ourselves to orbits with $\theta=\pi/2$ and $\dot{\theta}=0$. Then (3) yields three constants of motion:

$$l \equiv \left(1-\frac{2M}{r}\right)^{a} \dot{t},\ \ h \equiv \left(1-\frac{2M}{r}\right)^{(1-a-b)} r^2\,\dot{\phi},\ \ k \equiv \left(1-\frac{2M}{r}\right)^{b} \dot{\psi}. \tag{4}$$

The first two of these are associated with conventional energy and angular momentum, but the third is new and measures the "velocity" of a test particle in the extra dimension. When the latter quantity is small and $b\to 0$, (2) shows that $a\to 1$ and the metric (1) gives back the physics associated with the 4D Schwarzschild solution. From the perihelion advance of Mercury and other data, this appears to be the situation in our own solar system (Lim and Wesson [1992], Will [1992] and Kalligas et al. [1995]). However, it is apparent that any departure of b from zero forces the other constant a away from unity by (2), so the dominant or first part of the metric (1) will give non-standard physics which can in principle be looked for in other astronomical systems.

An interesting non-standard effect turns out to be the deflection of light from distant astronomical sources by objects in space with metric (1). The equation of motion for a photon can be obtained from (3) by putting $L=0$, using the constants (4), and introducing (as in other problems of celestial mechanics) the variable $u \equiv 1/r$. Then we obtain

$$\left(\frac{du}{d\phi}\right)^2 - \left(1-\frac{2M}{r}\right)^{(2-2a-b)}\frac{l^2}{h^2} + \left(1-\frac{2M}{r}\right)^{(2-a-2b)}\frac{k^2}{h^2} + \left(1-\frac{2M}{r}\right)u^2 = 0. \tag{5}$$

In the weak-field limit, defined by $M/r = Mu << 1$, we can solve this to first order. The solution is most conveniently couched in terms of a new constant $p \equiv (2-a-2b)(k^2/h^2) - (2-2a-b)(l^2/h^2)$ and the distance of closest approach r_o of the photon to the source. The angle of deflection, with conventional units now restored, is given by

$$\omega = \frac{4GM}{c^2 r_o} + \frac{2GMpr_o}{c^2}. \tag{6}$$

The first term here is the usual one of 4D general relativity. The second term is new, and arises indirectly because of the presence of the extra potential in the metric (1), which modifies the motion of a photon passing near a massive object. The new term decreases with distance r_o like the first term (because p is proportional to h^{-2} which by the geometry of the orbit is proportional to r^{-2}), but its size depends on the constants a and b that describe the perturbing object.

We have made a study of how ω depends on a and b (details of which will appear elsewhere), and have found a surprising result: ω can be negative. This is the opposite sign to that found in general relativity, and means light repulsion. To see that this is possible, we note that for $k \to 0$ (negligible velocity in the fifth dimension), the constant p introduced above is $p \simeq -(2-2a-b)l^2/h^2 \simeq -(2-2a-b){r_o}^{-2}$ and (6) becomes

$$\omega \simeq (4a+2b)\frac{GM}{c^2 r_o}. \tag{7}$$

Values of a and b to be inserted here will depend on the properties of the matter of the object causing the deflection, and for purposes of illustration we choose simple but realistic values. [See Gross and Perry [1983], Davidson and Owen [1985], Wesson [1992], Wesson et al. [1993] and Ponce de Leon [1988]: in the induced-matter interpretation of Kaluza-Klein theory, wherein geometry in 5D manifests itself as matter in 4D, we need to ensure that the density and gravitational mass of the source are positive; in this theory the solar system is close to Schwarzschild with $a \simeq 1$ and $b \simeq 0$ because it is close to vacuum.] Thus we take $a = 0.5$, whence from (2) we find $b \simeq -1.15$ is allowed. Then $(4a+2b) \simeq -0.3$, so ω of (7) is negative.

This effect is in principle observable. For example, if light from a quasar passes by an object described by metric (1) with appropriate values of a and b, then we should observe repulsion or "anti-lensing". A similar effect should be observable for light from remote astronomical sources that passes near dark-matter objects associated with the Milky Way. We suggest therefore a re-examination of the statistics of quasar lensing and Galactic microlensing, to see if light repulsion is indicated (Kochanek [1993]). By this, and other classical tests (Kalligas et al. [1995]), we can hope to confirm or deny the existence of an extra dimension.

References

Collins, P. D. B., Martin, A. D. and Squires, E. J. [1989] *Particle Physics and Cosmology,* Wiley, New York.

Davidson, A. and Owen, D. A. [1985] *Black Holes as Windows to Extra Dimensions,* Phys. Lett. B **155**, 247–250.

Gross, D. J. and Perry, M. J. [1983] *Magnetic Monopoles in Kaluza-Klein Theories,* Nucl. Phys. B **1226**, 29–48.

Kalligas, D., Wesson, P. S. and Everitt, C. W. F. [1995] *The Classical Tests in Kaluza-Klein Gravity,* Astrophys. J. **439**, 548–557.

Kaluza, T. [1921] *Zum Unitätsproblem der Physik,* Sitz. Preuss. Akad. Wiss. **33**, 966–972.

Klein, O. [1926] *Quantentheorie und fünf-dimensionale Relativitätstheorie,* Z. Phys. **37**, 895–906.

Kochanek, C. S. [1993] *The Analysis of Gravitational Lens Surveys, II, Maximum Likelihood Models and Singular Potentials,* Astrophys. J. **419**, 12–29.

Lim, P. and Wesson, P. S. [1992] *The Perihelion Problem in Kaluza-Klein Gravity,* Astrophys. J. **397**, L91–L94.

Ponce de Leon, J. [1988] *Weyl Curvature Tensor in Static Spherical Sources,* Phys. Rev. D **37**, 309–317.

Wesson, P. S. [1992] *The Effective Properties of Matter of Kaluza-Klein Solitons,* Phys. Lett. B **276**, 299–302.

Wesson, P. S., Ponce de Leon, J., Lim, P. and Liu, H. [1993] *Physical Properties of Matter Derived from Geometry of Kaluza-Klein Theory,* Int. J. Mod. Phys. D **2**, 163–170.

Will, C. M. [1992] *The Confrontation Between General Relativity and Experiment: A 1992 Update,* Int. J. Mod. Phys. D **1**, 13–68.

Fields Institute Communications
Volume **15**, 1997

A No-Go Theorem in String Cosmology

Nemanja Kaloper
Department of Physics
McGill University
Montréal, Québec Canada H3A 2T8
kaloper@hep.physics.mcgill.ca

Abstract. A no-go theorem pertaining to the graceful exit problem in Pre-Big-Bang inflation is reviewed. It is shown that dilaton selfinteractions and string fluid sources fail to facilitate branch changing necessary to avoid singularities. A comment on the failure of the higher genus corrections to induce graceful exit is also included.

Inflationary paradigm is one of the cornerstones of modern cosmology (Linde [1990] and Olive [1990]). It states that there must have occurred a stage of rapid expansion of the Universe, in order to account for such large scale properties as homogeneity and isotropy, flatness, absence of topological relics etc. A simple argument illustrating the need for inflation is the so called horizon problem. Namely, we cannot reconcile the observed high degree of correlation in the background radiation (with the angular fluctuations in black-body temperature of the order $\delta T/T \propto 10^{-6}$) with the evolution of the Universe dictated by Einstein relativity and dominated by fluid sources. The Universe must have evolved out of roughly 10^5 causally disconnected domains at the time of decoupling of radiation from matter in order to achieve its present size. As the radiation did not interact from that time on, it is hard to see how the present correlation is attained. Inflationary scenario saves the day by postulating an era of rapid expansion prior to decoupling, producing a huge causally correlated region from which our Universe evolved.

We still need to construct a fully self-consistent dynamics of inflation, in agreement with observations and free of internal inconsistencies. A possibility may be found in string theory (Green et al. [1987] and Lust and Theisen [1989]). Up to date, it is the only candidate for a unified theory of interactions. If we accept it for its promise, it is important to see if the inflationary scenario, dictated by observations, can be incorporated naturally in string cosmology. This in fact could be viewed as a test of string theory. It turns out that this is not an easy task. Typically there appear new difficulties, such as the rolling scalar fields and induced

---

1991 *Mathematics Subject Classification.* Primary 83E30; Secondary 81T30, 83F05.
The author wishes to thank R. Madden and K. A. Olive for a fruitful collaboration and many helpful discussions. This report was supported in part by NSERC of Canada and in part by an NSERC postdoctoral fellowship.

© 1997 American Mathematical Society

variation of particle masses and couplings (Campbell et al. [1991] and Campbell and Olive [1995]). In response, an alternative approach towards string-driven inflation, dubbed the Pre-Big-Bang inflation, has been suggested (Gasperini and Veneziano [1992], Brustein and Veneziano [1994] and Gasperini [1994]). It strives to induce inflation deriving from genuinely stringy mechanisms, relying precisely on the rolling dilaton. The scenario is defined in the string world-sheet frame, where there exist two branches of solutions related by string scale factor duality. They are characterized by superexponential, pole-driven inflation for $t < 0$ and a milder, power-law expansion for $t > 0$. If the jump (branch change) at (or near) $t = 0$ could be made from the superinflationary phase to the power-law one, the dreaded singularity may be avoided. However, the scenario does not work in its simplest form (Brustein and Veneziano [1994] and Kaloper et al. [1995]). The desired branch changes cannot be catalyzed by dilaton self-interactions and/or string fluid sources. Here we will review this recent result, and also will comment on the failure of the higher genus corrections introduced by Damour and Polyakov [1994] to induce such branch changes (Kaloper et al. [1995]). This review is based on the joint work of R. Madden, K. A. Olive and the author (Kaloper et al. [1995]), where a more detailed account can be found.

We start with the simplest case, defined by the action

$$S = \int d^4x \sqrt{g} e^{-2\phi} \Big\{ \frac{R}{2} + 2\partial_\mu \phi \partial^\mu \phi - \Lambda(\phi) \Big\}. \tag{1}$$

Here ϕ is the dilaton and $\Lambda(\phi)$ is the dilaton potential. We choose to work with the string world-sheet frame since it is the most transparent for the analysis; moreover, the conformal transformation relating it to the Einstein frame is just a field redefinition, and as such cannot affect the conclusions about the physics of this system. We have ignored the contribution of the axion since during expansion an axion dominated universe can be expected to quickly evolve to one dominated by the dilaton. Assuming a spatially flat Robertson-Walker geometry,

$$ds^2 = -dt^2 + a(t)^2 d\vec{x}^2 \tag{2}$$

we can write the two independent equations of motion in a relatively simple first order form:

$$\dot{\phi} = (3h \pm \sqrt{3h^2 + 2\Lambda(\phi)})/2 \tag{3}$$

$$\dot{h} = \pm h\sqrt{3h^2 + 2\Lambda(\phi)} - \Lambda'(\phi)/2 \tag{4}$$

with the $\pm$ sign chosen for both equations simultaneously. The variable h is just the standard Hubble parameter in string frame, defined as the logarithmic time derivative of the scale factor. These equations are easily solved when $\Lambda = 0$ resulting in four different solutions, two for each of the two branches corresponding to the choice of $(+)$ and $(-)$ sign. The $(+)$ branch is defined in the domain $t < 0$ and the $(-)$ branch in $t > 0$. The solutions are (Brustein and Veneziano [1994])

$$a = a_0 |t|^{\mp \frac{1}{\sqrt{3}}}, \; h = \mp \frac{1}{\sqrt{3}t}, \; \phi = \phi_0 + \frac{\mp\sqrt{3} - 1}{2} \ln|t| \text{ for } t < 0 \text{ ((+) branch)}$$

$$a = a_0 t^{\pm \frac{1}{\sqrt{3}}}, \; h = \pm \frac{1}{\sqrt{3}t}, \; \phi = \phi_0 + \frac{\pm\sqrt{3} - 1}{2} \ln t \text{ for } t > 0 \text{ ((−) branch)}. \tag{5}$$

We will focus here on the expanding solutions. The expanding solution for $t < 0$ begins in the weak coupling regime (ϕ large and negative) and evolves toward the strong coupling region (ϕ positive), compatible with the form of the action which represents a weak-coupling truncation of the full effective action of string theory. The expanding solution for $t > 0$ can be matched to a standard post-inflationary Robertson-Walker cosmology, by turning on dilaton self-interactions. We repeat here that for $t < 0$ we have a pole-driven expansion, reaching the singularity at $t = 0$, and for $t > 0$ we get a power-law expanding universe emerging from the singularity. If the singularity at the pole $t = 0$ were removed and the two branches joined sm oothly, the resulting solution would represent a completely nonsingular cosmology. In the region of the order of Planck scale the superinflating branch would metamorphose to a cooling, expanding universe which can be joined onto our own as $t \to \infty$ (Brustein and Veneziano [1994]). We note that the evolution should eventually lead to the decoupling of the dilaton by trapping it in a fixed point (potential well) in order to match the solution to a late time cosmology. The high curvature region around $t = 0$ would resemble the Big Bang and therefore, in addition to possibly solving the problems usually assigned to inflation, it would also give an elegant resolution to the question of initial singularity.

The mechanism of branch changing should be responsible for graceful exit. We now turn to it. It is obvious from the equations of motion that the two branches can never connect smoothly in the regions where the potential is positive (cf. Equations (3) and (4)) since we must match derivatives at the point of contact. This requires that the potential is negative in a certain region. If we represent the dynamics by the phase space portrait in the phase plane (ϕ, h), the regions where branch changes can occur are closed curves symmetric around the ϕ-axis, given by $3h^2 + 2\Lambda = 0$. They were conveniently named the "eggs" because of their concave shape in the regions containing a single negative minimum of Λ. Unfortunately, in our case although eggs lead to branch changes, there is no graceful exit since branch changes always occur in pairs. To see it, we define the egg function: $e = \sqrt{3h^2 + 2\Lambda(\phi)}$, which satisfies $\dot{e} = \pm(2h\dot{\phi} - \dot{h})$. By dividing both sides by $\dot{\phi}$ and integrating the result over ϕ along a trajectory, we obtain:

$$\pm(e(t_1) - e(t_0)) + h(t_1) - h(t_0) = 2 \int_{\phi(t_0)}^{\phi(t_1)} h d\phi. \tag{6}$$

The integral here is to be understood as a line integral along the path of the system between $\phi(t_0)$ and $\phi(t_1)$. In fact this is the equation of phase trajectory in integral form. The positivity of the integral in relevant cases is the input needed to obtain the no-go theorem.

This comes about as follows. From looking at the flows in the phase space, we see that any trajectory hitting the top of the egg must have come from the left, and any trajectory hitting below the egg must have come from the right. A case when a trajectory flows around the egg without hitting it for "half" a cycle is also consistent with the above remark. Then we can show that 1) no $(-)$ branch trajectory, originating from anywhere on the upper side of the egg can escape over the right end of the egg but must rehit it, and 2) no $(+)$ trajectory coming from the right and flowing below the egg hits the egg below, or on, the ϕ-axis. From this, we conclude that any $(+)$ branch entering an egg region from the left must go over the top of the egg, possibly experiencing several branch changes, and must exit the region of the egg while still on the $(+)$ branch. No $(+)$ branch entering an egg region from the right can hit below, and it must remain $(+)$ while flowing under the egg. Thus any $(+)$ trajectory entering the egg region cannot leave on a $(-)$ branch, and there is no "graceful exit". The egg can only convert $(-)$ to $(+)$. Clearly, multiple eggs cannot change this conclusion.

To prove the above statements, one notices that for the first case the integral formula applied for the $(-)$ bounce between the hit point (t_0) and the terminus of the egg (t_1) rules this option out. In this region the flow is to the right and $h \geq 0$. Therefore, the integral is equal to the area between the segment and the ϕ-axis and hence strictly positive $\int_{\phi(t_0)}^{\phi(t_1)} h d\phi = \mathcal{A} > 0$. Substituting the corresponding parameters in the integral formula we arrive at the sought contradiction:

$$0 < 2\mathcal{A} = -(\sqrt{3} - 1)h(t_1) - h(t_0) \leq 0. \tag{7}$$

Therefore, the $(-)$ bounce emerging from the upper side of the egg must rehit it, as we claimed. A similar contradiction shows the $(+)$ trajectory cannot hit below the egg. Furthermore, there may be several complications regarding the pathological trajectories hitting the singular egg points, corresponding to inflections of the potential or local maxima at $\Lambda = 0$. Although these are not covered by the above arguments, they do not lead to graceful exit either (Kaloper et al. [1995]).

As we have mentioned before, this no-go theorem can be generalized to the case when stringy fluid sources are present. In this case, the phase space of the model should be extended to three dimensions, the third coordinate being the energy density of the fluid ρ. The associated equations of motion are given by the following generalization of (3–4) (Brustein and Veneziano [1994] and Kaloper et al. [1995]):

$$\begin{aligned}
\dot{\phi} &= (3h \pm \sqrt{3h^2 + 2\Lambda(\phi) + \rho \exp{(2\phi)}})/2 \\
\dot{h} &= \pm h\sqrt{3h^2 + 2\Lambda(\phi) + \rho \exp{(2\phi)}} - \Lambda'(\phi)/2 + \frac{\gamma}{2}\rho \exp{(2\phi)} \\
\dot{\rho} &= -3(1+\gamma)h\rho.
\end{aligned} \tag{8}$$

Here $\gamma = p/\rho \in (-1/3, 1/3)$ is a constant representing the fluid equation of state. The physical restriction $\rho \geq 0$ is consistent with the equations of motion, as the ρ flow terminates at the $\rho = 0$ plane, which is like a potential barrier. The trajectories completely confined in this plane are governed by our previous theorem, so there is no graceful exit for them. Now we look at the fully three-dimensional trajectories. The egg function is given by $e = \sqrt{3h^2 + 2\Lambda(\phi) + \rho \exp{(2\phi)}}$ and the modified integral formula when sources are present is

$$\pm(e(t_1) - e(t_0)) + h(t_1) - h(t_0) = 2\int_{\phi(t_0)}^{\phi(t_1)} h d\phi + \frac{1+\gamma}{2}\int_{t_0}^{t_1} \rho e^{2\phi} dt. \tag{9}$$

This equation differs from (6) only in the presence of the last term, which is a nonnegative quantity for all trajectories. The other relevant characteristics of the phase space are also easily generalized. Without delving into the details, we note that qualitatively the picture remains the same: the egg is now a two-dimensional compact surface cut by the plane $\rho = 0$. Trajectories flow around the egg along helical paths, which if projected onto the $\rho = 0$ plane turn clockwise. We then need to consider trajectories crossing the cylindrical surface enclosing the egg, obtained by translating the curve representing the boundary of the egg in the $h = 0$ plane vertically upwards. We can see that the first integral on the LHS, representing the area enclosed by the projection of the trajectory onto the $\rho = 0$ plane, remains positive for all such trajectories. The second integral is always nonnegative, as we mentioned above. Thus all the arguments for the sourceless case extend to this case, again preventing favorable branch changes from occurring. As no new pathologies appear, we conclude that the no-go theorem must hold for this case too.

We note in passing that dilaton-dependent corrections to couplings, coming from the higher genus terms (Damour and Polyakov [1994], Damour and Vilenkin [1995]), also fail to induce the desired graceful exit (Kaloper et al. [1995]). In that case, the failure is caused solely by the flows around the eggs which preclude even the possibility for the needed hits on the egg. Thus the only option to consistently incorporate Pre-Big-Bang cosmology which remains open is to resort to higher derivative terms in the α' expansion (Kiritsis and Kounnas [1994a,b]). However this approach must consider systematically all the terms in the α' expansion, and thus should be implemented via the conformal field theory approach. At this moment, it appears that this goal is still beyond our means.

References

Brustein, R. and Veneziano, G. [1994] *The Graceful Exit Problem in String Cosmology*, Phys. Lett. B **329**, 429–434.

Campbell, B. A., Linde, A. D. and Olive, K. A. [1991] *Does String Theory Lead to Extended Inflation?*, Nucl. Phys. B **355**, 146–161.

Campbell, B. A. and Olive, K. A. [1995] *Nucleosynthesis and the Time Dependence of Fundamental Couplings*, Phys. Lett. B **345**, 429–434.

Damour, T. and Polyakov, A. M. [1994] *The String Dilaton and a Least Coupling Principle*, Nucl. Phys. B **423**, 532–558.

Damour, T. and Vilenkin, A. [1995] *String Theory and Inflation*, preprint IHES-P-95-26/hep-th/9503149.

Gasperini, M. and Veneziano, G. [1992] *Pre-Big Bang in String Cosmology*, Astroparticle Phys. **1**, 317–339.

Gasperini, M. [1994] *The Inflationary Role of the Dilaton in String Cosmology*, preprint DFTT-29/94/astro-ph/94 07026.

Green, M., Schwarz, J. and Witten, E. [1987] *Superstring Theory*, Cambridge U.P., Cambridge.

Kaloper, N., Madden, R. and Olive, K. A. [1995] *Towards a Non-singular Inflationary Universe?*, preprint McGill 95-17/UMN-TH-1333/95/hep-th/9506027.

Kiritsis, E. and Kounnas, C. [1994a] *Dynamical Topology Change in String Theory*, Phys. Lett. B **331**, 51–62.

——— [1994b] *Dynamical Topology Change, Compactification and Waves in a Stringy Early Universe*, preprint CERN-TH-7328-94/hep-th/9407005.

Linde, A. D. [1990] *Particle Physics and Inflationary Cosmology*, Harwood, Chur.

Lust, D. and Theisen, S. [1989] *Lectures on String Theory*, Springer, Berlin.

Olive, K. A. [1990] *Inflation*, Phys. Rep. **190**, 307–403.

Fields Institute Communications
Volume **15**, 1997

MapleTensor: A New System for Performing Indicial and Component Tensor Calculations by Computer

Masoud Kavian
Department of Applied Mathematics
University of Waterloo
Waterloo, Ontario, Canada N2L 3G1
mkavian@daisy.uwaterloo.ca

R. G. McLenaghan
Department of Applied Mathematics
University of Waterloo
Waterloo, Ontario, Canada N2L 3G1
rgmclenaghan@daisy.uwaterloo.ca

K. O. Geddes
Department of Computer Science
University of Waterloo
Waterloo, Ontario, Canada N2L 3G1
kogeddes@daisy.uwaterloo.ca

1 Introduction

MapleTensor (Kavian et al. [1995]) is a system for symbolic tensor computations which uses the facilities of the Maple computer algebra system. It is designed to perform both indicial and component tensor calculations in a natural and straightforward way. Some of the capabilities of MapleTensor are similar to the existing tensor packages, ITENSOR (Bogen and Pavelle [1977]) in MACSYMA, STENSOR (Hornfeldt [1988]) in SHEEP and MATHTENSOR (Christensen and Parker [1990]) in MATHEMATICA.

MapleTensor allows the operator to introduce user defined tensors with all their associated properties. For example the Riemannian tensor R_{abcd} is represented by **R(abcd)** in input. Covariant indices are represented by lower case latin letters while contravariant indices are represented by capital latin letters. For example the contravariant Riemannian tensor $R_{ab\ c}^{\ \ a}$ is entered as **R(abAc)** and MapleTensor returns R_{bc}.

All other tensor operations for example symmetrizations, the tensor product, contractions and covariant derivative are represented in the same way as they

1991 *Mathematics Subject Classification.* Primary 68Q40; Secondary 53A45.

One of the authors (M. Kavian) wishes to acknowledge NSERC and Waterloo Maple Software for the award of an Industrial NSERC Scholarship. This work was supported in part by NSERC Research Grants (K. O. Geddes & R. G. McLenaghan).

© 1997 American Mathematical Society

would be handwritten or typeset. In contrast with other systems, all operations in MapleTensor are performed using only one function called `tensor`. The following statements illustrate the use of this function.

Any input to the package will be processed using a large database of identities and substitution rules either to reduce the expression to canonical form or to derive other related algebraic expressions. For example the input **R(adbc)** is transformed to $-R_{abcd} - R_{acdb}$ by use of the cyclic identity.

In contrast to current systems MapleTensor gives considerable flexibility to the user in calculating the components of tensorial objects and invariants.

In order to permit the user to identify the optimal sequence of rules required to simplify a given tensor expression an expert system which operates in a non-deterministic way is under development. For this purpose a special programming language is being written. To increase the efficiency of MapleTensor, the package is mainly being written in C++.

2 Some Basic Tensorial Operations in MapleTensor

> # Covariant Differentiation of the Riemannian tensor.

> expr := tensor(R('adcE;b'),convertToPartials);

$$R_{adc\ ,b}^{\ \ \ e} - R_{pdc}^{\ \ \ e}\Gamma_{ab}^{\ \ p} - R_{apc}^{\ \ \ e}\Gamma_{bd}^{\ \ p} - R_{adp}^{\ \ \ e}\Gamma_{bc}^{\ \ p} + R_{adc}^{\ \ \ p}\Gamma_{bp}^{\ \ e}$$

> # Multivariate Covariant Differentiation of the Maxwell Tensor.

> expr := tensor(F('ab;ce'), multiCovDiff) ;

$$F_{ab,ce} - F_{pb,c}\Gamma^{p}{}_{ae} - \Gamma^{p}{}_{be}F_{ap,c} - \Gamma^{p}{}_{ce}F_{ab,p} - \Gamma^{p}{}_{ac}F_{bp,e} + \Gamma^{p}{}_{ac}\Gamma^{q}{}_{be}F_{pq} - \Gamma^{p}{}_{ac,e}F_{bp} + \Gamma^{p}{}_{qc}\Gamma^{q}{}_{ae}F_{bp}$$
$$+\Gamma^{p}{}_{aq}\Gamma^{q}{}_{ce}F_{bp} - \Gamma^{p}{}_{bc}F_{ap,e} - \Gamma^{p}{}_{bc,e}F_{ap} + \Gamma^{p}{}_{ae}\Gamma^{q}{}_{bc}F_{pq} + \Gamma^{p}{}_{qc}\Gamma^{q}{}_{be}F_{ap} + \Gamma^{p}{}_{bq}\Gamma^{q}{}_{ce}F_{ap}$$

> # Contraction with the metric tensor.

> expr := tensor(g(FE)*g(CB)*Riem(abcd)*Riem(AefD));

$$R_{ad}R^{ad}$$

> # Antisymmetrization of the Riemannian tensor.

> Riem('d[bac]');

$$-1/3R_{acbd} + 1/3R_{adbc} + 1/3R_{abcd}$$

> # Changing the metric to the Schwarzschild metric.

> Metric := Schwarzschild;

> # Computing the Kretschmann scalar.

> tensor(R(abcd)* R(ABCD), components) ;

$$48G^2M^2/r^6$$

3 Definition of Tensors in MapleTensor

A tensor in MapleTensor can be defined by writing to a file all the properties of the tensor. A compiler with the help of a pattern matcher is then able to compile and interpret each line of the file. The following example illustrates the concept in the case of the Riemannian tensor.

> # Defining the Riemannian tensor.

> tensor(defineTensor,tensorNames= { R,Riem,Riemannian });

The file Riemannian.def was successfully compiled.

> # Printing the content of the file Riemannian.def.

> # show('Riemannian.def');

```
# symmeries of the Riemannian tensor.
groupName = SYMMETRIES;
SYM1::R(bacd)= -R(abcd); sensitives= {a,b};
SYM2::R(abdc)= -R(abcd); sensitives= {c,d};
SYM3::R(cdab)= R(abcd) ; sensitives= {a,b,c,d };
# Symmetric AntiSymmetric Product.
# SYM_ANTISYM:R(abcd)*R(AB)=0;
# Cyclic Identity
CYC::R(adbc)= -R(abcd)-R(acdb);
# Bianchi Identity
BIANCHI::R(abde;c)=-R(abec;d)-R(abcd;e); sensitives={b,e,c,d};
# Contraction of indicies.
groupName = CONTRACTION
CONTR1::R(acCb)=-R(ab);
CONTR2::R(acCA)=-R;
# Covariant Derivatives
groupName = COVARIANT
COV1: R(abcd;A)=-R(bc;d)+R(bd;c);
COV2: R(abBc;Cd)=-R(;ad)/2;
COV3: R(abBc;dC)=R(;ad)/2+R(pa)*R(dP)-R(pq)*R(aQdP);
```

Having defined and compiled the Riemannian tensor, the system is able to browse recursively through the above set of rules or definitions with the help of the pattern matcher and can simplify a given expression containing the Riemannian tensor. The following examples show some of the capabilities of the system in such a calculation.

> # Automatic application of COV1 and CONTR1.

> R(abBd;A);

$$-R_{;d} + R_{bd;}{}^{b}$$

> **# Automatic application of SYM_ANTISYMMETRIC, CONTR1**
>
> **# and the simplification procedure.**
>
> **expr := tensor(R(abcd)*R(ACBD));**

$$0$$

4 SUPER-SIMPLIFICATION OF ALGEBRAIC EXPRESSIONS IN MAPLETENSOR.

Super-simplification of algebraic expressions containing tensorial objects is an integral part of the system which uses pattern matching and several other methods including the relabeling and reordering of indices. Attaining a high speed in super-simplification of tensorial expressions is very important since the super-simplification is itself used again in the pattern matcher which is the most frequently called procedure in a typical tensor algebraic operation.
We implemented the basic version of a super-simplification routine and we were able to attain a speed of at least one thousand times faster than MATHTENSOR (Christensen and Parker [1990]) for similar examples.

> **# Super-Simplification of the following polynomial of degree four**
>
> **# in the Riemannian tensor.**
>
> \# $R_{d}{}^{c}{}_{h}{}^{g} R^{h}{}_{g}{}^{fe} R^{d}{}_{acb} R_{fe}{}^{ba} + R^{c}{}_{d}{}^{g}{}_{h} R^{d}{}_{c}{}^{ab} R_{feab} R^{feh}{}_{g} - R^{g}{}_{h}{}^{e}{}_{f} R_{g}{}^{h}{}_{c}{}^{d} R_{d}{}^{b}{}_{a}{}^{c} R^{f}{}_{e}{}^{a}{}_{b}$
>
> **expr := tensor(R(dChG)*R(HgFE)*R(Dacb)*R(feBA) +**
> **R(CdGh)*R(DcAB)*R(feab)*R(FEHg) -**
> **R(GhEf)*R(gHcD)*R(dBaC)*R(FeAb),superSimplify) ;**

$$0$$

References

Bogen, R. and Pavelle, R. [1977] *Indicial Tensor Manipulation on Macsyma*, Lett. Math. Phys. **2**, 55–59.

Christensen, S. and Parker, L. [1990] *MathTensor: A System for Doing Tensor Analysis by Computer*, Mathematica Journal **1**, 51–61.

Hornfeldt, L. [1988] *STENSOR Reference Manual, Version 2.3*, Univ. of Stockholm, Dept. of Theor. Phys.

Kavian, M., McLenaghan, R. G. and Geddes, K. O. [1995] *Proceeding of the 7th Marcel Grossman Conference*, World Scientific, Singapore, to appear.

Fields Institute Communications
Volume **15**, 1997

Low-Energy Scattering of Extreme Black Holes in String Theory

Ramzi R. Khuri
Department of Physics
McGill University
Montréal, PQ, Canada H3A 2T8
and
CERN
CH-1211, Geneva 23, Switzerland

Robert C. Myers
Department of Physics
McGill University
Montréal, PQ, Canada H3A 2T8
rcm@hep.physics.mcgill.ca

Abstract. We discuss the low-energy dynamics of generalized extremal higher membrane black hole solutions of string theory and higher membrane theories following Manton's prescription for multi-soliton solutions. A flat metric is found for those solutions which possess κ-symmetry on the worldvolume.

1 Introduction

The construction of soliton and black hole solutions of string theory and their connections with various dualities in string theory have been the subject of much recent activity (see Duff et al. [1995] and references therein). The soliton solutions typically arise as extremal limits of two-parameter charged black hole solutions. These extremal black holes saturate a Bogomol'nyi bound between their ADM mass and charge, thus ensuring their stability, in analogy with the extremal Reissner-Nordstrom black holes of Einstein-Maxwell theory. In both the string and Einstein-Maxwell cases, the saturation of this bound is associated with the existence of a "zero-force" condition, which allows for the existence of multiple extreme black holes in a static configuration. In the string context, the saturation of the Bogomol'nyi bound is also associated with the existence of spacetime supersymmetry. A further

1991 *Mathematics Subject Classification.* Primary 54C40, 14E20; Secondary 46E25, 20C20.
This research was supported by NSERC of Canada and Fonds FCAR du Québec.
Talk given by R. Khuri.

© 1997 American Mathematical Society

generalization inherent in the string scenario is that these black hole solutions may have higher membrane structure, arising in a higher-dimensional spacetime.

Most of the solutions discussed in the string context are associated with an underlying "worldvolume" supersymmetry of a higher-membrane theory (generalizing worldsheet supersymmetry in string theory), called κ-symmetry, which ensures that the correct number of degrees of freedom arise in the theory. It turns out that these κ-symmetric solutions are also associated with the preservation of the maximal amount of spacetime supersymmetry in various embeddings.

In this work, we start with the generalized extremal black holes of Shiraishi [1993a], and note that these include both κ-symmetric and non-κ-symmetric solutions. We then generalize these solutions further to arbitrary higher-membrane black holes embedded in arbitrary higher dimensions and discuss their connections with the various dualities in string and higher membrane theories.

In Section 3 we follow Manton's method (Manton [1982]) for the computation of the low-velocity scattering of both the κ-symmetric and non-κ-symmetric generalized extreme higher-membrane black holes and find that this scattering is trivial (i.e., zero dynamical force) precisely for the κ-symmetric solutions.

Finally, we discuss our results in Section 4 as well as the intriguing possibility that some of the extremal black holes we discuss might correspond to elementary states in string theory.

2 Generalized Extreme Black Hole Solutions

Consider the action

$$I_4(1) = \int d^4x\sqrt{-g}\left(R - \frac{1}{2}(\partial\phi)^2 - \frac{1}{4}e^{-a\phi}F_2^2\right), \tag{2.1}$$

where g_{MN} is the metric, ϕ the scalar dilaton and $F_2 = dA_1$ the Maxwell field strength. a is an undetermined, constant. The generalized multi-extreme black hole solution is given by (Shiraishi [1993a])

$$\begin{aligned} ds^2 &= -F(x)^{-2/(1+a^2)}dt^2 + F(x)^{2/(1+a^2)}d\vec{x}\cdot d\vec{x}, \\ \phi &= -\frac{2a}{1+a^2}\ln F, \\ A_0 &= \pm\sqrt{\frac{4}{1+a^2}}F, \\ F(x) &= 1 + \sum_{i=1}^{N}\frac{k_i}{|\vec{x}-\vec{x}_i|}, \end{aligned} \tag{2.2}$$

where $\vec{x}_i$ are the locations of the N black holes. The mass $m_i = k_i/(1+a^2)$ and electric charge $e_i = 2k_i/\sqrt{1+a^2}$ of each black hole saturate the Bogomol'nyi bound $m_i^2 \geq e_i^2/4(1+a^2)$, thus ensuring the stability of the multi-black hole configuration.

The single black hole solution represents the extremal limit of a two parameter family of charged black hole solutions in which this bound is not saturated. One can equally well find magnetically charged solutions, where the magnetic charge is related to the electric charge via the Dirac quantization condition $eg = 2\pi n$, where n is an integer. For the specific value of $a = \sqrt{3}$, the action couples to a sigma-model of a theory with a fermionic worldline symmetry called κ-symmetry, which ensures that the nonphysical degrees of freedom in the worldline theory are projected out (see Duff et al. [1995] and references therein).

It is possible to generalize the above solutions in two ways. In the first step, we can consider point-like black holes in arbitrary higher-dimensional spacetime D. The four-dimensional action (2.1) is then replaced by a D-dimensional action via $\int d^4x \to \int d^Dx$ and $I_4(1) \to I_D(1)$. Note that the argument 1 in the action represents the dimension of the worldline swept out by the point-like black holes. In this case, the κ-symmetric solutions correspond to $a^2 = 2(D-1)/(D-2)$.

Furthermore[1], we can generalize the point-like black holes themselves to higher-membrane objects, with, say, $d-1$ spatial directions (called a $(d-1)$-brane) by coupling to an antisymmetric tensor A_d with d indices. Here the replacement takes the form $\frac{1}{4}F_2^2 \to \frac{1}{2\,d!}F_{d+1}^2$, where $F_{d+1} = dA_d$ is the field strength associated with A_d in further generalizing the action via $I_D(1) \to I_D(d)$.

A useful dimension to define is the "dual" dimension $\tilde{d} \equiv D - d - 2$. Then for a $(d-1)$-brane in D dimensions with "electric" charge e_d arising from the antisymmetric tensor, the natural solitonic object is a $(\tilde{d}-1)$-brane carrying topological "magnetic" charge $g_{\tilde{d}}$ (Duff et al. [1995]). These again are related by a (higher-dimensional) Dirac quantization condition. In this most general case, the κ-symmetric solutions correspond to $a^2 = 4 - 2d\tilde{d}/(D-2)$.

The existence of these generalized dual solitons is in fact the basis for the various dualities in string and higher-membrane theories. An especially interesting example of a duality which is of current interest is string/string duality in $D=6$ $(d = \tilde{d} = 2)$. An interesting consequence of this duality on reduction to four dimensions is the interchange of two previously studied dualities: the target-space T duality, which generalizes the compactification scale size duality in string theory, and the strong/weak coupling S duality (Duff and Khuri [1994]). An important implication of the latter lies in the application of perturbative techniques in the strong coupling region of string theory, which is of special interest in the attempt to understand string theory as a theory of quantum gravity. Furthermore, the existence of these dualities is likely to point to a reformulation of string theory in which these dualities are manifest.

3 Metric on Moduli Space

In the absence of exact solutions for time-dependent multi-soliton or multi-extreme black hole solutions, Manton's method for the computation of the metric on moduli space for two or more solitons represents a good low-velocity approximation to the exact dynamics.

Manton's prescription for the study of soliton scattering may be summarized as follows: One begins with a static multi-soliton solution, and gives the moduli characterizing this configuration a time-dependence. One then finds $O(v)$ corrections to the fields by solving the constraint equations of the system with time-dependent moduli. The resultant time-dependent field configuration only satisfies the full time-dependent field equations to lowest order in the velocities, but provides an initial data point for the fields and their time derivatives. Another way of saying this is that the initial motion is tangent to the set of exact static solutions. An effective action describing the motion of the solitons is determined by replacing

[1]The procedure is only outlined here. For full details see Khuri and Myers [1996]

the solution to the constraints into the field theory action. The kinetic action so obtained defines a metric on the moduli space of static solutions, and the geodesic motion on this metric determines the dynamics of the solitons (Manton [1982]).

A calculation of the metric on moduli space for the scattering of extreme Reissner-Nordstrom black holes and a description of its geodesics was worked out by Ferrell and Eardley [1987]. Shiraishi [1993b] then extended this work to black hole solutions of (2.1) for arbitrary a.

Following Ferrell and Eardley's technique, it is possible to compute this metric for the generalized, higher-membrane extreme black holes of the previous section (Khuri and Myers [1996]) . It turns out that precisely for the κ-symmetric solutions, this metric is flat, and the kinetic lagrangian describes noninteracting extreme black holes.

For the more general case, the direct computation of the metric is also possible, but rather tedious. A tremendous simplification occurs when one realizes that the low-velocity dynamics of an arbitrary $(d-1)$-brane extremal black hole embedded in D dimensions are simply related to that of the dimensionally reduced point-like black hole in $D-d+1$ dimensions (Khuri and Myers [1996]), essentially due to the fact that in both cases the dual dimension is the same and given by, $\tilde{d} = D-d-2$. This can be seen schematically via

$$(D, d) \to (D-1, d-1) \to \ldots \to (D-d+1, 1), \tag{3.1}$$

where the first number in braces indicates the spacetime dimension and the second the the dimension of the worldvolume swept out by the higher-membrane black hole, and where the arrows indicate dimensional reduction of both dimensions. Notice that the last pair of dimensions represents a point-like black hole.

As the dynamics of the point-like solutions have already been worked out by Shiraishi [1993b], we obtain our answer quite easily, taking care to account for the proper volume terms for the higher-membrane black holes. Note that, in particular, the κ-symmetric solutions can be expanded or reduced only to other κ-symmetric solutions, since only these solutions scatter trivially.

As an example of nontrivial scattering for non-κ-symmetric extreme black holes, consider the case of two point-like black holes in $D=4$ with $a=1$ in (2.1). Then the low-velocity Lagrangian is given following Manton's method by (Shiraishi [1993b])

$$L = -M + \frac{1}{2}MV^2 + \frac{1}{2}\mu v^2\left(1+\frac{2M}{r}\right), \tag{3.2}$$

where $M = m_1 + m_2$ is the total mass, $\vec{V} = \vec{v}_1 + \vec{v}_2$ is the center-of-mass velocity, $\mu = m_1 m_2/M$ is the reduced mass, $\vec{v} = \vec{v}_2 - \vec{v}_1$ is the relative velocity and $\vec{r} = \vec{x}_2 - \vec{x}_1$ is the relative separation of the two black holes. This interaction in fact yields Rutherford scattering.

4 Discussion

The flat metric and consequent trivial dynamics for the κ-symmetric solutions is a somewhat surprising result, and is probably connected with the existence of flat directions in the superpotentials associated with the underlying κ-symmetric theories. Another possibility is that these solutions also possess the maximal amount of spacetime supersymmetry in certain embeddings, and this may also constrain the dynamics considerably. For example, if we embed the four-dimensional black holes in $N = 8$, $D = 4$ supergravity, only the κ-symmetric $a = \sqrt{3}$ black hole preserves four of the spacetime supersymmetries.

Another interesting idea connected with the dynamics of these solutions is the possibility that the point-like extremal black holes might correspond to massive elementary states in the string spectrum. Particles with mass greater than the Planck mass possess a Schwarzschild radius greater than their Compton wavelength, and so presumably may display an event horizon even in a quantum-mechanical framework. In the absence of a theory of quantum gravity, however, this notion is hard to test. Since string theory is a possible theory of quantum gravity and predicts the existence of massive states at the Planck scale, it provides a framework in which to test this conjecture. Duff and Rahmfeld [1995] argued that certain massive string states in fact correspond to extremal black hole solutions of (2.1). More recently, dynamical evidence for this conjecture was found by the present authors (Khuri and Myers [1995]), where the dynamics of the black holes according to the Manton method and as discussed in this paper were compared with the string-theoretic scattering amplitudes of the corresponding string states. Remarkably, these two seemingly disparate methods yield the same long-range scattering. These results then promise to shed light on the nature of string theory as a theory of quantum gravity.

References

Duff, M. J. and Khuri, R. R. [1994] *Four-dimensional string/string duality* Nucl. Phys. B **411**, 473–486.

Duff, M. J., Khuri, R. R. and Lu, J. X. [1995] *String solitons* Phys. Reports **259**, 213–326.

Duff, M. J. and Rahmfeld, J. [1995] *Massive string states and extreme black holes* Phys. Lett. B **345**, 441–445.

Ferrell, R. C. and Eardley, D. M. [1987] *Slow-motion scattering and coalescence of maximally charged black holes* Phys. Rev. Lett. **59**, 1617–1620.

Khuri, R. R. and Myers, R. C. [1995] *Dynamics of extreme black holes and massive string states* Phys. Rev. D **52**, 6988–6996.

—— [1996] *Rusty Scatter Branes* Nucl. Phys. B **466**, 60–74.

Manton, N. S. [1982] *A remark on the scattering of BPS monopoles* Phys. Lett. B **110**, 54–56.

Shiraishi, K. [1993a] *Multicentered solution for maximally charged dilaton black holes in arbitrary dimensions* J. Math. Phys. **34**, 1480–1486.

—— [1993b] *Moduli space metric for maximally charged dilaton black holes* Nucl. Phys. B **B402**, 399–410.

Fields Institute Communications
Volume **15**, 1997

Entropy of Eternal Black Holes

Erik A. Martinez
Theoretical Physics Institute
University of Alberta
Edmonton, Alberta, Canada T6G 2J1
Current address: Center for Gravitational Physics and Geometry
Department of Physics
The Pennsylvania State University
University Park, PA USA 16802-6300
martinez@phys.psu.edu

Abstract. The entropy of a quantum-statistical system which is classically approximated by a general stationary eternal black hole is studied by means of a microcanonical functional integral. This requires computing the microcanonical action for a spatially bounded spacetime region when its two disconnected timelike boundary surfaces are located in different wedges of the Kruskal diagram. This approach opens the possibility of including explicitly the internal degrees of freedom of a physical black hole in path integral descriptions of its thermodynamical properties. If the functional integral is interpreted as the density of states of the system, the corresponding entropy equals $S = A_H/4 - A_H/4 = 0$ in the semiclassical approximation, where A_H is the area of the black hole horizon. The functional integral reflects the properties of a pure state.

1 Introduction

The relationship between the entropy of a physical black hole and its internal degrees of freedom remains a subject of active research. A natural question to ask in this regard is: can these degrees of freedom be effectively included in a functional integral description of black hole entropy? In an attempt to give an affirmative answer to this question, we investigate in this paper a microcanonical functional integral when applied to a quantum self-gravitating statistical system that includes spacetimes whose topology and boundary conditions coincide with the ones of (either distorted or Kerr-Newman) eternal black holes.

A proposal for the density of states of a gravitational system defined in terms of a microcanonical functional integral has been suggested recently in Brown and York [1993]. This integral is defined as a formal sum over Lorentzian geometries. The black hole density of states is obtained from this functional integral when the

1991 *Mathematics Subject Classification.* Primary 83C57; Secondary 51M05, 83C45.
I am grateful to Valeri Frolov and Werner Israel for helpful comments. This work was supported by the Natural Sciences and Engineering Research Council of Canada.

© 1997 American Mathematical Society

latter is approximated semiclassically by using a complex metric whose boundary data at its single boundary surface coincide with the boundary data of a Lorentzian stationary black hole. The density of states defined accordingly equals the exponential of one fourth of the area of the black hole horizon. This proposal opens the possibility of determining the thermodynamical properties of black hole systems starting from a sum over real Lorentzian geometries. However, several problems remain in this approach. First, a spacelike hypersurface Σ that describes the initial data of a Lorentzian black hole has to cross necessarily the event horizon and eventually intersect the interior singularity. Second, a Lorentzian stationary black hole is not an extremum of the microcanonical action for a spacetime region with a single timelike boundary surface. This implies that the black hole density of states whose boundary data correspond to the ones of a Lorentzian stationary black hole cannot be approximated semiclassically by using the same Lorentzian metric that motivates its boundary conditions. It is therefore necessary to use complex metrics to evaluate the Lorentzian functional integral in a steepest descents approximation. This procedure yields the correct result for the entropy but conceals its origin: the interior of the Lorentzian black hole literally disappears by virtue of this procedure, leaving effectively only a periodically identified Euclidean version of the "right" wedge region of a Kruskal diagram. The properties of the black hole interior become encoded in a set of conditions at the so-called "bolt" of the complex geometry. In this approach the statistical origin of entropy and its relationship to the internal degrees of freedom of a black hole remain obscure.

The problems mentioned above and the role of internal degrees of freedom in functional integral descriptions of black hole thermodynamics can be addressed by explicitly considering the eternal version of a black hole (Martinez [1995]). The excitations of the physical black hole can be associated with the deformations of an initial global Cauchy surface Σ of the eternal black hole. In general, the spatial slices Σ that foliate an eternal black hole are (deformed) Einstein-Rosen bridges with wormhole topology $R^1 \times S^2$. The spacetime is composed of two wedges M_+ and M_- located in the right and left sectors of a Kruskal diagram. Internal and external degrees of freedom of the black hole can be easily identified in this approach since the hypersurfaces Σ are naturally divided in two parts Σ_+ and Σ_- by a bifurcation two-surface S_0 where the lapse function N vanishes. While the "external" degrees of freedom of the original black hole are naturally given by the initial data at Σ_+, its "internal" degrees of freedom can be identified with initial data defined at Σ_- (Frolov and Martinez [1996]).

2 Microcanonical Action and Functional Integral

Consider a spacetime region whose three-dimensional timelike boundary surface B consists of two disconnected parts B_+ and B_-. The microcanonical action S_m is the action appropriate to a variational principle in which the fixed boundary conditions at the timelike boundaries $B_\pm$ are not the spacetime three-geometry but the surface energy density ε, surface momentum density j_a, and boundary two-metric σ_{ab} (Brown et al. [1991], Brown and York [1993]). The covariant form of the microcanonical action for a general spacetime whose timelike surfaces B_+ and B_- are located in the regions M_+ and M_- respectively has been presented in Martinez [1995]. Its Hamiltonian form is easily obtained under a $3+1$ spacetime split by recognizing that there exists a direction of time at the boundaries $B_\pm$ (Martinez [1995]). By introducing the momentum P^{ij} conjugate to the three-metric h_{ij} of Σ

for the so-called "tilted" foliation (Frolov and Martinez [1996]) and integrating the kinetic part of the volume integral, the action becomes $S_m = \int_M d^4x[P^{ij}\dot{h}_{ij} - N\mathcal{H} - V^i\mathcal{H}_i]$, where the dot denotes differentiation with respect to time, V^i denotes the shift vector, and the gravitational contributions to the Hamiltonian and momentum constraints are given by the usual expressions. Observe that the action vanishes identically for stationary solutions of the vacuum Einstein equations describing stationary eternal black holes. In this case $\dot{h}_{ij} = 0$, the constraint equations are satisfied, and no boundary terms remain in the action.

Consider now the microcanonical functional integral for a gravitational system whose timelike boundary surfaces $B_\pm$ are located in $M_\pm$. The path integral

$$\nu[\varepsilon_+, j_+, \sigma_+; \varepsilon_-, j_-, \sigma_-] = \sum_M \int \mathcal{D}H \exp(iS_m) \tag{1}$$

is a functional of the energy density $\varepsilon_\pm$, momentum density $j_\pm$, and two-metric $\sigma_\pm$ at the boundaries B_+ and B_-. The sum over M refers to a sum over manifolds of different topologies whose boundaries have topologies $B_+ \equiv S_+ \times S^1 = S^2 \times S^1$ and $B_- \equiv S_- \times S^1 = S^2 \times S^1$. The element S^1 is due to the periodic identification in the global time direction at the boundaries when the initial and final hypersurfaces are identified. The integral is a sum over periodic Lorentzian metrics that satisfy the boundary conditions at B_+ and B_-.

The eternal black hole functional integral ν_* is obtained when the boundary data $(\varepsilon_+, j_+, \sigma_+)$ and $(\varepsilon_-, j_-, \sigma_-)$ of the geometries summed over correspond to the boundary data of a general Lorentzian stationary eternal black hole. The boundary data of this solution can be determined at S_+ and S_- for each slice Σ. Observe that by virtue of the gravitational constraint equations, these data determine uniquely the size of the black hole horizon (Brown et al. [1990]) and are such that the two-metric is continuous at this horizon.

We evaluate now the functional integral in the semiclassical approximation. This requires finding a four-metric that extremizes the action S_m and satisfies the boundary conditions $(\varepsilon_+, j_+, \sigma_+)$ at S_+ and $(\varepsilon_-, j_-, \sigma_-)$ at S_-. Since the classical Lorentzian eternal black hole metric can be periodically identified and placed on a manifold whose two spatial boundaries have the desired topologies, the resulting metric can be used to approximate the path integral. The periodic identification alters neither the constraint equations nor the boundary data. In the semiclassical approximation the functional integral ν_* becomes (Martinez [1995])

$$\nu_*[\varepsilon_+, j_+, \sigma_+; \varepsilon_-, j_-, \sigma_-] \approx \exp\left(iS_m[\tilde{N}, \tilde{V}, \tilde{h}]\right) \approx \exp\left(0\right) , \tag{2}$$

since the microcanonical action $S_m[\tilde{N}, \tilde{V}, \tilde{h}]$ evaluated at the periodically identified geometry vanishes identically if the stationarity condition and the constraints are satisfied.

It is illustrative to consider now a complex four-metric which also extremizes the microcanonical action for eternal black hole boundary conditions and which can be used to reevaluate the path integral (1) in a steepest descent approximation. This alternative approximation of the quantity ν_* is useful in understanding the relationship of the result (2) with the density of states for an ordinary (that is, non-eternal) black hole (Brown and York [1993]). The complex metric can be obtained from the Lorentzian eternal black hole metric by a complexification map Ψ defined by $\Psi(N) = -iN$, $\Psi(V^i) = -iV^i$. This map preserves the reflection symmetry and the canonical variables h_{ij} and P^{ij} of the Lorentzian solution. The complex geometry consists of two complex sectors $\bar{M}_+$ and $\bar{M}_-$ which join at the locus of points at which the lapse vanishes. This geometry is also a solution of Einstein equations if one requires that conical singularities do not exist at that locus for every Σ. To do this, it is necessary to puncture each complex sector and to close smoothly the geometry at the inner boundaries ${}^3H_\pm = {}^2H_\pm \times S^1$ of $\bar{M}_\pm$, where ${}^2H_\pm$ denotes the intersection of the slices $\Sigma_\pm$ with the black hole horizon for the Lorentzian metric (Martinez [1995]). After imposing these regularity conditions, the topology of each sector $\bar{M}_\pm$ becomes $R^2 \times S^2$. However, each element ${}^3H_+$ and ${}^3H_-$ does contribute a term to the microcanonical action for the complex geometry.

The regular complex metric is not included in the sum over Lorentzian geometries $\bar{\nu}_*$ in (1) but can be used to approximate it by distorting the contours of integration for both lapse and shift into the complex plane (Brown and York [1993]). In this approximation the path integral becomes $\bar{\nu}_* \approx \exp(iS_m[-i\tilde{N}, -i\tilde{V}, \tilde{h}])$, where $S_m[-i\tilde{N}, -i\tilde{V}, \tilde{h}]$ is the action of the complex metric when the smoothness of the geometries at ${}^3H_+$ and ${}^3H_-$ is enforced. This action turns out to be $S_m[-i\tilde{N}, -i\tilde{V}, \tilde{h}] = -iA_+/4 + iA_-/4$, where A_+ and A_- denote the surface area of the elements ${}^2H_+$ and ${}^2H_-$ (Martinez [1995]). Since the periodic identification and the complexification Ψ do not alter the boundary data nor the gravitational constraint equations, the area A_+ of ${}^2H_+$ coincides with the area A_- of ${}^2H_-$: $A_+(\varepsilon_+, j_+, \sigma_+) = A_-(\varepsilon_-, j_-, \sigma_-) \equiv A_H$. This implies that, in agreement with (2), the eternal black hole functional integral is $\bar{\nu}_*[\varepsilon_+, j_+, \sigma_+; \varepsilon_-, j_-, \sigma_-] \approx \exp(A_H/4 - A_H/4) = \exp(0)$ in the "zero-loop" approximation.

If the microcanonical functional integral (1) is interpreted as the density of states of the statistical system, it is possible to express $\bar{\nu}_*$ approximately as

$$\bar{\nu}_*[\varepsilon_+, j_+, \sigma_+; \varepsilon_-, j_-, \sigma_-] \approx \exp(\mathcal{S}[\varepsilon_+, j_+, \sigma_+; \varepsilon_-, j_-, \sigma_-]), \tag{3}$$

where $\mathcal{S}$ represents the total entropy of the system. The above result implies that the entropy for the system in the semiclassical approximation is

$$\mathcal{S} \approx \frac{1}{4}A_H - \frac{1}{4}A_H = 0 , \tag{4}$$

where A_H is the area of the horizon of the physical eternal black hole solution that classically approximates the system (Martinez [1995]). The total entropy is given formally by the subtraction $\mathcal{S} = \mathcal{S}_+[\varepsilon_+, j_+, \sigma_+] - \mathcal{S}_-[\varepsilon_-, j_-, \sigma_-]$, where $\mathcal{S}_+$ and $\mathcal{S}_-$ can be interpreted as the semiclassical entropies associated with the external (M_+) and internal (M_-) regions respectively of the eternal black hole system.

3 Conclusions

The functional integral (2) refers to a quantum-statistical system which is classically approximated by a general stationary, eternal black hole solution of Einstein equations within a region bounded by two timelike surfaces B_+ and B_-. Its semiclassical value is a consequence of the choice of boundary data, the gravitational constraint equations, and the vanishing of the microcanonical action for the four-geometries that satisfy the boundary conditions and approximate the path integral. The calculation presented above applies to any distorted black hole in the strong gravity regime. It indicates that a pure state (of zero entropy) can be defined not only for matter fields perturbations propagating in the spacetime of an eternal black hole but also for the gravitational field itself. This is physically appealing: the initial data for the eternal black hole specified at the spacelike hypersurface Σ contain all the information required for the evolution of both the exterior and interior parts of a physical black hole. The entropy associated with Σ must therefore equal zero. Since in a microcanonical description it seems natural to relate the external and internal degrees of freedom of a black hole with the boundary data at the surfaces B_+ and B_- respectively (Frolov and Martinez [1996]), we believe that the microcanonical functional integral for eternal black hole systems opens the possibility of extending path integral formulations of black hole thermodynamics to situations when internal degrees of freedom are present and allows the study of gravitational statistical properties in terms of a single pure state (Martinez [1995]). These conclusions are in complete agreement with thermofield dynamics descriptions of quantum processes and, in particular, with the application of this approach to black hole thermodynamics developed originally by Israel [1976] for small perturbations. They strongly suggest that the thermofield dynamics description of quantum field processes in a curved background can be extended beyond perturbations to the gravitational field itself of distorted eternal black holes.

References

Brown, J. D., Comer, G. L., Martinez, E. A., Melmed, J., Whiting, B. F. and York, J. W. [1990] *Thermodynamic ensembles and gravitation*, Class. Quantum Grav. **7**, 1433–1444.

Brown, J. D., Martinez, E. A. and York, J. W., Jr. [1991] *Complex Kerr-Newman geometry and black hole thermodynamics*, Phys. Rev. Lett. **66**, 2281–2284.

Brown, J. D. and York, J. W., Jr. [1993] *Microcanonical functional integral for the gravitational field*, Phys. Rev. D **47**, 1420–1431.

Frolov, V. and Martinez, E. A. [1996] *Action and Hamiltonian for eternal black holes*, Class. Quantum Grav. **13**, 481–495.

Israel, W. [1976] *Thermo-field dynamics of black holes*, Phys. Lett. A **57**, 107–110.

Martinez, E. A. [1995] *Microcanonical functional integral and entropy for eternal black holes*, Phys. Rev. D **51**, 5732–5741.

Fields Institute Communications
Volume **15**, 1997

Nonexistence of Petrov Type III Space-Times on which Weyl's Neutrino Equation and Maxwell's Equations Satisfy Huygens' Principle

R. G. McLenaghan
Department of Applied Mathematics
University of Waterloo
Waterloo, Ontario, Canada N2L 3G1
rgmclena@daisy.uwaterloo.ca

Fernando D. Sasse
Department of Applied Mathematics
University of Waterloo
Waterloo, Ontario, Canada N2L 3G1
fsasse@barrow.uwaterloo.ca

Abstract. We show that there exist no Petrov type III space-times on which the Weyl neutrino equation or Maxwell's equations satisfy Huygens' principle. The necessary conditions for the validity of the principle are expressed in terms of a system of polynomial equations. Although the size of the system is large, Maple's package `grobner` can be applied with success, by dividing the problem into smaller ones.

1 Introduction

This paper is devoted to the solution of Hadamard's problem for the Weyl neutrino equation and Maxwell's equations. These equations can be written as

$$\nabla^{A\dot{A}}\varphi_A = 0\,, \tag{1.1}$$

and

$$\nabla^{A\dot{A}}\varphi_{AB} = 0\,, \tag{1.2}$$

respectively. Here we use the Newman-Penrose spin coefficient formalism. The notation and conventions are those of Carminati and McLenaghan [1988] (from now on referred to as CM). Huygens' principle is valid for (1.1) and (1.2) if and only if for every Cauchy initial value problem and every $x_0 \in V_4$, the solution depends only on the Cauchy data in an arbitrarily small neighbourhood of $S \cap C^-(x_0)$, where S denotes the initial surface and $C^-(x_0)$ the past null conoid from x_0 (Günther [1965] and Wünsch [1978]). Hadamard's problem for equations (1.1) and (1.2), originally posed only for scalar equations, is that of determining *all* space-times for which Huygens' principle is valid for a particular equation.

1991 *Mathematics Subject Classification.* 83-08.

© 1997 American Mathematical Society

We prove the following theorem:

Theorem 1 *There exist no Petrov type III space-times on which Weyl's equation (1.1) or Maxwell's equations (1.2) satisfy Huygens' principle.*

2 Necessary Conditions and Proof of the Theorem

In this paper we make use of the necessary condition determined by Alvarez and Wünsch [1991]:

$$T^{(1)}{}_{abcde} + \sigma_1 T^{(2)}{}_{abcde} + \sigma_2 T^{(3)}{}_{abcde} = 0\,, \tag{2.1}$$

where σ_1 and σ_2 are fixed real numbers, and

$$\begin{aligned} T^{(1)}{}_{abcde} = TS[&4{}^*C^k{}_{ab}{}^l C^u{}_{del;ukc} - 6{}^*C^k{}_{bc}{}^l{}_{;a}\, C^u{}_{del;uk} + 26{}^*C^u{}_{ab}{}^k{}_{;u}\, C^v{}_{dek;vc} + \\ &{}^*C^k{}_{ab}{}^l{}_;{}^n\, C_{kdel;nc} + 5{}^*C^k{}_{ab}{}^l\, C^n{}_{del;c} L_{kn} + 4{}^*C^k{}_{ab}{}^l C^n{}_{cdl;k} L_{en} \\ &+4{}^*C^k{}_{ab}{}^l\, C^u{}_{klc;u}\, L_{de} - 21{}^*C^k{}_{ab}{}^l\, C^u{}_{cdk;u}\, L_{el}]\,, \end{aligned} \tag{2.2}$$

$$T^{(2)}{}_{abcde} := TS[-12{}^*C^k{}_{ab}{}^l\, C_{kcd}{}^n\, C^u{}_{lne;u} - C_k{}^{nh}{}_a C_{lnhb}{}^*C^k{}_{de}{}^l{}_{;c}]\,, \tag{2.3}$$

$$T^{(3)}{}_{abcde} := TS[-8{}^*C^k{}_{ab}{}^l C_{kcd}{}^n C^u{}_{lne;u} +{}^*\, C_k{}^{nh}{}_a C_{lbnh}\, C^k{}_{de}{}^l{}_{;c}]\,, \tag{2.4}$$

$${}^*C_{abcd} := \frac{1}{2}\epsilon_{ab}{}^{ef} C_{efcd}\,, \tag{2.5}$$

$$C_{abcd} := R_{abcd} - 2g_{[a[d}L_{b]c]}\,, \tag{2.6}$$

$$L_{ab} := -R_{ab} + \frac{1}{6} R g_{ab}\,. \tag{2.7}$$

In the above C_{abcd} denotes the Weyl tensor and $TS[\ ]$ the operator which takes the trace free symmetric part of the enclosed tensor. We note that, in our conventions, the Riemann tensor, Ricci tensor and the Ricci scalar have opposite sign to those used by Alvarez and Wünsch [1991].

The explicit form of this necessary condition is obtained by first converting the tensorial expressions to the spinor form, then to dyad form and finally contracting them with appropriate products of o^A and ι^A and their complex conjugates. These computations required, for the case of the five-index condition (2.1), the use of the package `NPspinor` in Maple (Czapor et al. [1992]).

It was shown in CM that, for Petrov type III space-times, there exists a dyad o_A, ι_A and a conformal transformation such that

$$\kappa = \sigma = \rho = \tau = \epsilon = 0\,, \tag{2.8}$$

$$\Psi_0 = \Psi_1 = \Psi_2 = \Psi_4 = 0, \quad \Psi_3 = -1\,, \tag{2.9}$$

$$\Phi_{00} = \Phi_{01} = \Phi_{02} = \Lambda = 0\,, \tag{2.10}$$

$$D\alpha = D\beta = 0\,, \tag{2.11}$$

$$\Phi_{11} = c = \text{const.}\,, \tag{2.12}$$

$$D\pi = 0\,. \tag{2.13}$$

Also, from CM, we can determine the following Pfaffians:

$$\delta\beta = -\beta(\overline{\alpha} + \beta)\,, \tag{2.14}$$

$$D\Phi_{12} = 2\overline{\pi}\Phi_{11}\,, \tag{2.15}$$

$$D\Phi_{22} = -2(\beta + \overline{\beta}) + 2\Phi_{21}\overline{\pi} + 2\Phi_{12}\pi\,, \tag{2.16}$$

$$\delta\Phi_{12} = 2\overline{\alpha} + 4\overline{\pi} + 2\overline{\lambda}\Phi_{11} - 2\overline{\alpha}\Phi_{12}\,, \tag{2.17}$$

$$\overline{\delta}\Phi_{12} = -2\beta + 2\overline{\mu}\Phi_{11} - 2\overline{\beta}\Phi_{12}\,, \tag{2.18}$$

$$\begin{aligned}\delta\Phi_{22} &= \Delta\Phi_{12} + 2\overline{\gamma} + 4\overline{\mu} - 2\overline{\nu}\Phi_{11} + 2\overline{\lambda}\Phi_{21} \\ &\quad +2\Phi_{12}\mu - 2\Phi_{22}\beta - 2\Phi_{22}\overline{\alpha} + 2\overline{\gamma}\Phi_{12}\,,\end{aligned} \tag{2.19}$$

$$D\gamma = \overline{\pi}\alpha + \beta\pi + \Phi_{11}\,, \tag{2.20}$$

$$\delta\overline{\pi} = D\overline{\lambda} - \overline{\pi}^2 - \overline{\pi}\overline{\alpha} + \overline{\pi}\beta\,, \tag{2.21}$$

$$D\overline{\nu} = \Delta\overline{\pi} + \overline{\pi}\overline{\mu} + \overline{\lambda}\pi + \overline{\pi}\overline{\gamma} - \overline{\pi}\gamma - 1 + \Phi_{12}\,, \tag{2.22}$$

$$\delta\alpha = \overline{\delta}\beta + \alpha\overline{\alpha} + \beta\overline{\beta} - 2\beta\alpha + \Phi_{11}\,, \tag{2.23}$$

$$\delta\pi = D\mu - \overline{\pi}\pi + \pi\overline{\alpha} - \beta\pi\,, \tag{2.24}$$

$$\delta\overline{\beta} = -2\Phi_{11} - \overline{\beta}\overline{\alpha} - 4\overline{\beta}\overline{\pi} - 2D\overline{\mu} - \beta\overline{\beta} + 2\overline{\pi}\pi\,, \tag{2.25}$$

$$62\overline{\beta}\pi + 40\overline{\beta}\alpha + 6\pi\alpha + 3\alpha^2 + \overline{\delta}\alpha + 2\overline{\delta}\pi + \overline{\delta\beta} + \overline{\beta}^2 = 0\,, \tag{2.26}$$

$$\delta(2\overline{\pi} + \overline{\alpha}) = -62\overline{\pi}\beta - 39\beta\overline{\alpha} - 6\overline{\pi\alpha} - 3\overline{\alpha}^2\,, \tag{2.27}$$

$$\delta(2\pi + \alpha) = 2\pi\overline{\alpha} + \alpha\overline{\alpha} - 6\beta\pi - 3\beta\alpha - \Phi_{11}\,, \tag{2.28}$$

$$\begin{aligned}D\mu = \frac{1}{152}&\big(-1520\alpha\beta\pi + 208\alpha\overline{\alpha}\pi - 152\overline{\beta}\overline{\pi}\pi + 1040\beta\overline{\beta}\alpha \\ &+ 1968\overline{\beta}\beta\pi + 760\overline{\beta}\alpha\overline{\alpha} + 1216\overline{\beta}\pi\overline{\alpha} - 1228\pi\Phi_{11} \\ &+ 1688\alpha\pi\overline{\pi} - 739\alpha\Phi_{11} + 80\alpha^2\overline{\alpha} + 380\overline{\beta}\Phi_{11} + 2496\overline{\pi}\pi^2 \\ &- 2432\beta\pi^2 + 128\pi^2\overline{\alpha} + 80\overline{\pi}\alpha^2\big)/(-\overline{\beta} + 5\alpha + 8\pi)\,,\end{aligned} \tag{2.29}$$

$$\begin{aligned}D\overline{\mu} = \frac{1}{152}&\Big(776\alpha\overline{\beta}\overline{\alpha} - 304\overline{\beta}^2\overline{\pi} + 208\overline{\beta}^2\beta + 84\overline{\beta}\Phi_{11} - 1216\overline{\pi}\pi^2 \\ &+ 1536\alpha\overline{\pi}\overline{\beta} - 760\alpha\overline{\pi}\pi + 2744\overline{\pi}\pi\overline{\beta} + 1332\pi\overline{\beta}\overline{\alpha} \\ &+1140\pi\Phi_{11}\big)\,/(\overline{\beta} - 5\alpha - 8\pi)\,.\end{aligned} \tag{2.30}$$

Converting the necessary condition (2.1) to spinor form in the dyad basis, and contracting with $\iota^{AB}o^{CDE}\overline{\iota}^{\dot{A}\dot{B}\dot{C}\dot{D}\dot{E}}$, gives

$$\begin{aligned}&-14\overline{\alpha}\,\delta(\overline{\pi}) - 12\overline{\alpha}^3 - \delta(\delta(\overline{\alpha} + 2\overline{\pi})) - 21\beta\overline{\alpha}^2 + 7\overline{\lambda}\beta^2 + 14\beta^2\overline{\pi} \\ &-7\beta\delta(\overline{\alpha} + 2\overline{\pi}) - 42\beta\overline{\alpha}\overline{\pi} - (10\overline{\alpha} + 6\overline{\pi})\delta(\overline{\alpha}) - 24\overline{\alpha}^2\overline{\pi} = 0\,.\end{aligned} \tag{2.31}$$

We observe that the terms (2.3) and (2.4) do not contribute to this component.

Substituting (2.27) and (2.14) in (2.31) we have

$$\delta\overline{\alpha} = 192\beta\overline{\pi} - 3\overline{\alpha}^2 + 121\beta\overline{\alpha}\,, \tag{2.32}$$

and

$$\delta\overline{\pi} = -127\overline{\pi}\beta - 80\beta\overline{\alpha} - 3\overline{\pi}\overline{\alpha}\,. \tag{2.33}$$

The integrability conditions provided by the NP commutation relations can now be used. It is important to note that the expressions for $D\mu$, $D\overline{\mu}$, $\overline{\delta}\beta$, $\delta\beta$, etc. have been chosen such that they all have the same denominator.

Consider the NP commutator $[\overline{\delta}, \delta]\alpha$. Using the Pfaffians calculated previously, and solving for Φ_{11}, we obtain

$$\begin{aligned}\Phi_{11} := &-8\overline{\beta}\,(8\,\pi + 5\,\alpha)(172736\,\overline{\pi}\,\pi^2 - 7776\,\pi^2\,\overline{\alpha} - 13294\,\alpha\,\pi\,\overline{\alpha} + 7866\,\overline{\beta}\,\pi\,\overline{\alpha}\\ &+ 211556\,\alpha\,\overline{\pi}\,\pi + 9568\,\overline{\beta}\,\beta\,\pi - 22572\,\overline{\beta}\,\overline{\pi}\,\pi + 5330\,\overline{\beta}\,\beta\,\alpha + 4845\,\overline{\beta}\,\alpha\,\overline{\alpha}\\ &- 1805\,\overline{\beta}^2\,\overline{\alpha} - 2470\,\beta\,\overline{\beta}^2 - 13110\,\overline{\pi}\,\alpha\,\overline{\beta} - 5290\,\alpha^2\,\overline{\alpha} + 65010\,\overline{\pi}\,\alpha^2)\,/\\ &(-772320\,\pi^2\,\overline{\beta} + 2048352\,\alpha\,\pi^2 + 1158240\,\pi\,\alpha^2 - 4085\,\alpha\,\overline{\beta}^2 - 10868\,\pi\,\overline{\beta}^2\\ &- 335985\,\alpha^2\,\overline{\beta} - 1020276\,\alpha\,\pi\,\overline{\beta} + 214700\,\alpha^3 + 1191680\,\pi^3)\,. \end{aligned} \tag{2.34}$$

On the other hand, the commutator $[\overline{\delta}, \delta](\alpha + \overline{\beta})$, gives the following expression:

$$\begin{aligned}\Phi_{11} := &-8\overline{\beta}(-20672\,\overline{\beta}\,\overline{\pi}\,\pi - 12920\,\overline{\pi}\,\alpha\,\overline{\beta} + 8056\,\overline{\beta}\,\pi\,\overline{\alpha} + 172736\,\overline{\pi}\,\pi^2\\ &- 7776\,\pi^2\,\overline{\alpha} + 9568\,\overline{\beta}\,\beta\,\pi + 5035\,\overline{\beta}\,\alpha\,\overline{\alpha} - 5290\,\alpha^2\,\overline{\alpha} + 65010\,\overline{\pi}\,\alpha^2\\ &- 13294\,\alpha\,\pi\,\overline{\alpha} + 211556\,\alpha\,\overline{\pi}\,\pi + 5330\,\overline{\beta}\,\beta\,\alpha)\;/(-13047\,\overline{\beta}\,\alpha\\ &+ 42940\,\alpha^2 + 162944\,\pi\,\alpha + 10412\,\overline{\beta}^2 - 18564\,\pi\,\overline{\beta} + 148960\,\pi^2)\,. \end{aligned} \tag{2.35}$$

Using the fact that $\overline{\delta}(\Phi_{11}) = 0$, we obtain, from (2.35), a third expression for Φ_{11}, which has the same form as (2.35), but it is too long to be given here.

We suppose that spin coefficients α, β, π are different from zero. It was shown in CM that if one of these spin coefficients is equal to zero, the others must also be zero, and Huygens' principle is violated. Using this fact, we can reduce the number of complex variables from three to two, by defining $x_1 := \alpha/\pi$ and $x_2 := \beta/\overline{\pi}$. We subtract (2.34) from (2.35), take the numerator and divide it by $(8 - x_2 - 5x_1)(152\pi\overline{\pi})$, to get

$$\begin{aligned}N_1 := &\,178100\,x_1\,x_2\,\overline{x_2}^2 + 284960\,x_2\,\overline{x_2}^2 + 208240\,\overline{x_2}^2\,\overline{x_1} + 130150\,\overline{x_1}\,\overline{x_2}^2\,x_1\\ &+ 109825\,\overline{x_1}\,\overline{x_2}\,x_1^2 + 252850\,x_2\,\overline{x_2}\,x_1^2 + 523744\,x_2\,\overline{x_2} + 3451480\,x_1\,\overline{x_2}\\ &+ 341900\,\overline{x_2}\,x_1\,\overline{x_1} + 265888\,\overline{x_2}\,\overline{x_1} + 735800\,x_2\,\overline{x_2}\,x_1 + 2915264\,\overline{x_2}\\ &+ 1018400\,\overline{x_2}\,x_1^2 - 879008\,\overline{x_1} + 18263488 - 408050\,x_1^3\,\overline{x_1} + 4864450\,x_1^3\\ &+ 35335248\,x_1 + 22731900\,x_1^2 - 2101032\,x_1\,\overline{x_1} - 1622550\,x_1^2\,\overline{x_1} = 0\,. \end{aligned} \tag{2.36}$$

By subtracting (2.35) from $\overline{\delta}(\Phi_{11})$, with Φ_{11} being given by (2.35), taking the numerator and dividing by $(8 - x_2 - 5x_1)(152\pi\overline{\pi})$, we get

$$\begin{aligned}N_2 := &-11651821200\,x_1^3 - 26531539120\,x_1^2 - 10132263424 - 26800626944\,x_1\\ &+ 242619584\,\overline{x_2} + 593671488\,\overline{x_2}^2 + 2256829184\,\overline{x_1} - 35155250\,\overline{x_2}^2\,x_2\,x_1{}^2\\ &+ 21550100\,\overline{x_2}^3\,x_2\,x_1 - 128589500\,\overline{x_2}\,x_2\,x_1^3 + 11036720\,\overline{x_2}^3\,\overline{x_1}\\ &+ 21755825\,\overline{x_2}^2\,x_1^2\,\overline{x_1} + 130839250\,x_1^3\,\overline{x_2} - 216631750\,\overline{x_2}\,x_1^3\,\overline{x_1}\\ &- 17700400\,\overline{x_2}^3\,x_1 + 372958500\,x_1^4\,\overline{x_1} + 220600700\,\overline{x_2}^2\,x_1^2 + 34480160\,x_2\,\overline{x_2}^3\\ &- 1915591500\,x_1^4 - 28320640\,\overline{x_2}^3 + 6897950\,\overline{x_2}^3\,x_1\,\overline{x_1} + 724005800\,\overline{x_2}^2\,x_1\\ &- 112640320\,x_1\,x_2\,\overline{x_2}^2 - 90878112\,x_2\,\overline{x_2}^2 + 59839936\,\overline{x_2}^2\,\overline{x_1}\\ &+ 72209280\,\overline{x_1}\,\overline{x_2}^2\,x_1 - 1002142990\,\overline{x_1}\,\overline{x_2}\,x_1^2 - 585808600\,x_2\,\overline{x_2}\,x_1^2\\ &- 448045312\,x_2\,\overline{x_2} + 632690016\,x_1\,\overline{x_2} - 1546809184\,\overline{x_2}\,x_1\,\overline{x_1}\\ &- 887976960\,x_2\,\overline{x_2}\,x_1 + 509836460\,\overline{x_2}\,x_1^2 + 2328634300\,x_1^3\,\overline{x_1}\\ &+ 5728848896\,x_1\,\overline{x_1} + 5470002280\,x_1^2\,\overline{x_1} - 796702240\,\overline{x_2}\,\overline{x_1} = 0\,. \end{aligned} \tag{2.37}$$

At this point we shall consider x_1, x_2, $\overline{x_1}$, $\overline{x_2}$ as independent variables. We now apply the package **grobner**, from Maple, to the polynomial system defined the equations $N_1 = 0$ and $N_2 = 0$. In particular, we use the procedure **gsolve**, which computes a collection of reduced (lexicographic) Gröbner bases corresponding to a set of polynomials. The result is a list of reduced subsystems whose roots are those of the original system, but whose variables have been successively eliminated and separated as far as possible. In the present case we need to apply **gsolve** also to the resulting subsystems, after changing some equations by their complex conjugates or using further integrability conditions. We then get the result that x_1 and x_2 are constants. The same conclusion follow in the cases in which each denominator in the different expressions for Φ_{11} is zero. Eventually we can show that this implies that the necessary conditions for the validity of Huygens' principle are not satisfied.

3 Conclusions

The Theorem is proved here only for the case of Maxwell's equations. The proof is similar for the Weyl neutrino case where only numerical coefficients have to be changed. The fact that one of the components of Alvarez-Wünsch five-index condition turned out to be relatively simple, allowed us to construct the needed set of Pfaffians. The use of the package **NPpsinor** in Maple was essential for the conversion of the five-index tensorial form to components in a dyad basis. The resulting polynomial system could be solved using the the package **grobner** in Maple. A "divide and conquer" strategy was necessary in this step, due to the large size of the polynomials involved.

References

Alvarez, M. and Wünsch, V. [1991] *Zur Gültigkeit des huygensschen Prinzips bei der Weyl-Gleichung und den homogenen Maxwellschen Gleichungen für Metriken vom Petrov-typ N*, Wiss. Zeitschr. Päd. Hochsch. Erfurt/Mühlhausen, Math.-Naturwiss. **27**, 77–91.

Carminati, J. and McLenaghan, R. G. [1988] *An explicit determination of space-times on which the conformally invariant scalar wave equation satisfies Huygens' principle. Part III: Petrov type III space-times*, Ann. Inst. Henri Poincaré, Phys. Théor. **48**, 77–96.

Czapor, S. R., McLenaghan, R. G. and Carminati, J. [1992] *The automatic conversion of spinor equations to dyad form in MAPLE*, Gen. Rel. Grav. **24**, 911–928.

Günther, P. [1965] *Ein Beispiel einer nichttrivialen huygesschen Differentialgleichungen mit vier unabhängigen Variablen*, Arch. Rational Mech. Anal. **18**, 103–106.

Wünsch, V. [1978] *Cauchy-Problem und huygenssches Prinzip bei einigen Klassen spinorieller Feldleichungen I*, Beitr. zur Analysis **12**, 47–76.

Fields Institute Communications
Volume **15**, 1997

Symmetry Operators for the Weyl Neutrino Equation on Curved Spacetime

R. G. McLenaghan
Department of Applied Mathematics
University of Waterloo
Waterloo, Ontario, Canada N2L 3G1
rgmclena@daisy.uwaterloo.ca

David M. Walker
Department of Mathematics
University of Western Australia
Nedlands WA 6009, Australia
dmw@maths.uwa.edu.au

Abstract. We derive the necessary and sufficient conditions that a second order covariant differential operator be a symmetry operator of the Weyl-neutrino equation in a general curved spacetime background. It is found that such operators are naturally formulated in terms of conformal Killing vectors, tensors and spinors. Operators of this type play a role in the solution of the neutrino equation by separation of variables in type D vacuum spacetimes.

1 Introduction

A symmetry operator of a relativistic wave equation on curved spacetime is a linear differential operator that maps the space of solutions into itself. The most general first order symmetry operator for the massless Dirac equation is expressible in terms of conformal Killing-Yano tensors of valences one, two and three (Kamran and McLenaghan [1984a]). The separability property of the massless Dirac equation on vacuum type D spacetimes may be understood by the existence of a valence two Killing-Yano tensor on these spacetimes (Kamran and McLenaghan [1984b]). However, Fels and Kamran [1990] have demonstrated that a complete theory of separation of variables for the Dirac equation will involve symmetry operators that are of *higher* order than the Dirac operator and not related to the isometrics of the metric. The purpose of this paper, motivated by the above result, is to determine the form of the general second order symmetry operator of the Weyl-neutrino equation. We shall perform the calculations using the standard 2-component spinor formalism (Penrose and Rindler [1984]), which by the Weyl representation, is equivalent to

1991 *Mathematics Subject Classification.* Primary 83C60; Secondary 83C20.

The authors would like to thank Niky Kamran for enlightening discussions and his continued interest in the problem. This work has been supported in part by a Research Grant from the Natural Sciences and Engineering Research Council of Canada (R. G. McL.).

© 1997 American Mathematical Society

the 4-component formalism. The method is similar to that used by Kalnins et al. [1992] to determine the second order symmetry operators of Maxwell's equations.

2 Symmetry Operators

The Weyl-neutrino field equation has the form

$$\nabla^{AA'}\phi_A = 0. \tag{2.1}$$

We shall determine the form of the most general second-order symmetry operator of (2.1). One observes that there are two possible types of such operators, viz., those which map solutions ϕ_B into solutions $\tilde{\phi}_A$ and those which map solutions ϕ_B into solutions $\tilde{\chi}_{A'}$. We shall refer to these operators as operators of the first and second type respectively.

The above symmetry operators have the following general forms

$$Q_A{}^B = K_A{}^{BEE'FF'}\nabla_{EE'}\nabla_{FF'} + L_A{}^{BEE'}\nabla_{EE'} + M_A{}^B, \tag{2.2}$$

$$Q_{A'}{}^B = K_{A'}{}^{BEE'FF'}\nabla_{EE'}\nabla_{FF'} + L_{A'}{}^{BEE'}\nabla_{EE'} + M_{A'}{}^B. \tag{2.3}$$

They must satisfy the defining equations

$$\nabla^A{}_{A'}Q_A{}^B\phi_B = 0 \tag{2.4}$$

and

$$\nabla_A{}^{A'}Q_{A'}{}^B\phi_B = 0 \tag{2.5}$$

where ϕ_B is any solution of (2.1).

We begin the calculation by removing from the operators (2.2) and (2.3) any operators which are trivial as a result of being 'multiples' of (2.1) or are expressions of the Ricci identities. This result is achieved for (2.2) by imposing the following necessary symmetries on the coefficient spinors:

$$K_{ABEFE'F'} = K_{A(BEF)(E'F')}. \tag{2.6}$$

$$L_{ABEE'} = L_{A(BE)E'} \tag{2.7}$$

Similarly, for the operator (2.3) the coefficients must satisfy

$$K_{A'E'F'BEF} = K_{A'(E'F')(BEF)}. \tag{2.8}$$

$$L_{A'E'BE} = L_{A'E'(BE)}\,. \tag{2.9}$$

3 Operators of the First Type

The form of the general symmetry operator of the first type is given by the following theorem:

Theorem 3.1 *The most general second-order covariant differential symmetry operator for the Weyl-neutrino equation which maps solutions ϕ_B into solutions $\tilde{\phi}_A$ is of the form*

$$\begin{aligned}\tilde{\phi}_A = \; & K^{EFE'F'}\nabla_{(E(E'}\nabla_{FF')}\phi_{A)} + J_A{}^{BEE'}\nabla_{(EE'}\phi_{B)} \\ & + L^{EE'}\nabla_{(EE'}\phi_{A)} + M_A{}^B\phi_B\end{aligned} \tag{3.1}$$

where the coefficients $K^{EFE'F'}$, $J_A{}^{BEE'}$, $L^{EE'}$ and $M_A{}^B$ are any spinor fields which satisfy

$$\nabla_{(B(A'}K_{EF)E'F')} = 0, \tag{3.2}$$

$$J_{ABCD'} = \nabla_{(AK'}K_{BC)}{}^{K'}{}_{D'}, \tag{3.3}$$

$$\begin{aligned}\nabla_{(B(A'}L_{E)E')} = &\ \tfrac{8}{15}\nabla_{(BK'}\nabla_{|L|(A'}K_{E)}{}^{L}{}_{E')}{}^{K'} - \tfrac{8}{5}\overline{\Psi}_{K'A'E'M'}K_{BE}{}^{M'K'} \\ &-\tfrac{11}{15}\Psi_{ABEM}K^{MA}{}_{A'E'} - \tfrac{58}{15}\Phi_{A(B(A'|M'|}K_{E)}{}^{A}{}_{E')}{}^{M'} \\ &+\tfrac{56}{15}\Lambda K_{EBA'E'},\end{aligned} \tag{3.4}$$

$$\begin{aligned}M_{(EB)} = &\ -\tfrac{1}{3}\nabla_{(B(L'}\nabla_{|K|K')}K_{E)}{}^{KK'L'} + \tfrac{1}{2}\nabla_{(BL'}L_{E)}{}^{L'} \\ &+\tfrac{2}{3}\Phi_{L'K'M(B}K_{E)}{}^{MK'L'},\end{aligned} \tag{3.5}$$

and the consistency condition

$$\begin{aligned}&\nabla^{A}{}_{A'}M_{AB} + \tfrac{4}{3}K^{EFE'F'}\nabla_{EE'}\Phi_{A'F'FB} - \tfrac{4}{3}K^{E}{}_{B}{}^{E'}{}_{A'}\nabla_{EE'}\Lambda \\ &-\tfrac{1}{3}K^{EFE'}{}_{A'}\nabla^{K}{}_{E'}\Psi_{KEFB} + \tfrac{1}{3}J_{KEFA'}\Psi^{EFK}{}_{B} - L_{FF'}\Phi_{A'}{}^{F'F}{}_{B} \\ &+3\Lambda L_{BA'} = 0.\end{aligned} \tag{3.6}$$

Corollary 3.2 *The most general first-order covariant differential symmetry operator for the Weyl-neutrino equation which maps solutions ϕ_B into solutions $\tilde{\phi}_A$ is of the form*

$$\tilde{\phi}_A = L^{EE'}\nabla_{(EE'}\phi_{A)} + M_A{}^B\phi_B \tag{3.7}$$

where the various quantities appearing satisfy

$$\nabla_{(B(A'}L_{E)E')} = 0, \tag{3.8}$$

$$M_{(EB)} = \frac{1}{2}\nabla_{(BL'}L_{E)}{}^{L'}, \tag{3.9}$$

and the consistency condition

$$\nabla^{A}{}_{A'}M_{AB} - L_{FF'}\Phi_{A'}{}^{FF'}{}_{B} + 3\Lambda L_{BA'} = 0. \tag{3.10}$$

We see from equation (3.2) that $K_{EFE'F'}$ satisfies the conformal Killing tensor equation and from equation (3.8) that $L_{EE'}$ satisfies the conformal Killing vector equation. We also notice that the condition (3.4) on $L_{EE'}$ is invariant under the addition of a conformal Killing vector to $L_{EE'}$.

4 Operators of the Second Type

The form of the general symmetry operator of the second type is given by the following theorem:

Theorem 4.1 *The most general second-order covariant differential symmetry operator for the Weyl-neutrino equation which maps solutions ϕ_A into solutions $\tilde{\chi}_{A'}$ is of the form*

$$\begin{aligned}\tilde{\chi}_{A'} = &\ K^{F'BEF}\nabla_{(E(A'}\nabla_{FF')}\phi_{B)} + J_{A'}{}^{BEE'}\nabla_{(EE'}\phi_{B)} \\ &+L^{EB}\nabla_{(EA'}\phi_{B)} + M_{A'}{}^{B}\phi_B\end{aligned} \tag{4.1}$$

where the coefficients $K^{F'BEF}$, $J_{A'}{}^{BEE'}$, L^{EB} and $M_{A'}{}^{B}$ are any spinor fields which satisfy

$$\nabla_{(A(E'}K_{F')BEF)} = 0, \tag{4.2}$$

$$J_{A'B'AB} = \frac{3}{4}\nabla_{L(A'}K_{B')}{}^{L}{}_{(AB)}, \tag{4.3}$$

Fields Institute Communications
Volume **15**, 1997

Two-Dimensional Dilaton Black Holes

Guy Michaud
Department of Physics
McGill University
Montréal, Québec, Canada H3A 2T8
gmichaud@hep.physics.mcgill.ca

Robert C. Myers
Department of Physics
McGill University
Montréal, Québec, Canada H3A 2T8
rcm@hep.physics.mcgill.ca

Abstract. The two-dimensional CGHS model provides an interesting toy-model for the study of black hole evaporation. For this model, a quantum effective action, which incorporates Hawking radiation and backreaction, can be explicitly constructed. In this paper, we study a generalization of this effective action. In our extended model, it is possible to remove certain curvature singularities arising for the original theory. We also find that the flux of Hawking radiation is identical to that encountered in other two-dimensional models.

1 Introduction

In 1992, Callan, Giddings, Harvey and Strominger (CGHS) presented an interesting two-dimensional toy-model (Callan et al. [1992]) for the study of black hole evaporation (Hawking [1975]). Much greater analytic progress can be made in studying such a two-dimensional model because of the fewer number of degrees of freedom and reduced complexity as compared to gravity in four dimensions. However in two dimensions, one cannot use the Einstein action because it leads to trivial equations of motion. In the CGHS model, an extra scalar field, the dilaton, is included to produce a nontrivial theory of gravity. The "classical" CGHS model is based on the string-inspired action:

$$S_0 = \frac{1}{2\pi}\int d^2x\sqrt{-g}\left\{e^{-2\phi}[R+4(\nabla\phi)^2+4\lambda^2] - \frac{1}{2}\sum_{i=1}^{N}(\nabla f_i)^2\right\} . \tag{1.1}$$

1991 *Mathematics Subject Classification.* Primary 83C57; Secondary 83C75.

Talk presented by Guy Michaud.

This research was supported by NSERC of Canada and Fonds FCAR du Québec.

© 1997 American Mathematical Society

This theory couples the two-dimensional metric, g_{ab}, and the dilaton, ϕ, to N massless scalar fields, f_i. It also includes a cosmological constant, λ^2.

In fact, the general solution for this theory may be constructed analytically (Callan et al. [1992]). Usually, the solutions are written in the conformal gauge where the metric is given in diagonal form as:

$$g_{\pm\pm} = 0 \qquad g_{+-} = -\frac{1}{2}e^{2\rho}$$

where we use the null coordinates defined by $x^{\pm} = t \pm x$. The general solution includes many black hole solutions. The simplest of these is the eternal black hole, in which all of the matter fields are set to zero. In Kruskal gauge with $\rho = \phi$ (Callan et al. [1992]), one has:

$$e^{-2\phi} = e^{-2\rho} = \frac{M}{\lambda} - \lambda^2 x^+ x^-$$

where M is the mass of the black hole as it can be seen by a computation of the Bondi mass (Callan et al. [1992]). This solution has the same causal structure as the extended Schwarzschild black hole, with future and past spacelike singularities. With $M = 0$, one is left with the linear dilaton vacuum. In this case, the metric describes flat two-dimensional Minkowski space while (in appropriate coordinates) the dilaton increases linearly in the spatial direction.

Other solutions describe the formation of a black hole from the collapse of a shell of matter. One example, which contains an infinitely thin shell (or shock wave) collapsing along $x^+ = x_0^+$, has:

$$e^{-2\phi} = e^{-2\rho} = -\lambda^2 x^+ x^- - m\,(x^+ - x_0^+)\,\Theta(x^+ - x_0^+)\ .$$

This solution is divided in two different regions as we see from the step function $\Theta(x^+ - x_0^+)$. In the first region below the infall line ($x^+ < x_0^+$), the solution corresponds to the linear dilaton vacuum. The region above the infall line ($x^+ > x_0^+$) is a portion of the eternal black hole with an event horizon located at $x^- = -m/\lambda^2$. Thus it contains the future spacelike singularity.

1.1 One-loop effective action The previous black hole solutions are classical, and so they do not include Hawking radiation. In order to study the latter, we must include quantum effects. One approach is to define the quantum theory with the following functional integral:

$$\mathcal{Z} = \int \mathcal{D}g\,\mathcal{D}\phi\,\mathcal{D}f_i\ e^{iS_{DG}(g,\phi)+iS_M(g,f_i)} \tag{1.2}$$

where S_{DG} and S_M denote the pure dilaton-gravity and the matter contributions to the action S_0, respectively. Actually, we cannot perform this functional integral completely, but we can compute the matter functional integral which is a simple Gaussian. One finds:

$$\int \mathcal{D}f_i\ e^{iS_M(g,f_i)} = e^{iS_1(g)}$$

where $S_1(g)$ is the Polyakov action, which may be written in a covariant non-local form (Polyakov [1981]):

$$S_1 = -\frac{\kappa}{8\pi}\int d^2x\sqrt{-g}\,R\frac{1}{\nabla^2}R \tag{1.3}$$

where $\kappa \equiv N/12$, and $1/\nabla^2$ is the Green's function for the scalar d'Alembertian ∇^2. This action can also be written in a local form in conformal gauge. This one-loop contribution combined with S_0 yields a quantum effective action for the CGHS theory. Solutions of this effective action incorporate Hawking radiation and backreaction at least to leading order in a $1/N$ expansion for large N. Unfortunately, it is apparently not possible to solve the resulting equations of motion exactly, although some qualitative results have been produced.

2 Generalized Model

The measure used in the functional integral (1.2) is not uniquely defined, and one can define alternative theories with different measures. Explicitly, this means that we can build new theories by adding local, covariant counterterms to the Polyakov action S_1. Here, we exploit this freedom to build a generalization of the previous quantum effective action, and look at the effects of the new interactions on black hole physics. Our extended action will be $S = S_0 + S_1 + S_2 + S_3$, where S_0 is the classical CGHS action (1.1) and S_1 is the Polyakov action (1.3). The remaining terms are:

$$S_2 = -\frac{\kappa}{8\pi}\int d^2x\sqrt{-g}\left[\alpha\phi R + \beta(\nabla\phi)^2\right]$$

$$S_3 = -\frac{\kappa}{8\pi}\int d^2x\sqrt{-g}\sum_{n=2}^{K}\left[a_n\phi^n R + b_n\phi^{n-1}(\nabla\phi)^2\right]$$

where α, β, a_n and b_n are coupling constants, and $K \geq 2$ is some integer. One of our objectives in selecting the coupling constants will be to produce an exactly soluble theory. A condition which will achieve this result is requiring the simple current equation (Russo et al. [1992] and Kazama et al. [1995]):

$$\partial_+\partial_-(\rho - \phi) = 0 \ .$$

The difference between T_{+-} and the dilaton equation of motion yields the above if we set:

$$\begin{aligned} \beta &= 4 - 2\alpha \\ b_n &= -2n\, a_n \ . \end{aligned}$$

Note that further setting $\alpha = 2$ and $a_n = 0$ (and hence $\beta = 0 = b_n$) produces the model of Russo, Susskind and Thorlacius [1992].

2.1 Liouville theory With the above simple current conditions and in conformal gauge, our action S is simplified by performing the field redefinition:

$$\chi = \rho - \frac{\alpha}{4}\phi + \frac{1}{\kappa}e^{-2\phi} - \frac{1}{4}\sum_{n=2}^{K} a_n\phi^n$$

$$\Omega = \left(1 - \frac{\alpha}{4}\right)\phi + \frac{1}{\kappa}e^{-2\phi} - \frac{1}{4}\sum_{n=2}^{K} a_n\phi^n \ .$$

With these new fields, the action takes a Liouville form:

$$S = \frac{1}{\pi}\int d^2x\left\{-\kappa\,\partial_+\chi\,\partial_-\chi + \kappa\,\partial_+\Omega\,\partial_-\Omega + \lambda^2 e^{2(\chi-\Omega)} + \frac{1}{2}\sum_{i=1}^{N}\partial_+ f_i\,\partial_- f_i\right\} \ .$$

In terms of the fields χ and Ω, we can write the Ricci curvature scalar as:

$$\begin{aligned} R &= 8e^{-2\rho}\partial_+\partial_-\rho \\ &= 8e^{-2\rho(\chi,\Omega)}\frac{1}{\Omega'}\left[\partial_+\partial_-\chi - \frac{\Omega''}{\Omega'}\partial_+\Omega\,\partial_-\Omega\right] \end{aligned}$$

where $\rho(\chi, \Omega)$ is the conformal factor as an implicit function of the new fields and the prime ($'$) denotes a derivative with respect to the field ϕ. Hence the curvature diverges whenever Ω' vanishes. Such extrema of the function $\Omega(\phi)$ lead to timelike singularities, even for the vacuum solution in which the dilaton always increases along the spatial direction. However, in the extended model, we can avoid these singularities by properly constraining the coupling constants. For example, $\Omega(\phi)$ has no extrema if $\alpha > 4$, $a_n > 0$ for odd n, and $a_n = 0$ for even n.

2.2 Solution We can further simplify the solutions with another field redefinition:

$$\begin{aligned} U &= \frac{\kappa}{2}(\chi + \Omega) \\ V &= \chi - \Omega\,. \end{aligned}$$

Then the metric equations of motion become:

$$\begin{aligned} T_{\pm\pm} &= -2\partial_\pm U \partial_\pm V + \frac{\kappa}{2}\partial_\pm^2 V + \partial_\pm^2 U + \frac{1}{2}\sum_{i=1}^{N}(\partial_\pm f_i)^2 + t_\pm(\sigma^\pm) = 0 \\ T_{+-} &= -\frac{\kappa}{2}\partial_+\partial_- V - \partial_+\partial_- U - \lambda^2 e^{2V} = 0 \end{aligned}$$

where the functions $t_\pm$ arise because of zero-mode ambiguities in defining the matter Green's function in the Polyakov action (1.3). We consider the solution for a collapsing shock wave for which the matter configuration is:

$$T^f_{++} = \frac{1}{2}\sum_{i=1}^{N}(\partial_+ f_i)^2 = m\,\delta(\sigma^+ - \sigma_0^+) \qquad T^f_{--} = \frac{1}{2}\sum_{i=1}^{N}(\partial_- f_i)^2 = 0$$

where m is the amplitude of the shock wave. We also choose the $t_\pm$-functions to have the form (Russo et al. [1992]):

$$t_\pm(x^\pm) = -\frac{\kappa}{4}\frac{1}{(x^\pm)^2}\,.$$

Now before presenting the solution, we transform to asymptotically Minkowskian coordinates for which at large radius $ds^2 \simeq -d\sigma^+ d\sigma^-$. With these choices, the solution is:

$$\begin{aligned} U &= e^{\lambda(\sigma^+ - \sigma^-)} - \frac{m}{\lambda}\left(e^{\lambda(\sigma^+ - \sigma_0^+)} - 1\right)\Theta(\sigma^+ - \sigma_0^+) - \frac{\kappa}{4}\ln\left[1 + \frac{m}{\lambda}e^{\lambda\sigma^-}\right] \\ V &= \frac{\lambda}{2}(\sigma^+ - \sigma^-)\,. \end{aligned} \tag{2.1}$$

This solution is the analogue of the classical shock wave. However, the above solution now includes the effects of Hawking radiation. The black hole evaporation can be examined by looking at the evolution of the Bondi mass.

3 Bondi Mass

The Bondi mass is defined on the future null infinity $\mathcal{J}_R^+$ and in the asymptotically Minkowskian coordinates, $\sigma^{\pm}$. It is expressed in terms of the linear variations δT_{ab} of the stress-energy tensor around some reference solution. As a reference, we take the "vacuum solution" which is given by setting $m = 0$ in the shock wave solution (2.1). Thus, the Bondi mass (de Alwis [1992]):

$$M(\sigma^-) = -\int^{\mathcal{J}_R^+} d\sigma^+ (\delta T_{++} + \delta T_{+-})$$

for the shock wave solution is:

$$M(\sigma^-) = m - \frac{\lambda}{4}\kappa \left\{ \ln\left[1 + \frac{m}{\lambda} e^{\lambda\sigma^-}\right] + \frac{m}{\lambda e^{-\lambda\sigma^-} + m} \right\} .$$

The flux of Hawking radiation may be obtained by differentiating this expression with respect to σ^-. Note that our results are completely independent of any of the coupling constants α and a_n. Hence the radiation in these solutions is essentially the same as in other two-dimensional models (Callan et al. [1992], Russo et al. [1992]). The radiation goes to zero in the far past $\sigma^- \to -\infty$, while it approaches a constant in the far future $\sigma^- \to \infty$. The latter constant flux means that the black holes never stop Hawking radiating even when their mass reaches zero! Thus our generalized model does not avoid this unphysical behavior found in other models.

4 Conclusion

We have presented a new generalization for the effective quantum action derived by CGHS for the study of Hawking radiation in two dimensions. We have shown that our generalization gives the possibility of removing the timelike curvature singularity arising in other models. Hence we may avoid the related boundary problems (Das and Mukherji [1994], Russo et al. [1993] and Strominger and Thorlacius [1994]). Also, this model is exactly solvable in the sense that we can find the general solution analytically. By studying the solution for the collapsing shock wave, we found that black hole evaporation proceeds in essentially the same way as in other dilaton gravity models. In particular, the Hawking radiation in our generalized model never stops! Thus we must conclude that the physics of this generalized theory remains incomplete, and we are still unable to determine the end-state of the black hole evaporation.

References

Callan, C. G., Giddings, S. B., Harvey, J. A. and Strominger, A. [1992] *Evanescent black holes,* Phys. Rev. D **46**, R1005–R1009.

Das, S. R., and Mukherji, S. [1994] *Boundary dynamics in dilaton gravity,* Mod. Phys. Lett. A **9**, 3105–3118.

de Alwis, S. P. [1992] *Quantum black holes in two dimensions,* Phys. Rev. D **46**, 5429–5438.

Hawking, S. W. [1975] *Particle creation by black holes,* Comm. Math. Phys. **43**, 199–220.

Kazama, Y., Satoh, Y. and Tsuchiya, A. [1995] *A unified approach to solvable models of dilaton gravity in two-dimensions based on symmetry,* Phys. Rev. D **51**, 4265–4276.

Polyakov, A. M. [1981] *Quantum geometry of bosonic strings,* Phys. Lett. B **103**, 207–210.

Russo, J. G., Susskind, L. and Thorlacius, L. [1992] *End point of Hawking radiation,* Phys. Rev. D **46**, 3444–3449.

——— [1993] *Cosmic censorship in two-dimensional gravity,* Phys. Rev. D **47**, 533–539.

Strominger, A. and Thorlacius, L. [1994] *Conformally invariant boundary conditions for dilaton gravity,* Phys. Rev. D **50**, 5177–5187.

Fields Institute Communications
Volume **15**, 1997

Gravitationally–Induced Three–Flavor Neutrino Oscillations as a Possible Solution to the Solar Neutrino Problem

Jonas R. Mureika
Department of Physics
University of Waterloo
Waterloo, Ontario, Canada N2L 3G1
newt@avatar.uwaterloo.ca

Robert B. Mann
Department of Physics
University of Waterloo
Waterloo, Ontario, Canada N2L 3G1
mann@avatar.uwaterloo.ca

Abstract. Neutrinos can undergo flavor–oscillations if they possess flavor–dependent couplings to the surrounding gravitational field (the VEP mechanism). The neutrino fields can be massless, in accord with the Minimal Standard Model, but at the expense of the Einstein Equivalence Principle. We show that it is possible to explain the observed Solar Neutrino data from the various experiments using the VEP solution in a realistic three–generation framework, and further note how the three–flavor model can offer larger allowed regions of parameter space over the two–flavor models.

1 Introduction

Various explanations of the Solar Neutrino Problem (SNP) are based on the assumption that neutrinos possess two non–degenerate eigenbases in which they can be described. One of these is the flavor eigenbasis $|\nu\rangle_W$, relevant to electroweak phenomena, while the other is diagonal in the quantum mechanical equations of motion. The first of such models was proposed by Mikheyev and Smirnov [1985] and Wolfenstein [1978], and is therefore dubbed the MSW mechanism. In this model, the second eigenstate must be a massive one, which implies that the neutrinos must have non–trivial masses. Such a solution saves the Standard Solar Model (Bahcall and Pinsonneault [1992]), but requires an extension to the Minimal Standard Model of Particle Physics, i.e., massive neutrinos.

This compromise of the Standard Model can be saved, as was pointed out in 1988 (Gasperini [1988]). Instead of having mass, if each neutrino couples differently to the (solar) gravitational field $\phi(r)$, then the same oscillation mechanism can be

1991 *Mathematics Subject Classification.* Primary 81V25; Secondary 83B05.

This work was supported in part by the Natural Sciences and Engineering Research Council of Canada.

© 1997 American Mathematical Society

obtained (up to the form of the energy dependence, the primary distinction between the massive and gravitational oscillation models). That is, each eigenstate ν_i has a different coupling $G_i = (1 + f_i)G$, with G Newton's constant, and each $f_i \ll 1$ a dimensionless "violation parameter". For first generation neutrinos, we take $f_i = 0$, i.e., $G_1 = G$.

The two eigenbases are related by a matrix $V_{N_g} \in SU(N_g)$. For three flavors, $N_g = 3$; discounting CP violations in the neutrino sector, V_3 becomes an orthogonal three–parameter $(\theta_{12}, \theta_{13}, \theta_{23})$ matrix. With $|\nu\rangle_W = V_3|\nu\rangle_{M,G}$, the neutrino states evolve according to the equations of motion

$$\begin{aligned} & i\frac{d}{dr}\,|\nu\rangle_{M,G} = H_{M,G}|\nu\rangle_{M,G} \\ \Longrightarrow \quad & i\frac{d}{dr}\,|\nu\rangle_W = H'_{M,G}|\nu\rangle_W. \end{aligned} \tag{1.1}$$

Here, $H_{M,G}$ are the diagonal Hamiltonians for MSW and VEP respectively in the mass/gravitational eigenbasis,

$$H_M = \frac{1}{2E}\,\mathrm{diag}\left\{m_1^2, m_2^2, m_3^2\right\} \tag{1.2}$$

$$H_G = 2E|\phi(r)|\,\mathrm{diag}\left\{f_1, f_2, f_3\right\} \tag{1.3}$$

while $H' = V_3^{-1} H V_3 + A$ is the corrected version for the electroweak interactions ($A = \mathrm{diag}\left\{\sqrt{2}G_F N_e, 0, 0\right\}$). It is the off–diagonal nature of H' which induces flavor oscillations, and the presence of A creates parameter–dependent resonances. We can subtract a total factor of unity $\mathbb{1} \cdot H_{11}$ from H, since this yields only an unobservable phase, and hence deal only with eigenvalue differences $2E|\phi(r)|\Delta f_{21,31}$ ($\Delta m^2_{21,31}/2E$ for MSW).

Re–diagonalization of H' by a matter–enhanced matrix V_3^m creates a new eigenbasis $|\nu\rangle_{MAT.}$ in which we can describe the evolution, with $|\nu\rangle_W = V_3^m|\nu\rangle_{MAT.}$. For an electron neutrino ν_e created in the solar core, the averaged probability that it reaches the Earth as a ν_e is found to be (Baldini and Giudice [1987])

$$\begin{aligned} \langle P(\nu_e \to \nu_e)\rangle &= \sum_{i,j=1}^{3} |(V_3)_{1i}|^2\, |(P_{LZ})_{ij}|^2\, |(V_3^m)_{1j}|^2 \\ &= c^2_{m12}c^2_{m13}\left\{(1-P_1)c^2_{12}c^2_{13} + P_1 s^2_{12}c^2_{13}\right\} + \\ &\quad s^2_{m12}c^2_{m13}\left\{P_1(1-P_2)c^2_{12}c^2_{13} + (1-P_1)(1-P_2)s^2_{12}c^2_{13} + P_2 s^2_{13}\right\} \\ &\quad + s^2_{m13}\left\{P_1P_2c^2_{12}c^2_{13} + P_2(1-P_1)s^2_{12}s^2_{13} + (1-P_2)s^2_{13}\right\}\ . \end{aligned} \tag{1.4}$$

Here, $s_{ij}, c_{ij} \equiv \sin\theta_{ij}, \cos\theta_{ij}$ and $s_{mij}, c_{mij} \equiv \sin\theta^m_{ij}, \cos\theta^m_{ij}$ are the parameters of V_3 and V_3^m, respectively. The matrix P_{LZ} has elements $P_{1,2}$ which describe the probability of non–adiabatic level crossing between $\nu_e \to \nu_\mu$, $\nu_e \to \nu_\tau$ (hereafter 12– and 13–transitions).

2 Behavior of $\langle P(\nu_e \to \nu_e)\rangle$ for a Double Resonance

If the ν_e's are created at electron densities higher than the corresponding resonance density for either 12– or 13–transitions, then the matter–enhanced mixing

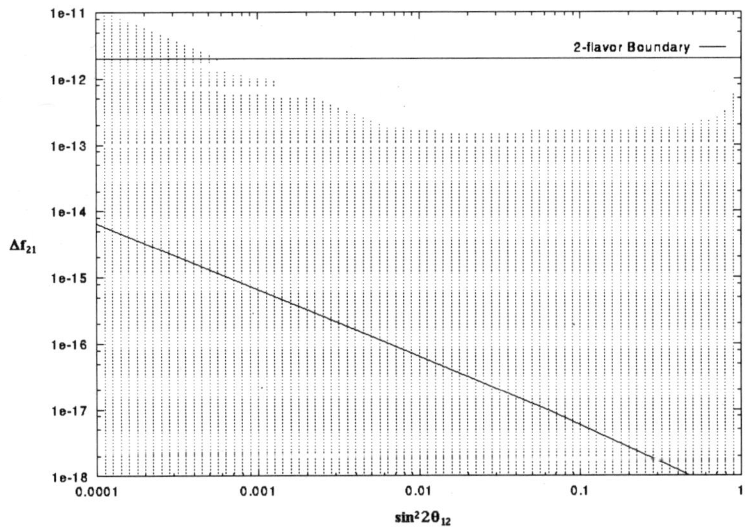

Figure 1 3σ overlap (see Bahcall et al. [1995]), with $\Delta f_{31} = 10^{-13}$, $s_{13}^2 = 0.4$

angles approach the value $\theta_{12}^m, \theta_{13}^m \to \frac{\pi}{2}$. Hence, the survival probability reduces to the simpler form (Mureika [1995])

$$\langle P(\nu_e \to \nu_e)\rangle = P_1 P_2 c_{12}^2 c_{13}^2 + P_2(1 - P_1) s_{12}^2 s_{13}^2 + (1 - P_2) s_{13}^2 \ . \tag{2.1}$$

To see how this is affected by the addition of the third flavor ν_τ, we can examine its limiting form for small and large θ_{13}. In the former case, we have

$$\langle P(\nu_e \to \nu_e)\rangle = c_{12}^2 P_1 P_2 \ , \tag{2.2}$$

which shows energy dependence through both 12– and 13–transitions in the $P_{1,2}$ terms. Solutions to the solar neutrino problem in this case are similar to the small–angle solution in the two–flavor limit, and are of questionable statistical validity (Mureika [1995]). However, we note that for large θ_{13}, the term $P_2 \to 0$ (adiabatic approximation for 13–transition), and so the above expression further reduces to

$$\langle P(\nu_e \to \nu_e)\rangle = s_{13}^2 \ . \tag{2.3}$$

This limiting form has interesting implications, as it suggests that the ν_e suppression is not only energy–independent (as is usually the case with large–angle oscillation solutions), but that it is also independent of the 12–transition. Figure 1 shows the allowed parameter–space overlap for the most recent solar neutrino experiment data in the large θ_{13} case. Clearly, the addition of a third flavor greatly broadens the regions from the much smaller two–flavor results (see Bahcall et al. [1995], Butler et al. [1993] and Pantaleone et al. [1993] for these).

3 Comparison of ^8B Neutrino Fluxes

As previously mentioned, the difference between the two oscillation mechanisms resides in their energy dependence. A study of the spectrum of ^8B neutrinos incident on terrestrial detectors can help shed light on which suppression mechanism, if any,

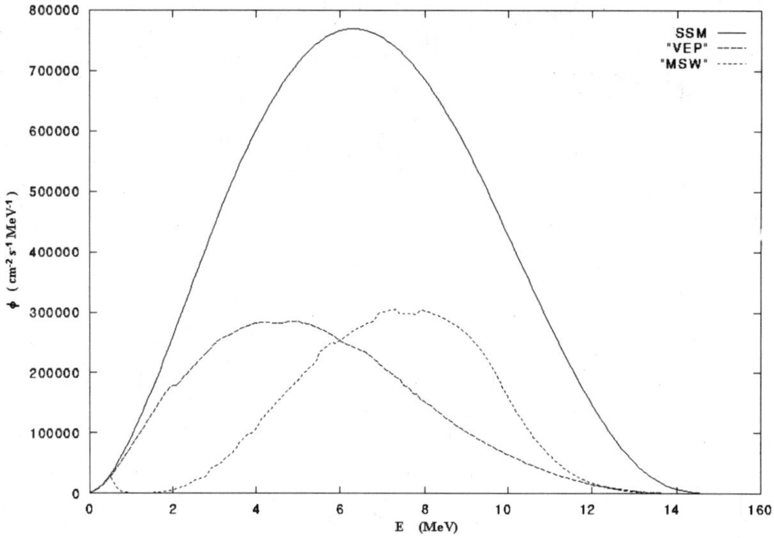

Figure 2 MSW and VEP reduced fluxes yielding counting rate $R = 2.00$ SNU, with both 12– and 13–resonances allowed.

is at work. In the previous section, the large θ_{13} form was shown to be energy independent, while the small θ_{13} case is a function of both 12– and 13–transitions. So, we should expect to see some type of spectral distortion in the ^8B neutrino flux in this small angle case (Mureika [1995], Mureika and Mann [1996]). Figure 2 shows that the attenuation of ν_es is indeed affected quite differently by each model: MSW suppresses low–energy neutrinos, while VEP suppresses higher–energy ones. Since detectors such as Kamiokande II, or SNO (when it comes online) can detect subtle variations in the high energy portion of the flux spectrum, we would expect such behavior as that in Figure 2 to be a major clue as to the solution of the SNP.

References

Bahcall, J. N., Krastev, P. I. and Leung, C. N. [1995] *Solar Neutrinos and the Principle of Equivalence*, Phys. Rev. D **52**, 1770-1779.

Bahcall, J. N. and Pinsonneault, M. H. [1992] *Solar models and neutrinos*, Rev. Mod. Phys. **64**, 885–926.

Baldini, A. and Giudice, G. F. [1987] *Oscillations of Three Generations of Neutrinos and the Solar Neutrino Problem*, Phys. Lett. B **186**, 211–216.

Butler, M. N. et al. [1993] *Gravitationally induced neutrino oscillations*, Phys. Rev. D **47**, 2615–2618.

Gasperini, M. [1988] *Testing the principle of equivalence with neutrino oscillations*, Phys. Rev. D **38**, 2635–2637.

Mikheyev, S. P. and Smirnov, A. Yu. [1985] *Resonance enhancement of oscillations in matter and solar neutrino spectroscopy*, Sov. J. Nucl. Phys. **42**, 913–917.

Mureika, J. [1995] *Gravitationally-Induced Three-Flavor Neutrino Oscillations as a Possible Explanation of the Solar Neutrino Problem*, M.Sc. Thesis, University of Waterloo, pp. 1–100.

Mureika, J. and Mann, R. B. [1996] *Mass or Gravitationally Induced Neutrino Oscillations? – A Comparison of ^8B Neutrino Flux Spectra in a Three–Generation Framework*, Phys. Lett. B **368**, 112–118.

Pantaleone, J., Halprin, A. and Leung, C. N. [1993] *Neutrino mixing due to a violation of the equivalence principle*, Phys. Rev. D **47**, R4199–R4202.

Wolfenstein, L. [1978] *Neutrino oscillations in matter*, Phys. Rev. D **17**, 2369–2374.

Fields Institute Communications
Volume **15**, 1997

Differential Invariants and Regularity

Peter Musgrave
Department of Physics
Queen's University
Kingston, Ontario, Canada K7L 3N6
musgrave@astro.queensu.ca

Kayll Lake
Department of Physics
Queen's University
Kingston, Ontario, Canada K7L 3N6
lake@astro.queensu.ca

Abstract. We explore the role of differential invariants on the regularity of spacetime. Specifically, we consider spherically symmetric, static spacetimes which have regular curvature invariants at the origin, but a diverging higher order differential invariant. Exact "solutions" which meet this criteria and satisfy all energy conditions are constructed. We then examine the physical source of the divergence in the differential invariants.

Scalar invariants such as $R_{abcd}R^{abcd}$ are often used to examine the "scalar polynomial" singular structure of spacetimes and recognize coordinate singularities. If the complete set of curvature invariants is non-singular we often assume the spacetime is (scalar polynomial) regular. In this paper we look for examples of spacetimes which are scalar polynomial regular (i.e., have regular second order differential invariants), but have diverging third order differential invariants. We then examine the physical source of the divergence in the differential invariants.

Recall that there are 14 independent scalar invariants for a four dimensional Riemannian manifold. The number of differential invariants grows rapidly (Kramer et al. [1980]).

# derivatives	# invariants
0	14
1	74
2	200
3	424

1991 *Mathematics Subject Classification.* 83C99.
This work was supported in part by the Natural Sciences and Engineering Research Council of Canada and the School of Graduate Studies, Queen's University. We would like to thank Denis Pollney for many helpful comments and Roy Maartens for pointing out to us Schmidt [1966].

© 1997 American Mathematical Society

We will restrict ourselves to those that are currently built into GRTensorII (Musgrave et al. [1994]):

Object Name	Definition
diRiem	$R_{abcd;e}R^{abcd;e}$
d2Riem	$R_{abcd;ef}R^{abcd;ef}$
d3Riem	$R_{abcd;efg}R^{abcd;efg}$
d4Riem	$R_{abcd;efgh}R^{abcd;efgh}$
dR	$R_{;e}R^{;e}$
d2R	$R_{;ef}R^{;ef}$
d3R	$R_{;efg}R^{;efg}$
d4R	$R_{;efgh}R^{;efgh}$

If the metric functions are only C^2 then clearly the differential invariants may diverge, but this is a *coordinate dependent* statement. The invariants give us a definition of spacetime continuity independent of coordinates. We will label a spacetime as $\mathcal{C}^2$ if all algebraic (i.e., second order) invariants are regular, $\mathcal{C}^3$ if the third order differential invariants are regular and so on.

Bonnor [1994] has recently reexamined the Kinnersley "photon rocket"

$$\begin{aligned} ds^2 &= (1 - a^2r^2\sin^2\theta - 2ar\cos\theta - 2mr^{-1})du^2 + 2du\,dr \\ &\quad -r^2(d\theta^2 + \sin^2\theta\,d\phi^2) - 2ar^2\sin\theta\,d\theta\,du \end{aligned}$$

$$-\infty < u < \infty,\ 0 < r < \infty,$$
$$0 \le \theta \le \pi,\ 0 \le \phi \le 2\pi$$

where $m = m(u)$ is the mass of a particle at the origin and $a = a(u)$ its acceleration along the negative direction of the polar axis. One of the issues he studies is whether $a(u)$ is physical or a consequence of the coordinate system. A first step to resolving this issue is to consider the curvature scalars. Using GRTensorII we can determine that the only non-vanishing invariants are:

$$\begin{aligned} C_{ab}^{\ \ cd}C_{cd}^{\ \ ab} &= 48\frac{m^2}{r^6} \\ C_{ab}^{\ \ cd}C_{cd}^{\ \ ef}C_{ef}^{\ \ ab} &= -96\frac{m^3}{r^9} \end{aligned}$$

and indeed $a(u)$ does not appear. But, if we consider the differential Riemann invariant

$$R_{abcd;e}R^{abcd;e} = \frac{144m}{r^9}\left(-2r^2m' - 5rm + 6amr^2\cos\theta + 10m^2\right) \tag{1}$$

where $' \equiv d/dr$, $a(u)$ does appear and so cannot be transformed away (this result has been obtained independently by M. A. H. MacCallum; see Bonnor [1994]).

This led us to ask the following question: Can we construct a "reasonable" spacetime with regular curvature invariants and diverging differential invariants? The simplest possibility is a spherically symmetric static spacetime and so we consider the metric

$$ds^2 = b(r)dr^2 + r^2d\theta^2 + r^2\sin^2\theta\,d\phi^2 - a(r)dt^2. \tag{2}$$

The conditions for the regularity of all invariants polynomial in the Riemann tensor at the origin $r = 0$ are (Schmidt [1966], Maartens and Maharaj [1990] and

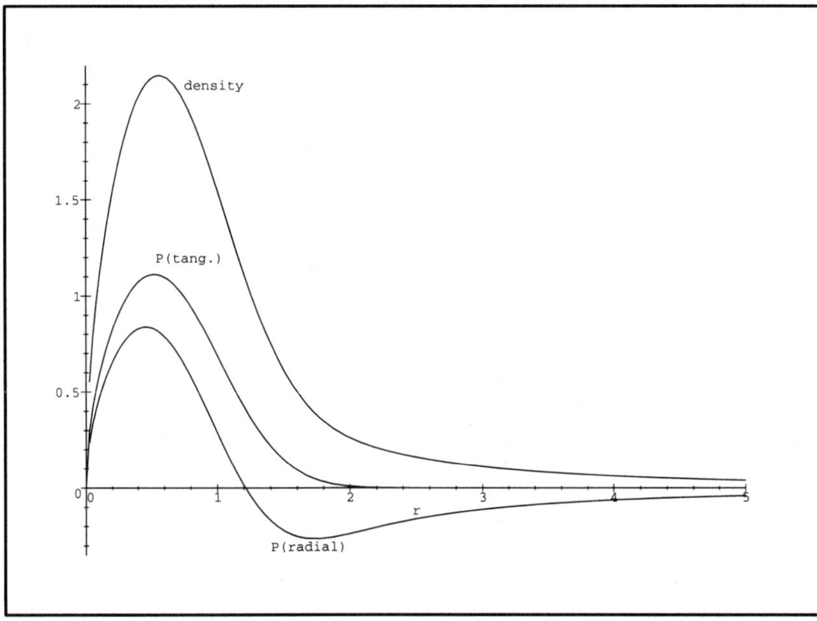

Figure 1 Density and pressures

Lake and Musgrave [1994])

$$a(0) = 1,\ b(0) = 1,\ a'(0) = 0,\ b'(0) = 0,\ ('\equiv d/dr), \tag{3}$$

with $a(r), b(r) \in C^2$.

If we examine $R_{abcd;e}R^{abcd;e}$ at $r = 0$ given the above restrictions we determine:

$$R_{abcd;e}R^{abcd;e} = \frac{5}{2}a'''(0)^2 + \frac{13}{18}b'''(0)^2 \tag{4}$$

and if we consider the strong energy condition for $a(r) = b(r)$ we get:

$$\frac{2a'}{r} + a'' \geq \frac{(a')^2}{a}. \tag{5}$$

Hence we consider:

$$a(r) = b(r) = 1 + r^{n+3/2}, n = 1, 2, 3 \ldots \tag{6}$$
$$a(r) = b(r) = \exp(r^{n+3/2}), n = 1, 2, 3 \ldots. \tag{7}$$

which clearly satisfy (3,5).

We now restrict our attention to (6) for the remainder of this paper.

The pressures and density for (6) with $n = 1$ are shown in Figure 1. The Einstein tensor for this spacetime is

$$\begin{aligned} G_r^r &= \frac{r^{n-1/2}(1+2n-2r^{n+3/2})}{2(1+r^{n+3/2})^2} \equiv P(r)/8\pi \\ G_\theta^\theta = G_\phi^\phi &= \frac{r^{n-1/2}(3+8n+4n^2-(6+4n)r^{n+3/2})}{8(1+r^{n+3/2})^3} \\ &\equiv P_T(r)/8\pi \\ G_t^t &= -\frac{r^{n-1/2}(5+2n+2r^{n+3/2})}{2(1+r^{n+3/2})^2} \equiv -\rho(r)/8\pi \end{aligned}$$

and the mass function is given by:

$$\frac{r}{2m(r)} = \frac{1+r^{n+3/2}}{r^{n+3/2}}, \tag{8}$$

and so $r > 2m(r)$ for $0 \le r < \infty$. This is a consequence of the regularity conditions.

Are these "solutions" physically reasonable? Since any spherically symmetric static spacetime can be interpreted as an anisotropic fluid we must check energy conditions. The weak, dominant and strong energy conditions are satisfied for $n > 0$.

The Ricci scalar can be expressed as:

$$R = 8\pi(\rho - P - 2P_T) \tag{9}$$

and evaluation of $R_{;\alpha}R^{;\alpha}$ for $n = 1$ gives

$$\frac{(7-254r^{5/2}+146r^5+32r^{15/2})^2}{64(1+r^{5/2})^9 r}$$

which diverges as $r \to 0$ like $1/r$. Note also that the gradients of ρ, P and P_T all diverge like $1/r^{1/2}$ as $r \to 0$.

Is the solution regular in the sense of Baumgarte and Rendall [1993]? Baumgarte and Rendal have shown that with isotropic pressure ($P = P_T$) if $\rho > 0$ and $\rho \in C^0$ with $P(0) = P_0$ where $0 < P_0 < \infty$ then $r > 2m(r)$ and $P \in C^1$ for $P > 0$. In the anisotropic pressure case they prove a similar theorem (with $P(0) = P_T(0)$) but require $P_T \in C^1$ in a closed neighborhood of $r = 0$. For the case $n = 1$ P_T is *not* C^1 at $r = 0$. This then can be considered as the "physical source" of the singularity in the differential invariants.

The diverging differential invariant can be pushed "deeper". We have verified (explicitly for $p < 5$) that the Ricci invariant $R_{;a_1 \ldots a_p} R^{;a_1 \cdots a_p}$ diverges for the metric

$$ds^2 = (1+r^{n+3/2})dr^2 + d\Omega^2 - (1+r^{n+3/2})dt^2. \tag{10}$$

(Presumably this is true in general, but it is difficult to establish.) So, in our notation, (10) is $\mathcal{C}^{n+2}$. Solutions with a divergence in a higher order ($p > 1$) Ricci invariant will have a density and pressures which do meet the differentiability requirements of Baumgarte and Rendall and which satisfy $r > 2m(r)$.

References

Baumgarte, T. W. and Rendall, A. D. [1993] *Regularity of spherically symmetric static solutions to the Einstein equations*, Class. Quantum Grav. **10**, 327–332.

Bonnor, W. B. [1994] *The photon rocket*, Class. Quantum Grav. **11**, 2007–2012.

Kramer, D., Stephani, H., MacCallum, M. and Herlt, E. [1980] *Exact solutions of Einstein's field equations*, Cambridge University Press, Cambridge.

Lake, K. and Musgrave, P. [1994] *The regularity of static, spherically, cylindrically and plane symmetric spacetimes at the origin*, Gen. Rel. Grav. **26**, 917–925.

Maartens, R. and Maharaj, M. S. [1990] *Conformally symmetric static fluid spheres*, J. Math. Phys. **31**, 151–155.

Musgrave, P., Pollney, D. and Lake, K. [1994] GRTensorII software (available by anonymous ftp from astro.queensu.ca [130.15.26.39]). GRTensorII requires MapleV Release 3, available from Waterloo Maple Software, Waterloo, Ontario, Canada.

Schmidt, B. [1966] *Kugelsymmetrische statische Materielösungen der Einsteinschen Feldgleichungen*, Diplomarbeit, Hamburg University.

Fields Institute Communications
Volume **15**, 1997

GRTensorII: A Computer Algebra System for General Relativity

Peter Musgrave
Department of Physics
Queen's University
Kingston, Ontario, Canada K7L 3N6
musgrave@astro.queensu.ca

Denis Pollney
Faculty of Mathematical Studies
University of Southampton
Highfield, Southampton, S017 1BJ United Kingdom
dp@maths.soton.ac.uk

Kayll Lake
Department of Physics
Queen's University
Kingston, Ontario, Canada K7L 3N6
lake@astro.queensu.ca

Abstract. We describe a number of features of the GRTensorII computer algebra package emphasizing its input and output facilities, the ability to define new tensor objects, and the ability to use and switch between a number of different formalisms (classical tensor methods, general bases, and null tetrads).

GRTensorII is a package for the calculation and manipulation of components of tensors and related objects. Rather than focus upon a specific type or method of calculation, the program has been designed to operate efficiently on a wide range of applications allowing use of a number of different formalisms. The standard object libraries make use of the most efficient known algorithms in carrying out calculations. Additionally, the package allows for customization and expansion with the ability to define new objects, user-defined algorithms, and add-on libraries.

We have found that regardless of the algorithm or formalism used, it is often the case that only certain simplifications applied at a crucial stages can make some problems tractable. For this reason, GRTensorII has been designed to provide full control of the calculation path and the simplifications to be performed at each

---

1991 *Mathematics Subject Classification.* Primary 83-08; Secondary 53-04.

The authors would like to thank Bob McKellar for indicating some useful clarifications of the output. This work was supported in part by grants (to K. L.) from the Natural Sciences and Engineering Research Council of Canada and the Advisory Research Committee of Queen's University. P. M. and D. P. would like to thank the Ministry of Colleges and Universities of Ontario for financial support through Ontario Graduate Scholarships.

© 1997 American Mathematical Society

stage. Once a calculation is completed, a variety of commands are available for the manipulation and simplification of results.

Emphasis has also been placed on the interface, allowing simple user input of calculations and tensor definitions, as well as presenting readable output. Metrics and basis vectors are easily defined and can be saved for later use. Calculations are specified in an intuitive manner using a minimal number of commands.

GRTensorII has been available on all platforms supporting MapleV.3 since October, 1994. It has since been ported to Mathematica. The software and complete documentation is distributed free of charge via ftp[1].

Basics

Tensors are specified by their name and index configuration, using the abbreviations `dn` and `up` to indicate covariant and contravariant indices, respectively. For instance, the Ricci tensor could be referenced

`R(dn,dn)` $(= R_{ab})$, `R(up,dn)` $(= R^{a}{}_{b})$, `R(bup,bdn)` $(= R^{(a)}{}_{(b)})$,

where the last example uses the labels `bup` and `bdn` to reference the components of the Ricci tensor in terms of a basis rather than coordinates.

Three commands are used most frequently within a GRTensorII session:

grcalc: calculates the components of tensors;
grdisplay: displays the components;
gralter: applies simplification routines to tensors.

Thus, to calculate the covariant Ricci tensor for the predefined Kerr metric, we might use the command sequence

```
> qload ( kerr ):
> grcalc ( R(dn,dn) ):
> grdisplay ( R(dn,dn) ):
```

At this point, since no simplifications of the results have been applied, the components of R_{ab} might not obviously be zero as we would expect. In Boyer-Lindquist coordinates, for instance, the following commands are needed:

```
> gralter ( R(dn,dn), trig ):
> grdisplay ( R(dn,dn) ):
```

The first command applies trigonometric simplification to the components of `R(dn,dn)`. Upon using `grdisplay`, we see that the the simplification has indeed reduced the components to zero.

Note that for a more complicated metric, it may be the case that it is better to carry out the computation in stages. That is, we can first calculate the Christoffel symbols and apply simplifications, then the Ricci tensor, and finally, for instance, the Ricci scalar.

A number of alternate simplification routines exist within `gralter`, including trigonometric, radical, expansion and factorization techniques. The command `grmap` can also be used to apply *any* Maple function to the components of a tensor.

[1]Both GRTensorII software and its documentation can be obtained directly from our ftp site at `astro.queensu.ca` in the `/pub/grtensor` directory, or from our WWW page at `http://astro.queensu.ca/~grtensor/GRHome.html`.

Input of Spacetimes

Spacetimes are specified within GRTensorII through use of the makeg facility. They can be entered in the following forms:

Metric, g_{ab}: A $n \times n$ symmetric matrix;
Line element: An equation of the standard form $ds^2 = \ldots$;
Non-holonomic basis: A set of basis vectors (covariant or contravariant) with an inner product;
Null tetrad: As described by the Newman-Penrose formalism.

As an illustrative (but computationally trivial) example, we consider the twisting Einstein-Maxwell solution described on p. 135 of Kramer et al. [1980],

$$k\,ds^2 = -2\left(x^a du - \frac{dy}{(a+1)x}\right)\left[dt + \frac{a(tdx+dy)}{x} + f(t)\left(x^a du - \frac{dy}{(a+1)x}\right)\right]$$
$$+ (dx^2+dy^2)(t^2+1)/2x^2,$$

We can enter this metric directly in this form using makeg:

```
> makeg ( sm ):
Makeg 2.0:  GRTensorII metric/basis entry utility
To quit makeg, type 'exit' at any prompt.
Do you wish to enter a 1) metric [g(dn,dn)],
                       2) line element [ds],
                       3) non-holonomic basis [e1...e4], or
                       4) null tetrad [l,n,m,mbar]?

> 2:
Enter coordinates as a LIST (e.g.  [r,theta,phi,t]):
> [ t, x, y, u ]:

Enter the line element using 'd[coord]' to indicate differentials.
(for example, r^2*(d[theta]^2 + sin(theta)^2*d[phi]^2)
[Type 'exit' to quit makeds]
ds^2 =

> 1/k*(-2*(x^a*d[u] - d[y]/((a+1)*x))*(d[t] + a*(t*d[x]
+ d[y])/x + f(t)*(x^a*d[u]-d[y]/((a+1)*x))) + (d[x]^2 +
d[y]^2)*(t^2+1)/(2*x^2)):
```

Note that in the above expression the coordinate differentials are labelled using the variables d[*coordinate*], allowing us to enter the line element in exactly the form that it is given in Kramer et al. [1980]. The output from makeg is:

$$ds^2 = \;\; 2\frac{dt\;dy}{k(a+1)x} - 2\frac{x^a dt\;du}{k} + \left(\frac{1}{2}\frac{t^2}{kx^2} + \frac{1}{2}\frac{1}{kx^2}\right)dx^2 + 2\frac{a\;t\;dx\;dy}{k(a+1)x^2}$$
$$- 2\frac{x^a a\;t\;dx\;du}{kx} + \left(2\frac{a}{k(a+1)x^2} - 2\frac{f(t)}{k(a+1)^2x^2} + \frac{1}{2}\frac{t^2}{kx^2} + \frac{1}{2}\frac{1}{kx^2}\right)dy^2$$
$$+ 2\left(-\frac{x^a a}{kx} + 2\frac{x^a f(t)}{k(a+1)x}\right)dy\;du - 2\frac{(x^a)^2 f(t) du^2}{k}$$

```
makeg completed.
```

The parameter values can now be set for the particular Type-III case studied by MacCallum and Siklos [1992] (referred to in Kramer et al. [1980] on p. 135):

```
> a := 1/2:  k := -32/(78*Lambda):  f(t) := (13*t^2+17)/32:
```

For this simple metric, it is not difficult to calculate the (rather uninteresting, in this case) components of the Ricci tensor:

```
> grcalc ( R(dn,dn) ):
> grdisplay ( _ ):
```

For the sm metric:

$$R_{ab} = \begin{bmatrix} 0 & 0 & -\frac{13}{8}\frac{1}{x} & \frac{39}{16}\sqrt{x} \\ 0 & -\frac{39}{32}\frac{t^2+1}{x^2} & -\frac{13}{16}\frac{t}{x^2} & \frac{39}{32}\frac{t}{\sqrt{x}} \\ -\frac{13}{8}\frac{1}{x} & -\frac{13}{16}\frac{t}{x^2} & -\frac{65}{192}\frac{5+t^2}{x^2} & -\frac{13}{128}\frac{5+13t^2}{\sqrt{x}} \\ \frac{39}{16}\sqrt{x} & \frac{39}{32}\frac{t}{\sqrt{x}} & -\frac{13}{128}\frac{5+13t^2}{\sqrt{x}} & \frac{39}{256}x(13t^2+17) \end{bmatrix}$$

The Newman-Penrose Formalism

A more complicated problem is that of determining the Petrov type of this metric. In order to do this, the metric must be cast in the form of a null tetrad. Since no tetrad obviously presents itself from the form of the metric given above, we can use the command **nptetrad** to generate one. This command accepts as input a space-like or time-like vector (not necessarily normalized), and performs a Gram-Schmidt orthonormalization process to determine a null tetrad. Naturally, this tetrad may not be optimal for further calculation, but is often a good start to which further simplifications can be applied.

```
> nptetrad ( [0,0,0,1], -1 ):
    The null tetrad has been stored as e(bdn,up).
> grmap ( e(bdn,up) , radsimp, `x` ):
    Applying routine radsimp to e(bdn,up)
> grdisplay ( e(bdn,up) ):
```

$${e_{(a)}}^b =$$

$$\begin{bmatrix} -\frac{1}{78}\frac{I\sqrt{3}\sqrt{13}\sqrt{13t^2+17}\sqrt{2}}{\sqrt{\Lambda}}, 0, 0, 0 \\ \frac{1}{78}\frac{I\sqrt{3}\sqrt{13}\sqrt{13t^2+17}\sqrt{2}}{\sqrt{\Lambda}}, 0, 0, -\frac{16}{39}\frac{I\sqrt{2}\sqrt{13}\sqrt{3}}{\sqrt{13t^2+17}\sqrt{x}\sqrt{\Lambda}} \\ -\frac{2}{39}\frac{(t+I)\sqrt{13}\sqrt{3}}{\sqrt{-t^2-1}\sqrt{\Lambda}}, \frac{4}{39}\frac{x\sqrt{3}\sqrt{13}}{\sqrt{-t^2-1}\sqrt{\Lambda}}, \frac{4}{39}\frac{I\sqrt{3}\sqrt{13}x}{\sqrt{-t^2-1}\sqrt{\Lambda}}, \frac{8}{117}\frac{I\sqrt{13}\sqrt{3}}{\sqrt{-t^2-1}\sqrt{x}\sqrt{\Lambda}} \\ -\frac{2}{39}\frac{(t-I)\sqrt{13}\sqrt{3}}{\sqrt{-t^2-1}\sqrt{\Lambda}}, \frac{4}{39}\frac{x\sqrt{3}\sqrt{13}}{\sqrt{-t^2-1}\sqrt{\Lambda}}, -\frac{4}{39}\frac{I\sqrt{3}\sqrt{13}}{\sqrt{-t^2-1}\sqrt{\Lambda}}, -\frac{8}{117}\frac{I\sqrt{13}\sqrt{3}}{\sqrt{-t^2-1}\sqrt{x}\sqrt{\Lambda}} \end{bmatrix}$$

From this point we can proceed to determine the Petrov type of our example metric by first calculating the spin coefficients (**NPSpin** and their conjugates **NPSpinbar**) and the Weyl scalars (**WeylSc**).

```
> grcalc ( NPSpin, NPSpinbar ):
> grcalc ( WeylSc ):
> grdisplay ( _ ):
```

$$\Psi 0 = 0$$
$$\Psi 1 = 0$$
$$\Psi 2 = 0$$

$$\Psi 3 = \frac{20}{13} \frac{\sqrt{2}(t - I)}{\Lambda (t^2+1)^{3/2} \sqrt{13t^2+17}}$$

$$\Psi 4 = -\frac{40}{39} \frac{13It^3 + 23t^2 + It + 11}{(t^2+1)^2 \Lambda (13t^2+17)}$$

Note that the calculation of NPSpin and NPSpinbar did not need to be specified in a separate step, since this step is implicit in the calculation of WeylSc. However, in general a good rule of thumb is to calculate these quantities first and simplify them fully (not required in this case) before proceeding to the curvature scalars.

```
> grcalc ( Petrov ):
> grdisplay ( _ ):
```

For the sm spacetime:
Petrov Type = III (or simpler)

Note the caveat 'or simpler' in the final line of output. The calculation of the Petrov type involves determining whether a number of intermediate quantities can be evaluated to zero (Letniowski and McLenaghan [1988]). It is sometimes the case that complicated ways of expressing zero do not, in fact, reduce to zero within the computer algebra system. In such cases it will return a Petrov type which is unnecessarily specialized. To determine the reliablility of the computer's evaluation, the report() command is provided, which details the intermediate calculations performed in the determination of the Petrov type.

```
> report():
```

The conclusion 'Petrov type = III (or simpler)' for the sm metric
was based on the following results:
Weyl scalar Psi0 = 0
Weyl scalar Psi1 = 0
Weyl scalar Psi2 = 0
Weyl scalar Psi3 could not be evaluated to zero.
Weyl scalar Psi4 could not be evaluated to zero.
---> Therefore the metric is Petrov III (or simpler).

We see from the previous output that the indeed the calculations have been carried out correctly (all of the terms which the computer believes to be non-zero are actually non-zero).

Defining New Objects

Finally, we describe a facility which exists within GRTensorII for defining tensors which are not part of the standard object libraries. Here we define the Bel-Robinson tensor,

$$T_{cdef} := C_{acdb} C^{a}{}_{ef}{}^{b} + C^{*}_{acdb} C^{*a}{}_{ef}{}^{b},$$

as well as its trace and its divergence using the command grdefine:

```
> grdefine(`T{c d e f}`,{},`C{a c d b}*C{^a e f^b}
    + Cstar{a c d b}*Cstar{^a e f^b}`):
> grdefine(`TT{c d}`,{},`T{^a a c d}`):
> grdefine(`TC{b c d}`,{},`T{^a b c d ;a}`);
```

Note that in the strings defining these tensors, contravariant indices are indicated by preceding them with a caret, '^', while covariant differentiation is indicated as usual, with a semi-colon, ';'. We can now load the Kerr-Newman metric and calculate these objects:

```
> qload(newkn):
```

Coordinates:

$$x^a = [\ r\ u\ \phi\ t\ ]$$

Line element

$$\begin{aligned} ds^2 = & \frac{(r^2+u^2)dr^2}{r^2-2mr+a^2+Q^2} + \frac{(r^2+u^2)du^2}{a^2-u^2} \\ & + \frac{(a^2-u^2)\left(r^2+a^2+\dfrac{(a^2-u^2)(2mr-Q^2)}{r^2+u^2}\right)d\phi^2}{a^2} \\ & - \frac{(a^2-u^2)(2mr-Q^2)d\phi dt}{a(r^2+u^2)} + \left(-1+\frac{2mr-Q^2}{r^2+u^2}\right)dt^2 \end{aligned}$$

Kerr–Newman Solution in Boyer–Lindquist coordinates $(u = a * \cos(\theta))$

```
> grcalc(T(dn,dn,dn,dn)):
```

We display a component (e.g., T_{rrrr}):

```
> grcomponent(T(dn,dn,dn,dn),[r,r,r,r]):
```

$$6\frac{m^2r^2-2mrQ^2+m^2u^2+Q^4}{(r^2+u^2)^2(r^2-2mr+a^2+Q^2)^2}$$

We now check that $T^a{}_{acd} = 0$ and that $\nabla_a T^a{}_{bcd} = 0$ in vacuum:

```
> grcalc(TT(dn,dn)):
> grdisplay( _ ):
```

TT(dn,dn) = *All components are zero*

```
> Q:=0:
> grcalc(TC(dn,dn,dn)):
> grdisplay( _ ):
```

TC(dn,dn,dn) = *All components are zero*

References

Kramer, D., Stephani, H., MacCallum, M. and Herlt, E. [1980] *Exact Solutions of Einstein's Field Equations*, Cambridge University Press, Cambridge.

Letniowski, F. W. and McLenaghan, R. G. [1988] *An Improved Algorithm for Quartic Equation Classification and Petrov Classification*, Gen. Rel. Grav. **20**, 463–483.

MacCallum, M. A. H. and Siklos, S. T. C. [1992] *Algebraically-special hypersurface-homogeneous Einstein spaces in general relativity*, J. Geom. Phys. **8**, 221–242.

Fields Institute Communications
Volume **15**, 1997

Exact Solution for Two Charged Spherically Symmetric Sources in Equilibrium

G. P. Perry
Department of Physics and Astronomy
University of Victoria
P. O. Box 3055
Victoria, B. C., Canada V8W 3P6
perry@uvphys.phys.uvic.ca

Fred I. Cooperstock
Department of Physics and Astronomy
University of Victoria
P. O. Box 3055
Victoria, B. C., Canada V8W 3P6
cooperstock@uvphys.phys.uvic.ca

Abstract. Up to the present time, the problem of gravitational-electrostatic balance of two bodies in general relativity without an intervening strut has been solved exactly only for critically charged masses, $m_i^2 = q_i^2$. Approximate solutions for arbitrary point sources have led to contradictory arguments in the literature with respect to the condition of balance. In the present paper, an exact electrostatic solution of the Einstein-Maxwell equations representing the exterior field of two arbitrary charged essentially spherically symmetric (Reissner-Nordström) bodies in equilibrium is given. This solution generalizes the Weyl class solution for which the gravitational and electrostatic equipotential surfaces overlap. Properties of the solution, including the balance issue without tension or strut, will be discussed.

It is of considerable interest to compare the phenomena of classical (Newtonian) physics and general relativity. The classical condition for the electrostatic balance of two spherically charged masses is

$$m_1 m_2 = q_1 q_2, \tag{1}$$

in appropriately chosen units. This balance occurs for arbitrary separation of the sources.

To study the conditions for balance in general relativity, one must first obtain appropriate physically meaningful solutions to the Einstein-Maxwell equations. Since directly solving the Einstein-Maxwell equations is often an insurmountable task, several researchers have employed techniques which greatly simplify the field

---

1991 *Mathematics Subject Classification.* Primary 83C15; Secondary 83C20, 83C22.

This research was supported, in part, by a grant from the Natural Sciences and Engineering Research Council of Canada and an NSERC Postgraduate Scholarship (GPP).

© 1997 American Mathematical Society

equations. One such class of solution is the Weyl class (Cooperstock and de la Cruz [1979]). It is characterized by the metric function $w = w(\Phi)$ in the line element

$$ds^2 = e^w dt^2 - e^{v-w}\left(d\rho^2 + dz^2\right) - \rho^2 e^{-w} d\phi^2, \tag{2}$$

where Φ is the electrostatic potential. For asymptotically flat boundary conditions, the unique functional relationship between w and Φ is

$$e^w = 1 - 2\frac{m_T}{q_T}\Phi + \Phi^2, \tag{3}$$

where m_T and q_T are the total mass and charge respectively. Furthermore, it is found that the constraint (Cooperstock and de la Cruz [1979])

$$m_1 q_2 - m_2 q_1 = 0 \tag{4}$$

must also hold. In this class, balance without a Weyl strut singularity for two Reissner-Nordström sources can only be achieved under the more restrictive condition that each mass be "critically charged", i.e.,

$$m_i = |q_i| \qquad i = 1, 2. \tag{5}$$

Several authors have studied the electrostatic balance problem using approximation techniques. Kimura and Ohta [1977] concluded that, in the post-post Newtonian approximation, the necessary and sufficient condition for balance is that each mass be critically charged. Barker and O'Connell [1977], using a different approximation method, claimed that in the post-Newtonian approximation, the equation

$$(m_1 q_2 - m_2 q_1)(q_1 - q_2) = 0 \tag{6}$$

had to be satisfied in addition to condition (1). Recently, Bonnor [1993] determined the equilibrium conditions for a charged test particle in the field of a Reissner-Nordström source. He found that while condition (5) was sufficient but not necessary for electrostatic balance, condition (1) was neither necessary nor sufficient.

The following electrostatic solution represents the exterior field of two arbitrary charged non-linearly superposed Reissner-Nordström solutions in equilibrium. It is obtained with the aid of Sibgatullin's method (Manko and Sibgatullin [1992] and [1993]) for constructing the complex Ernst potentials (Ernst [1968]). It is a special case of the general mathematical solution given by Ernst [1994]. Our parametrization incorporates the Weyl-class double Reissner-Nordström solution and generalizes it in the simplest possible manner when $m_1 q_2 \neq m_2 q_1$. As a result, the gravitational and electrostatic equipotential surfaces no longer overlap when Equation (4) is not satisfied.

In Sibgatullin's method, it is required that the Ernst potentials along the z-axis be specified. Our choice was

$$\mathcal{E}(\rho = 0, z) \equiv e(z) = 1 - \frac{2(m_1(z+z_2) + m_2(z+z_1))}{(z+z_1+m_1)(z+z_2+m_2) - q_1 q_2},$$

$$\Psi(\rho = 0, z) \equiv F(z) = \frac{q_1(z+z_2) + q_2(z+z_1)}{(z+z_1+m_1)(z+z_2+m_2) - q_1 q_2}. \tag{7}$$

(Note: the extension to include spin (double Kerr-Newman) can be achieved by replacing $z_n \to z_n - ia_n$, $n = 1, 2$, and then applying Sibgatullin's method.) The full Ernst potentials $\mathcal{E}(\rho, z)$ and $\Psi(\rho, z)$ for the solution, expressed in terms of the cylindrical coordinates $(\rho,\ z)$, are found to be

$$\mathcal{E} = \frac{A-B}{A+B}, \quad \Psi = \frac{C}{A+B}, \tag{8}$$

where

$$A \equiv \sum_{i<j}^{4} a_{ij} r_i r_j, \ B \equiv \sum_{i=1}^{4} b_i r_i, \ C \equiv \sum_{i=1}^{4} c_i r_i, \ r_n \equiv \sqrt{\rho^2 + (z-\alpha_n)^2}, \quad (n = 1 \to 4).$$

The constants α_n in Equation (8) are the roots of the equation

$$e(z) + [F(z)]^2 = 0. \tag{9}$$

The remaining constants a_{ij}, b_i and c_i are defined as follows:

$$a_{ij} \equiv (-1)^{i+j+1} s_i s_j t_i t_j (s_i t_j - s_j t_i) \begin{vmatrix} s_k v_k & s_l v_l \\ t_k u_k & t_l u_l \end{vmatrix},$$

$$(i < j;\ k < l;\ k, l \neq i, j;\ \ i, k = 1 \to 3;\ j, l = 2 \to 4);$$

$$b_i \equiv (-1)^i s_i t_i (s_i - t_i) \begin{vmatrix} s_k^2 t_k^2 & s_l^2 t_l^2 & s_m^2 t_m^2 \\ s_k v_k & s_l v_l & s_m v_m \\ t_k u_k & t_l u_l & t_m u_m \end{vmatrix},$$

$$(k < l < m;\ k, l, m \neq i;\ \ i = 1 \to 4;\ k = 1, 2;\ l = 2, 3;\ m = 3, 4);$$

$$c_i \equiv (-1)^{i+1} s_i t_i (s_i - t_i)(K_3 G_i + K_4 H_i),$$

$$G_i \equiv \begin{vmatrix} s_k t_k^2 & s_l t_l^2 & s_m t_m^2 \\ s_k v_k & s_l v_l & s_m v_m \\ t_k u_k & t_l u_l & t_m u_m \end{vmatrix}, \qquad H_i \equiv \begin{vmatrix} s_k^2 t_k & s_l^2 t_l & s_m^2 t_m \\ s_k v_k & s_l v_l & s_m v_m \\ t_k u_k & t_l u_l & t_m u_m \end{vmatrix}, \tag{10}$$

$$(k < l < m;\ k, l, m \neq i;\ \ i = 1 \to 4;\ k = 1, 2;\ l = 2, 3;\ m = 3, 4);$$

$$s_i \equiv \beta_1 - \alpha_i, \quad t_i \equiv \beta_2 - \alpha_i,$$

$$u_i \equiv K_1 s_i t_i + K_3^2 t_i + K_3 K_4 s_i, \quad v_i \equiv K_2 s_i t_i + K_4^2 s_i + K_3 K_4 t_i,$$

$$K_1 \equiv \frac{m_1 z_2 + m_2 z_1 + (m_1 + m_2)\beta_1}{\beta_1 - \beta_2}, \quad K_2 \equiv \frac{m_1 z_2 + m_2 z_1 + (m_1 + m_2)\beta_2}{\beta_2 - \beta_1},$$

$$K_3 \equiv \frac{q_1 z_2 + q_2 z_1 + (q_1 + q_2)\beta_1}{\beta_1 - \beta_2}, \quad K_4 \equiv \frac{q_1 z_2 + q_2 z_1 + (q_1 + q_2)\beta_2}{\beta_2 - \beta_1},$$

$$\beta_1 \equiv -\frac{1}{2}\left(z_1 + m_1 + z_2 + m_2 - \sqrt{(z_1 - z_2 + m_1 - m_2)^2 + 4 q_1 q_2}\right),$$

$$\beta_2 \equiv -\frac{1}{2}\left(z_1 + m_1 + z_2 + m_2 + \sqrt{(z_1 - z_2 + m_1 - m_2)^2 + 4 q_1 q_2}\right),$$

where all of the subsequent quantities introduced are constants ultimately defined in terms of m_i, q_i, z_i, $i = 1, 2$, which specify the character and locations of the sources.

The expressions for $\mathcal{E}$ and Ψ are in Kinnersley's form (Kinnersley [1973]) and this permits one to write the corresponding metric functions as

$$e^w = \frac{A\bar{A} - B\bar{B} + C\bar{C}}{(A+B)(\bar{A}+\bar{B})}, \quad e^v = \frac{A\bar{A} - B\bar{B} + C\bar{C}}{K_0 r_1 r_2 r_3 r_4}, \tag{11}$$

where

$$K_0 = \left(\sum_{i<j}^{4} a_{ij}\right)\left(\sum_{i<j}^{4} \bar{a}_{ij}\right) \tag{12}$$

and a bar denotes complex conjugation. The electrostatic potential for a static metric is related to the Ernst potential simply by $\Phi = \Psi$, which completes the solution.

Between the bodies, the condition for elementary flatness (or removal of the strut singularity) requires that the metric function $v = 0$ for $\rho = 0$, $\mathrm{Re}(\alpha_2) > z > \mathrm{Re}(\alpha_3)$, where the α_n are chosen such that $\alpha_1 \geq \alpha_2 > \alpha_3 \geq \alpha_4$ when we have two Weyl class undercharged ($m_i > |q_i|$, $i = 1, 2$) Reissner-Nordström bodies. This yields the equation

$$K \equiv \frac{a_{12}\left(\bar{a}_{13} + \bar{a}_{14}\right) + \bar{a}_{12}\left(a_{13} + a_{14}\right)}{K_0} = 0 \tag{13}$$

which must be satisfied in order to have electrostatic balance.

Since our solution contains the Weyl class, we know that Equation (5) satisfies Equation (13). Assuming that the above is a faithful general representation of the individual masses and charges of the composite system, numerical analysis of Equation (13) has revealed an additional *distance dependent* balance condition which does not satisfy conditions (1) or (5). For example, given $m_1 = 9$, $q_1 = 3$, $q_2 = 8$, $z_1 = -l_1 - d$, $z_2 = d$, with $l_1 \equiv \sqrt{m_1^2 - q_1^2}$ and $d = 30$ (note: the values of z_1 and z_2 are chosen such that the solution merges with the Weyl class solution (Cooperstock and de la Cruz [1979]) under the appropriate conditions), Equation (13) is satisfied for $m_2 \simeq 2.47$. If Equation (1) were the balance condition, then we would have $m_2 = 2.\dot{6}$. Hence, we obtain balance for $m_1 m_2 < q_1 q_2$. As the separation parameter d increases, $m_2 \to 2.\dot{6}$ asymptotically in order to satisfy Equation (13). This result parallels Bonnor's test particle analysis (Bonnor [1993]).

Both the mass and charge Simon relativistic multipole moments of our solution are, in general, dependent upon the mass and charge of each source and the separation of the bodies. However, as the balance condition approaches Equation (1) (with increasing separation), the form of the relativistic multipoles approaches that which we expect for separated monopoles in Newtonian physics. When the bodies are critically charged, the resulting relativistic multipole structure is identical to that in Newtonian physics for any separation and strength of the sources, at least up to 16-pole order. This behaviour of the multipole structure of our solution is absent in other works (see Manko et al. [1994] and references therein) and substantiates the essentially spherical nature of the sources.

References

Barker, B. M. and O'Connell, R. F. [1977] *Conditions for Static Balance for the Post-Newtonian Two-Body Problem with Electrostatic Charge in General Relativity*, Phys. Lett. A **61**, 297-298.

Bonnor, W. B. [1993] *The Equilibrium of a Charged Test Particle in the Field of a Spherical Charged Mass in General Relativity*, Class. Quantum Grav. **10**, 2077-2082.

Cooperstock, F. I. and de la Cruz, V. [1979] *Static and Stationary Solutions of the Einstein-Maxwell Equations*, Gen. Rel. Grav. **10**, 681–697.

Ernst, F. J. [1968] *New Formulation of the Axially Symmetric Gravitational Field Problem. II*, Phys. Rev. **168**, 1415–1417.

——— [1994] *Fully Electrified Neugebauer Spacetimes*, Phys. Rev. D **50**, 6179–6189.

Kimura, T. and Ohta, T. [1977] *On the Conditions for Static balance in the Post-Post-Newtonian Approximation*, Phys. Lett. A **63**, 193–195.

Kinnersley, W. [1973] *Generation of Stationary Einstein-Maxwell Fields*, J. Math. Phys. **14**, 651–653.

Manko, V. S., Martín, J. and Ruiz, E. [1994] *Metric of Two Arbitrary Kerr-Newman Sources Located on the Symmetry Axis*, J. Math. Phys. **35**, 6644–6657.

Manko, V. S. and Sibgatullin, N. R. [1992] *Metric of a Rotating, Charged, Magnetized Mass*, Phys. Lett. A **168**, 343–347.

——— [1993] *Kerr-Newman Metric Endowed with Magnetic Dipole Moment*, J. Math. Phys. **34**, 170–177.

Fields Institute Communications
Volume **15**, 1997

The Equation of Motion in Higher-Dimensional Gravity

Jaime Ponce de Leon
Department of Physics
University of Puerto Rico
P.O. Box 23343, Rio Piedras
Puerto Rico USA 00931

Paul S. Wesson
Department of Physics
University of Waterloo
Waterloo, Ontario, Canada N2L 3G1
wesson@astro.uwaterloo.ca

Abstract. We show that in 5D gravity where the metric is allowed to depend on the extra coordinate there are fifth-force like terms in the equation of motion.

The equation of motion of a particle in 5D has been studied by a number of people (Leibowitz and Rosen [1973], Gegenberg and Kunstatter [1984], Ferrari [1989] and Mashhoon et al. [1994]), because it provides in principle a way of testing whether there are extra dimensions to spacetime of the sort first proposed by Kaluza [1921] and Klein [1926]. However, much previous work was based on the older version of Kaluza-Klein theory, wherein the extra dimension was assumed to be compactified to an unobservably small size and there was no dependency of the metric on the extra (or internal) coordinate. These restrictions have been known for a while to lead to problems, such as large values of particle masses (the hierarchy problem) and the energy density of the vacuum (the cosmological constant problem). These problems can be avoided, and at the same time new physics obtained, in the newer version of Kaluza-Klein theory wherein the fifth dimension is introduced on exactly the same footing as the usual four. There is now a considerable literature on this form of the theory (see Wesson et al. [1993] for a review). In it, the metric is allowed to depend on the extra coordinate, a condition which means that Einstein's 4D equations of general relativity with matter can always be regarded as a subset of the Kaluza-Klein 5D equations for vacuum (Wesson and Ponce de Leon [1992]). In other words, matter in 4D can be regarded as induced by the fifth

1991 *Mathematics Subject Classification.* 83E15.

© 1997 American Mathematical Society

dimension. This version of Kaluza-Klein theory is in agreement with the classical tests of relativity (Kalligas et al. [1995]), and includes as a special case the 5D "black holes" or solitons that have been extensively studied on the basis of the older theory (Wesson and Ponce de Leon [1994]). However, the newer or induced-matter version of Kaluza-Klein theory clearly implies the need to rederive the equation of motion in the case where the metric can depend on the extra coordinate and where compactification is not necessarily assumed for the extra dimension. In the present note we will therefore derive the general equation of motion, leaving the question of its testing to other work (Wesson [1995] and Wesson and Ponce de Leon [1995]).

We let capital English letters run 01234 and small Greek letters run 0123. The signature of the 5D space is $+---\epsilon$, where $\epsilon = \pm 1$ is free to choose. We absorb the speed of light c and the gravitational constant G by a choice of units that renders them unity. Then the 5D interval in terms of the metric tensor γ_{AB} and coordinates x^A is given by

$$dS^2 = \gamma_{AB} dx^A dx^B \tag{1}$$

where a repeated index indicates summation as usual. As is well know, we can always decompose (1) into a 4D part associated with gravity and another part associated with a scalar field and electromagnetism:

$$dS^2 = ds^2 + \epsilon \Phi^2 (dx^4 + A_\mu dx^\mu)^2. \tag{2}$$

Here the gravitational potentials are defined via $ds^2 = g_{\mu\nu} dx^\mu dx^\nu$, and the scalar and electromagnetic potentials via

$$\Phi^2 \equiv \epsilon \gamma_{44}, \quad A_\mu \equiv \frac{\gamma_{\mu 4}}{\gamma_{44}}. \tag{3}$$

From the above we can obtain some relations that will be useful in what follows:

$$\gamma_{\alpha\beta} = g_{\alpha\beta} + \epsilon \Phi^2 A_\alpha A_\beta, \quad \gamma^{4\mu} = -A^\mu, \quad \gamma^{\mu\nu} = g^{\mu\nu}. \tag{4}$$

The last means we can raise or lower indices using either the 5D metric or the 4D metric.

The equations of motion are contained in the 5D geodesic equation:

$$\frac{d^2 x^A}{dS^2} + K^A_{BC} \frac{dx^B}{dS} \frac{dx^C}{dS} = 0, \tag{5}$$

where the five-dimensional Christoffel symbol of the second kind is

$$K^A_{BC} = \frac{\gamma^{AD}}{2} \left(\frac{\partial \gamma_{BD}}{\partial x^C} + \frac{\partial \gamma_{CD}}{\partial x^B} - \frac{\partial \gamma_{BC}}{\partial x^D} \right). \tag{6}$$

The equation (5) has 5 components, which it is convenient to study in order 4, 0123.

The 4-component of (5) can be written using the 5D Christoffel symbol of the first kind as

$$\gamma_{4A} \frac{d^2 x^A}{dS^2} + K_{4,CD} \frac{dx^C}{dS} \frac{dx^D}{dS} = 0. \tag{7}$$

It may be verified by expansion, using (4), that (7) is equivalent to

$$\frac{dB}{dS} = \frac{1}{2} \frac{\partial \gamma_{CD}}{\partial x^4} \frac{dx^C}{dS} \frac{dx^D}{dS}, \tag{8}$$

where B is a scalar function:

$$B \equiv \epsilon\Phi^2 \left(\frac{dx^4}{dS} + A_\alpha \frac{dx^\alpha}{dS} \right). \tag{9}$$

From (8) we see that if the five-dimensional metric γ_{CD} were independent of x^4, then B would be a constant of the motion. This is the case studied in the old version of Kaluza-Klein theory (Kaluza [1921], Klein [1926], Leibowitz and Rosen [1973], Gegenberg and Kunstatter [1984], Ferrari [1989] and Mashhoon et al. [1994]), but the noted condition does not in general hold in the new version (Wesson and Ponce de Leon [1992] and Wesson et al. [1993]). We also note from (2) and (9) that

$$dS = \frac{ds}{(1 - \epsilon B^2/\Phi^2)^{1/2}}, \tag{10}$$

which gives the 5D interval in terms of its 4D part.

The 0123 components of (5), which include the equation of motion in ordinary space, require the following Christoffel symbols [which can be obtained from (6) using (4)]:

$$K^\mu_{44} = -\epsilon\Phi\Phi^{;\mu} + \epsilon\Phi \left(A^\mu \frac{\partial\Phi}{\partial x^4} + \Phi g^{\mu\lambda} \frac{\partial A_\lambda}{\partial x^4} \right) \tag{11}$$

$$K^\mu_{4\nu} = -\epsilon\Phi\Phi^{;\mu} A_\nu - \frac{\epsilon\Phi^2}{2} F^\mu_\nu + \frac{g^{\mu\lambda}}{2} \frac{\partial\gamma_{\lambda\nu}}{\partial x^4} \tag{12}$$

$$K^\mu_{\alpha\beta} = \Gamma^\mu_{\alpha\beta} - \epsilon\Phi\Phi^{;\mu} A_\alpha A_\beta - \frac{\epsilon\Phi^2}{2} \left(A_\alpha F^\mu_\beta + A_\beta F^\mu_\alpha \right) + \frac{A^\mu}{2} \frac{\partial\gamma_{\alpha\beta}}{\partial x^4}. \tag{13}$$

Here a semicolon as in $\Phi^{;\mu} = g^{\mu\nu}\Phi_{;\nu} = g^{\mu\nu}(\partial\Phi/\partial x^\nu)$ involves the usual 4D covariant derivative, $\Gamma^\mu_{\alpha\beta}$ is the usual 4D Christoffel symbol defined by a relation analogous to (6), and $F_{\mu\nu}$ is the usual 4D antisymmetric tensor:

$$F_{\mu\nu} \equiv \frac{\partial A_\nu}{\partial x^\mu} - \frac{\partial A_\mu}{\partial x^\nu} = A_{\nu;\mu} - A_{\mu;\nu}. \tag{14}$$

We have introduced this because Kaluza-Klein theory can be interpreted as a unified theory of gravity and electromagnetism whenever $\gamma_{4\alpha}$ are finite. However, to bring out both the gravitational and electromagnetic parts of the geodesic (5) in a recognizable form, we need to substitute for (11)-(14) and use the definition (9) for the quantity B. After considerable algebra, we find that (5) with $A = \mu$ can be written

$$\begin{aligned} \frac{d^2x^\mu}{dS^2} + \Gamma^\mu_{\alpha\beta} \frac{dx^\alpha}{dS}\frac{dx^\beta}{dS} &= BF^\mu_\alpha \frac{dx^\alpha}{dS} + \epsilon B^2 \frac{\Phi^{;\mu}}{\Phi^3} - A^\mu \frac{dB}{dS} \\ &\quad - g^{\mu\lambda} \frac{dx^4}{dS} \left(B \frac{\partial A_\lambda}{\partial x^4} + \frac{\partial g_{\lambda\nu}}{\partial x^4} \frac{dx^\nu}{dS} \right). \end{aligned} \tag{15}$$

If we replace total derivatives with respect to S by their counterparts with respect to s using (10), we obtain

$$\begin{aligned} \frac{d^2x^\mu}{ds^2} + \Gamma^\mu_{\alpha\beta} \frac{dx^\alpha}{ds}\frac{dx^\beta}{ds} &= \frac{B}{(1 - \epsilon B^2/\Phi^2)^{1/2}} \left[F^\mu_\nu \frac{dx^\nu}{ds} - \frac{A^\mu}{B}\frac{dB}{ds} - g^{\mu\lambda} \frac{\partial A_\lambda}{\partial x^4} \frac{dx^4}{ds} \right] \\ &\quad + \frac{\epsilon B^2}{(1 - \epsilon B^2/\Phi^2)\Phi^3} \left[\Phi^{;\mu} + \left(\frac{\Phi}{B}\frac{dB}{ds} - \frac{d\Phi}{ds} \right) \frac{dx^\mu}{ds} \right] - g^{\mu\lambda} \frac{\partial g_{\lambda\nu}}{\partial x^4} \frac{dx^\nu}{ds}\frac{dx^4}{ds}. \end{aligned} \tag{16}$$

This is the fully general equation of motion in Kaluza-Klein theory, and for $\mu = 123$ shows how a test particle moves in ordinary space.

Equation (16) is a remarkable relation that tells us a lot about the nature of Kaluza-Klein theory. The left-hand side of (16) is the usual 4D relation for straight general relativity, and if equal to zero would express geodesic motion under pure Einstein gravity. However, the terms on the right-hand side modify the motion. Of these, the three inside the first bracket are to do with electromagnetism. All three vanish if the electromagnetic potentials A_μ are zero, i.e., if $\gamma_{4\mu} = 0$ by (3). Of these three terms, the last two would vanish independently if the metric did not depend on x^4 (since then B would be a constant of the motion). This leaves the first term, which we recognize as the usual Lorentz force, provided we identify the charge-to-mass ratio of the test particle as $q/m = B(1 - \epsilon B^2/\Phi^2)^{-1/2}$. This identification is straightforward in the case where the 4D part of the metric does not depend on x^4 and the extra fifth part is flat ($\gamma_{44} = \epsilon\Phi^2 = constant$). But we should be wary of using it in the more general case, since it appears to lead to problems in cosmological metrics that we will investigate elsewhere (Wesson and Ponce de leon [1995]). Here we proceed to look at the other terms in (16), and note that the three inside the second bracket have to do with the scalar field Φ. Of these, the last two show that the 4D acceleration involved with Φ depends also on the four-velocity dx^μ/ds. Specifically, the acceleration of a test particle in ordinary space depends on its velocity in ordinary space. Passing now to the final term in (16) we note that it does not depend explicitly on the scalar field Φ, but exists if the *four*-dimensional metric depends on the extra coordinate. This term could be significant if the physics of 5D Kaluza-Klein theory were to be "compressed" into the 4D Einstein metric by a choice of coordinates that render $\gamma_{4\alpha} = 0$ and $\gamma_{44} = constant$. It should be noted that the last term in (16), and others in it that involve dependency on x^4, are absent in work on Kaluza-Klein dynamics by Leibowitz and Rosen [1973], Gegenberg and Kunstatter [1984] and Ferrari [1989]. However, their work agrees with ours in the appropriate limit.

In summary, we see that the general equation of motion (16) in 5D Kaluza-Klein theory consists of a left-hand side that is identical to that of pure Einstein gravity and a right-hand side that includes parts due to electromagnetism and parts due to a scalar field. The non-electromagnetic parts of (16) could, from the viewpoint of general relativity, be ascribed to the action of a kind of fifth force. Whether these terms should be regarded as giving rise to a violation of the Equivalence Principle is to a certain extent a matter of definition. The important thing as far as we can see is that they are in principle measurable, and we investigate ways to test the general equation of motion in other work (Wesson [1995] and Wesson and Ponce de Leon [1995]). However, it is apparent already that the equation of motion (16) provides in principle a way of deciding if the world has 4 or 5 dimensions.

References

Ferrari, J. A. [1989] *On an Approximate Solution for a Charged Object and the Experimental Evidence for the Kaluza-Klein Theory,* Gen. Rel. Grav. **21**, 683–695.

Gegenberg, J. and Kunstatter, G. [1984] *The Motion of Charged Particles in Kaluza-Klein Space-Time,* Phys. Lett. A **106**, 410–414.

Kalligas, D., Wesson, P. S. and Everitt, C. W. F. [1995] *The Classical Tests in Kaluza-Klein Gravity,* Astrophys. J. **439**, 548–557.

Kaluza, T. [1921] *Zum Unitätsproblem der Physik,* Sitz. Preuss. Acad. Wiss. **33**, 966–972.

Klein, O. [1926] *Quantentheorie und fünf-dimensionale Relativitätstheorie,* Z. Phys. **37**, 895–906.

Leibowitz, E. and Rosen, N. [1973] *Five-Dimensional Relativity,* Gen. Rel. Grav. **4**, 449–474.

Mashhoon, B., Liu, H. and Wesson, P. S. [1994] *Particle Masses and the Cosmological Constant in Kaluza-Klein Theory,* Phys. Lett. B **331**, 305–312.

Wesson, P. S. [1995] *Possible Violations of the Equivalence Principle in Kaluza-Klein Theories,* Proc. Conference on Satellite Test of the Equivalence Principle (Reinhardt, J., ed.), to appear, E. S. A.

Wesson, P. S. and Ponce de Leon, J. [1992] *Kaluza-Klein Equations, Einstein's Equations and an Effective Energy-Momentum Tensor,* J. Math. Phys. **33**, 3883–3887.

______ [1994] *The Physical Properties of Kaluza-Klein Solitons,* Class. Quant. Grav. **11**, 1341–1351.

______ [1995] *The Equations of Motion in Kaluza-Klein Cosmology and Its Implications for Astrophysics,* Astron. Astrophys. **294**, 1–7.

Wesson, P. S., Ponce de Leon, J., Lim, P. and Liu, H. [1993] *Physical Properties of Matter Derived from Geometry of Kaluza-Klein Theory,* Int. J. Mod. Phys. D **2**, 163–170.

Fields Institute Communications
Volume **15**, 1997

The Nature of Cosmic Time

David S. Salopek
Department of Physics & Astronomy
University of British Columbia
Vancouver, B.C., Canada V6T 1Z1
and
Department of Physics
University of Alberta
Edmonton, Alberta, Canada T6G 2J1

Abstract. Hamilton-Jacobi theory provides a natural starting point for a covariant description of the gravitational field. Using a spatial gradient expansion, one may solve for the phase of the wavefunction by using a line-integral in superspace. Each contour of integration corresponds to a particular choice of time-hypersurface, and each yields the same answer. In this way, one can describe all time choices simultaneously. In an interesting application to observational cosmology, I compute large-angle microwave background anisotropies and the galaxy-galaxy correlation function associated with the scalar and tensor fluctuations of power-law inflation.

1 Introduction

Hamilton-Jacobi (HJ) theory is a cornerstone of modern theoretical physics. It may be profitably applied to numerous problems in cosmology. Since a full quantum theory is lacking, a semi-classical analysis provides our best understanding of the gravitational field.

HJ theory provides an elegant formalism for computing density perturbations as well as microwave background fluctuations arising from the inflationary scenario (Halliwell and Hawking [1985], Salopek [1995a] and Salopek and Stewart [1995]). It has also been successfully employed in deriving the Zel'dovich approximation (Zel'dovich [1970]) (which describes the formation of sheet-like structures in the Universe) from general relativity (Croudace et al. [1994], Salopek et al. [1994]). Numerous researchers have employed HJ methods in an attempt to recover the inflaton potential from cosmological observations (Copeland et al. [1993]). Lastly, HJ techniques can be used to construct inflationary models that yield non-Gaussian primordial fluctuations (Salopek [1991] and [1992a]); such models could possibly

1991 *Mathematics Subject Classification.* Primary 85A40; Secondary 83F05.
I thank K. M. Croudace, J. Parry and J. M. Stewart for a fruitful collaboration on Hamilton-Jacobi topics. The author also received partial support from the Natural Sciences and Engineering Research Council of Canada.

© 1997 American Mathematical Society

resolve the problems of large scale structure in the Universe (Moscardini et al. [1993]).

I will focus on one particularly attractive feature of HJ theory: it provides a covariant formulation of the gravitational field (Parry et al. [1994]). In the semi-classical theory, the answer to the question of time is clear: time is arbitrary. HJ theory enables one to consider all such time choices simultaneously. I will now consider a simple analogy from potential theory which illuminates the general technique.

2 Potential Theory

The fundamental problem in potential theory is: given a force field $g^i(u_k)$ which is a function of n variables u_k, what is the potential $\Phi \equiv \Phi(u_k)$ (if it exists) whose gradient returns the force field,

$$\frac{\partial \Phi}{\partial u_i} = g^i(u_k) \, ? \tag{2.1}$$

Not all force fields are derivable from a potential. Provided that the force field satisfies the integrability relation,

$$0 = \frac{\partial g^i}{\partial u_j} - \frac{\partial g^j}{\partial u_i} = \left[\frac{\partial}{\partial u_j}, \frac{\partial}{\partial u_i} \right] \Phi \, , \tag{2.2}$$

(i.e., it is curl-free), one may find a solution which is conveniently expressed using a line-integral

$$\Phi(u_k) = \int_C \sum_j dv_j \, g^j(v_l) \, . \tag{2.3}$$

If the two endpoints are fixed, all contours return the same answer. In practice, one employs the simplest contour that one can imagine: a line connecting the origin to the observation point u_k. Using s, $0 \leq s \leq 1$, to parameterize the contour, the line-integral may be rewritten as

$$\Phi(u_k) = \sum_{j=1}^{n} \int_0^1 ds \, u_j \, g^j(s u_k) \, . \tag{2.4}$$

Similarly, in solving for the phase of the wavefunctional, one utilizes a line-integral in *superspace.*

3 Solving the Hamilton-Jacobi Equation for General Relativity

The Hamilton-Jacobi equation for general relativity is derived using a Hamiltonian formulation of gravity. One first writes the line element using the ADM 3+1 split,

$$ds^2 = (-N^2 + \gamma^{ij} N_i N_j) \, dt^2 + 2 N_i \, dt \, dx^i + \gamma_{ij} \, dx^i dx^j \, , \tag{3.1}$$

where N and N_i are the lapse and shift functions, respectively, and γ_{ij} is the 3-metric. Hilbert's action for gravity interacting with a scalar field becomes

$$\mathcal{I} = \int d^4x \left(\pi^\phi \dot{\phi} + \pi^{ij} \dot{\gamma}_{ij} - N\mathcal{H} - N^i \mathcal{H}_i \right) . \tag{3.2}$$

The lapse and shift functions are Lagrange multipliers that imply the energy constraint $\mathcal{H}(x) = 0$ and the momentum constraint $\mathcal{H}_i(x) = 0$.

The object of chief importance is the generating functional $\mathcal{S} \equiv \mathcal{S}[\gamma_{ij}(x), \phi(x)]$. For each universe with field configuration $[\gamma_{ij}(x), \phi(x)]$ it assigns a number which can be complex. The generating functional is the 'phase' of the wavefunctional in the semi-classical approximation: $\Psi \sim e^{i\mathcal{S}}$. The probability functional, $\mathcal{P} \equiv |\Psi|^2$, is given by the square of the wavefunctional.

Replacing the conjugate momenta by functional derivatives of $\mathcal{S}$ with respect to the fields,

$$\pi^{ij}(x) = \frac{\delta \mathcal{S}}{\delta \gamma_{ij}(x)} \, , \quad \pi^{\phi}(x) = \frac{\delta \mathcal{S}}{\delta \phi(x)} \, , \tag{3.3}$$

and substituting into the energy constraint, one obtains the Hamilton-Jacobi equation (Peres [1962] and Misner et al. [1973]),

$$\begin{aligned} \mathcal{H}(x) &= \gamma^{-1/2} \frac{\delta \mathcal{S}}{\delta \gamma_{ij}(x)} \frac{\delta \mathcal{S}}{\delta \gamma_{kl}(x)} \left[2\gamma_{il}(x)\gamma_{jk}(x) - \gamma_{ij}(x)\gamma_{kl}(x) \right] \\ &\quad + \frac{1}{2}\gamma^{-1/2} \left(\frac{\delta \mathcal{S}}{\delta \phi(x)} \right)^2 + \gamma^{1/2} V(\phi(x)) \\ &\quad - \frac{1}{2}\gamma^{1/2} R + \frac{1}{2}\gamma^{1/2}\gamma^{ij}\phi_{,i}\phi_{,j} = 0 \, , \end{aligned} \tag{3.4}$$

which describes how $\mathcal{S}$ evolves in superspace. R is the Ricci scalar associated with the 3-metric, and $V(\phi)$ is the scalar field potential. In addition, one must also satisfy the momentum constraint

$$\mathcal{H}_i(x) = -2 \left(\gamma_{ik} \frac{\delta \mathcal{S}}{\delta \gamma_{kj}(x)} \right)_{,j} + \frac{\delta \mathcal{S}}{\delta \gamma_{lk}(x)} \gamma_{lk,i} + \frac{\delta \mathcal{S}}{\delta \phi(x)} \phi_{,i} = 0 \, , \tag{3.5}$$

which legislates gauge invariance: $\mathcal{S}$ is invariant under reparametrizations of the spatial coordinates (Peres [1962] and Misner et al. [1973]). (Units are chosen so that $c = 8\pi G = \hbar = 1$.) Since neither the lapse function nor the shift function appears in Equations (3.4) and (3.5) the temporal and spatial coordinates are *arbitrary*: HJ theory is *covariant*.

As a first step in solving Equations (3.4) and (3.5), I will expand the generating functional

$$\mathcal{S} = \mathcal{S}^{(0)} + \mathcal{S}^{(2)} + \mathcal{S}^{(4)} + \ldots \, , \tag{3.6}$$

in a series of terms according to the number of spatial gradients that they contain. The invariance of the generating functional under spatial coordinate transformations suggests a solution of the form,

$$\mathcal{S}^{(0)}[\gamma_{ij}(x), \phi(x)] = -2 \int d^3x \, \gamma^{1/2} H[\phi(x)] \, , \tag{3.7}$$

for the zeroth order term $\mathcal{S}^{(0)}$. The function $H \equiv H(\phi)$ satisfies the separated HJ equation of order zero (Salopek [1991] and [1992a]),

$$H^2 = \frac{2}{3} \left(\frac{\partial H}{\partial \phi} \right)^2 + \frac{1}{3} V(\phi) \, , \tag{3.8}$$

which is an ordinary differential equation. Note that $\mathcal{S}^{(0)}$ contains no spatial gradients.

In order to compute the higher order terms, one introduces a change of variables, $(\gamma_{ij}, \phi) \to (f_{ij}, u)$:

$$u = \int \frac{d\phi}{-2\frac{\partial H}{\partial \phi}} \, , \quad f_{ij} = \Omega^{-2}(u)\, \gamma_{ij} \, , \tag{3.9}$$

where the conformal factor $\Omega \equiv \Omega(u)$ is defined through

$$\frac{d\ln\Omega}{du} \equiv -2\frac{\partial H}{\partial\phi}\frac{\partial\ln\Omega}{\partial\phi} = H\ . \tag{3.10}$$

in which case the equation for $\mathcal{S}^{(2m)}$ becomes

$$\left.\frac{\delta\mathcal{S}^{(2m)}}{\delta u(x)}\right|_{f_{ij}} + \mathcal{R}^{(2m)}[u(x), f_{ij}(x)] = 0\ . \tag{3.11}$$

The remainder term $\mathcal{R}^{(2m)}$ depends on some quadratic combination of the previous order terms (i.e., it may be written explicitly (Parry et al. [1994])). For example, for $m = 1$, it is

$$\mathcal{R}^{(2)} = \frac{1}{2}\gamma^{1/2}\gamma^{ij}\phi_{,i}\phi_{,j} - \frac{1}{2}\gamma^{1/2}R. \tag{3.12}$$

Equation (3.11) has the form of an infinite dimensional gradient. It may be integrated using a line integral analogous to Equation (2.4):

$$\mathcal{S}^{(2m)} = -\int d^3x \int_0^1 ds\ u(x)\ \mathcal{R}^{(2m)}[su(x), f_{ij}(x)]\ . \tag{3.13}$$

Typically, $\mathcal{S}^{(2m)}$ is an integral of terms which contain the Ricci tensor and derivatives of the scalar field (Parry et al. [1994]).

The integrability condition for the HJ equation (Moncrief and Teitelboim [1972]) follows from the Poisson bracket of the energy constraints evaluated at spatial points x and x',

$$\{\mathcal{H}(x^k), \mathcal{H}(x^{k'})\} = [\gamma^{ij}(x^k)\mathcal{H}_j(x^k) + \gamma^{ij}(x^{k'})\mathcal{H}_j(x^{k'})]\ \frac{\partial}{\partial x^i}\delta^3(x^k - x^{k'})\ . \tag{3.14}$$

In fact, alternative contours replacing the line-integral Equation (3.13) will correspond to different time-hypersurface choices. Provided that the generating functional is invariant under reparametrizations of the spatial coordinates, (e.g., $\mathcal{H}_i$ vanishes in the right-hand-side of Equation (3.14)), different time-hypersurface choices will lead to the same generating functional. Hypersurface invariance is closely related to gauge invariance.

4 Computing Large-Angle Microwave Background Fluctuations and Galaxy Correlations

In order to describe the fluctuations arising during the inflationary epoch, it is necessary to sum an infinite subset (Salopek [1995a], Salopek and Stewart [1995]) of the terms $\mathcal{S}^{(2m)}$. In this case, one considers all terms which are quadratic in the Ricci tensor $\tilde{R}_{ij}$ of the conformal 3-metric $f_{ij}(x)$ defined in Equation (3.9). Once again, no explicit choice of time hypersurface is made.

However, when one compares theory with observations, there are indeed preferred gauges. The phase transition of photon-decoupling occurs essentially on a uniform temperature slice, $T \sim 4000K$, when protons combine with electrons to form neutral hydrogen. For adiabatic perturbations at large wavelengths, this slice is the same as a comoving, synchronous time hypersurface which Sachs and Wolfe [1967] used in the computation of large-angle microwave background anisotropies.

The power-law inflationary model (Lucchin and Matarrese [1985]) provides an excellent example of HJ techniques. For this model, the scalar factor of the Universe evolves as $a \sim t^p$ which describes an inflationary epoch provided $p > 1$. The scalar field potential has an exponential form

$$V(\phi) = V_0 \exp\left(-\sqrt{\frac{2}{p}}\phi\right) . \tag{4.1}$$

Power-law inflation is of high interest for observational cosmology because it may produce copious amounts of primordial gravitational radiation (Davis et al. [1992] and Salopek [1992b] and [1995b]), which is in essence a quantum gravitational effect.

Figure 1 illustrates the observed power spectrum,

$$\mathcal{P}_\delta(k) \equiv \frac{k^3}{2\pi^2} \int d^3x \, e^{-i\vec{k}\cdot\vec{x}} < \frac{\delta\rho(x)}{\rho} \frac{\delta\rho(0)}{\rho} >, \tag{4.2}$$

for the linear density perturbation at the present epoch; here k is the comoving wavenumber. The data points were compiled using eight galaxy surveys (Peacock and Dodds [1994]). Also shown are the power spectra arising from power-law inflation for various values of p. After the inflationary epoch, I have assumed that the evolution of the fluctuations is described by the cold-dark-matter transfer function (Peebles [1982] and Bond and Efstathiou [1984]) where the present Hubble parameter is taken to be $H_0 = 50$ km s^{-1}Mpc^{-1}. With a correction for gravitational waves, the theoretical power spectra for density perturbations have been normalized using the 2-year DMR data set (Bennett et al. [1994]) of the Cosmic Background Explorer (COBE) satellite: $\sigma_{sky}(10^0) = 30.5 \pm 2.7\mu K$ (68% confidence level).

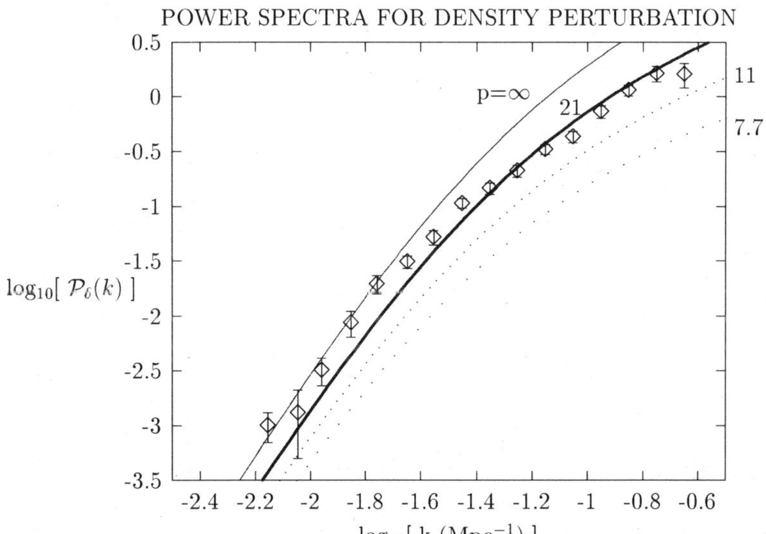

Figure 1 For the present epoch, the power spectra for the linear density perturbation $\delta\rho/\rho$ is shown. The data points are the observed power spectrum derived from galaxy surveys. The curves are theoretical predictions of the power-law inflationary model for several values of p: $p = \infty$ is the standard cold-dark-matter model; $p = 21$ provides the best fit.

The discrepancy between the galaxy surveys and the standard cold-dark-matter model ($p = \infty$) is quite severe at short length scales, $k > 10^{-1.4}$ Mpc^{-1}. The bold line, $p = 21$, provides the best fit to the observed data. The agreement is excellent at short scales. At longer scales, the theoretical model under-predicts the observed power but the deficit is not very severe. For $p = 21$, gravitational waves contribute 35% to σ_{sky}^2, the square of COBE's microwave anisotropy.

5 Summary

The question of time choice in general relativity is a difficult one, particularly for the quantum theory (Kuchař [1992]). For semi-classical problems of interest to observational cosmology, one may construct a covariant formalism which treats all time choices on an equal footing. Power-law inflation with $p = 21$ yields a better fit to cosmological data than the standard cold-dark-matter model.

References

Bennett, C. L., Kogut, A., Hinshaw, G., Banday, A. J., Wright, E. L., Gorski, K. M., Wilkinson, D. T., Weiss, R., Smoot, G. F., Meyer, S. S., Mather, J. C., Lubin, P., Loewenstein, K., Lineweaver, C., Keegstra, P., Kaita, E., Jackson, P. D. and Cheng, E. S. [1994] *Cosmic Temperature Fluctuations from 2 Years of COBE Differential Microwave Radiometers Observations*, Astrophys. J. **436**, 423–442.

Bond, J. R. and Efstathiou, G. [1984] *Cosmic Background Radiation Anisotropies in Universes Dominated by Nonbaryonic Dark Matter*, Astrophys. J. Lett. **285**, L45–L48.

Copeland, E. J., Kolb, E. W., Liddle, A. R. and Lidsey, J. E. [1993] *Reconstructing the Inflation Potential, in Principle and in Practice*, Phys. Rev. D **48**, 2529–2547.

Croudace, K. M., Parry, J., Salopek, D. S. and Stewart, J. M. [1994] *Applying the Zel'dovich Approximation to General Relativity*, Astrophys. J. **423**, 22–32.

Davis, R. L., Hodges, H. M., Smoot, G. F., Steinhardt, P. J. and Turner, M. S. [1992] *Cosmic Microwave Background Probes Models of Inflation*, Phys. Rev. Lett. **69**, 1856–1859.

Halliwell, J. J. and Hawking, S. W. [1985] *The Origin of Structure in the Universe*, Phys. Rev. D **31**, 1777–1791.

Kuchař, K. V. [1992] *Time and Interpretations of Quantum Gravity*, Proc. of 4th Canadian Conference on GR and Relativistic Astrophys., May 16-18 1991 (Kunstatter, G., Vincent, D. E. and Williams, J. G., eds.), World Scientific, Singapore, pp. 211–314.

Lucchin, F. and Matarrese, S. [1985] *Power-Law Inflation*, Phys. Rev. D **32**, 1316–1322.

Misner, C., Thorne, K. S. and Wheeler, J. A. [1973] *Gravitation*, Freeman, New York, p. 1185.

Moncrief, V. and Teitelboim, C. [1972] *Momentum Constraints as Integrability Conditions for the Hamiltonian Constraint in General Relativity*, Phys. Rev. D **6**, 966–969.

Moscardini, L., Borgani, S., Coles, P., Lucchin, F., Matarrese, S., Messina, A. and Plionis, M. [1993] *Large Scale Angular Correlations in Cold Dark Matter Models*, Astrophys. J. **413**, L55–L58.

Parry, J., Salopek, D. S. and Stewart, J. M. [1994] *Solving the Hamilton-Jacobi Equation for General Relativity*, Phys. Rev. D **49**, 2872–2881.

Peacock, J. A. and Dodds, S. J. [1994] *Reconstructing the Linear Power Spectrum of Cosmological Mass Fluctuations*, Mon. Not. Roy. Astron. Soc. **267**, 1020–1034.

Peebles, P. J. E. [1982] *Large-Scale Background Temperature and Mass Fluctuations Due to Scale-Invariant Primeval Perturbations*, Astrophys. J. **263**, L1-L5.

Peres, A. [1962] *On Cauchy's Problem in General Relativity, II*, Nuovo Cim. **26**, 53–62.

Sachs, R. K. and Wolfe, A. M. [1967] *Perturbations of a Cosmological Model and Angular Variations of the Microwave Background*, Astrophys. J. **147**, 73–90.

Salopek, D. S. [1991] *Nonlinear Solutions of Long-Wavelength Gravitational Radiation*, Phys. Rev. D **43**, 3214–3233.

——— [1992a] *Cold Dark Matter Cosmology with NonGaussian Fluctuations from Inflation*, Phys. Rev. D **45**, 1139–1157.

——— [1992b] *Consequences of the COBE Satellite Results for the Inflationary Scenario*, Phys. Rev. Lett. **69**, 3602–3605.

——— [1995a] *Characteristics of Cosmic Time*, Phys. Rev. D **52**, 5563–5575.

——— [1995b] *Cosmological Inflation, Microwave Background Anisotropies and Large Scale Structure of the Universe*, Proc. of the Intern. School of Astrophys., "D. Chalonge", Third Course, Erice, Italy, September 4-16, 1994 (Sanchez, N., ed.), Kluwer, Dordrecht, pp. 179-204.

Salopek, D. S. and Stewart, J. M. [1995] *Hypersurface Invariant Approach to Cosmological Perturbations*, Phys. Rev. D **51**, 517–535.
Salopek, D. S., Stewart, J. M. and Croudace, K. M. [1994] *The Zel'dovich Approximation and the Relativistic Hamilton-Jacobi Equation*, Mon. Not. Roy. Astr. Soc. **271**, 1005–1016.
Zel'dovich, Y. B. [1970] *Gravitational Instability: An Approximate Theory for Large Density Perturbations*, Astron. Astrophys. **5**, 84–89.

Fields Institute Communications
Volume **15**, 1997

Numerical Quantum Gravity by Dynamical Triangulation

Kristin Schleich
Department of Physics
University of British Columbia
Vancouver, B. C., Canada V6T 1Z1
schleich@physics.ubc.ca

Donald Witt
Department of Physics
University of British Columbia
Vancouver, B. C., Canada V6T 1Z1
donwitt@noether.physics.ubc.ca

Abstract. Recently an alternate technique for numerical quantum gravity, dynamical triangulation, has been developed. In this method, the geometry is varied by adding and subtracting equilateral simplices from the simplicial complex. This method overcomes certain difficulties associated with the traditional approach in Regge calculus of varying geometry by varying edge lengths. However additional complications are introduced: three of the four moves in dynamical triangulation can violate the simplicial nature of the complex. Simulations indicate that the rate of these violations is significant. Thus additional conditions must be placed on the dynamical triangulation moves to ensure that the simplicial complex and its topology are preserved.

1 Introduction

Numerical evaluations of path integral expressions for quantum amplitudes can address many issues not tractable to analytic or perturbative methods. It is thus natural to formulate such methods for the study of quantum gravity and quantum cosmology. A standard approach to numerically computing quantum amplitudes for gravity is Regge calculus. In this method, the topology of a history in the path integral is given in terms of a simplicial complex. Its geometry is specified by fixing the edge lengths of all simplices in the complex. The geometry is varied by varying the edge lengths. A sum over all geometries is carried out by evaluating the contribution of these varied geometries to the path integral. This approach has been used by several workers with some interesting results (See the review by Williams and Tuckey [1992], Section 5). However, the requirement that simplices remain

1991 *Mathematics Subject Classification.* Primary 83C27; Secondary 83C45.

The first author is supported in part by an NSERC grant.

© 1997 American Mathematical Society

nondegenerate under variation of edge length and the necessity of recomputing the curvature after each variation are factors that make this technique somewhat involved.

An alternate approach to numerical computation of quantum gravity, dynamical triangulation, has been used recently by several workers. This work has concentrated on searching for phase transitions and characterizing their behavior; recent results have found some evidence for such a phase transition in four dimensions (Agishtein and Migdal [1992], Ambjorn and Jurkiewicz [1992], Varsted [1992] and Brugmann and Marinari [1993]). In the dynamical triangulation approach, the topology is again given in terms of a simplicial complex. However, now all simplices in the complex are fixed to be equilateral. The geometry is then varied by changing the number of simplices in the complex. Simplices are thus always nondegenerate and additionally always produce the same contribution to the curvature which allegedly allows for more rapid numerical evaluation.

It is key to the use of a simplicial complex in dynamical triangulation that any changes in geometry preserve both the simplicial nature of the complex and the topology[1]. Surprisingly, as described in detail in this paper, this is not the case for the standard moves for three dimensional dynamical triangulation: three of the four moves used to change the number of simplices in the complex can violate its simplicial nature. Therefore restrictions on these moves must be enforced if the simplicial nature of the complex is to be preserved by dynamical triangulation. Fortunately, as detailed below, such restrictions can be enforced in 3 dimensions.

2 Problems with Dynamical Triangulation

A discretization of histories appropriate for quantum gravity is given by a simplicial complex[2]. First, a *n-simplex* is the convex hull of $(n+1)$ affinely independent points in $\mathbb{R}^{n+1}$. A 0-simplex is a point, a 1-simplex is an edge, a 2-simplex is a triangle, a 3-simplex is a tetrahedron and so on. A simplex that is the convex hull of a subset of the vertices of an n-simplex is a *face* of the n-simplex. Then

Definition *A simplicial complex K is a topological space $|K|$ and a collection of simplices K such that*

1. *$|K|$ is a closed subset of some finite dimensional euclidean space,*
2. *If F is a face of a simplex in K, then F is also contained in K,*
3. *If B,C are simplices in K, then $B \cap C$ is a face of both B and C.*

The topological space $|K|$ is the union of all simplices in K.

Simplicial complexes can describe very general topological spaces; however those relevant for the current discussion are manifolds. First, the *star* of vertex v is the complex consisting of all simplices containing v. The *link* of v is the complex consisting of the subset of simplices in the star of v that do not contain v itself. Then a closed combinatorial n-manifold is a simplicial complex for which the link of every vertex is a combinatorial $(n-1)$-sphere. The key element in the definition of simplicial complex is that all elements are uniquely determined by the vertices they contain; no two elements in the complex are specified by the same vertices. Were elements of the complex not uniquely specified by their vertices, additional

[1] Indeed it is this requirement that restricts edge lengths in the Regge approach.

[2] For a more detailed presentation of this introductory material, see Schleich and Witt [1993].

information would be needed to differentiate the elements. Moreover, it is precisely this feature that allows the geometry of a given space to be modeled by filling each simplicial complex with flat space and fixing all edge lengths. As the geometry is flat on the interior of the simplices, the curvature is no longer distributed over the space but is carried on $(n-2)$-simplices. In dynamical triangulation, one chooses all n-simplices to have equal edge lengths. An advantage of this choice is that the curvature at each $(n-2)$-simplex is determined very simply by counting the number k of n-simplices containing it: $R = 2\pi - k\delta$ where δ is the deficit angle between adjacent faces of an equilateral n-simplex. For three dimensions $\delta = \cos^{-1}(\frac{1}{3})$. The Euclidean action for a closed 3-manifold then reduces to especially simple form, $I = \kappa_3 N_3 - \kappa_1 N_1$ where N_i is the number of i-simplices in the manifold and κ_i are constants depending on δ, G, Λ and the edge length. Thus changes in the action can be computed directly from changes in the number of simplices in the complex.

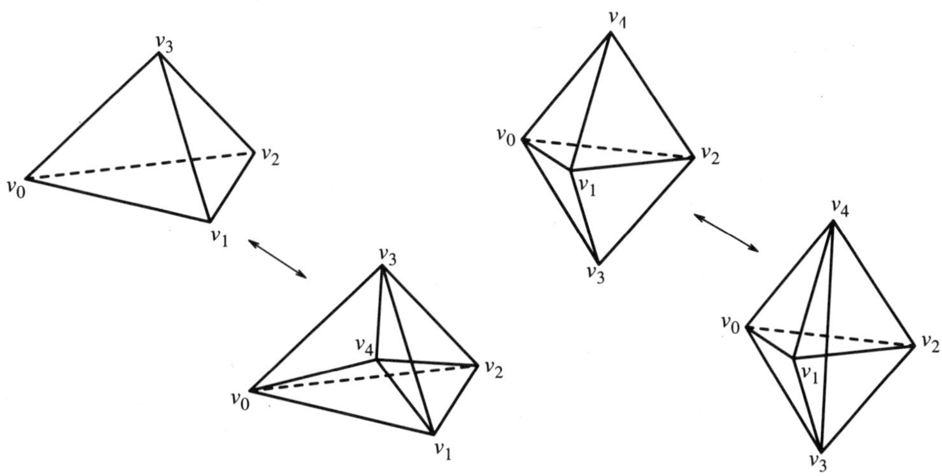

Figure 1 The dynamical triangulation moves in three dimensions. The first picture illustrates the subdivision move; the original tetrahedron is replaced with 4 new tetrahedra sharing a common vertex. Observe that the boundary of the subdivision is the same as that of the original tetrahedron. The second illustrates the flip move; two tetrahedra sharing a common triangle with vertices v_0, v_1, v_2 are replaced with three tetrahedra sharing an edge v_3, v_4. Again the boundary of the region is preserved.

Clearly changing the number of 3-simplices adjacent to a given edge will change the curvature of the space and correspondingly the geometry of the combinatorial manifold. Such changes are carried out by a special set of moves. In three dimensions, there are four moves as shown in Figure 1; a subdivision of one tetrahedron into four tetrahedra, a flip move in which two tetrahedra sharing a triangle are taken into three tetrahedra sharing an edge, and their inverses (Godfrey and Gross [1991] and Ambjorn and Varsted [1992]). Note that a necessary condition for performing the inverse flip move is that the edge is in precisely three tetrahedra. Similarly a necessary condition for the inverse subdivision move is that the vertex is in precisely four tetrahedra. The subdivision changes the number of edges and tetrahedra in the complex by $\Delta N_1 = 4$, $\Delta N_3 = 3$, and the flip move by $\Delta N_1 = 1$, $\Delta N_3 = 1$. Clearly the inverse moves change the counting by the opposite amount.

At first glance, it would appear that as these moves preserve the boundary of the region, they preserve the simplicial nature of the complex. This is not true; three of the four moves can violate the simplicial properties of the 3-manifold. This fact is not noted in the literature on three dimensional dynamical triangulation. It is useful to demonstrate how such moves can break the complex in two dimensions, where we can clearly illustrate this. The flip move in two dimensions is illustrated at the top of Figure 2. Now consider performing the flip move on the 2-manifold at the bottom of Figure 2; by the nature of the curvature at the flip location, the edge to be added into the complex is already present. Thus the simplicial nature of the complex is broken at this location and the standard assignment of curvature in Regge calculus cannot be used on the resulting space.

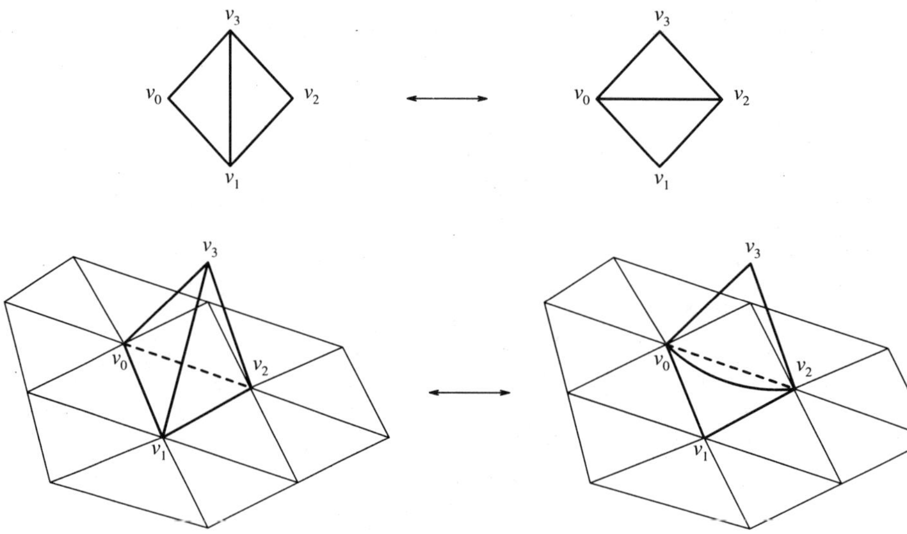

Figure 2 Two dimensional example of a failed move. The top picture illustrates the flip move in two dimensions. The bottom picture illustrates the consequences of this move for a vertex v_3 of high curvature. Note edges specified by v_0, v_1, v_2 are not the boundary of a triangle in the complex. Before the move, all edges are uniquely specified by their vertices. After the move there are two edges with vertices v_0, v_2. To distinguish them, the new edge has been curved. Curvature is thus no longer carried only at vertices and the Regge formulas for it are no longer valid.

In three dimensions analogous violations occur: The flip move can attempt to add an edge that is already in the simplicial 3-manifold. The inverse flip move can similarly attempt to add a triangle that is already present. Finally the inverse subdivision can attempt to add a tetrahedron that is already in the complex. Although it is hard to illustrate pictorially, examples of all of these violations occur for the smallest combinatorial 3-sphere consisting of the surface of a 4-simplex, $K = \{v_0, v_1 \ldots, v_4, e_0, e_1 \ldots, e_9, t_0, t_1 \ldots, t_9, q_0, q_1 \ldots q_9\}$ with the higher dimensional simplices containing the following vertices:

$e_0 \supset v_0, v_1$	$e_5 \supset v_1, v_3$	$t_0 \supset v_2, v_3, v_4$	$t_5 \supset v_0, v_2, v_4$	$q_0 \supset v_0, v_1, v_2, v_3$
$e_1 \supset v_0, v_2$	$e_6 \supset v_1, v_4$	$t_1 \supset v_1, v_3, v_4$	$t_6 \supset v_0, v_2, v_3$	$q_1 \supset v_0, v_1, v_2, v_4$
$e_2 \supset v_0, v_3$	$e_7 \supset v_2, v_3$	$t_2 \supset v_1, v_2, v_4$	$t_7 \supset v_0, v_1, v_4$	$q_2 \supset v_0, v_1, v_3, v_4$

$e_3 \supset v_0, v_4$	$e_8 \supset v_2, v_4$	$t_3 \supset v_1, v_2, v_3$	$t_8 \supset v_0, v_1, v_3$	$q_3 \supset v_0, v_2, v_3, v_4$
$e_4 \supset v_1, v_2$	$e_9 \supset v_3, v_4$	$t_4 \supset v_0, v_3, v_4$	$t_9 \supset v_0, v_1, v_2$	$q_4 \supset v_1, v_2, v_3, v_4$

All triangles are in two tetrahedra (this is true for all closed 3-manifolds) and are therefore candidates for a flip. Consider t_0; it is in q_3 and q_4 and the vertices that appear in the new edge will be v_0, v_1. However, this edge is already in the complex; it is e_0. Therefore this move cannot be performed. Similarly, all edges are contained in precisely 3-tetrahedra (this is not generic to 3-manifolds, but true for this special case) and are thus candidates for an inverse flip move. Consider e_5; it is in q_0, q_2, q_4 and thus would add a triangle with vertices v_0, v_2, v_4. But again this triangle is already there; it is t_5. Finally, the inverse subdivision move at vertex v_0 attempts to replace the four tetrahedra q_0, q_1, q_2, q_3 with one containing vertices v_1, v_2, v_3, v_4. Again this is already there; it is q_5. Therefore, all of these moves fail. Note that these violations can occur for more general 3-manifolds; it is clear that this 3-sphere can be joined to any more general space to produce a region with edges, triangles and vertices that have similar properties.

These violations of the simplicial nature of the complex are serious. Fortunately, they are local. Therefore, the simplicial nature of the complex can be preserved so long as one tests for these violations before performing a move. Such a test is easy to design: Check to see that all new elements to be added to the complex are not currently present. Finally, after performing the move, verify that the manifold topology is preserved by an algorithm. Such a violation testing code has been developed for three dimensional dynamical triangulation by the authors. This code is based on a data structure that encodes the topology of the simplicial complex in hierarchical form. Moves are only implemented after testing for violations of the simplicial nature of the complex. Data is kept on the frequency of attempts to perform a move that violates the simplicial nature of the complex. This code has been run on small scale simulations. Results of these simulations indicate that the violation rate of the flip move, defined as the percentage of successful flips on a random sample of all triangles is about 60 percent. The violation rate of the inverse flip move defined as the percentage of successful inverse flips on a random sample of all edges that are contained in precisely three tetrahedra is around 40 percent. These high violation frequencies can be intuitively understood in terms of the subdivision move; it adds 4 edges to the simplex that contain precisely 3 tetrahedra, but these edges also contain a vertex that contains only 4 tetrahedra. Therefore, these edges cannot be used in an inverse flip move. It also adds 6 triangles that cannot be used in a flip move as the new vertex contains only 4 tetrahedra. As the subdivision move can be performed at any tetrahedra in the complex, one expects a continued violation rate for arbitrarily sized complexes. Finally, the inverse subdivision move has a very low violation rate, 0.2 percent. Clearly the violation rates for the flip move and inverse flip move are significant.

3 Conclusions

Any application of dynamical triangulation should preserve the topology and simplicial nature of the complex. It has been demonstrated here that additional conditions on the neighborhoods of the elements subject to moves must be enforced to do so. Preliminary simulations indicate that failing to enforce these conditions leads to a high violation rate. Thus enforcement is important if dynamical triangulation simulations are to implement a discretized version of three dimensional gravity. Furthermore, these results have implications for the four dimensional case as the problems discussed in this paper will appear in four dimensions. These difficulties will be magnified by the lack of an easily implementable algorithm for testing for a 4-manifold. Thus it may be difficult to verify the additional conditions on the moves needed to preserve the simplicial nature of the complex are sufficient.

References

Agishtein, M. and Migdal, A. [1992] *Critical behavior of dynamically triangulated quantum gravity in four dimensions,* Nucl. Phys. B **385**, 395–412.

Ambjorn, J. and Jurkiewicz, J. [1992] *Four dimensional simplicial quantum gravity,* Phys. Lett. B **278**, 42–50.

Ambjorn, J. and Varsted, S. [1992] *Three dimensional simplicial quantum gravity,* Nucl. Phys. B **373**, 557–577.

Brugmann, B. and Marinari, E. [1993] *Four dimensional simplicial quantum gravity with a non-trivial measure,* Phys. Rev. Lett. **70**, 1908–1911.

Godfrey, N. and Gross, M. [1991] *Simplicial quantum gravity in more than two dimensions,* Phys. Rev. D **43**, R1749–R1753.

Schleich, K. and Witt, D. [1993] *Generalized sums over histories for quantum gravity II: Simplicial conifolds,* Nucl. Phys. B **402**, 469–530.

Varsted, S. [1992] *Four dimensional quantum gravity by dynamical triangulations,* Nucl. Phys. B **412**, 406–414.

Williams, R. and Tuckey, P. [1992] *Regge calculus: a brief review and bibliography,* Class. Quantum Gravity **9**, 1409–1422.

Fields Institute Communications
Volume **15**, 1997

Conformal Motions in Spherical and Plane Symmetric Spacetimes

Brian O. J. Tupper
Department of Mathematics and Statistics
University of New Brunswick
Fredericton, N. B., Canada E3B 5A3
brian@math.unb.ca

Abstract. It is shown how all conformal motions of those spherical and plane symmetric spacetimes which are not conformally flat may be found easily from the isometries and homotheties of $2+2$ decomposable spacetimes.

1 Introduction

In this article we are concerned with spacetimes admitting conformal symmetries which are described by a global conformal Killing vector (CKV) field, ξ^a, defined by

$$\mathcal{L}_\xi g_{ab} = 2\psi\, g_{ab}, \tag{1.1}$$

where $\mathcal{L}_\xi$ signifies the Lie derivative along ξ^a and $\psi(x^a)$ is the conformal scalar. Our particular interest is in proper CKV, i.e., CKV for which $\psi_{;ab} \neq 0$, in which case the presence of the unknown scalar function $\psi(x^a)$ renders Equation (1.1) rather more difficult to solve than in the special cases of Killing vectors (KV, $\psi = 0$), homothetic vectors (HV, $\psi_{,a} = 0$, $\psi \neq 0$), and special CKV (SCKV, $\psi_{;ab} = 0$, $\psi_{,a} \neq 0$).

If a spacetime is conformally flat (CF) one can, in principal, find the 15 CKV of the spacetime by transforming the metric to the form

$$ds^2 = e^{2\chi}(-dt^2 + dx^2 + dy^2 + dz^2), \tag{1.2}$$

where $\chi = \chi(t, x, y, z)$, so that the CKV are just the CKV (10 KV, 1 HV, 4 SCKV) of the underlying Minkowski spacetime but, in general, are 15 proper CKV because of the effect of the conformal factor, $e^{2\chi}$, on the conformal scalar, $\psi(x^a)$. However, it may be more difficult to transform the metric into the form (1.2) than to directly solve Equation (1.1). For non-CF spacetimes there is no 'in principal' method for finding the proper CKV, if any, that are admitted by the spacetime other than by direct integration of Equation (1.1). In this article we present a straightforward method of calculating the CKV, particularly proper CKV, of a large class

1991 *Mathematics Subject Classification.* Primary 83C20; Secondary 53B21.

A longer version of this paper appears as: Tupper, B. O. J., *Conformal symmetries of conformal-reducible spacetimes with non-zero Weyl tensor*, Class. Quantum Grav. **13**, Institute of Physics (1996), 1679–1687.

© 1997 American Mathematical Society

of spacetimes which are not CF but are conformal to $2+2$ decomposable spacetimes. Among the spacetimes in this category are all non-CF spherical symmetric and plane symmetric spacetimes; we focus our attention exclusively on these two classes of spacetime.

The basis of the method is a result proved by Coley and Tupper [1992], namely that non-CF $2+2$ decomposable spacetimes admit no proper CKV nor, from Coley and Tupper [1989], any SCKV. Also any KV of either of the constituent 2-spaces is a KV of the entire $2+2$ spacetime, while a HV exists iff both 2-spaces admit a HV (Hall and da Costa [1988]). It follows that, in order to find the CKV (proper or improper) of a non-CF spacetime which is conformal to a $2+2$ decomposable spacetime we need only find the KV and HV of each 2-space of the underlying $2+2$ spacetime and to calculate the effect of the conformal factor on the conformal scalar ψ.

More precisely, we start with the metric of a non-CF spacetime M of the form

$$ds^2 = e^{2\chi(x^a)} d\Sigma^2, \tag{1.3}$$

where $e^{2\chi}$ is the conformal factor and $d\Sigma^2$ is the metric of $2+2$ decomposable spacetime N where

$$d\Sigma^2 = g_{AB}(x^C)dx^A dx^B + g_{\alpha\beta}(x^\gamma)dx^\alpha dx^\beta, \tag{1.4}$$

with $A, B, C = 0, 1$ and $\alpha, \beta, \gamma = 2, 3$ and the 2-spaces with metric tensors g_{AB}, $g_{\alpha\beta}$ are labelled N_1, N_2 respectively. We know that each 2-space admits, locally, three KV iff it is a space of constant curvature, otherwise it admits at most one KV (Eisenhart [1949]) and all KV of N_1 and N_2 are KV of N. Also a 2-space of constant curvature admits a HV iff the constant curvature is zero, so if either of N_1, N_2 is of constant curvature, κ_1, κ_2 respectively, with either $\kappa_1 \neq 0$ or $\kappa_2 \neq 0$, then N cannot admit a HV. Thus, for a spherical symmetric spacetime, N cannot admit a HV since $\kappa_2 = 1$ whereas, for a plane symmetric spacetime, it is possible for N to admit a HV since $\kappa_2 = 0$. Furthermore N and, consequently, M are CF iff $\kappa_1 + \kappa_2 = 0$; we avoid this case. Once we have found all KV and HV of N, these are then KV, HV or CKV of M (Defrise-Carter [1975]) the exact nature of the symmetries depending on the effect of the conformal factor $e^{2\chi}$.

2 Spherical Symmetric Spacetimes

The metric of a general spherical symmetric spacetime, M, is

$$ds^2 = -e^{2v}dt^2 + e^{2w}dr^2 + R^2(d\theta^2 + \sin^2\theta\, d\phi^2), \tag{2.1}$$

where v, w and R are each functions of t and r. The metrics of the 2-spaces N_1, N_2 are, respectively:

$$\begin{aligned} d\sigma_1{}^2 &= R^{-2}(-e^{2v}dt^2 + e^{2w}dr^2), & (2.2)\\ d\sigma_2{}^2 &= d\theta^2 + \sin^2\theta\, d\phi^2, & (2.3) \end{aligned}$$

and the conformal factor $e^{2\chi} = R^2$. The 2-space N_2 is of constant curvature, $\kappa_2 = 1$, and so admits three KV which are thus KV of N. These are the usual KV associated with spherical symmetry, i.e.,

$$\xi^a_{(1)} = \cos\phi\,\partial_\theta - \sin\phi\cot\theta\,\partial_\phi,\ \ \xi^a_{(2)} = \sin\phi\,\partial_\theta + \cos\phi\cot\theta\,\partial_\phi,\ \ \xi^a_{(3)} = \partial_\phi, \tag{2.4}$$

and, on restoring the conformal factor R^2, these three vector fields are also KV of M. Since N_2 admits no HV any further symmetries of M must arise from the KV of N_1. If N_1 is of constant curvature κ_1 ($\neq -1$ else N and M are CF) then the following condition holds

$$\begin{aligned} &e^{-2v}(R^{-1}R_{tt} - R^{-2}{R_t}^2 - v_tR^{-1}R_t + w_tR^{-1}R_t - w_{tt} - w_t^2 + v_tw_t) \\ &-e^{-2w}(R^{-1}R_{rr} - R^{-2}{R_r}^2 + v_rR^{-1}R_r - w_rR^{-1}R_r - v_{rr} - v_r^2 + v_rw_r) \\ &\qquad +\kappa_1R^{-2} = 0. \end{aligned} \tag{2.5}$$

and N_1 admits three KV. If (2.5) does not hold then we check whether or not N_1 admits a single KV. Hence, N admits a further three, one or zero KV and we restore the conformal factor R^2 to determine the corresponding symmetry type for M. Thus a non-CF spherical symmetric spacetime always admits the three KV given by (2.4) and can admit three, one or zero further symmetries which may be KV, HV or CKV.

Example 2.1 The Schwarzschild solution

$$M: \; ds^2 = -(1-2mr^{-1})dt^2 + (1-2mr^{-1})^{-1}dr^2 + r^2(d\theta^2 + \sin^2\theta\, d\phi^2) \tag{2.6}$$
$$N_1: \; d{\sigma_1}^2 = -r^{-2}(1-2mr^{-1})dt^2 + r^{-2}(1-2mr^{-1})dr^2. \tag{2.7}$$

N_1 is not of constant curvature; it admits one KV only, $\xi^a_{(4)} = \partial_r$, and the corresponding symmetry of M is KV. Thus M admits four KV, no HV or CKV. (This is a well-known result, but the above shows how quickly the result is obtained.)

Example 2.2 The Vaidya solution

$$M: \quad ds^2 = -[1-2m(u)r^{-1}]du^2 - 2du\,dr + r^2d\Omega^2 \tag{2.8}$$
$$N_1: \quad d{\sigma_1}^2 = -r^{-2}[1-2m(u)r^{-1}]du^2 - 2r^{-2}du\,dr. \tag{2.9}$$

N_1 is not of constant curvature and, in general, admits no KV, so that M admits no symmetries other than (2.4). However, in the special case $m(u) = \alpha u$ (α constant) N_1, and thus N, admits the KV.

$$\xi^a_{(4)} = u\,\partial_u + r\,\partial_r \tag{2.10}$$

and the corresponding symmetry of M is HV with $\psi = 1$. Thus M with $m = \alpha u$ admits three KV and one HV.

Example 2.3 The LRS solution

$$M: \quad ds^2 = -dt^2 + t^{2a}dr^2 + t^2(d\theta^2 + \sin^2\theta\, d\phi^2) \tag{2.11}$$
$$N_1: \quad d\sigma_1^2 = -t^{-2}dt^2 + t^{2a-2}dr^2. \tag{2.12}$$

N_1 is of constant positive curvature $\kappa_1 = (a-1)^2$ (so that M is not CF) and thus N admits three additional KV

$$\begin{aligned} \xi^a_{(4)} &= \partial_r, \qquad \xi^a_{(5)} = t\,\partial_t + (1-a)r\,\partial_r, \\ \xi^a_{(6)} &= \begin{cases} t\,r\,\partial_t + \frac{1}{2}[(1-a)^{-1}t^{2(1-a)} + (1-a)r^2]\partial_r & (a \neq 1) \\ t\,r\,\partial_t + \log t\,\partial_r & (a = 1). \end{cases} \end{aligned} \tag{2.13}$$

The corresponding symmetries of M are: $\xi^a_{(4)}$ is a KV, $\xi^a_{(5)}$ is a HV with $\psi_{(5)} = 1$, and $\xi^a_{(6)}$ is a CKV with $\psi_{(6)} = r$.

3 Plane Symmetric Spacetimes

The metric of a general plane symmetric spacetime, M is

$$ds^2 = -e^{2v}dt^2 + e^{2w}dx^2 + R^2(dy^2 + dz^2) \tag{3.1}$$

where v, w and R are each function of t and x. The metrics of the 2-spaces N_1 and N_2 are

$$d\sigma_1{}^2 = R^{-2}(-e^{2v}dt^2 + e^{2w}dx^2) \tag{3.2}$$

$$d\sigma_2{}^2 = dy^2 + dz^2 \tag{3.3}$$

and the conformal factor $e^{2\chi} = R^2$. The 2-space N_2 is of constant curvature $\kappa_2 = 0$ and so admits three KV which are thus KV of N. These are

$$\xi^a_{(1)} = \partial_y, \ \xi^a_{(2)} = \partial_z, \ \xi^a_{(3)} = z\,\partial_y - y\,\partial_z \tag{3.4}$$

and, on restoring the conformal factor R^2, these three vector fields are KV of M. Also N_2 admits a HV $\eta^\alpha = y\,\partial_y + z\,\partial_z$ and this will give rise to a HV iff N_1 admits a HV in which case N_1 cannot be of constant curvature since, for a HV, the constant curvature must be zero which implies that M is CF. Thus for a non-CF plane symmetric spacetime, N admits a HV only if N_1 is not of constant curvature. If N_1 is of constant curvature κ_1 the condition (2.5) holds (with the differentiating suffix r replaced by x) and N_1 admits three KV. If N_1 is not of constant curvature then M may admit zero, one or two further symmetries depending on whether or not N_1 admits a KV and/or a HV.

Example 3.1 The plane symmetric Bianchi I spacetime

$$M: \quad ds^2 = A^2(t)[-dt^2 + dx^2 + \sinh^2 at(dy^2 + dz^2)] \tag{3.5}$$

$$N_1: \quad d\sigma_1{}^2 = A^2(t)\,\mathrm{csch}^2 at(-dt^2 + dx^2), \tag{3.6}$$

where $A(t)$ is arbitrary. N_1 is of constant curvature $\kappa_1 = a^2$ and admits the three KV

$$\xi^a_{(4)} = \partial_x, \ \xi^a_{(5)} = \cosh ax\, \sinh at\,\partial_t + \sinh ax\, \cosh at\,\partial_x,$$
$$\xi^a_{(6)} = \sinh ax\, \sinh at\,\partial_t + \cosh ax\, \cosh at\,\partial_x \tag{3.7}$$

which are also KV of N. Restoring the conformal factor $\sinh^2 at$ we find that $\xi^a_{(4)}$ is a KV of M while $\xi^a_{(5)}$ and $\xi^a_{(6)}$ are CKV of M with

$$\psi_{(5)} = \cosh ax\,(a\cosh at + A_t A^{-1}\sinh at) \tag{3.8}$$

$$\psi_{(6)} = \sinh ax\,(a\cosh at + A_t A^{-1}\sinh at) \tag{3.9}$$

which are proper CKV unless $A = \mathrm{csch}\, at$. Thus M admits four KV and two proper CKV in general.

Example 3.2 The plane symmetric Bianchi I spacetime

$$M: \quad ds^2 = A^2(t)\,[-dt^2 + dx^2 + t^{2m}(dy^2 + dx^2)] \tag{3.10}$$

$$N_1: \quad d\sigma_1{}^2 = A^2(t)\,t^{-2m}(-dt^2 + dx^2) \tag{3.11}$$

where $m \neq 1$ and $A(t)$ is arbitrary. N_1 is not of constant curvature but admits one KV $\xi^a_{(4)} = \partial_x$, which is also a KV of M. However, N_1 admits a HV and so N admits the HV

$$\xi^a_{(5)} = t\,\partial_t + x\,\partial_x + (1-m)(y\,\partial_y + z\,\partial_z) \tag{3.12}$$

with $\psi_{(5)} = 1 - m$. The corresponding symmetry of M is a CKV with $\psi_{(5)} = 1 + A^{-1}A_t\, t$, which is a proper CKV if $A \neq t^n$ for any n. Hence the spacetime M with $m \neq 1$, $A \neq t^n$ admits four KV and one proper CKV.

4 Summary

We have shown how the conformal symmetries of non-CF spherical and plane symmetric spacetimes may be found by investigating the KV and HV of a 2-space and that the maximum number of conformal symmetries admitted by such spacetimes is six.

References

Coley, A. A. and Tupper, B. O. J. [1989] *Special conformal Killing vector spacetimes and symmetry inheritance*, J. Math. Phys. **30**, 2616–2625.

Coley, A. A. and Tupper, B. O. J. [1992] *Affine conformal vectors in spacetime*, J. Math. Phys. **33**, 1754–1764.

Defrise-Carter, L. [1975] *Conformal groups and conformally equivalent isometry groups*, Comm. Math. Phys. **40**, 273–282.

Eisenhart, L. P. [1949] *Riemannian Geometry*, Princeton University Press, Princeton.

Hall, G. S. and da Costa, J. [1988] *Affine collineations in spacetime*, J. Math. Phys. **29**, 2465–2472.

Fields Institute Communications
Volume **15**, 1997

Dissipative Cosmology

Robert van den Hoogen
Department of Mathematics, Statistics and Computing Science
Dalhousie University
Halifax, Nova Scotia, Canada B3H 3J5
hoogen@cs.dal.ca

Abstract. Dissipative cosmological models are analyzed and compared by using dynamical systems techniques. The equations of state used to complete the system are chosen so that the equilibrium points of the dynamical system represent self-similar cosmological models.

1 Dissipative Processes

Why study dissipative processes in the universe? Weinberg [1972] stressed that it is an attractive hypothesis that the high entropy per baryon we currently observe in the universe resulted from dissipative processes in the early universe. Misner [1967], [1968] studied dissipative processes and investigated whether they could account for the currently observed highly isotropic matter distribution. Dissipative processes are of interest because they provide an alternative mechanism for inflation (Lima and Portugal [1988]). In addition, dissipative processes can affect the nature of the initial singularity and in some instances the initial singularity can be avoided (Coley and van den Hoogen [1994a], Murphy [1973]).

Unfortunately, in most of the previous work concerning dissipative cosmology, it was assumed that the dissipative processes could be described by the first order theory of irreversible thermodynamics proposed by Eckart [1940]. However, it was later shown that the equations describing the Eckart theory were parabolic and hence disturbances in the fluid could propagate with infinite speeds (Hiscock and Salmonson [1991]). This acausal behaviour in the first order theory led people like Israel and Stewart to develop a second order theory of irreversible thermodynamics (Israel and Stewart [1979a,b]). The Israel-Stewart theory of irreversible thermodynamics has been shown to be causal, stable, and satisfy a set of reasonable hyperbolic differential equations (Hiscock and Lindblom [1983]).

The Israel-Stewart theory of irreversible thermodynamics has recently been divided into two classes, the *Truncated* theory and the *Full* or *Non-truncated* theory. The *Truncated* theory results from assuming that certain terms involving the gradients of some of the parameters multiplying second order quantities can be

1991 *Mathematics Subject Classification.* Primary 83F05; Secondary 85A40.
The author was supported by a Killam Scholarship.

© 1997 American Mathematical Society

considered small (Israel and Stewart [1979b]) and therefore these terms have been dropped resulting in a truncated version of the equations. The *Full* theory of irreversible thermodynamics is described by the following equations (assuming that the model is spatially homogeneous with the matter flowing orthogonal to the spatial hypersurfaces):

$$\begin{aligned}
\Pi &= -\zeta\left\{\theta + \beta_0\dot{\Pi} - \alpha_0 q^c{}_{;c} + \frac{\epsilon}{2}\Pi T\left(\frac{\beta_0 u^c}{T}\right)_{;c}\right\},\\
q_a &= -\kappa T h_a{}^b\left\{\frac{T_{;b}}{T} - \alpha_0\Pi_{;b} - \alpha_1\Pi^c{}_{b;c} + \frac{\epsilon}{2}q_b T\left(\frac{\beta_1 u^c}{T}\right)_{;c}\right\}, \qquad (1)\\
\Pi_{ab} &= -2\eta\left\{\sigma_{ab} + \beta_2\dot{\Pi}_{ab} - \alpha_1\langle q_{b;a}\rangle + \frac{\epsilon}{2}\Pi_{ab} T\left(\frac{\beta_2 u^c}{T}\right)_{;c}\right\},
\end{aligned}$$

where

$$\langle A_{ab}\rangle \equiv \frac{1}{2}h_a{}^c h_b{}^d(A_{cd} + A_{dc}) - \frac{1}{3}h_{ab}h^{cd}A_{cd}.$$

The following interpretations can be made about the parameters in the theory. The quantity Π represents the bulk viscous stress, q_a is the heat conduction vector, Π_{ab} is the shear viscous stress, and T is the temperature as measured by an observer moving with the fluid. If u^a is the four-velocity of the fluid flow, then θ is the expansion and σ_{ab} is the shear of the fluid. The Eckart theory results by setting $\beta_i = 0$ and $\alpha_j = 0$ [where $(i = 0,1,2)$ and $(j = 0,1)$], and in this case we can interpret ζ as the bulk viscosity coefficient, κ as the thermal conductivity and η as the shear viscosity coefficient. The so-called *Truncated* theory is described by the above equations by setting $\epsilon = 0$. In this case there are the three parameters, β_i, and the two coupling parameters α_j. Therefore, in the *Truncated* theory we need equations of state for ζ, η, κ, β_0, β_1, β_2, α_0 and α_1. In the *Full* theory, we need equations of state for the above quantities as well as an equation of state for the temperature T. Two questions now arise: 1) What equations of state should be employed?, and 2) What theory ought to be used, *Truncated* or *Full*?

2 Equations of State

Early investigations of OSH (orthogonal spatially homogeneous) cosmological models discovered that for a large class of OSH models, the past and future asymptotic solutions are approximated by exact power-law solutions. It was then later shown that these exact power-law solutions admit a homothetic vector and hence are *Self-Similar*. Using dynamical systems techniques, Wainwright and his collaborators (Hewitt and Wainwright [1993], Wainwright and Hsu [1989]) illustrated that the equilibrium points of the dynamical system (describing the OSH Bianchi type A and type B perfect fluid models) represented self-similar cosmological models. It was shown by Coley and van den Hoogen [1994b] that the property of having self-similar asymptotic limit points is dependent upon the equations of state that are chosen. If the system of ordinary differential equations describing the OSH models is rewritten in terms of dimensionless variables (Coley [1990b], Coley and van den Hoogen [1994b], Wainwright and Hsu [1989]) and if the equations of state are physically reasonable functions of the dimensionless variables then self-similar limit points are to be expected (for exceptions, see Coley and van den Hoogen [1994b]). It can be shown that the barotropic equation of state for the pressure,

p, in the perfect fluid models is precisely of the form required to have self-similar limit points. (E.g., if $p = \alpha\rho^2$ then one would not expect self-similar limit points to the future and to the past.)

When the fluid is no longer a perfect fluid and we include dissipative effects such as heat conduction and viscosity, the need for more equations of state for the various viscosity parameters arises. These equations of state can be calculated (in a few instances) from kinetic theory; however, these are often too complicated and cumbersome to use. Therefore, phenomenological equations of state must be assumed. If we make the requirement that the asymptotic limit points of the imperfect fluid models be self-similar (analogous to the perfect fluid case) then the equations of state needed to ensure this requirement are precisely the dimensionless equations of state mentioned earlier (Coley [1990a], Coley and van den Hoogen [1994b], [1995]).

3 A Simple Viscous Model

The Einstein field equations for the Friedmann-Robertson-Walker (FRW) viscous fluid models are (Burd and Coley [1994], Coley and van den Hoogen [1995])

$$\begin{aligned} \dot{\theta} &= -\frac{1}{3}\theta^2 - \frac{1}{2}(\rho + 3p + 3\Pi), \\ \dot{\rho} &= -\theta(\rho + p + \Pi), \\ \theta^2 &= 3\rho - \frac{1}{3}\,{}^3R, \end{aligned} \tag{2}$$

where θ is the expansion, ρ is the energy density, p is the thermodynamic pressure, Π is the bulk viscous pressure, and ${}^3R$ is the curvature of the homogeneous hypersurfaces. The evolution of Π is governed by equation (1).

We shall employ the following dimensionless variables,

$$x \equiv \frac{3\rho}{\theta^2}, \quad y \equiv \frac{9\Pi}{\theta^2} \quad \text{and a new time variable} \quad \frac{d\Omega}{dt} = -\frac{\theta}{3}.$$

As described above, we impose the condition that the asymptotic limit points be self-similar and hence we choose appropriate dimensionless equations of state of the form,

$$\frac{p}{\theta^2} = \frac{1}{3}(\gamma - 1)x, \qquad \frac{3}{\beta_0\theta^2} = ax, \qquad \frac{\zeta}{\theta} = \zeta_0 x^m = \frac{a}{b}x^m.$$

In addition, an equation of state for the temperature T is required. There is no known form for T, but it is commonly assumed that $T = T_0\rho^r \equiv T_0 x^r \theta^{2r}$ (Maartens [1995], Zakari and Jou [1993]). We shall adopt the same convention here.

The dynamical system in terms of the new dimensionless variables and employing the dimensionless equations of state becomes:

$$\begin{aligned} x' &= (1 - x)\left[(3\gamma - 2)x + y\right] \\ y' &= -y[2 + y + (3\gamma - 2)x] + 9ax^m + bx^{1-m}y \\ &\qquad + \frac{\epsilon}{2}y\left\{3 + (r+1)(3\gamma + y/x)\right\}. \end{aligned} \tag{3}$$

The invariant set $x = 1$ is of particular interest since it represents the zero curvature FRW models. In order to illustrate the differences between the various thermodynamical theories, we hereafter assume $x = 1$. In the set $x = 1$, the two dimensional system (3) reduces to

$$y' = y^2\left(\frac{\epsilon}{2}(r+1)-1\right)+y\left(b-3\gamma+\frac{3\epsilon}{2}[1+\gamma(r+1)]\right)+9a.$$

3.1 Eckart: If $\beta_0 = 0$, then equations (1) describe the Eckart theory which was studied by Burd and Coley [1994]. In the Eckart theory we have an algebraic equation for y. For the flat models y remains a constant, i.e., $y = -9\zeta_0$.

3.2 Truncated: If $\epsilon = 0$ and $\beta_0 \neq 0$, then we have the truncated Israel-Stewart theory. This situation was studied thoroughly by Coley and van den Hoogen [1995]. We give a short analysis below. The evolution of y is given by,

$$y' = -y^2 + (b-3\gamma)y + 9a.$$

The equilibrium points are at

$$y^{\pm} = \frac{(b-3\gamma)\pm\sqrt{(b-3\gamma)^2+36a}}{2}.$$

It is easy to see that the qualitative behaviour is of the form seen in Figure 1. The solution represented by the equilibrium point $y = y^+$ represents the past asymptotic behaviour while the solution represented by the point $y = y^-$ represents the future asymptotic behaviour.

3.3 Full theory: If $\epsilon = 1$, then the evolution equation for y has the form

$$y' = \frac{(r-1)}{2}y^2 + \frac{1}{2}[3+2b+3\gamma(r-1)]y + 9a.$$

The qualitative behaviour of a simple quadratic differential equation is characterized by the coefficients

$$A = \frac{(r-1)}{2}, \qquad B = \frac{1}{2}[3+2b+3\gamma(r-1)], \qquad C = 9a.$$

See Figures 1–5. The qualitative behaviour changes significantly when $r = r*$, where r^* is the positive root of the equation

$$9\gamma^2(r-1)^2 + [6\gamma(3+2b)-72a](r-1)+(3+2b)^2 = 0.$$

See Figures 3 and 5.

4 Summary

As we see, the qualitative behaviour of the system depends upon the parameter r (equation of state for T) and ϵ (the particular theory). If $r < 1$, then the qualitative behaviour of the flat models is the same in the *Truncated* and *Full* theories. In particular, models evolve from a visco-elastic singularity at $y = y^+$ to models at infinity or asymptotically approach the perfect fluid FRW model at $y = y^-$. If $1 < r < r^*$, then models evolve from a state with infinitely negative bulk viscous pressure at $y = y_1^-$ to models at infinity or asymptotically approach the model at the point $y = y_2^-$. If $r^* < r$, then there are no equilibrium points at finite values. Models evolve from a state with infinite viscous pressure and after a finite time start to inflate, after-which they evolve towards a model with infinitely negative bulk viscous pressure.

These preliminary results suggest that the truncated and the full theory have the same qualitative behaviours if $r < 1$, and radically different behaviours if $r > r^*$.

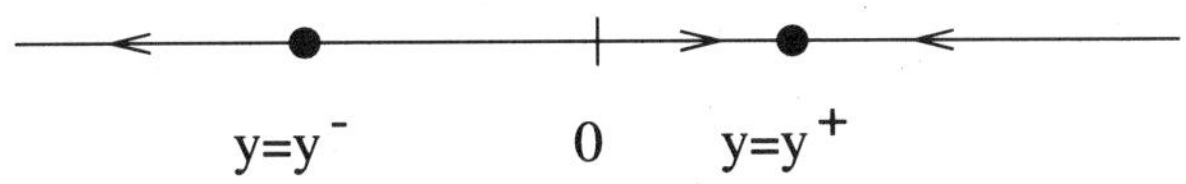

Figure 1 If $r < 1$, then $A < 0$ and $B^2 - 4AC > 0$. There are two equilibrium points, one positive $y = y^+$, and one negative $y = y^-$. One is a sink, the other is a source. (This phase portrait also represents the qualitative behaviour of the flat FRW models using the *Truncated* theory)

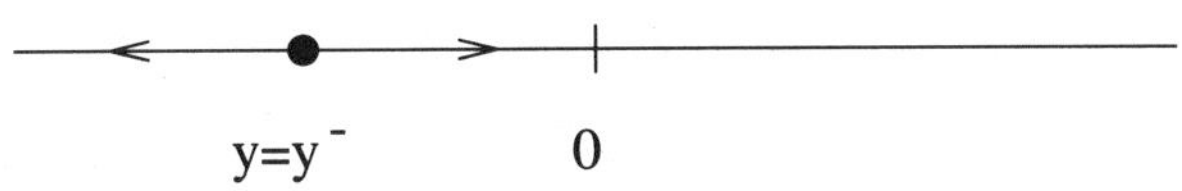

Figure 2 If $r = 1$, then $A = 0$. There is only one equilibrium point $y^- < 0$. It is a source.

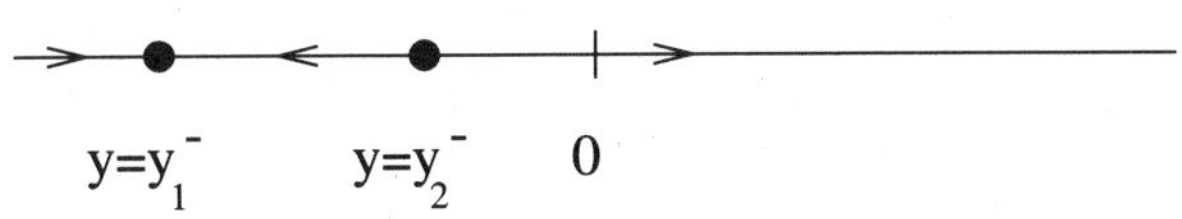

Figure 3 If $r^* > r > 1$, then $A > 0$ and $B^2 - 4AC > 0$. There are two equilibrium points that are less than zero. One is a sink, the other is a source.

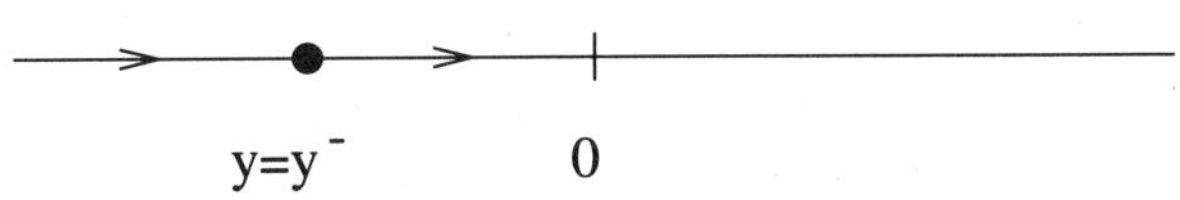

Figure 4 If $r* = r > 1$, then $A > 0$ and $B^2 - 4AC = 0$. There is only one equilibrium point. It is degenerate.

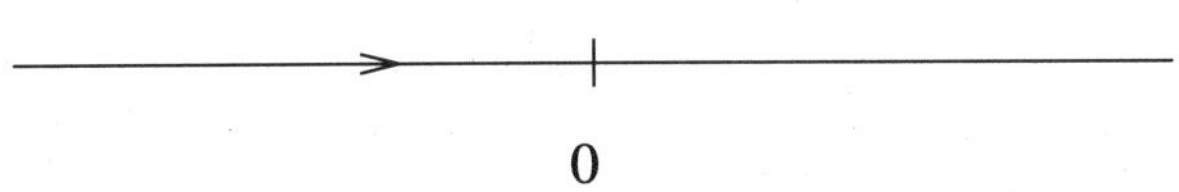

Figure 5 If $r > r^*$, then $A > 0$ and $B^2 - 4AC < 0$. There are no equilibrium points at finite values.

A complete analysis of the system will appear elsewhere (Coley et al. [1995]). Models more complex than the FRW models will also be analyzed, for instance the isotropic curvature models.

References

Burd, A. and Coley, A. [1994] *Viscous Fluid Cosmology*, Class. Quantum Grav. **11**, 83–105.

Coley, A. A. [1990a] *Bianchi V Imperfect Fluid Cosmology*, Gen. Rel. Grav. **22**, 3–18.

——— [1990b] *Equations of State and Plane-Autonomous Systems in Bianchi V Imperfect Fluid Cosmology*, J. Math. Phys. **31**, 1698–1703.

Coley, A. A. and van den Hoogen, R. J. [1994a] *Qualitative Analysis of Diagonal Bianchi Type V Imperfect Fluid Cosmological Models*, J. Math. Phys. **35**, 4117–4144.

——— [1994b] *Self-similar Asymptotic Solutions of Einstein's Equations*, Deterministic Chaos in General Relativity, NATO ASI 332B, Plenum, New York.

——— [1995] *Qualitative Analysis of Viscous Fluid Cosmological Models Satisfying the Israel-Stewart Theory of Irreversible Thermodynamics*, Class. Quantum Grav. **12**, 1977–1994.

Coley, A. A., van den Hoogen, R. J. and Maartens, R. [1997] *Qualitative Viscous Cosmology*, Physical Review D, **54**, 1393-1397.

Eckart, C. [1940] *The Thermodynamics of Irreversible Processes III: Relativistic Theory of the Simple Fluid*, Phys. Rev. **58**, 919–924.

Hewitt, C. G. and Wainwright, J. [1993] *A Dynamical Systems Approach to Bianchi Cosmologies: Orthogonal Models of Class B*, Class. Quantum Grav. **10**, 99–124.

Hiscock, W. A. and Lindblom, L. [1983] *Stability and Causality in Dissipative Relativistic Fluids*, Ann. Phys. **151**, 466–496.

Hiscock, W. A. and Salmonson, J. [1991] *Dissipative Boltzmann-Robertson-Walker Cosmologies*, Phys. Rev. D **43**, 3249–3258.

Israel, W. and Stewart, J. M. [1979a] *Transient Relativistic Thermodynamics and Kinetic Theory*, Ann. Phys. **118**, 341-372.

——— [1979b] *On Transient Relativistic Thermodynamics and Kinetic Theory II*, Proc. R. Soc. Lond. A **365**, 43–52.

Lima, J. A. S., Portugal, R. and Waga, I. [1988] *Bulk-Viscosity-Driven Asymmetric Inflationary Universe*, Phys. Rev. D **37**, 2755–2760.

Maartens, R. [1995] *Dissipative Cosmology*, Class. Quantum Grav. **12**, 1455–1465.

Misner, Charles W. [1967] *Transport Processes in the Primordial Fireball*, Nature **214**, 40-41.

——— [1968] *The Isotropy of the Universe*, Astrophys. J. **151**, 431–457.

Murphy, G. L. [1973] *Big-Bang Without Singularities*, Phys. Rev. D **8**, 4231–4233.

Wainwright, J. and Hsu, L. [1989] *A Dynamical Systems Approach to Bianchi Cosmologies: Orthogonal Models of Class A*, Class. Quantum Grav. **6**, 1409–1431.

Weinberg, Steven [1972] *Gravitation and Cosmology*, Wiley and Sons, New York.

Zakari, M. and Jou, D. [1993] *Equations of State and Transport Equations in Viscous Cosmological Models*, Phys. Rev. D **48**, 1597–1601.

Fields Institute Communications
Volume **15**, 1997

Counting Kinks in 1+1 Dimensions

J. G. Williams
Department of Mathematics and Computer Science
(and the Winnipeg Institute for Theoretical Physics)
Brandon University
Brandon, Manitoba, Canada R7A 6A9
williams@brandonu.ca

Peter D. Zvengrowski
Department of Mathematics and Statistics
University of Calgary
Calgary, Alberta, Canada T2N 1N4
zvengrow@acs.ucalgary.ca

Abstract. The Frenet formulas are used to derive an equation for the number of Finkelstein-Misner kinks on any closed one-surface Σ imbedded in a (1+1)-dimensional spacetime. The resulting kink number equation is given as an integral over Σ, where the integrand depends upon both the geodesic curvature of Σ and the nature of the vector field describing the tipping of the light cones on Σ. Kink numbers are calculated for two well known examples of (1+1)-dimensional kinked spacetimes.

Alty [1995] has recently presented a kink number formula in terms of an integral of certain differential forms that occur in the Gauss-Bonnet-Chern theorem (Gilkey [1995]). For any given general relativistic metric $g_{\mu\nu}$, taking this integral over a hypersurface Σ yields kink$(\Sigma, g_{\mu\nu})$, i.e., the number of Finkelstein-Misner kinks (Finkelstein and Misner [1959]) existing on the hypersurface. (See also Torre [1995].) The purpose of the present paper is to derive a similar counting number formula for general relativistic kinks in 1+1 dimensions (Dunn et al. [1992]). As will be seen, the derivation is much easier than in higher dimensions because, in 1+1 dimensions, use can be made of the Frenet equations and the fact that Gauss' second fundamental form becomes simply the geodesic curvature. The end result is equivalent to that of Alty [1995] reduced from 3+1 to 1+1 dimensions. (See also the kink counting procedure outlined in Dunn et al. [1991] which, although not generally valid, works under certain restricted circumstances.)

Consider a (1+1)-dimensional general relativistic metric $g_{\mu\nu}$ that is well defined throughout a (1+1)-dimensional spacetime manifold M. Let Σ denote (the image

---

1991 *Mathematics Subject Classification.* Primary 83C20; Secondary 83C05.

It is a pleasure to thank Lloyd Alty and Peter Gilkey for a number of helpful comments and to acknowledge the hospitality of the Aspen Center for Physics. This work was supported by grants from the Natural Sciences and Engineering Research Council of Canada.

© 1997 American Mathematical Society

of) a closed curve in M. Situations of interest include, but are not limited to, the case where Σ is the boundary of M ($\Sigma = \partial M$) or one component of the boundary of M or where Σ is a hypersurface of M (e.g., $M = \mathbf{R}^1 \times \Sigma$). The presence of kinks implies that Σ cannot be everywhere spacelike. Choose any *positive-definite* metric $h_{\mu\nu}$ on the spacetime manifold M and consider the equation

$$g_{\mu\nu} X^\nu \;=\; \lambda\, h_{\mu\nu} X^\nu.$$

Each "eigenvalue" λ will be a function of the spacetime coordinates and there will be exactly *one* negative eigenvalue. Let the corresponding eigenvector, normalized to unity with respect to $h_{\mu\nu}$, be denoted by V^ν. The eigenvector V^ν is only defined to within a $\pm$ sign. However, we shall assume that the spacetime is time-orientable (or in topological terms, that the line bundle in question is trivial – i.e., admits a nowhere zero section). It is then possible to make a consistent sign choice globally, and we shall henceforth assume that such a choice has been made. In this way, any time-orientable general relativistic metric $g_{\mu\nu}$ can be associated with a vector field. We have $h_{\mu\nu} V^\mu V^\nu = 1$ and $g_{\mu\nu} V^\mu V^\nu = \lambda < 0$. Intuitively, the "turning" of the timelike vector V^μ along the closed curve Σ keeps pace with the turning of the light cone and so is a measure of the kink number, $\mathrm{kink}(\Sigma, g_{\mu\nu})$. The latter can equally well be denoted by $\mathrm{kink}(\Sigma, V^\mu)$, or by $\mathrm{kink}(\Sigma)$ if the metric or vector field are understood to be given.

Before proceeding further, we establish some notation. Let x^0, x^1 denote spacetime coordinates on M and parametrize Σ by an affine parameter such as arc length,

$$\sigma \;:=\; \int_{t_0}^{t} \left[h_{\mu\nu} \frac{dx^\mu(t)}{dt} \frac{dx^\nu(t)}{dt} \right]^{1/2} dt.$$

Christoffel symbols $\Gamma^\mu_{\lambda\nu}$ are defined with reference to the positive-definite metric $h_{\mu\nu}$, as are the absolute and the covariant derivatives of a vector field $W^\mu(\sigma)$:

$$\frac{DW^\mu}{D\sigma} \;:=\; \frac{dW^\mu}{d\sigma} + \Gamma^\mu_{\lambda\nu} W^\lambda \frac{dx^\nu}{d\sigma} \;=\; W^\mu{}_{\|\nu} \frac{dx^\nu}{d\sigma},$$

$$W^\mu{}_{\|\nu} \;:=\; \frac{\partial W^\mu}{\partial x^\nu} + \Gamma^\mu_{\lambda\nu} W^\lambda.$$

Unit vectors tangent and normal to Σ are given, respectively, by $t^\mu := dx^\mu/d\sigma$ and $n^\nu := \varepsilon^{\mu\nu} t_\mu$, where $\varepsilon^{\mu\nu}$ is the (1+1)-dimensional completely antisymmetric tensor having the properties: $\varepsilon^{01} = h^{-1/2}$, $\varepsilon_{01} = h^{1/2}$, $\varepsilon^{\mu\nu}\varepsilon_{\mu\nu} = 2$, $\varepsilon^{\lambda\mu}\varepsilon_{\lambda\nu} = \delta^\mu_\nu$, $\varepsilon^{\mu\nu}\varepsilon_{\lambda\rho} = \delta^\mu_\lambda \delta^\nu_\rho - \delta^\mu_\rho \delta^\nu_\lambda$ and $\varepsilon^{\mu\nu}{}_{\|\lambda} = 0$. The geodesic curvature of Σ is defined by

$$\kappa_g \;=\; \varepsilon_{\mu\nu} \frac{dx^\mu}{d\sigma} \left(\frac{d^2x^\nu}{d\sigma^2} + \Gamma^\nu_{\lambda\rho} \frac{dx^\lambda}{d\sigma} \frac{dx^\rho}{d\sigma} \right) \;=\; n_\nu \frac{Dt^\nu}{D\sigma} \;=\; \varepsilon_{\mu\nu} n^\mu \frac{Dn^\nu}{D\sigma}.$$

Covariant differentiation of $h_{\mu\nu} t^\mu t^\nu = h_{\mu\nu} n^\mu n^\nu = 1$ yields the well known Frenet formulas:

$$\frac{Dt^\mu}{D\sigma} \;=\; \kappa_g\, n^\mu, \quad \frac{Dn^\mu}{D\sigma} \;=\; -\kappa_g\, t^\mu.$$

At each point of Σ, the unit vector V^μ, which was derived from the metric $g_{\mu\nu}$, can be decomposed into directions tangent to Σ and normal to Σ:

$$V^\mu \;=\; t^\mu \sin\alpha \;+\; n^\mu \cos\alpha.$$

Since V^μ lies inside the light cone, the angle α measures the tipping of the light cone at any spacetime point lying on Σ. The integral $(2\pi)^{-1}\oint_\Sigma (d\alpha/d\sigma) d\sigma$ is exactly what is meant by "the number of kinks on Σ." To find a more useful expression for $\text{kink}(\Sigma)$, we use the Frenet formulas to simplify the following jacobian-like quantity.

$$\begin{aligned}
\varepsilon_{\mu\nu}V^{\mu}V^{\nu}{}_{\|\lambda}\frac{dx^{\lambda}}{d\sigma} &= \varepsilon_{\mu\nu}V^{\mu}\frac{DV^{\nu}}{D\sigma} \\
&= \varepsilon_{\mu\nu}\left(t^{\mu}\sin\alpha + n^{\mu}\cos\alpha\right)\frac{D}{D\sigma}(t^{\nu}\sin\alpha + n^{\nu}\cos\alpha) \\
&= \varepsilon_{\mu\nu}\left(t^{\mu}\sin\alpha + n^{\mu}\cos\alpha\right) \\
&\quad \times\left(\kappa_g n^{\nu}\sin\alpha + t^{\nu}\cos\alpha\frac{d\alpha}{d\sigma} - \kappa_g t^{\nu}\cos\alpha - n^{\nu}\sin\alpha\frac{d\alpha}{d\sigma}\right) \\
&= -\kappa_g + \frac{d\alpha}{d\sigma}.
\end{aligned}$$

It follows that

$$\text{kink}(\Sigma) \equiv \frac{1}{2\pi}\oint_{\Sigma}\frac{d\alpha}{d\sigma}d\sigma = \frac{1}{2\pi}\oint_{\Sigma}\left(\varepsilon_{\mu\nu}V^{\mu}V^{\nu}{}_{\|\lambda}\frac{dx^{\lambda}}{d\sigma} + \kappa_g\right)d\sigma.$$

This is the desired formula for the kink number in 1+1 dimensions.

The above formula is related to the Gauss-Bonnet theorem if Σ happens to be the boundary ∂M of the (1+1)-dimensional spacetime manifold M. By Stokes' theorem, the kink counting number formula can then be written

$$\text{kink}(\partial M) = \frac{1}{2\pi}\int_M (\varepsilon^{\rho\lambda}\varepsilon_{\mu\nu}V^{\mu}V^{\nu}{}_{\|\lambda})_{\|\rho}\, d^2\Omega + \frac{1}{2\pi}\oint_{\partial M}\kappa_g\, d\sigma,$$

where $d^2\Omega$ represents, as usual, the 2-volume element on M. Since $V^{\mu}V_{\mu} = 1$, it follows that $V^{\mu}{}_{\|\rho}V_{\mu} = 0$ so that $\det\|V^{\mu}{}_{\|\rho}\| = 0$ and hence $\varepsilon^{\rho\lambda}\varepsilon_{\mu\nu}V^{\mu}{}_{\|\rho}V^{\nu}{}_{\|\lambda} = 0$, which implies

$$(\varepsilon^{\rho\lambda}\varepsilon_{\mu\nu}V^{\mu}V^{\nu}{}_{\|\lambda})_{\|\rho} = \varepsilon^{\rho\lambda}\varepsilon_{\mu\nu}V^{\mu}V^{\nu}{}_{\|\lambda\|\rho}.$$

The latter is precisely the Gaussian curvature K of the manifold M (Laugwitz [1965]) and it now follows that

$$\text{kink}(\partial M) = \frac{1}{2\pi}\int_M K\, d^2\Omega + \frac{1}{2\pi}\oint_{\partial M}\kappa_g\, d\sigma.$$

The right side of this equation occurs in the Gauss-Bonnet theorem (Laugwitz [1965]) and is equal to the Euler number $\chi(M)$. It was this relationship between kink number, Euler number and the Gauss-Bonnet theorem that Alty [1995] used in his derivation of a formula for $\text{kink}(\Sigma)$ in 3+1 dimensions.

To illustrate the use of the above (1+1)-dimensional counting number formula, consider the 1+1 analogue of the one-kink spacetime given by Gibbons and Hawking [1992]. The manifold M is the unit disc, $(x^0)^2 + (x^1)^2 \leq 1$, and the general relativistic metric is the Minkowski metric, $\|g_{\mu\nu}\| = \|\eta_{\mu\nu}\| := \text{diag}(-1, 1)$. A simple choice for $h_{\mu\nu}$ is $\|h_{\mu\nu}\| = \|\delta_{\mu\nu}\| := \text{diag}(1, 1)$. The eigenvalue equation is solved by $(V^0, V^1) = (1, 0)$ with $\lambda = -1$. The metric $h_{\mu\nu}$ is flat, so that all Christoffel symbols are zero and $V^{\nu}{}_{\|\lambda} = V^{\nu}{}_{,\lambda} = 0$. The circular boundary $\Sigma = \partial M$ can be parametrized in terms of its arc length σ by $x^0(\sigma) = \cos\sigma$, $x^1(\sigma) = \sin\sigma$, whence

$$\text{kink}(\Sigma) = \frac{1}{2\pi}\int_{\sigma=0}^{\sigma=2\pi}\varepsilon_{\mu\nu}\frac{dx^{\mu}}{d\sigma}\frac{d^2x^{\nu}}{d\sigma^2}d\sigma = 1.$$

A second example is provided by the general relativistic metric (Dunn et al. [1992])

$$\begin{aligned} ds^2 &\equiv g_{\mu\nu}\, dx^\mu dx^\nu \\ &= -\cos 2\alpha\, dt^2 - 2\sin 2\alpha\, dt\, dx + \cos 2\alpha\, dx^2, \end{aligned}$$

with $x^0 \equiv t$, $x^1 \equiv x$, $-\infty < t, x < \infty$ and $\alpha = \alpha(x)$. The angle α has the same meaning as before, namely α measures the angle of tilt of the light cones. The spacetime manifold is the whole tx-plane, $M = \mathbf{R}^2$, and the hypersurface Σ will be chosen to be the x-axis, $\Sigma = \{x^0(\sigma) = 0,\ x^1(\sigma) = \sigma \mid \sigma \in \mathbf{R}^1\}$. The hypersurface $\Sigma \approx \mathbf{R}^1$ is noncompact, but we impose the boundary conditions $\alpha(\pm\infty) = 0 \pmod{2\pi}$ which imply $g_{\mu\nu}(-\infty) = g_{\mu\nu}(+\infty)$ so that the results of the present analysis will be homotopically the same as the outcome obtained if M were replaced by $\mathbf{R}^1 \times S^1$. It follows from the boundary conditions on α that

$$\alpha(+\infty) - \alpha(-\infty) = 2\pi n,$$

where n is an integer. It is intuitively clear that n will equal the number of kinks. As in the previous example, it is convenient to choose $h_{\mu\nu} = \delta_{\mu\nu}$. The eigenvalue equation is solved by $(V^0, V^1) = (\cos\alpha, \sin\alpha)$ with $\lambda = -1$. Since Σ is a straight line in flat space, it follows that $\kappa_g = 0$. All covariant derivatives become ordinary derivatives and

$$\begin{aligned} \text{kink}(\Sigma) &= \frac{1}{2\pi}\int_{-\infty}^{\infty} \left(V^0 \frac{dV^1}{dx} - V^1 \frac{dV^0}{dx} \right) dx \\ &= \frac{1}{2\pi}\int_{-\infty}^{\infty} \frac{d\alpha}{dx}\, dx \\ &= \frac{1}{2\pi}[\alpha(\infty) - \alpha(-\infty)] \;=\; n. \end{aligned}$$

Finally, we remark that the formula for kink(Σ) can be rewritten

$$\text{kink}(\Sigma) = \oint_\Sigma \varepsilon_{\mu\nu} \left(V^\mu \frac{DV^\nu}{D\sigma} + n^\mu \frac{Dn^\nu}{D\sigma} \right) d\sigma,$$

or, in terms of differential forms, $\text{kink}(\Sigma) = \oint_\Sigma \mathbf{V} \wedge d\mathbf{V} + \mathbf{n} \wedge d\mathbf{n}$. The two expressions in the integrand correspond to Alty's $\Theta_2(t)$ and $\Theta_2(n)$ (Alty[1995], Eq. (4.10)). Alty's $\Theta_1(t)$ and $\Theta_1(n)$ are not present in 1+1 dimensions.

References

Alty, L. J. [1995] *The generalized Gauss-Bonnet-Chern theorem*, J. Math. Phys. **36**, 3094–3105.

Dunn, K. A., Harriott, T. A. and Williams, J. G. [1991] *Kink number in general relativity*, J. Math. Phys. **32**, 476–479.

——— [1992] *Toy model for gravitational kinks*, J. Math. Phys. **33**, 1437–1444.

Finkelstein, D. and Misner, C. W. [1959] *Some new conservation laws*, Ann. Phys. (NY) **6**, 230–243.

Gibbons, G. W. and Hawking, S. W. [1992] *Kinks and topology change*, Phys. Rev. Lett. **69**, 1719–1721.

Gilkey, P. B. [1995] *Invariance Theory, the Heat Equation, and the Atiyah-Singer Index Theorem*, CRC Press, Boca Raton, FL, p. 201.

Laugwitz, D. [1965] *Differential and Riemannian Geometry*, Academic Press, New York, pp. 207–211.

Torre, C. G. [1995] *Some remarks on gravitational analogues of magnetic charge*, Class. Quantum Grav. **12**, L43–L50.

Fields Institute Communications
Volume **15**, 1997

Toward a Definition of Chaos for General Relativity

Donald Witt
Department of Physics
University of British Columbia
Vancouver, B. C., Canada V6T 1Z1
donwitt@noether.physics.ubc.ca

Kristin Schleich
Department of Physics
University of British Columbia
Vancouver, B. C., Canada V6T 1Z1
schleich@physics.ubc.ca

Abstract. General relativity exhibits a unique feature not represented in standard examples of chaotic systems; it is a spacetime diffeomorphism invariant theory. Thus many characterizations of chaos do not work. It is therefore necessary to develop a definition of chaos suitable for application to general relativity. This presentation will present results towards this goal.

1 Introduction

Recent work in chaos has lead to quantitative studies of chaotic behavior in different mathematical and physical systems. Such results have led to useful insight into the nature of these systems. Clearly, a characterization of whether or not general relativity exhibits chaos would be valuable. However, there are well known difficulties in applying the current characterizations of chaos to general relativity; they are not diffeomorphism invariant. Thus when applied to general relativity, they give different results in different coordinate systems (see for example, summary and references in Rugh [1994]). In particular, one test for chaos commonly used by physicists is to search for sensitive dependence to initial conditions in a system $\dot{x}(t) = F(x)$ by computing Liapunov exponents via $\lambda = \lim_{t\to\infty} \frac{1}{t} \ln |z(t)|$ where $z(t)$ is the linearized perturbation of $x(t)$. Positive Liapunov exponent is taken as indicating chaotic behavior. Application of this test to general relativity is fundamentally flawed; a coordinate change rescales a positive Liapunov exponent to zero. For example, consider computing the Liapunov exponents for the one parameter model corresponding to expanding de Sitter spacetime; one finds $\lambda = H$

1991 *Mathematics Subject Classification.* Primary 70K50; Secondary 83C05.
The second author is supported in part by an NSERC grant.

© 1997 American Mathematical Society

in the coordinate chart $ds^2 = -dt^2 + a^2(t)(dx^2 + dy^2 + dz^2)$ whereas $\lambda = 0$ in the chart $ds^2 = -\frac{dt^2}{H^2t^2} + a^2(t)(dx^2 + dy^2 + dz^2)$. Clearly, as the physics of the solution is independent of coordinate choice, there is a problem with this characterization.

It is clear that a coordinate independent definition of chaos is needed for the study of general relativity. This paper discuss work toward developing such a definition. A longer version of this paper with details is in preparation (Schleich and Witt [1995]). The starting point will be a discussion of a rigorous definition of chaos for standard dynamical systems. Next crucial aspects of general relativity that distinguish it from these standard systems will be outlined. Finally a generalized definition of chaos that addresses these crucial aspects will be proposed.

2 Chaos in Standard Dynamical Systems

The literature contains many characterizations of chaos, often adapted to the type of the system being studied. However, it is well known that testing for chaos simply by testing for positive Liapunov exponent is flawed. For example, the system with evolution $\dot{x} = cx$ where c is a positive constant has positive Liapunov exponent. Clearly, this system is not chaotic. Rigorous characterizations of chaos require ergodicity in addition to tests of sensitive dependence on initial conditions such as Liapunov exponents. (See for example, Wiggins [1990], p. 608 and Pollicott [1993] in the context of non-uniformly hyperbolic diffeomorphisms.)

While developing the machinery for a rigorous definition of chaos, it is useful to keep in mind two standard examples of dynamical systems: One corresponds to two linear coupled oscillators; for fixed initial momenta, this system is isomorphic to the translational flow on the flat torus $T^2 = \mathbb{R}^2/\mathbb{Z}^2$, $f_t : T^2 \to T^2$ given by $f_t(x_1, x_2) = (x_1 + \omega_1 t, x_2 + \omega_2 t) + K$, $K \in \mathbb{Z}^2$. The other is the cat map on the flat torus T^2 given by $f(x_1, x_2) = (2x_1 + x_2, x_1 + x_2) + K$, $K \in \mathbb{Z}^2$.

A general dynamical system is a measurable map: A measure space $(X, \mathcal{B})$ consists of a set X and a σ-algebra $\mathcal{B}$. A map $T : (X_1, \mathcal{B}_1) \to (X_2, \mathcal{B}_2)$ between two measure spaces is called measurable if and only if $T^{-1}(B) \in \mathcal{B}_1$ for all $B \in \mathcal{B}_2$. A measure m on $(X, \mathcal{B})$ is a function $m : \mathcal{B} \to \mathbb{R}^+ \cup \{\infty\}$ which satisfies $m(\emptyset) = 0$ and $m(\cup_k B_k) = \sum_k m(B_k)$ for any countable disjoint collection $\{B_k\}$. Then

Definition 2.1 *A measure m on $(X, \mathcal{B})$ is a probability measure if $m(X) = 1$.*

Note that X need not be a compact space; for example the measure induced by $\frac{1}{\sqrt{\pi}} \exp(-x^2)dx$ is a probability measure on $\mathbb{R}$. For simplicity of notation, the term measure will refer to probability measure in this paper. Next

Definition 2.2 *Given a measurable map T on $(X, \mathcal{B})$, m is T-invariant if and only if $T^*m(B) = m(B)$ for all $B \in \mathcal{B}$ where $T^*m(B) = m(T^{-1}(B))$.*

When the map is understood, a T-invariant measure is called an invariant measure. In general, a map T on a given space may have more than one invariant measure. Especially note that m need not have support on the entire space; for example $m = \delta(x)$ is an invariant measure for the map $T : \mathbb{R} \to \mathbb{R}$ given by $T(x) = ax$. At this point a key concept can be introduced:

Definition 2.3 *An invariant measure m is called an ergodic measure if whenever $T^{-1}B = B$ for some $B \in \mathcal{B}$, then either $m(B) = 0$ or $m(B) = 1$.*

Note that in general, the space of ergodic measures is not equal to the space of invariant measures. One can show that when T is a homeomorphism, there exists at least one ergodic measure on any compact space. Both examples have ergodic measures; the measure induced by $dx_1 dx_2$ is the unique invariant and ergodic measure for the cat map and for the translational flow when $\frac{\omega_1}{\omega_2} \notin \mathbb{Q}$. For $\frac{\omega_1}{\omega_2} \in \mathbb{Q}$, there are an *infinite* number of ergodic measures for the translational flow; these correspond to delta functions that concentrate support on one periodic orbit. A rigorous definition of chaos is then

Definition 2.4 *A system $T : (X, \mathcal{B}) \to (X, \mathcal{B})$ is chaotic if and only if it has an ergodic measure m and exhibits sensitive dependence on initial conditions with respect to this measure*[1]*.*

When applied to the examples, (2.4) distinguishes chaotic behavior. The translational flow is not chaotic for any values of the parameters, even incommensurate ones; it is easy to show that trajectories do not diverge by any test of sensitive dependence as they are simply straight lines. In contrast, the cat map is chaotic as it has positive Liapunov exponent; it illustrates the importance of sensitive dependence in distinguishing chaotic behavior.

Note especially that the use of ergodic measure implies that chaotic behavior can be concentrated on a subset of the measure space. This feature is important as dissipative systems such as the Henon map exhibit chaotic behavior on such a subset. Thus the use of the ergodic measure make (2.4) a very general definition of chaos. Moreover, other characterizations of chaos such as self-similarity and dense sets of periodic orbits can be recovered under certain conditions. It is this definition that will be the starting point for a more general definition of chaos suitable both to the standard examples and to general relativity.

3 Important Characteristics of General Relativity

There are two issues that must be addressed by any formulation of chaos appropriate for application to general relativity: its diffeomorphism invariance and its lack of a physically motivated choice of measure.

The most discussed difficulty associated with diffeomorphism invariance is concentrated in time reparameterization invariance; time is no longer an absolute quantity but is tied to the coordinate system. The coordinate systems given for the de Sitter space model clearly indicate this fact. Note that one can always associate an observer (though not necessarily a freely falling observer) to any choice of coordinate system; thus any attempt to tie chaos to observers is simply making physical a particular coordinate choice.

This problem is closely linked to the fact that diffeomorphism invariance is a form of gauge invariance; the metric degrees of freedom are not all physical. However, it is key to the correct identification of chaos that it be carried out on physical degrees of freedom alone. For example, consider the lifts of the translational flow

[1]One test for sensitive dependence is now given by positive Liapunov exponents *subject* to computation in the ergodic measure m (See for example Pollicott [1993] for application to nonuniformly hyperbolic diffeomorphisms). This definition differs significantly from the standard physicist's use in that Liapunov exponents defined with respect to an ergodic measure are constant over the space. Alternately Wiggins [1990] presents a definition that does not presume exponential divergence.

and the cat map from T^2 to the covering space $\mathbb{R}^2$. The measure induced by $dx_1 dx_2$ is not ergodic for either map on $\mathbb{R}^2$; there are obviously invariant sets for both the translational flow and the cat map. Therefore neither system is ergodic on the covering space in this measure. Clearly the global topology of the space plays a key role in the nature of the dynamical system!

Such global topology also arises in general relativity; removing the large gauge transformations will change the global structure of the phase space. An example is provided by Bianchi I (Bhushan et al. [1995]). Spatial gauge is fixed locally by writing the metric as $ds^2 = -N^2(t)dt^2 + a^2(t)dx^2 + b^2(t)dy^2 + c^2(t)dz^2$. Residual transformations corresponding to interchanges of spatial coordinates are not fixed. Removing these modifies the nature of the flow; the fully reduced phase space is no longer $\mathbb{R}^2$ but a wedge and the flows are no longer straight lines but include bounces at the boundaries of this wedge. This change in behavior in a such a simple model stresses the need to fully reduce to physical degrees of freedom.

Unfortunately, this problem is not confined to large gauge transformations; note that the usual Hamiltonian formulation of gravity in terms of superspace is also *not* in terms of physical degrees of freedom. Although the spatial diffeomorphisms have been removed, those corresponding to timelike diffeomorphisms have not. Therefore the study of flows on this space includes redundant information unless a further reduction is carried out. In particular as different time reparameterizations correspond to different gauge choices, studying the dynamics of the system without further reduction imparts physical meaning to different gauges. Thus gauge effects can mask the underlying physics of the system. Unfortunately there are no known ways to reduce to physical degrees of freedom explicitly; one obvious approach is to isolate a preferred timelike Killing vector or conformal Killing vector for superspace. However, Kuchař [1981] has shown no such vectors exist in general. It is thus unlikely that explicit reductions can be carried out.

The second issue is the lack of a physically motivated choice of measure for general relativity. It is well known that there is no preferred metric on superspace. Therefore, one cannot begin with the corresponding measure as a candidate for an invariant or ergodic measure. This problem is also complicated by the fact that superspace is not in terms of physical degrees of freedom. Thus it is not clear what role any choice of metric would play in defining any such measure.

4 Toward a Generalized Definition of Chaos

The above issues clearly demonstrate the futility of testing for chaos in general relativity by (2.4) by studying flows on superspace. However, what should this definition be replaced by? First observe that general relativity is a hamiltonian system in terms of its physical degrees of freedom, even if these variables cannot be identified explicitly. It is therefore natural to first ask the question in the context of the fully reduced theory, even if it cannot be explicitly constructed. Given such a definition on the physical degrees of freedom, one can then attempt to parameterize it to be applicable to more tractable forms of the theory.

Observe that a key element in (2.4) is ergodicity, intuitively that orbits in the phase space are dense in some measurable set. Topological generalizations of ergodicity are well known (see for example Peterson [1990]). Therefore, a natural approach is to generalize the condition of ergodic measure to a topological condition.

Given any invariant measure m and a map T one can define measurable entropy

$0 \leq h_{\mathrm{meas}}(T, m)$ (see Peterson [1990] Sect. 5.1). The measurable entropy is related to the (rigorous) Liapunov exponents by the Pesin-Reulle inequality under certain restrictive conditions on the map and measure $h_{\mathrm{meas}} \leq \sum_{\lambda_i > 0} \lambda_i$ where the sum is over positive Liapunov exponents (Pollicott [1993] p. 46). Therefore $h_{\mathrm{meas}} > 0$ implies chaos for ergodic measures. Next

Definition 4.1 *The topological entropy is* $h_{\mathrm{top}}(T) = \sup_m h_{\mathrm{meas}}(T, m)$ *where* m *is an invariant measure.*

Clearly if there is any measure with $h_{\mathrm{meas}} > 0$, then $h_{\mathrm{top}} > 0$. This is independent of whether or not the measure has support on the entire space. The utility of this definition of entropy is given in the following theorem

Theorem 4.2 *If the topological entropy of a map* T *is positive, then there exists an ergodic measure such that the measurable entropy is positive.*

Thus topological entropy encompasses both ergodicity and sensitive dependence in one quantity. The translational flow has zero topological entropy by explicit calculation. The cat map has positive topological entropy; it is a special case of a family of chaotic systems with positive topological entropy given by maps on a 2-manifold of genus > 1, $T_* : H_1(M^2; \mathbb{R}) \to H_1(M^2; \mathbb{R})$ such that T_* has all eigenvalues > 1. The difficulty with topological entropy has been in its application. However, numerical estimates of this quantity are possible (Gribble [1995]).

Given the above, a natural generalized definition of chaos is

Definition 4.3 *A dynamical system is chaotic if and only if* $h_{\mathrm{top}} > 0$.

This definition is especially suited to systems without known ergodic measure such as general relativity.

5 Conclusions

Definition (4.3) can be taken as a working definition of chaos for general relativity; it is not a final definition but a plausible starting point. Its difficulty is that it must be applied to the theory in terms of physical degrees of freedom. A key task is to see if it can be recast into parameterized form. Such a hope is not unreasonable given that topological entropy is closely related to partition functions which have obvious analogs in gauge theories. Furthermore quantization of gauge theories in general proceeds along a very similar track of reparameterization of quantities formally given in terms of physical degrees of freedom. Finally, infinite dimensional systems such as general relativity only have finite evolutions, that is singularities and caustics can arise in the future evolution of regular initial data. These effects do not appear in finite dimensional systems. Although such effects appear in other infinite dimensional theories, they are an essential part of the nature of general relativity. Thus it must be verified that (4.3) correctly handles these features.

References

Bhushan, V., Schleich, K. and Witt, D. [1995] *Global Topology and Ergodicity in Bianchi I*, in preparation.

Gribble, S. [1995] *Topological Entropy as Practical Tool for the Indentification and Characterization of Chaotic Systems*, Senior Thesis, Univ. British Columbia.

Kuchař, K. [1981] *General relativity: dynamics without symmetry,* J. Math. Phys. **22**, 2640–2654.

Petersen, K. [1983] *Ergodic Theory,* Cambridge University Press, Cambridge.

Pollicott, M. [1993] *Lectures on Ergodic Theory and Pesin Theory on Compact Manifolds*, Cambridge University Press, Cambridge.

Rugh, S. [1995] *Chaos in the Einstein equations - characterization and importance?*, Deterministic Chaos in General Relativity (Hobill, D., ed.), Plenum Press, New York, pp. 359–422.

Schleich, K. and Witt, D. *Chaos in General Relativity*, in preparation.

Wiggins, S. [1990] *Introduction to Applied Nonlinear Dynamical Systems and Chaos*, Springer-Verlag, New York.

Fields Institute Communications
Volume **15**, 1997

Causal Precedence Formalism

Eric Woolgar
Department of Mathematics
University of Saskatchewan
Saskatoon, SK, Canada S7N 0W0
woolgar@math.usask.ca

Abstract. In this talk, I describe recent work in collaboration with R. D. Sorkin. On lorentzian manifolds, we introduce a new causal relation, called the precedence relation, which is the smallest closed and transitive relation that contains the chronology relation I^+ as a subset. This new relation allows us to extend the techniques of global causal analysis to metrics which may not be differentiable or, in some cases, even invertible. We apply this relation, together with standard order-theoretic arguments, to give a simple proof of the familiar theorem that the space of causal curves joining fixed compact sets of a globally hyperbolic manifold is itself compact.

In riemannian manifolds, metrical properties play an important role not only at the obvious geometric level, but also at the topological level and therefore at the level of global structure as well. Consider for example the Hopf-Rinow theorem (Beem and Ehrlich [1981]), that metric completeness is equivalent to geodesic completeness and implies existence of minimal-length curves joining any pair of points. However, in the case of lorentzian manifolds, the topology and the global structure are in certain ways more directly related to the causal order of points than to the full lorentzian metric. With this in mind, consider the following theorem, which is fundamental to the field of global causal analysis and does lead to something of an analogue of Hopf-Rinow (namely, the existence of "fastest causal curves" in globally hyperbolic manifolds).

Theorem 1 *Let $\mathcal{A}$ and $\mathcal{B}$ be compact subsets of a globally hyperbolic lorentzian manifold $\mathcal{M}$. Then the space $C(\mathcal{A},\mathcal{B})$ of causal curves joining $\mathcal{A}$ to $\mathcal{B}$ is bi-compact (compact and Hausdorff).*

The proof usually given for this theorem is arguably not conceptually simple[1]. Perhaps the reason for this is that the proof involves techniques of *differential topology*, to borrow from the title of Penrose [1972], which are often more suited to (and often borrowed from) the riemannian case.

1991 *Mathematics Subject Classification.* Primary 53C50; Secondary 06A06.

[1]Roger Penrose, in his SIAM monograph (Penrose [1972]), commences his proof of this theorem (for what we call C^2-lorentzian manifolds) with the words, "I do not have a nice simple argument, but I feel sure that one must exist (exercise: find one!)."

© 1997 American Mathematical Society

In the formulation of global causal analysis, one may thus feel that it should be possible to replace certain differential topological constructs by notions which are more appropriate to lorentzian manifolds, and so more elementary. Here we have in mind order-theoretic notions which capture the causal order of a lorentzian manifold. R. D. Sorkin and myself, in recent work (Sorkin and Woolgar [1996]), have reformulated certain basic concepts in causal theory using order-theoretic ideas instead of differential-topological ones. In accordance with the idea that the order-theoretic approach may be more elementary, we have been able to use this formulation to give a simple and direct proof of the compactness theorem quoted above, which is valid even in cases where the lorentz metric may be only C^0 (an extension beyond standard proofs, such as that given in Hawking and Ellis [1973]). In what follows, I will attempt to provide an outline of the key ideas and a sketch of the logic entering into the compactness proof, although details cannot be provided for reasons of space limitations. It is intended that the interested reader will refer to Sorkin and Woolgar [1996] for details and for proofs.

Herein, $\mathcal{M}$ is taken to be a C^0-lorentzian manifold — a manifold is C^k-lorentzian iff it is time-orientable, has a C^{k+1} differentiable structure, and possesses a C^k lorentzian metric. We make use of the concept of *relations* on $\mathcal{M}$; these are just subsets of $\mathcal{M} \times \mathcal{M}$. An example is the chronology relation, which contains the ordered pair (p, q) iff $q \in I^+(p)$. Here we say that there is a timelike curve from p to q, and we write $q \in I^+(p)$ or $p < q$, iff there is a C^0, piecewise C^1 curve with everywhere future-timelike tangent that begins at p and ends at q.

We introduce a new relation on $\mathcal{M}$, which we call the *precedence relation.* We define this relation as the smallest closed and transitive relation which contains the chronology relation as a subset. If (p, q) is in this relation, we write $q \in K^+(p)$ or $p \prec q$ (and we say that p *precedes* q — there is an analogous definition for $K^-(p)$ as well). The closure condition means that if $p_n \prec q_n$ for convergent sequences $p_n \to p$, $q_n \to q$, then $p \prec q$, while transitivity is the statement that $x \prec y$ and $y \prec z \Rightarrow x \prec z$. Just as there is a "relative" chronology relation (which we denote by $<_{\mathcal{O}}$, so $p <_{\mathcal{O}} q \Leftrightarrow q \in I^+(p, \mathcal{O})$), one may also define "precedence relative to the subset $\mathcal{O} \subseteq \mathcal{M}$." Specifically, we define this relation to be the smallest closed and transitive relation on $\mathcal{O}$ that contains the $<_{\mathcal{O}}$ relation, and if (p, q) is in this relation, we write $p \prec_{\mathcal{O}} q$ or $q \in K^+(p, \mathcal{O})$. Based on this, we now define a *causal curve* to be the image of a continuous, locally increasing map (called a *causal path*)[2] $\gamma : [0, 1] \to \mathcal{M}$. By locally increasing, we mean that for every open $\mathcal{O} \ni \gamma(t)$, there is a neighbourhood of t in $[0, 1]$ within which $t' < t'' \Rightarrow \gamma(t') \prec_{\mathcal{O}} \gamma(t'')$.

An open set $\mathcal{O} \subseteq \mathcal{M}$ is said to be K-*causal* if $\prec$ induces a partial order on $\mathcal{O}$; that is, if the relation $\prec$ is *asymmetric* on $\mathcal{O}$ (so that, if both $p \prec q$ and $q \prec p$, then $p = q$). This is a causality condition, quite analogous to the familiar strong causality condition (but note that K-causality is stronger, and can be shown to imply strong causality)[3]. A key point is that on K-causal manifolds, the definition of causal curves can be considerably simplified, by using standard results from order theory. Namely, these standard results give

[2]It is convenient here that the parameter of the curve should take values in $[0, 1]$. For other applications, it may be convenient to consider open or half-open domains instead.

[3]Is K^+ equivalent to Seifert's J_s^+? Is K-causality equivalent to stable causality? These open questions and arguments that might suggest affirmative answers were brought to the author's attention by R. Low in private communications.

Standard Theorem 2 *Let Γ be: (a) a compact, connected, topological space containing a countable dense subset; (b) a linearly ordered set (with some order $\leq$), containing both a minimum and a maximum element with respect to the order; and (c) such that the order-closed intervals $\{y|x \leq y \leq z\}$ are topologically closed for all $x, z \in \Gamma$. Then Γ is simultaneously order- and topologically isomorphic to the real interval* $[0, 1]$.

(A set is linearly ordered by $\leq$ iff for every p, q in it either $p \leq$ q or $q \leq p$.) Using this lemma and the available partial order $\prec$, we may show that our definition of causal curves can be simplified on K-causal manifolds. Although the following is actually a lemma that can be derived by applying Standard Theorem 2 to our general definition of causal curves (see Sorkin and Woolgar [1996]), for purposes of this note we take it as a definition:

"Definition" 3 On a K-causal manifold, a *causal curve* is a compact connected set linearly ordered by $\prec$.

We will say that a set $\mathcal{O}$ is *globally hyperbolic* iff it is K-causal and for every pair $p, q \in \mathcal{O}$, the set $K^+(p) \cap K^-(q)$ is compact and contained in $\mathcal{O}$. In Sorkin and Woolgar [1996] it is shown that in the case of C^2-lorentzian manifolds, this definition is equivalent to the definition used in standard references (such as Hawking and Ellis [1973]), but the definition given here represents an extension that includes all C^0-lorentzian manifolds.

Let $\mathcal{A}$ and $\mathcal{B}$ be compact subsets of a globally hyperbolic C^0-lorentzian manifold $\mathcal{M}$. Let $C(\mathcal{A}, \mathcal{B})$ be the space of all causal curves ("Definition" 3 suffices here since the space is automatically K-causal) from $\mathcal{A}$ to $\mathcal{B}$. We topologise this space by first considering the Vietoris topology on the spaces $2^{\mathcal{K}}$ of *all* closed non-empty subsets of the compact sets $\mathcal{K} := K^+(x) \cap K^-(y)$, for $x, y \in \mathcal{M}$. The open sets of the Vietoris topology are those generated by taking arbitrary unions of the sets $\mathcal{V}(\mathcal{O}; \mathcal{O}_1, \ldots, \mathcal{O}_n)$, where $\mathcal{O}, \mathcal{O}_1, \ldots, \mathcal{O}_n$ are open subsets of $\mathcal{M}$ and where $\gamma \in \mathcal{V}(\mathcal{O}; \mathcal{O}_1, \ldots, \mathcal{O}_n)$ iff γ is a closed subset of $\mathcal{M}$ that is contained in $\mathcal{O}$ and meets each of $\mathcal{O}_1, \ldots, \mathcal{O}_n$. We refer to limits of nets taken with respect to this topology as *Vietoris limits*. The advantage of this approach is that we may use the following standard theorem.

Standard Theorem 4 *The space $2^{\mathcal{X}}$ of closed subsets of the bi-compact (i.e., compact and Hausdorff) space $\mathcal{X}$ is bi-compact.*

In particular, let Γ_α be a net of curves joining compact sets $\mathcal{A}$ to $\mathcal{B}$. Let p (q) be an accumulation point of the initial (final) endpoints, and choose any $x \in I^-(p)$, $y \in I^+(q)$. Then $\mathcal{K} := K^+(x) \cap K^-(y)$ is compact (by global hyperbolicity) and eventually contains each curve in the net (or at least in a subnet). By Standard Theorem 4 applied to the space $2^{\mathcal{K}}$, the net of curves has a convergent subnet with Vietoris limit in $2^{\mathcal{K}}$. It is an easily proved result that this limit, call it Γ, is a compact connected set, since it is a Vietoris limit of a net whose elements are themselves compact connected sets (in particular, causal curves).

To establish that Γ is linearly ordered by $\prec$, let $r, s \in \Gamma$ and let $\mathcal{O}_1 \ni r$ and $\mathcal{O}_2 \ni s$ be open in $\mathcal{M}$. Since Γ is a Vietoris limit, we have $\Gamma_\alpha \in \mathcal{V}(\mathcal{M}; \mathcal{O}_1, \mathcal{O}_2)$ for α large enough. As the neighbourhoods $\mathcal{O}_1$ and $\mathcal{O}_2$ shrink down, we may extract convergent sequences $r_n, s_n \in \Gamma_n$ such that $r_n \to r$ and $s_n \to s$. If the Γ_α are curves in the K-causal manifold $\mathcal{M}$, then "Definition" 3 implies that either $r_n \prec s_n$ or vice versa (since each Γ_α is linearly ordered), whence by closure $r \prec s$ (or vice versa), whence Γ is linearly ordered by $\prec$. Furthermore, it can be shown that $p \in \mathcal{A}$ and

$q \in \mathcal{B}$ are the infimum and supremum with respect to this order on Γ. Combining all these results, we have that Γ is a compact, connected, linearly ordered set — thus, by "Definition" 3, a causal curve — joining $\mathcal{A}$ to $\mathcal{B}$, whence $\Gamma \in C(\mathcal{A}, \mathcal{B})$. Thus, we have shown that every net in $C(\mathcal{A}, \mathcal{B})$ has a convergent subnet, whence $C(\mathcal{A}, \mathcal{B})$ is compact. Lastly, it is Hausdorff, and therefore bi-compact, since it is immediate that if distinct curves can be separated by open sets in $2^{\mathcal{K}}$ (which is Hausdorff by Standard Theorem 4), they will also be so separated in $C(\mathcal{A}, \mathcal{B})$. This proves Theorem 1.

References

Beem, J. K. and Ehrlich, P. E. [1981] *Global Lorentzian Geometry*, Dekker, New York.

Hawking, S. W. and Ellis, G. F. R. [1973] *The Large Scale Structure of Spacetime*, Cambridge University Press, Cambridge.

Penrose, R. [1972] *Techniques of Differential Topology in Relativity*, SIAM, Philadelphia.

Sorkin, R. D. and Woolgar, E. [1996] *A Causal Order for Spacetimes with C^0 Lorentzian Metrics: Proof of Compactness of the Space of Causal Curves*, Class. Quantum Gravit., **13**, 1971–1993.

Appendix

Sixth Canadian Conference on General Relativity and Relativistic Astrophysics University of New Brunswick Fredericton, N. B.

List of Participants

Abolghasem, Hossein, Department of Mathematics, Statistics and Computing Science, Dalhousie University, Halifax, N. S., Canada B3H 3J5 *hossein@cs.dal.ca*
Acharya, Raghunath, Department of Physics, Arizona State University, Tempe, Arizona USA 85287-1504 *acharya@phyast.la.asu.edu*
Benoit, Patricia M., Department of Mathematics, Statistics and Computing Science, Dalhousie University, Halifax, N. S., Canada B3H 3J5 *pmbenoit@cs.dal.ca*
Billyard, Andrew, Department of Physics, Dalhousie University, Halifax, N. S., Canada B3H 1J3 *jaf@cs.dal.ca*
Bonanno, Alfio, Theoretical Physics Institute, Department of Physics, University of Alberta, Edmonton, Alberta, Canada T6G 2J1 *abonanno@phys.ualberta.ca*
Braham, Stephen P., Centre for Experimental & Constructive Mathematics, Simon Fraser University, Burnaby, B. C., Canada V5A 1S6 *warp@lizardo.cecm.sfu.ca*
Burgess, Cliff, Department of Physics, McGill University, Montreal, P. Q., Canada H3A 2K6 cliff@hep.physics.mcgill.ca
Burton, Howard S., Department of Physics, University of Waterloo, Waterloo, Ontario, N2L 3G1 *hsburton@physics.watstar.uwaterloo.ca*
Carot, Jaume, Dep. Física, Univ. Illes Balears (UIB), Cra. Valldemossa km 7.5, E-07071 Palma de Mallorca, Spain *dfsjcg0@ps.uib.es*
Chamblin, Andrew, Department of Applied Math & Theoretical Physics, Silver Street, Cambridge CB3 9EW, England *h.a.chamblin@damtp.cam.ac.uk*
Chan, Jim, Department of Applied Mathematics, University of Waterloo, Waterloo, Ontario, Canada N2L 3G1 *sfjchan@jeeves.uwaterloo.ca*
Chan, Kevin C.K., Department of Physics, University of Waterloo, Waterloo, Ontario, Canada N2L 3G1 *kckchan@avatar.uwaterloo.ca*
Coley, Alan A., Department of Mathematics, Statistics & Computing Science, Dalhousie University, Halifax, N. S., Canada B3H 3J5 *aac@cs.dal.ca*
Corkum, Krista, Department of Mathematics and Computer Science, Mount Saint Vincent University, Halifax, N. S., Canada B3M 2J6 *kcorkum@linden.msvu.ca*
Creighton, Jolien, Department of Physics, University of Waterloo, Waterloo, Ontario, Canada N2L 3G1 *jolien@avatar.uwaterloo.ca*
da Costa, Jose Manuel, Departamento de Matemática, Universidade da Madeira, Colegio dos Jesuitas, Largo do Municipio, P-9000 Funchal, Portugal *costa@dragoeiro.uma.pt*
Darian, Bahman, Department of Physics, University of Alberta, Edmonton, Alberta, Canada T6G 2J1 *darian@phys.ualberta.ca*
Davies, George, 254 Sydenham St., Kingston, Ontario, Canada K7K 3M5 *davies@astro.queensu.ca*
Demers, Jean-Guy, Department of Physics, McGill University, Montreal, P. Q., Canada H3A 2T8 *jgdemers@hep.physics.mcgill.ca*
Dunnett, Chris, Department of Mathematics & Statistics, University of New Brunswick, Fredericton, N. B., Canada E3B 5A3 *q296@unb.ca*

Fee, Greg, Department of Applied Mathematics, University of Waterloo, Waterloo, Ontario, Canada N2L 3G1 *gjfee@daisy.uwaterloo.ca*

Friedman, John L., Department of Physics, University of Wisconsin - Milwaukee, Milwaukee, WI USA 53201 *friedman@thales.phys.uwm.edu*

Gauthier, Claude, Département de Mathématiques, Université de Moncton, Moncton, N. B., Canada E1A 3E9 *gauthic@umoncton.ca*

Gautreau, Ronald, Department of Physics, New Jersey Institute of Technology, Newark, NJ USA 07102 *gautreaur@admin.njit.edu*

Gegenberg, Jack, Department of Mathematics and Statistics, University of New Brunswick, Fredericton, N. B., Canada E3B 5A3 *lenin@math.unb.ca*

Goode, Stephen M., Department of Mathematics, California State University, Fullerton, CA USA 92634 *stephen_goode@qmail2.fullerton.edu*

Gravel, Pierre, Department of Mathematics & Computer Science, Royal Military College of Canada, Kingston, Ontario, Canada K7K 5LO *gravel_p@banmath.rmc.ca*

Hall, Graham S., Department of Mathematics, University of Aberdeen, Edward Wright Building, Dunbar Street, Aberdeen AB9 2TY, Scotland *gsh@aberdeen.ac.uk*

Harriott, Tina, Mathematics Department, Mount Saint Vincent University, Halifax, N. S., Canada B3M 2J6 *tharriot@linden.msvu.ca*

Hobill, David, Department of Physics & Astronomy, University of Calgary, Calgary, Alberta, Canada T2N 1N4 *hobill@crag.ucalgary.ca*

Medrano, Jesus Ibáñez, Dept. Física Teórica, Universidad del País Vasco, Apdo. 644, E-48080, Bilbao, Spain *wtpibmej@lg.ehu.es*

Isenberg, Jim, Department of Mathematics, University of Oregon, Eugene, OR USA 97403-1222 *jim@newton.uoregon.edu*

Israel, Werner, Department of Physics, University of Alberta, Edmonton, Alberta, Canada T6G 2J1 *userudey@mts.ucs.ualberta.ca*

Kalligas, Dimitri, Department of Physics, University of Waterloo, Waterloo, Ontario, Canada, N2L 3G1

Kaloper, Nemanja, Department of Physics, McGill University, Montreal, P. Q., Canada H3A 2T8 *kaloper@hep.physics.mcgill.ca*

Kavian, Masoud, Department of Applied Mathematics, University of Waterloo, Waterloo, Ontario, Canada N2L 3G1 *mkavian@daisy.uwaterloo.ca*

Khuri, Ramzi, Department of Physics, McGill University, Montreal, P. Q., Canada H3A 2T8 *rrk@campy.physics.mcgill.ca*

Kunstatter, Gabor, Physics Department, University of Winnipeg, Winnipeg, Manitoba, Canada R3B 2E9 *gabor@theory.uwinnipeg.ca*

Lafrance, René, Department of Physics, McGill University, Montreal, P. Q., Canada H3A 2T8 *lafrance@physics.mcgill.ca*

Louis Martinez, Domingo J., Physics Department, University of Winnipeg, Winnipeg, Manitoba, Canada R3B 2E9 *martinez@theory.uwinnipeg.ca*

Martinez, Erik, Theoretical Physics Institute, 412 Avadh Bhatia Physics Laboratory, University of Alberta, Edmonton, Alberta, Canada T6G 2J1 *martinez@phys.ualberta.ca*

McKellar, R. J., Department of Mathematics & Statistics, University of New Brunswick, Fredericton, N. B., Canada E3B 5A3 *bob@math.unb.ca*

McLenaghan, R. G., Department of Applied Mathematics, University of Waterloo, Waterloo, Ontario, Canada N2L 3G1 *rgmclenaghan@daisy.uwaterloo.ca*

Michaud, Guy, Physics Department, McGill University, Montreal, P. Q., Canada H3A 2T8 *gmichaud@hep.physics.mcgill.ca*

Morsink, Sharon, Department of Physics, University of Alberta, Edmonton, Alberta, Canada T6G 2J1 *morsink@euclid.phys.ualberta.ca*

Mureika, Jonas, Department of Physics, University of Waterloo, Waterloo, Ontario, Canada N2L 3G1 *newt@avatar.uwaterloo.ca*

Musgrave, Peter, Department of Physics, Queen's University, Kingston, Ontario, Canada K7L 3N6 *musgrave@astro.queensu.ca*

Nezami, Reza, Department of Mathematics & Statistics, University of New Brunswick, Fredericton, N. B., Canada E3B 5A3 *reza@math.unb.ca*

Overduin, James, Department of Physics & Astronomy, University of Victoria, Victoria, B. C., Canada V8W 3P6 *overduin@otter.phys.uvic.ca*

Paul, Chris, P.O. Box 750, Sackville, N. B., Canada E0A 3C0 *medianet@nbnet.nb.ca*

Perry, Phil, Department of Physics, University of Victoria, Victoria, B. C., Canada V8W 3P6 *perry@sol.uvic.ca*

Poisson, Eric, Department of Physics, Washington University, St. Louis, MO USA 63130 *poisson@wurel.wustl.edu*

Pollney, Denis, Department of Physics, Queen's University, Kingston, Ontario, Canada K7L 3N6 *pollney@astro.queensu.ca*

Pullin, Jorge A., Center for Gravitational Physics & Geometry, Department of Physics, Pennsylvania State University, University Park, PA USA 16802 *pullin@phys.psu.edu*

Qayum, Hisham A., Department of Physics, University of California, Davis, CA USA 95616 *qayum@bethe.ucdavis.edu*

Salopek, David S., Department of Physics, University of Alberta, Edmonton, Alberta, Canada T6E 4S6 *salopek@phys.ualberta.ca*

Sasse, Fernando, Department of Applied Mathematics, University of Waterloo, Waterloo, Ontario, Canada N2L 3G1 *fsasse@barrow.uwaterloo.ca*

Schleich, Kristin, Department of Physics, University of British Columbia, Vancouver, B. C., Canada V6T 1Z1 *schleich@physics.ubc.ca*

Söler, François, Département de Mathématique, Université de Moncton, Moncton, N. B., Canada E1A 3E9 *solerf@umoncton.ca*

Steif, Alan, Physics Department, University of California, Davis, Davis, CA USA 95616, *steif@dirac.ucdavis.edu*

Swamy, Padmanabha, Department of Physics, Southern Illinois University, Edwardsville IL USA 62026–1654 *swamyp@eniac.ac.siue.edu*

Torre, Charles G., Department of Physics, Utah State University, Logan, Utah USA 84322-4415 *torre@cc.usu.edu*

Tupper, B. O. J., Department of Mathematics and Statistics, University of New Brunswick, Fredericton, N. B., Canada E3B 5A3 *brian@math.unb.ca*

van den Hoogen, Robert, Department of Mathematics, Statistics and Computing Science, Dalhousie University, Halifax, N. S., Canada B3H 3J5 *hoogen@cs.dal.ca*

Vera Mège, Rafael A., Departamento de Fisica, Universidad de Concepción. Chile, Casilla 4009. Concepción, Chile *rvera@buho.dpi.udec.cl*

Vincent, Dwight, Physics Department, University of Winnipeg, Winnipeg, Manitoba, Canada R3B 2E9 *vincent@uwpg02.uwinnipeg.ca*

Wald, Robert M., Enrico Fermi Institute, University of Chicago, Chicago, IL USA 60637-1433 *rmwa@midway.uchicago.edu*

Wesson, Paul, Department of Physics, University of Waterloo, Waterloo, Ontario, Canada N2L 3G1 *wesson@astro.uwaterloo.ca*

West, Michael J., Department of Astronomy and Physics, Saint Mary's University, Halifax, N. S., Canada B3H 3C3 *west@sisyphus.stmarys.ca*

Williams, Jeff, Department of Mathematics & Computer Science, Brandon University, Brandon, Manitoba, Canada R7A 6A9 *williams@brandonu.ca*

Witt, Donald, Department of Physics, University of British Columbia, Vancouver, B. C., Canada V6T 1Z1 *donwitt@noether.physics.ubc.ca*

Wong, Jeff, Faculty of Computer Science, University of New Brunswick, Fredericton, N. B., Canada E3B 5A3 *z3u7@unb.ca*

Woolgar, Eric, Department of Mathematics, University of Saskatchewan, Saskatoon, Sask., Canada S7N 0W0 *woolgar@math.usask.ca*

Zaidi, Haider, Department of Physics, University of New Brunswick, Fredericton, N. B., Canada E3B 5A3 *zaidi@unb.ca*